Cell Physiology

Cell Physiology

Editor: Gloria Doran

RCALLISTO REFERENCE

www.callistoreference.com

Callisto Reference,
118-35 Queens Blvd., Suite 400,
Forest Hills, NY 11375, USA

Visit us on the World Wide Web at:
www.callistoreference.com

ISBN: 978-1-63239-815-4 (Hardback)

The publisher's policy is to use permanent paper from mills that operate a sustainable forestry policy. Furthermore, the publisher ensures that the text paper and cover boards used have met acceptable environmental accreditation standards.

Trademark Notice: Registered trademark of products or corporate names are used only for explanation and identification without intent to infringe.

Printed in the United States of America.

Cataloging-in-publication Data

Cell physiology / edited by Gloria Doran.
    p. cm.
Includes bibliographical references and index.
ISBN 978-1-63239-815-4
1. Cell physiology. 2. Cytology. 3. Cytogenetics. I. Doran, Gloria.
QH631 .C45 2017
571.6--dc23

Table of Contents

Permissions

List of Contributors

Index

Preface

This book traces the progress of cell physiology and highlights some of its key concepts and applications. It elucidates new techniques and their applications in a multidisciplinary approach. Cell physiology is concerned with the study of cell functions and cell structure. It also examines the processes performed by cells to function. The topics included in this book are of utmost significance and bound to provide incredible insights to readers. The topics included in this book are of utmost significance and bound to provide incredible insights to readers. It presents researches and studies performed by experts across the globe. This book will help the readers in keeping pace with the rapid changes in this field. Coherent flow of topics, student-friendly language and extensive use of examples make this text an invaluable source of knowledge.

The world is advancing at a fast pace like never before. Therefore, the need is to keep up with the latest developments. This book was an idea that came to fruition when the specialists in the area realized the need to coordinate together and document essential themes in the subject. That's when I was requested to be the editor. Editing this book has been an honour as it brings together diverse authors researching on different streams of the field. The book collates essential materials contributed by veterans in the area which can be utilized by students and researchers alike.

Each chapter is a sole-standing publication that reflects each author's interpretation. Thus, the book displays a multi-facetted picture of our current understanding of applications and diverse aspects of the field. I would like to thank the contributors of this book and my family for their endless support.

Editor

Membrane Topology and Cellular Dynamics of Foot-and-Mouth Disease Virus 3A Protein

Mónica González-Magaldi[1], Miguel A. Martín-Acebes[1], Leonor Kremer[2], Francisco Sobrino[1]*

1 Centro de Biología Molecular Severo Ochoa, Consejo Superior de Investigaciones Científicas-Universidad Autónoma de Madrid, Madrid, Spain, **2** Centro Nacional de Biotecnología, Consejo Superior de Investigaciones Científicas, Madrid, Spain

Abstract

Foot-and-mouth disease virus non-structural protein 3A plays important roles in virus replication, virulence and host-range; nevertheless little is known on the interactions that this protein can establish with different cell components. In this work, we have performed *in vivo* dynamic studies from cells transiently expressing the green fluorescent protein (GFP) fused to the complete 3A (GFP3A) and versions including different 3A mutations. The results revealed the presence of a mobile fraction of GFP3A, which was found increased in most of the mutants analyzed, and the location of 3A in a continuous compartment in the cytoplasm. A dual behavior was also observed for GFP3A upon cell fractionation, being the protein equally recovered from the cytosolic and membrane fractions, a ratio that was also observed when the insoluble fraction was further fractioned, even in the presence of detergent. Similar results were observed in the fractionation of GFP3ABBB, a 3A protein precursor required for initiating RNA replication. A nonintegral membrane protein topology of FMDV 3A was supported by the lack of glycosylation of versions of 3A in which each of the protein termini was fused to a glycosylation acceptor tag, as well as by their accessibility to degradation by proteases. According to this model 3A would interact with membranes through its central hydrophobic region exposing its N- and C- termini to the cytosol, where interactions between viral and cellular proteins required for virus replication are expected to occur.

Editor: Yi Li, Wuhan Bioengineering Institute, China

Funding: Work at the F.S. laboratory was supported by Spanish grant BIO2011–24351, by an institutional grant from Fundación Ramón Areces, and by the ICTS program from the Spanish Ministry of Science and Innovation. Work at the L. Kremer laboratory was supported by The Instituto de Salud Carlos III grant PI10/00594 and the CSIC grant 201120E007 from the Spanish Ministerio de Economía y Competitividad. MAMA is recipient of a JAE-Doc fellowship from CSIC. The funders had no role in study design, data collection and analysis, decision to publish, or preparation of the manuscript.

Competing Interests: The authors have declared that no competing interests exist.

* Email: fsobrino@cbm.csic.es

Introduction

Foot-and-mouth disease virus (FMDV) is an aphthovirus that belongs to the *Picornaviridae* family and the etiological agent of an extremely contagious disease of cloven-hoofed animals (FMD) that is responsible for high economic losses in affected countries [1,2]. FMDV RNA is a positive strand molecule of about 8500 nucleotides that encodes a single ORF [3]. The polyprotein resulting from its translation is processed by viral proteases to yield structural proteins as well as precursors and mature non-structural (NS) proteins [4]. The NS protein 3A is produced by cleavage of 3ABC precursor, reviewed in [5], and is one of the most variable viral proteins encoded by FMDV, being the variable residues preferentially accumulated at its C-terminus [6]. An 18 amino acids long hydrophobic region (HR, spanning residues 59 to 76) is predicted in the middle of the molecule [7,8,9]. In other picornaviruses this hydrophobic domain has been reported to target 3A to intracellular membranes [10,11] and could contribute to locate the viral replication complex within a membrane context [12,13,14,15], but the origin of the membranes involved in FMDV replication and the type of interactions they establish with viral proteins remain uncertain [16]. In cells transiently expressing FMDV 3A, about 50% of the cellular pool of the protein was recovered from the membrane fraction, suggesting an association of 3A with cellular membranes [8].

FMDV 3ABC region shows unique characteristics among picornaviruses, such as encoding 3 copies of viral genome-bound 3B protein [7,17] that serves as a primer for RNA replication [18]. The three copies of 3B are required for both optimal replication in cell culture [19] and for virulence in natural hosts [20]. In addition, the C-terminal fragment of FMDV 3A (up to the HR) is considerably longer than those of the other picornaviruses. On the other hand, 3A is not the responsible for blocking the endoplasmic reticulum (ER)-to-Golgi transport of proteins as occurs in poliovirus (PV), being this function carried out by 2B and 2BC [8]. FMDV 3A partially colocalizes with ER and Golgi markers [21,22] and recent evidences point to the involvement of ER exit sites for virus replication, supporting to the involvement of ER in virus replication [23].

On the other hand, 3A protein has been reported to play a role on FMDV host range, as a single amino acid replacement (Q44R) in this protein conferred FMDV the ability to cause vesicular lesions in guinea pigs [24] and deletions and mutations in the C-terminal region associate both to viral attenuation in cattle [25] and to decreased replication rates in bovine epithelial cells [26].

A molecular model of the N-terminal fragment of FMDV 3A protein, derived from the corresponding NMR structure of the PV 3A [27], predicted a hydrophobic interface composed of two α-helices spanning residues 25 to 44 as the main determinant for 3A dimerization. Replacements L38E and L41E, involving charge acquisition at residues predicted to contribute to the hydrophobic interface, reduced dimerization and led to production of infective viruses that replaced the acidic residues introduced (E) by non-polar amino acids, indicating that preservation of the hydrophobic interface is essential for virus replication [9].

To facilitate its study in transient expression assays we fused FMDV 3A wt and mutant versions of this protein − including different deletions, as well as point mutations at the dimerization interface and at the odd cysteine present in 3A − to the green fluorescent protein (GFP). Live cell imaging in combination with photobleaching can provide insights into the movement of proteins and on their interaction with cellular components [28,29]. Time-lapse microscopy revealed that the cytoplasmic mobility of GFP3A was spatially confined and the analysis of the fluorescence loss in photobleaching (FLIP) [30,31] supported the location of GFP3A in a continuous compartment in the cytoplasm. In addition, fluorescence recovery after photobleaching (FRAP) analyses [32] revealed the presence of a mobile fraction of GFP3A, which was shown increased in most of the mutants analyzed. On the other hand, biochemical analyses of transfected cells showed that about 60% of GFP3A protein interacted with cellular membranes. Membrane bounded fractions of GFP3A and its precursor GFP3ABBB were further analyzed by different biochemical treatments, resulting in interaction profiles different from that of an integral membrane protein. Further analysis of 3A interactions with membranes by glycosylation tagging experiments and by a biochemical protease protection assay allowed proposing a model of 3A membrane topology.

Materials and Methods

Cells and virus

Vero cells (African green monkey epithelial kidney cells; ATCC CCL-81), IBRS-2 (swine kidney cell line) [33], HeLa (human cervical epithelial cells) and BHK-21 cells (Baby hamster kidney cells; ATCC CCL-10) were grown at 37°C and maintained in Dulbecco's modified Eagle's medium (DMEM) (Gibco-BRL), prepared without phenol red for *in vivo* microscopy, supplemented with 5% fetal bovine serum (Gibco-BRL), 2 mM glutamine, 1 µg/ml streptomycin and 1 µg/ml penicillin. A viral stock from type C FMDV C-S8c1 isolate [34] was produced by amplification in BHK-21 cells.

Antibodies and reagents

Monoclonal Ab (MoAb) 2C2 to NS protein 3A (38), rabbit polyclonal Ab 346 and 479− directed to the C- and N- termini of 3A protein, respectively (9) −, rabbit polyclonal Ab to caveolin-1 (BD Transduction Laboratories), rabbit polyclonal Ab to calreticulin (Abcam) and a MoAb to GFP (Roche) were used.

Construction of fusion proteins

For *in vivo* experiments 3ABBB, 3A and its mutants (3AL38E, 3AL41E and 3AC65S) were fused to the C-terminus of GFP using plasmid pEGFP-C2 (Clontech). The sequences encoding 3A and 3ABBB wt proteins were amplified by PCR from the infectious clone pMT28 that encodes the genomic RNA of FMDV isolate C-S8c1 [35]. Primers 3A1/3A2 and 3A1/3A-3BBBr were used to amplify 3A wt and 3ABBB, respectively (Table 1). The resulting amplicons and pEGFP were digested with the corresponding restriction enzyme (New England BioLabs), indicated in Table 1, and ligated with DNA ligase T4 (Roche), as described [36]. Substitutions of selected amino acids were generated by site-directed mutagenesis [9]. Deletion mutants were obtained by PCR amplification of the selected 3A sequences: 3AΔHR-C-ter (K53-E153), 3AΔC-ter (R82-E153) and 3AΔN-ter (I1-L41), using primers 3A1/ΔHR-C-ter, ΔC-ter f/ΔC-ter r and ΔN-ter f/ΔN-ter r, respectively. The resulting amplicons were cloned into plasmid pEGFP following digestion with the corresponding restriction enzyme (Table 1). The correct orientation and sequence of the plasmids obtained were confirmed by sequencing with GFP primers.

N-glycosylation insertion mutagenesis and deglycosylation assays

For topologic analyses 3A protein was cloned in pcDNA3.1+ vector (Invitrogen) using 3A wt amplicons and the restriction enzymes BamHI and XbaI. To construct plasmids pcDNA3A-glyc, pcDNAglyc-3A, and pcDNAglyc-3AΔN-ter(I1-N58), a N-glycosylation acceptor site (Asn-Ser-Thr-Ser-Ala-Asn) (36) was fused in-frame to the C- or N-termini of 3A, or to the N-terminus of 3AΔN-ter. Amplicons were obtained by PCR using pcDNA3A as template, the sense primers Gly-Nter F, Gly-TMC-ter F, and the antisense primer C-ter-Gly R with the corresponding sense and antisense primers (Table 1). For the deglycosylation assay, Vero cells were grown in 35-mm dishes, transfected with 2 µg of DNA and 24 h post transfection (pt) lysed in 200 µl NBP [50 mM Tris-HCl, pH 7.5, 150 mM NaCl, 1% NP-40, 1% sodium deoxycholate, 0.1% SDS, 1 mM phenylmethylsulfonylfluoride, protease inhibitor cocktail 1x (Roche)], treated with 1 µl benzonase (Novagen). Lysed cells were centrifuged 10 min at 300×g and the supernatant centrifuged 30 min at 30000×g. Pellets were resuspended in NPB and deglycosylation was performed with PGNase F (New England Biolabs) as recommended by manufacturer. Laemmli sample buffer [37] was added and proteins were separated by SDS-PAGE and analyzed by Western blot. As positive controls pTM-DV-NS4A(1–150)-GFP-Glyc and pTM-DV-NS4A(1–100)-GFP-Glyc − expressing dengue virus NS4A full length (FL) and a C-terminal truncation of NS4A (amino acids 1–100), respectively [38] − were transfected. One h before transfection with these plasmids, cells were infected with vaccinia virus VTF7–3 expressing the T7 RNA polymerase to allow cytoplasmic transcription of the constructs [39]. At 20 h pt cells were lysed and processed as described above.

Biochemical treatment of cell lysates

Vero cells were grown in 60-mm dishes and transfected with 1–2 µg of plasmid DNA using Lipofectamine (Invitrogen). 24 h pt cells were lysed with PBS supplemented with protease inhibitor cocktail 1x (Roche) by five cycles of freezing in liquid nitrogen and thawing at 37°C. Lysed cells were centrifuged at 300×g for 10 min and then at 30000×g for 20 min twice. Pellets were resuspended in the following solutions: 0.1 M Na_2CO_3, 4 M Urea (Merck), 1 M KCl (Merck), or 0.5% Triton X-100 (Sigma). Samples were boiled, resolved on SDS-PAGE and immunoblotted, following addition of Laemmli sample buffer.

Biochemical protease protection assay

HeLa cells grown in 35-mm dishes, without or with coverslips (in case of immunofluorescence analysis), were transfected with pcDNA3A. This cell line was used because of it high level of transfection efficiency. 24 h pt cells were washed with KHM buffer (110 mM potassium acetate, 20 mM HEPES pH 7.2,

Table 1. Oligonucleotides used for construction of wt and mutant versions of 3A and 3ABBB fused to GFP.

Oligonucleotide	Sequence (5′→3′)	Genomic orientation	Restriction enzyme
3A1	TAGGGGATCCGTATCTCAATACCTTCC	S	BamHI
3A2	GCAGATCTTTATTCAGCTTGCGGTTG	A	BglII
3A-3BBB	GCAGATCTTTACTCAGTGACAATCAA	A	BglII
ΔHR-C-ter	GCAGATCTTTAAAAAGCACGTTTCAC	A	BglII
ΔC-ter f	GCGAAGCTTTCTAGAAATGATCTCAATACCTTCC	S	HindIII
ΔC-ter r	GCCGGATCCTTACTTGTGAGTCTCGC	A	BamHI
ΔN-ter f	CGGAGATCTGGATCCAACAAACTTCA	S	BglII
ΔN-ter r	GCAAGCTTTATTCAGCTTGCGGTTG	A	HindIII
GlyN-ter F	TTCGCGGATCCGACATGAATTCGACCTCGGCTACATCTCAATACCTTCCCAA	S	BamHI
GlyTMC-ter F	TTCGCGGATCCGACATGAATTCGACCTCGGCTACTTTGAAATTGTTGCACTG	S	BamHI
C-terGly R	TCGCCTCTAGACTAGTTAGCCGAGGTCGAATTTCAGCTTGCGGTTGCTC	A	Xba I

2 mM MgCl$_2$) and permeabilized with 50 μM digitonin in KHM buffer for 1 min at room temperature. Then, cells were washed in KHM and treated with 0.025% trypsin or 50 μM proteinase K for 5 min at room temperature. Finally, cells were washed and lysed, Laemmli sample buffer was added, and proteins were separated in a SDS-PAGE and analyzed by Western blotting with antibodies to the N- and C-termini of 3A and to calreticulin. For immunofluorescence analysis, cells in coverslips with the same treatment were fixed in 4% paraformaldehyde.

Western blot analysis

Vero cells grown on 35-mm dishes were transfected as described above with 1 μg of different plasmids. At 24 h pt, cells were scraped on ice into NP-40 lysis buffer (10 mM EGTA, 2.5 mM MgCl2, 1% NP-40, 20 mM HEPES pH 7.4) and sonicated. Equal volumes of each sample mixed with Laemmli sample buffer were boiled, separated by SDS-PAGE 12%, and transferred onto a nitrocellulose membrane. The membrane was blocked, and proteins were detected by incubation with the selected primary antibody and the corresponding horseradish peroxidase-coupled secondary antibody that was developed using a chemiluminescence kit (Perkin-Elmer).

Density gradient fractionation

The procedure for isolation of Triton X-100-insoluble membranes by centrifugation to equilibrium in sucrose density gradients was essentially as described [40]. Cells grown in 100-mm dishes were transfected and 24 h later washed tree times with cold PBS, scraped on 0.5 ml of 0.5% Triton X-100 in TNE buffer (25 mM Tris-HCl, 150 mM NaCl, 5 mM EDTA pH 7.4) and maintained for 30 min on ice. The lysate was passed through a 22-gauge needle, mixed with 70% sucrose in TNE buffer supplemented with 1 mM PMSF and protease inhibitor cocktail, and brought to 35% sucrose in a final volume of 4 ml beneath a 8 ml 5–30% linear sucrose gradient. Gradients were centrifuged at 4°C for 18 h at 180000×g in a SW40 rotor (Beckman) and 12 fractions of 1 ml, collected from top to bottom, were analyzed by Western blotting or stored at −80°C.

Immunofluorescence and confocal microscopy

Cells grown on glass cover slips were transfected or infected with FMDV C-S8c1 (moi = 5 PFU/ml). At 24 h pt or 4 h post infection (pi), cells were fixed in 4% paraformaldehyde for 15 min at room temperature, blocked, and permeabilized with PBTG buffer (0.1% Triton X-100, 1% bovine serum albumin (BSA), and 1 M glycine in PBS) for 15 min. Samples were incubated with the selected primary antibody diluted in 1% BSA in PBS for 1 h at room temperature, washed with PBS and incubated with the corresponding secondary antibody for 30 min. Finally, samples were mounted in Fluoromount G (Southern Biotech) and cells were observed with a Microradiance confocal (Biorad/Zeiss) microscope.

Time-lapse microscopy

Time-lapse was performed in Vero cells grown on glass bottom 35-mm dishes (MatTek) transfected with GFP3A, which 6 h pt were transferred to the microscope incubator previously warmed at 37°C. Images (5 different planes along Z axis at intervals of 5 min for 3 h) were acquired using an inverted Axiovert200 microscope (Zeiss) with a 63x/1.2 Water C-Apochromat Corr, coupled to a digital camera C9100–02 (Hamamatsu). Humidity, CO$_2$ and temperature (37°C) were controlled using the *In Vivo* Cell Observer system (Zeiss). Manual tracking of the fluorescence was performed using ImageJ plug-in: Manual tracking.

FRAP analysis

Vero cells grown on glass bottom 35-mm dishes were transfected with GFP3A. At 6 h pt, cells were observed using an *in vivo* system in an inverted Axio Observer Confocal laser scanning microscope (Zeiss), in triplicate experiments (n>10 cells/experiment). The 488-nm laser line was used to perform photobleaching of a defined circular 10 μm ø region of interest (ROI) at full laser power (100% laser power, 100 interactions). Recovery of the fluorescence was monitored by continuous scanning of a control ROI either in a neighbor cell or at a different region in the cytoplasm of the same cell, using a low laser power (1%) until the fluorescence of the bleached area reached a plateau. Cells were scanned six times before photobleaching to determine the maximum initial intensity of fluorescence. No additional photobleaching was observed during recovery. The images were captured with the Zenon (Zeiss) software. Fcalc program, Turu Centre for Biotechnology, Finland [41] was used to analyzed the FRAP data and to calculate the mobile fraction fitting an exponential curve to the corrected data using a least square fit: A1 (1-ek1t) +A2 (1-ek2t), where A1 and A2 represent the mobile fractions with a two function fit, and k the kinetic constant.

FLIP analysis

Cells grown on glass bottom 35-mm dishes were transfected with pEGFP3A or plgLdR1KDEL-RFP [32]. Twenty four h pt cells were observed as described before. The 488-nm laser line was used to carry out sequential photobleaching events of a selected ROI. This area was exposed to 15 interactions of 100% laser power for RFP and 60% for GFP every six scanners (7 s) for 150 repeats. Loss of fluorescence was monitored in a different ROI in the cytoplasm of the same repetitive bleached cell. Fluorescence intensity of a neighbor cell ROI was determined to estimate global photobleaching in the field. Images were analyzed with the Zenon program.

Data analysis

To probe statistical significance of the data, one-way analysis of the variance was performed with statistical package SPSS 19.0 (SPSS, Inc.) for Windows. For multiple comparisons, Bonferroni's correction was applied. The data are presented as means $\pm$ the standard deviations and statistically significant differences are indicated in the figures by an *.

Results

Characterization of 3A fluorescent protein in transfected cells

The analysis of protein dynamics *in vivo* requires fluorescent probes whose biophysical properties can be monitored to infer changes in cellular biochemistry [28,42]. In this work we have studied *in vivo* the properties of FMDV 3A protein by means of its fusion to GFP. A bioinformatic application (TMHMM) [43] that estimates the likelihood of membrane protein topology indicated that fusion of GFP to the N-terminus, but not to the C-terminus of 3A, maintained the topology of this viral protein. Therefore, GFP was cloned as fusion with the N-terminus of 3Awt, and expression of GFP3A confirmed by Western blot analysis (Fig. 1A). As a first step in the use of GFP3A as a tool to study the 3A *in vivo* distribution and dynamics, transfected cells were examined by confocal microscopy. Fluorescence of GFP and GFP3A in Vero cells is shown in Fig. 1B-i. While GFP fluorescence was observed throughout the whole cell including the nucleus, GFP3A fluorescence was restricted to the cytoplasm, including a perinuclear distribution similar to that found for 3A in FMDV-infected IBRS-2 cells (Fig. 1B-ii). As expected, fluorescence of GFP and 3A were shown to colocalize in transfected cells (Fig. 1B-iii).

Then, the cytoplasmic distribution of GFP3A was analyzed by time-lapse microscopy. Representative images of Vero cells 6 h pt with pEGFP3Awt are shown in Fig.1C. The *in vivo* record showed a pattern of fluorescence with puncta located predominantly in the perinuclear region as well as a diffuse fluorescence dispersed in the cytoplasm. Tracking of punctate structures revealed a confined movement pattern represented in the trajectories drawn in Fig. 1D, and the velocity of the fluorescence puncta ranged from 0.1 to 0.7 μm/s. This confined pattern is different from those described for other viral proteins associated to microtubules that usually move over longer distances [32,44].

Cellular dynamics of GFP3A protein

Time-lapse imaging of proteins with a restricted mobility has to face limitations imposed by long time and repetitive expositions of cells to the excitation light that are detrimental for cellular viability. Consequently, FRAP was used to further investigate the

movement of GFP3A in the cytoplasm of cells at different times pt. As shown in Fig. 2, a fraction of GFP3A recovered fluorescence after photobleaching indicating that, when transiently expressed, 3A protein shows a mobile (Mf) and a non- mobile fraction. From the images in Fig. 2A-B a Mf of 35±15% was determined at 24 h pt. Higher Mf values were found at shorter times pt, being of 71±11% at 6 h pt (Fig 2C). Most FRAP experiments are usually conducted between 16 to 36 h pt to ensure the correct expression of the transiently expressed protein, to have enough bright specimens, as well as to minimize overexpression artifacts [45]; for this reason further FRAP analysis were performed at 24 h pt.

As commented in the Introduction, FMDV 3A protein has been associated to the ER, so we analysed the localization of GFP3A by confocal microscopy. As reported for transiently expressed 3A [22], our fusion protein colocalized with the ER marker calreticulin (Fig. 3A), confirming the association of GFP3A with this organelle. It is well documented that membranes and luminal spaces of the ER are normally continuous throughout the cell and that rough and smooth ER form an interconnected membrane system [31,46,47], which has been confirmed by FLIP [30,48]. Therefore, we decided to repetitively photobleach a selected fluorescent area in the cytoplasm to measure the fluorescence loss in photobleaching in cells transfected with GFP3A. A mouse IgH leader sequence-derived ER targeted mRFP1 version, termed IgLdR1kdel [32], was used as control of a ER-resident protein. As expected for a protein located at a continuous compartment, repetitive photobleaching of a region in the cytoplasm led to extinction of IgLdR1kdel fluorescence in the whole cell (Fig. 3B–C). A similar behavior was found in cells expressing GFP3A (Fig. 3D–E), supporting that this protein is also placed in a continuous cytoplasmic compartment, which is compatible with its localization at the ER.

Analysis of different mutants of GFP3A protein

Based on the predicted structure of FMDV 3A [9], different mutations were introduced to analyze their effect on the properties of this protein (Fig. 4A–B). These mutations included: i) replacements L38E and L41E that had been shown to destabilize the hydrophobic interaction between residues involved in 3A dimerization, ii) substitution of the odd cysteine present in 3A and located in the HR (replacement C65S), iii) a truncated protein lacking the N-terminal region of 3A (ΔN-ter), and iv) truncated proteins lacking the C-terminal region and maintaining (ΔC-ter) or not (ΔHR-C-ter) the HR. The 3ABBBwt precursor was also included in this study. All constructs were fused to the C-terminus of GFP, and their expression was confirmed by Western blot analysis as a band of the expected electrophoretic mobility was found for each construction (Fig. 4C). In addition, no major differences in the distribution of the fusion proteins were observed by fluorescence microscopy of Vero cells transfected with each of the plasmids, with the exception of GFP3ABBB whose distribution was similar to that reported for 3ABBB [21], and of GFP3AΔHR-C-ter whose fluorescence was accumulated in the cytoplasm of cells that showed an altered morphology and pyknotic nucleus (Fig. 4D). For this reason GFP3AΔHR-C-ter was not included in the subsequent *in vivo* analysis. The trend of GFP3AΔC-ter to appear accumulated in the cytoplasm and the cell alterations associated to its expression were of lower magnitude that those observed for GFP3AΔHR-C-ter; therefore, the properties of GFP3AΔC-ter were further analyzed.

The different 3A mutants constructed were analyzed in FRAP experiments and their Mfs compared with that of GFP3A (Fig. 4E). With the exception of replacement L38E at the dimerization interface that did not alter the mobile fraction of

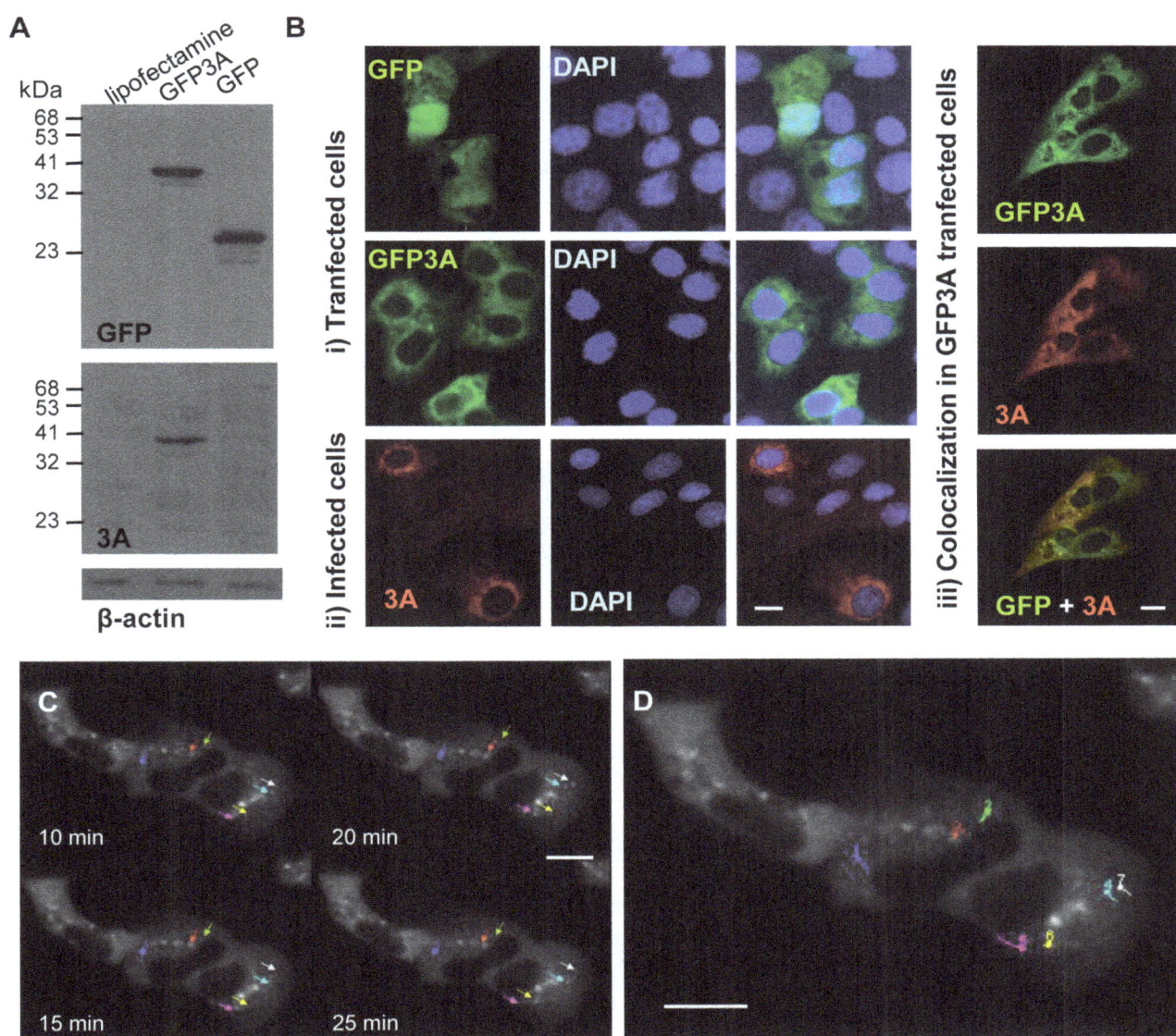

Figure 1. Expression of GFP3A and *in vivo* analyses of 3A protein in transfected cells. A) Vero cells were transfected with 1 µg of pEGFP3A. Proteins were detected by Western blotting with a polyclonal Ab to 3A (479) or a MoAb to GFP as primary antibodies. Blotting to β-actin was used as control of protein loading. Molecular weights are indicated in kDa. B) Fluorescence microscopy of: i) Vero cells transiently expressing GFP or GFP3A; ii) IBRS cells 4 h pi with FMDV; iii) Vero cells transiently expressing GFP3A (24 h pt) incubated with a polyclonal Ab to 3A (346) (red) or showing the autofluorescence of GFP (green). Co-localization is shown in the merge image. Cell nuclei were counterstained with DAPI. C) For time-lapse microscopy, Vero cells were transfected with pEGFP3A and 6 h later cells were scanned by 488-line laser every 5 min for 3 h, as described in Materials and Methods. Images at different times are shown. Colored arrows point to selected tracked dots. D) Manual tracking of the fluorescence of selected dots was performed using ImageJ plug-in. Scale bar, 20 µm.

the protein, the remaining single mutations resulted in a significant increase of Mfs values. Such increase was also observed in the deletion mutants GFP3AΔC-ter and GFP3AΔN-ter analyzed. Thus, different 3A mutations can alter the interactions responsible for the mobility observed for GFP3A.

To gain insight on the interactions established by 3A with cellular membranes, biochemical analyses were performed with cells transiently expressing the 3A fusion proteins. The solubility of the GFP3A, GFP3ABBB and the mutant fusion proteins was analyzed from supernatants (soluble fraction) and pellets (membrane-associated insoluble fraction) recovered after centrifugation of lysates from transfected cells. The presence of 3A was revealed

by immunoblotting using MoAb to 3A and to GFP. In Fig. 5A the relative percentage of the Ab staining intensity in the soluble and the insoluble fractions is represented for each fusion protein; in these analyses a non-fused 3A wt protein (pRSV3A), as well as GFP were included. A similar proportion, about 60%, of the cellular pools of FMDV GFP3A and 3A was detected in the soluble fraction while the remaining protein was found in the membrane fraction supporting the partial association of both proteins with intracellular membranes. As observed with the substitution of the odd cysteine C65S, replacements 3AL38E and 3AL41E that diminish 3A dimerization, did not alter the protein solubility. In the case of the precursor GFP3ABBB a slightly

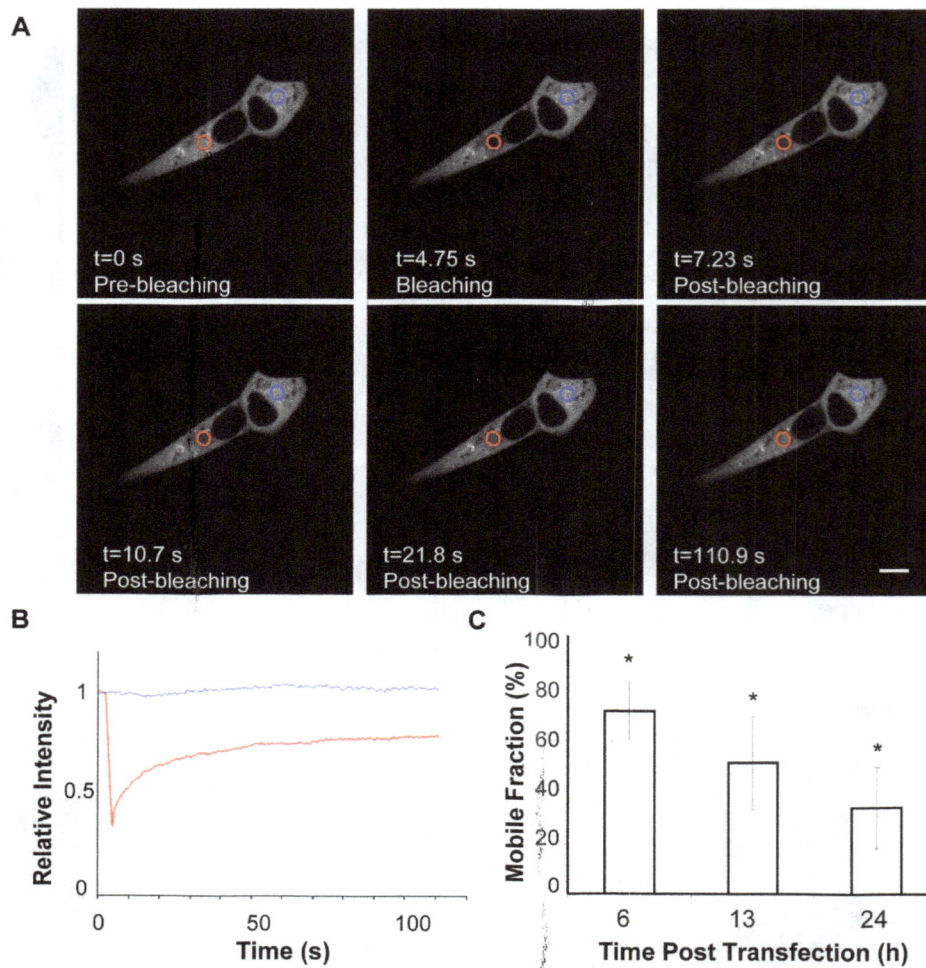

Figure 2. FRAP analysis. Vero cells were transfected with pEGFP3A. A) Images of transfected cells pre bleaching, bleaching and post bleaching. At different times pt a ROI of 10 μm Ø circular region (red circle) was photobleached. Recovery of the fluorescence was monitored by continuous scanning the whole cell (including ROI). The area selected as control in the neighbor cell is indicated by a blue circle. Experiments were done in triplicate (n>10). B) Relative intensity vs. time in FRAP determined in (A). C) Percentages of the GFP3A mobile fraction, determined as described in Materials and Methods, at different times pt. Data are presented as means ± the standard deviations. An asterisk denotes statistically significant differences (P≤0.005). Scale bar, 20 μm.

increase was observed in the proportion of protein recovered in the soluble fraction, although values were not statistically significant. Among the deletion mutants, GFP3AΔC-ter showed the highest decrease in solubility, while GFP3AΔN-ter, the 3A version with the N-terminus truncated in the predicted dimerization region, was the most soluble of the proteins analyzed. These results suggest that N-ter contributes to 3A insolubility although the single replacements that impair dimerization did not significantly affect the solubility of this protein.

Biochemical characterization of the interaction of 3A with cell membranes

Peripheral and integral membrane proteins differentially respond to treatments with high salt, high pH, or chaotropic reagents such as guanidine or urea that will dissociate peripheral membrane proteins from the lipid bilayer [49,50]. In contrast, interaction of integral membrane proteins with the lipid bilayer is much stronger than that of peripheral membrane proteins, requiring the use of detergents for its membrane extraction [15].

Given that about half of 3A was found associated to membranes in transfected cells, the subcellular membrane fractions including GFP3A and its precursor GFP3ABBB were subjected to further biochemical treatments to characterize its interactions with cell membranes (Fig. 5B). Pellets of lysed transfected cells were either dissolved in PBS or treated with mild chaotropic salt conditions (4 M urea), high pH (0.1 M Na_2CO_3) or high salt concentration (1 M KCl), as described [11,51], as well as with a non-ionic detergent (0.5% Triton-X 100), prior to a second fractionation by centrifugation. The effect of these treatments on calnexin was analyzed, as a control for the behavior of an integral membrane protein. As expected statistically significant increases in calnexin solubility were only observed upon treatment of pellets with triton X-100. The solubility of GFP3A and GFP3ABBB differed from that of calnexin. Upon centrifugation the solubility in PBS of GFP3A and GFP3ABBB, was of about 50%, a value similar to that obtain for 3A and GFP3A in the first fractionation, suggesting a dynamic equilibrium between membrane-associated and soluble GFP3A that pulls protein from the membrane when the initial soluble protein is removed. Furthermore, no statistically significant

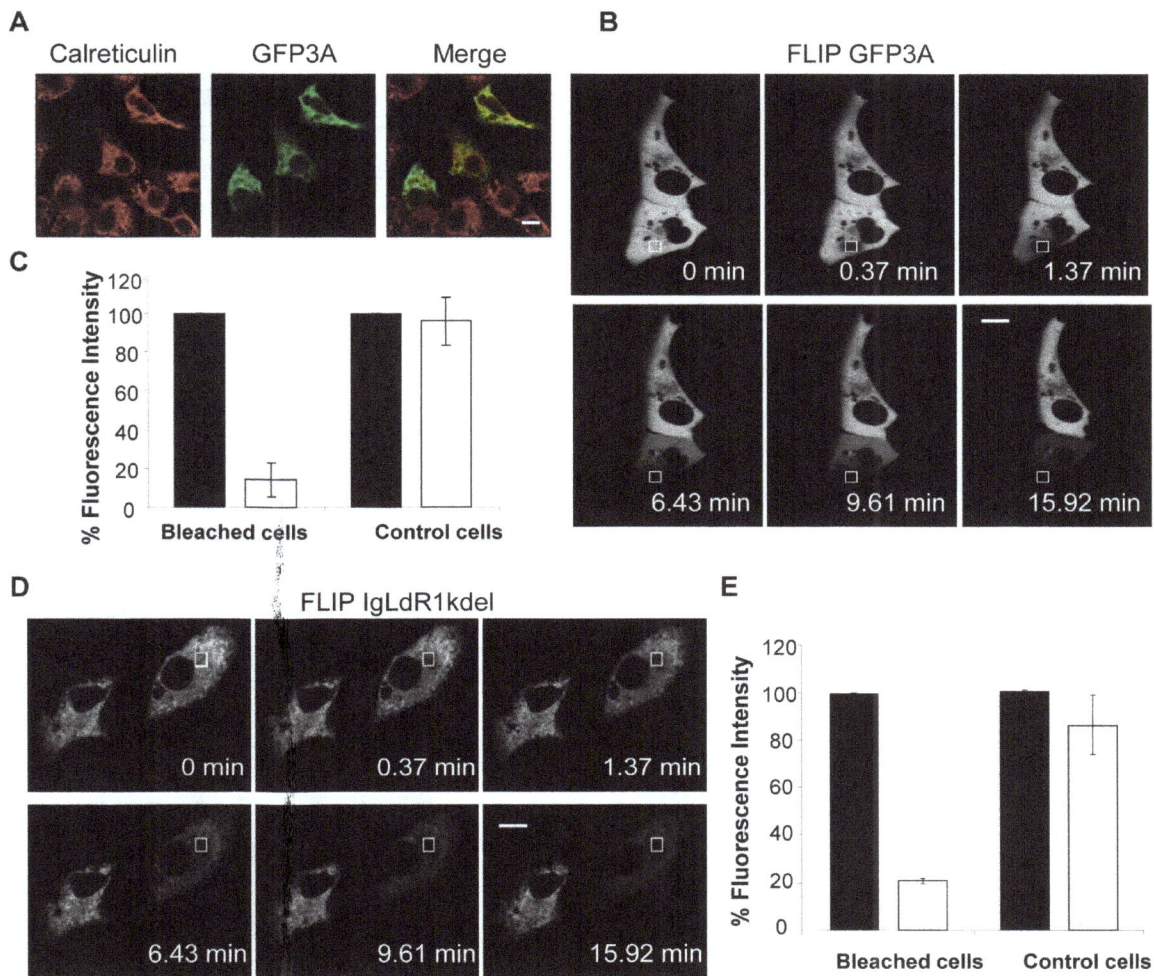

Figure 3. Distribution of GFP3A in the cytoplasm of transfected cells. A) Colocalization of GFP3A and calreticulin in Vero cells transfected with pEGFP3A. B) FLIP analysis; images of pre and post bleached cells. Vero Cells were transfected with pEGFP3A and 24 h pt the indicated area (white rectangle) was subjected to sequential photobleaching. An image of the field was acquired after each bleaching event to determine the loss of fluorescence in the cytoplasm of the cell. C) The percentage of the fluorescence intensity determined in B is represented for bleached and neighbor control (not bleached) cells. Mean fluorescence intensities of prebleached events (black bars) and after all the bleaching repeats (white bars) are indicated. FLIP analysis, as in (B), of cells transfected with pIgLdR1KDEL-RFP. E) Percentage of the fluorescence intensity determined in (D). Scale bar, 20 μm.

increases in solubility were observed with any of the treatment conditions tested. These results indicate that 3A and its precursor 3ABBB interacts with cellular membranes in a manner different from that of an integral membrane protein.

Lack of interaction of 3A with lipid rafts

Since the previous results showed that 3A protein could be associated with lipid membranes, the possible involvement in this interaction of cholesterol enriched micro domains of lipid rafts was analyzed. The association of different viral proteins to lipid rafts has been described for many viruses [52,53,54,55,56]. Detergent-resistant membrane (DRM) fractions were isolated by sucrose density gradient from Vero and BHK-21 cells transfected with pEGFP3A (Fig. 6). Western-blot analysis showed no overlapping between the fractions detected by the anti-GFP antibody and those stained with an anti-caveolin 1 antibody (present in lipid rafts), confirming that 3A is not associated with lipid rafts.

Membrane topology of 3A protein

Characterization of membrane topology of viral proteins contributes to understand the structural organization of viral replication complexes in infected cells. Glycosylation assays can provide information on membrane topology of proteins associated with the ER [57]. To address the membrane topology of FMDV 3A wt, a glycosylation acceptor site was fused in-frame to the C- or N-termini of 3A, as well as to the N-terminus of the HR in a construction with the N-terminus deleted (Fig. 7A). None of the proteins expressed were found glycosylated under the assay conditions used. As expected, glycosylation was observed for dengue virus NS4A protein carrying the same glycosylation acceptor, used as positive control [38] (Fig. 7B), supporting that both 3A protein termini are located towards cytosol.

To confirm the topology suggested by the glycosylation results, a biochemical protease protection assay [58] was performed. Thus, cells transiently expressing 3A were treated with trypsin or proteinase K and analyzed, using antibodies to the N- and the

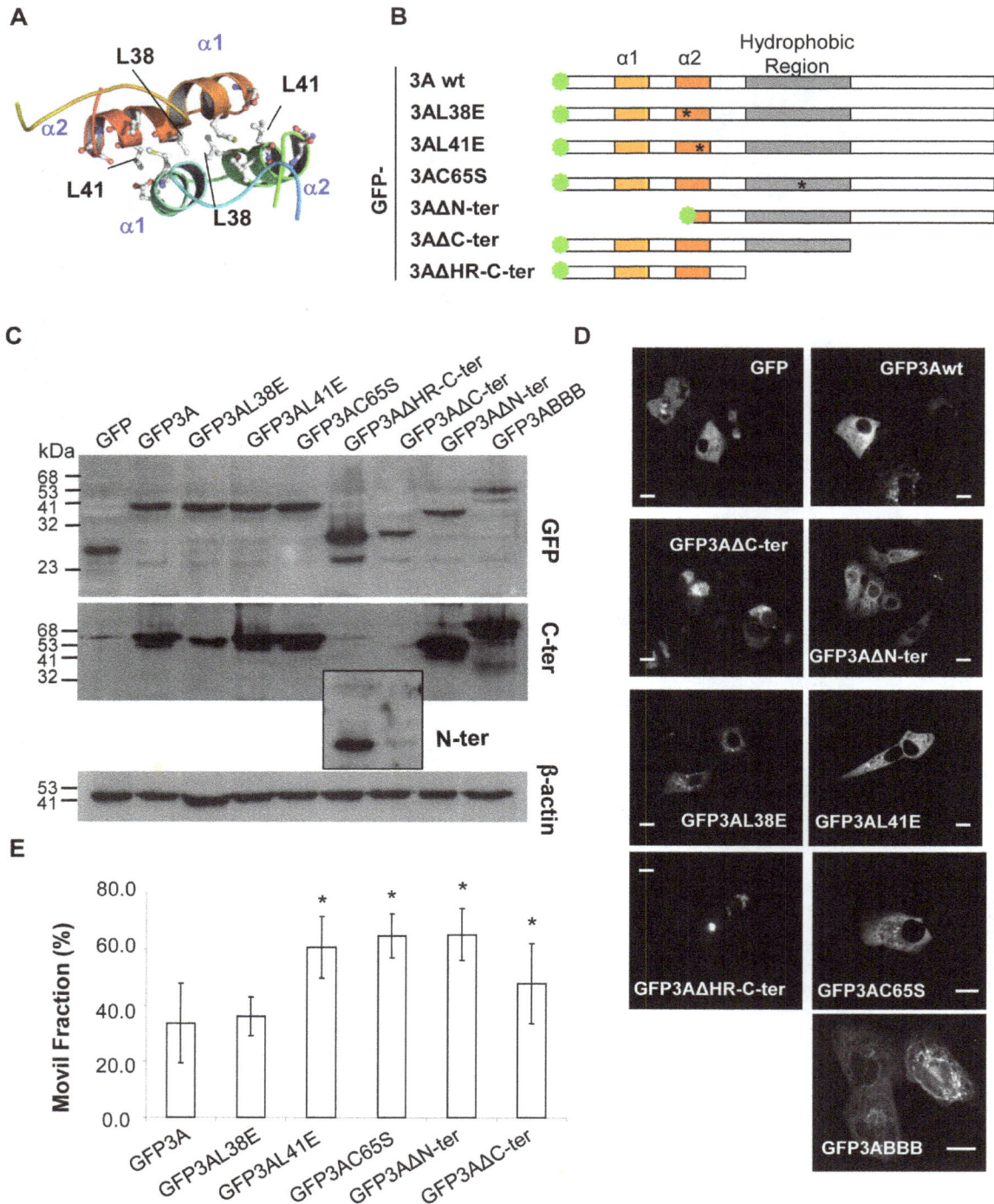

Figure 4. Analysis of GFP fusion proteins carrying mutations in 3A. A) Structural model for FMDV 3A protein dimer (9). Ribbons represent α-helixes 1 and 2. Leucines at positions 38 and 41 are indicated. B). Schematic representation of the fusion proteins analyzed in which GFP (green stars), α-helixes (orange boxes) and the hydrophobic region (gray boxes) are indicated. Substitutions (L38E, L41E and C65S at residues conserved in 99, 99 and 85% among the FMDV sequences from the NCBI database, respectively) and deletions − ΔN-ter (I1-L41), ΔC-ter (R82-E153) and ΔHR-C-ter (K53-E153) −, generated as described in Materials and Methods, are shown. Asterisks denote single replacements. An alignment of the 3A sequences spanning the different mutations constructed among different FMDV serotypes can be found at [9] C) Western blotting of cells transiently expressing fusion proteins. Vero cells were transfected with 1 μg of plasmids expressing the fusion proteins indicated. Proteins were detected by incubation with a primary polyclonal antibody to the C-terminus (346) − with the exception of GFPΔHR-C-ter and GFPΔC-ter (shown boxed) that were blotted with serum 443 to the N-terminus − or with a MoAb to GFP. Blotting to an anti-β-actin was used as control of protein loading. Molecular weights are indicated in kDa. D) Fluorescent pattern of different GFP fusion proteins. Vero cells were fixed 24 h pt with the plasmids expressing the fusion proteins indicated were fixed and processed for confocal microscopy as described in Materials and Methods. E) Comparison of mobile fractions in FRAP of GFP3A fusion proteins. Vero cells were transfected as in (D), and 24 h pt FRAP was determined as described in the legend of Fig. 5. Data are presented as means ± the standard deviations of triplicate experiments (n>10). Statistically significant differences relative to GFP3A percentage of mobile fraction are indicated by an asterisk (P≤0.001). Scale bar, 20 μm.

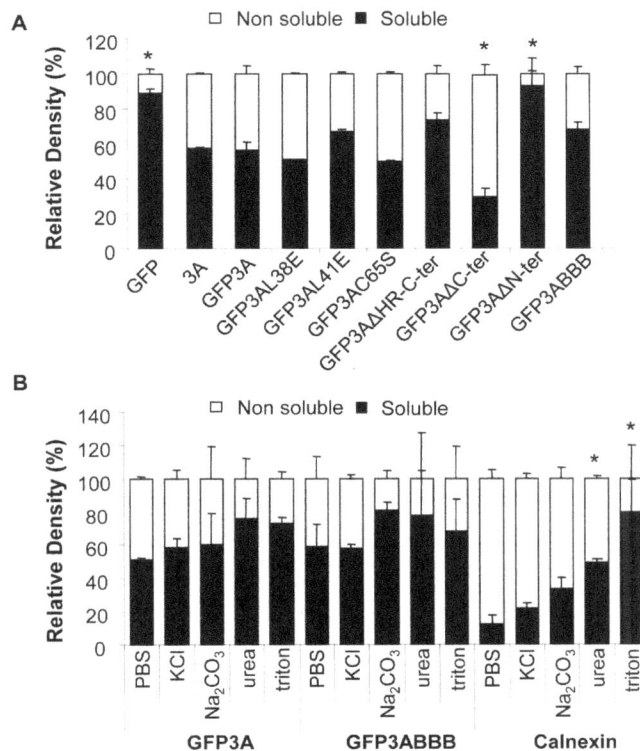

Figure 5. Solubility of GFP3A fusion proteins. A) Distribution of fusion proteins in soluble or insoluble fractions of transfected cells. Vero cells transfected with 1 µg of pEGFP, pRSV3A and the plasmids expressing the fusion proteins indicated, were lysed in PBS buffer by freeze–thawing and fractionated by centrifugation. Proteins in pellets and supernatants were resolved on a 12% SDS-PAGE, transferred to a membrane, and blotted with MoAb to GFP or 3A (2C2). Statistically significant differences, relative to GFP3A are indicated by * (P≤0.05). B) Insoluble fraction association of transiently expressed GFP3A and GFP3ABBB proteins. Vero cells transfected with 1 µg of pEGFP3A or pEGFP3ABBB, were processed as in (A). Pellets were further treated with: Na_2CO_3, Urea, KCl or Triton X-100 (as described in Materials and Methods) and their proteins blotted with MoAb to GFP or to calnexin. Plots represent the percentage of the relative intensity of the protein bands in the blot that were quantified by densitometry with ImageJ program. Statistically significant differences, relative to PBS treatment, are indicated by * (P≤0.05).

C-termini of the protein, by immunofluorescence (Fig. 7C) as well as by Western blotting (Fig. 7D) [57,58]. The results revealed that both termini of the protein were proteolyzed indicating that the N- and the C-termini of 3A are accessible to the enzymes and, therefore, oriented towards the cytosol. Indeed, no proteolysis was observed for calreticulin, a protein that resides in the protease-protected ER lumen.

The results obtained led us to propose a membrane topology model (Fig. 7E) in which 3A protein interacts with ER membranes through its hydrophobic stretch, while its N- and C-terminus face the cytosol being accessible to other viral proteins for viral replication.

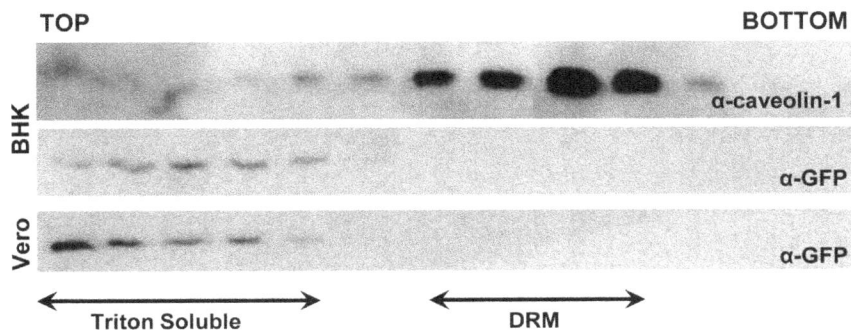

Figure 6. Lack of association of 3A with membranes rich in lipid rafts. Cells were transfected with pEGFP3A and 24 h later lysed with cold 0.5% Triton X-100 in TNE Buffer and the rafts were purified by density gradient fractionation, top and bottom are indicated. Detergent resistant membrane fractions (DRM) are indicated. Proteins in the different fractions were resolved on a 12% SDS-PAGE and blotted with a polyclonal antibody to caveolin-1 and a MoAb to GFP.

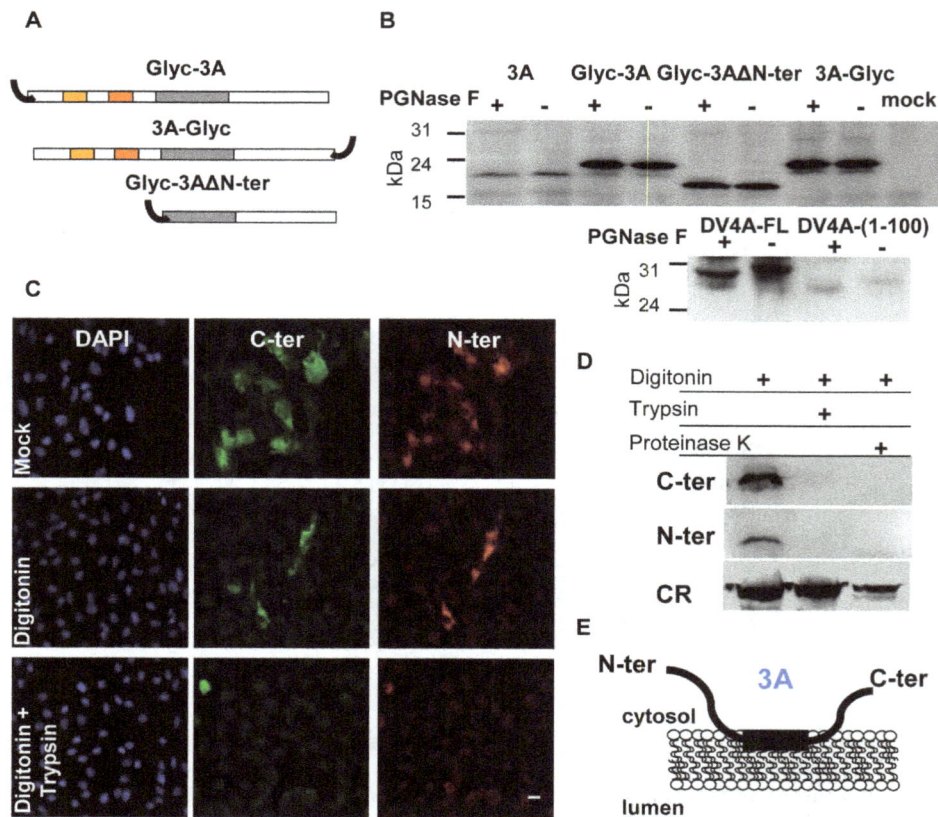

Figure 7. Membrane topology of 3A. A) Schematic representation of fusion proteins of the complete 3A and 3AΔN-ter, with the glycosylation acceptor site Asn-Ser-Thr-Ser-Ala-Asn (black curve lines). For 3A, α-helixes (orange boxes) and the hydrophobic region (gray boxes) are indicated. B) Deglycosylation assay of transiently expressed fusion proteins. Vero Cells were transfected with pcDNAGlyc-3A, 3A-Glyc or Glyc-3AΔN-ter, and 24 h later cells were lysed in NPB and PNGase F treated for 1 h at 37°C. Proteins were separated in 12% SDS-PAGE and blotted with a polyclonal antibody to the C-terminus of 3A (346). As positive control for glycosylation cells previously infected with vaccinia T7/F3 were transfected with pTM-DV4AFL(1–150)-eGFP-Glyc or with pTM-DV4A(1–100)-eGFP-Glyc and processed as before using a MoAb to GFP. C and D) Biochemical protease protection assay of transiently expressed 3A. C) Vero cells grown on coverslips were transfected with pcDNA3A and 24 h later were permeabilized with digitonin for 1 min, treated with trypsin for 5 min and fixed in PFA 4% after proteolysis. Cells were analyzed by immunofluorescence with a MoAb to the C-terminus (2C2) and a polyclonal Ab to the N-terminus (479) of 3A. Alexa fluor 488 anti-mouse and alexa fluor 555 anti-rabbit were used as secondary antibodies. D) Vero cells were transfected and processed as in (C). Cells were lysed and analyzed by Western blotting using polyclonal antibodies to the C- (346) and the N-termini (479) of 3A and to calreticulin (CR). E) Schematic representation of the model proposed for the membrane topology of 3A protein. Cytosol and lumen are indicated. Scale bar, 20 µm.

Discussion

Replication of positive strand RNA viruses is intimate associated with membranes, which confers advantages not only for viral replication but also in protecting viral RNA from sensing by cell pattern-recognition receptors and the subsequent triggering of innate immunity [59]. The FMDV NS protein 3A is involved in the host range, pathogenicity and virulence of the virus [24,25,26,60] and exhibits properties and characteristics different from those of other picornaviruses. It is thus interesting to gain insight on the function and properties of this "key-protein" involved in FMDV replication.

The picornavirus replication cycle occurs in the cell cytoplasm [61,62]. Replication complexes appear associated to virus-recruited membrane structures to which NS proteins anchor [63,64,65]. In this context, the data available for picornaviruses point to 3A as a multifunctional NS protein [66].

Characterization of protein dynamics in the cell may contribute to understand the different functional roles of viral proteins. To this end, we have *in vivo* studied the properties of GFP fusions with FMDV 3A wt protein and with mutant versions including point mutations that either destabilize dimer formation (L38E and L41E) or impair the establishment of disulfide intermolecular bonds in the odd cysteine residue present in 3A (C65S), as well as deletions corresponding to the N-terminal or the C-terminal regions of 3A protein.

FMDV GFP3A was correctly expressed and its fluorescence displayed a punctuated perinuclear distribution similar to that described for 3A wt in FMDV infected cells [21,22]. The movements of the GFP3A fluorescent puncta, revealed by time-lapse microscopy, showed a confined track in the cytoplasm. This pattern is different to that associated with microtubules and could be related to the location of the protein in association with the membranes involved in formation of the replication complex. Alterations in the distribution of microtubules and intermediate filaments components have been described in FMDV-infected cells, being 3C(pro) the only FMDV protein involved in these changes [67].

The study by FRAP analysis of the inner dynamics of the sites where GFP3A resides in the cytoplasm revealed that the mobile fraction of the protein was higher at early times pt (70%) and decreased at 24 h pt (about 35%); this later time was chosen for further analyses as it was considered to better reflect the interaction of mature GFP3A with the cell components. Most of the mutations studied, including deletions on the N- and C-termini and the single replacements L41E and C65S, resulted in an increase in protein mobility. Alterations in the mobility and fluorescent pattern of the C-ter deleted mutant could be related with the lack of the interaction domain of 3A with the cellular protein DCTN3 that has been implicated in the motility of viral proteins and whose deletion attenuates the disease in cattle [68]. The lack of effect of replacement 3AL38E on the protein mobility remains to be explained. Taken together, these results indicate that different mutations can alter the interactions responsible for the mobility observed for GFP3A, suggesting a remarkable complexity in the determinants of 3A cellular dynamics.

In FLIP analyses, fluorescence in one area of the cell is repeatedly bleached while images in a non-bleached region are collected. If fluorescent molecules from any other region of the cell can diffuse into the area being bleached, loss of fluorescence will occur in both ROIs, indicating that the regions are connected and the protein can diffuse between them [28]. FLIP experiments have been used to clarify the extent of continuity of various intracellular membrane systems [29,47]. Here, a monomeric red fluorescent protein targeted to the ER via an immunoglobulin leader sequence and retained in the ER lumen by a KDEL retention signal (IgLdR1kdel) was used as a control of a protein resident in a continuous compartment [32]. When an area in the cytoplasm of cells transiently expressing IgLdR1kdel or GFP3A was repeatedly photobleached, loss of fluorescence was observed to occur in the whole cell cytoplasm, indicating that the protein is located in a continuous compartment. These results, along with the colocalization observed between GFP3A and calreticulin, support an interaction of 3A with the ER, which is consistent with previous data [21,22].

The interactions of FMDV 3A with cell membranes are poorly understood. In this work we also performed a biochemical characterization GFP3A and of different point and deletions mutants of this protein. In cells transiently expressing 3Awt and GFP3A, these proteins were similarly found in the soluble (about 60%) and the insoluble (about 40%) membrane fractions. The partial association of 3A and GFP3A with cell membranes observed is in agreement with previous analysis of 3A in transfected cells [8]. Interestingly, a fraction of GFP3ABBB was also found in the insoluble fraction indicating that the presence of the 3 copies of 3B does not significantly alter ability for membrane interaction of 3A protein. Deletion of the N-terminus of 3A (GFP3AΔN-ter) significantly increased the solubility of GFP3A. Conversely, deletion of the C-terminus (GFP3AΔC-ter) decreased the solubility, which appeared to be associated to a tendency of this fluorescent protein to accumulate in define points of the cytoplasm (Fig. 4D). On the other hand, an increase in GFP3AΔC-ter mobility was found in FRAP. This apparent discrepancy between solubility and mobility could be due to a

bias introduced by the exclusion of cells with highly altered morphology from FRAP analyses.

None of the point mutations analyzed resulted in significant alterations of 3A solubility, suggesting that neither the potential establishment of intermolecular disulfide bridges, nor the efficient 3A dimerization are critical for 3A association to cellular membranes. These results indicate also that residues other than L38 and L41 are likely to be involved in the increase in solubility associated with deletion of the N-terminus of 3A.

Further analyses of the insoluble fraction of GFP3A and GFP3ABBB revealed that high ionic strength and high pH, conditions that favor solubilization of proteins whose binding to membranes mainly depend on electrostatic forces, slightly altered the solubility of GFP3A. The most stringent conditions tested (a chaotropic agent and a non-ionic detergent) enhanced GFP3A solubility, albeit in a non-statistically significant manner. Interestingly, the results obtained with the solubilization treatments demonstrated that GFP3A and GFP3ABBB are not integral membrane proteins, such as calnexin, albeit they can establish strong interactions with intracellular membranes. These results led us to investigate 3A topology by different approaches. The deglycosylation assay used indicated that 3A could display both N- and C-termini towards cytosol, an observation that was confirmed by the protease protection assay. Based on these results, we proposed a model for the 3A membrane topology in which both protein termini are exposed to the cytosol (Fig 7E). This model could be compatible with an infected cell context, where the mature protein and the 3AB precursors would face the cytosol where viral replication takes place and protein-protein interactions expected to occur.

In the model proposed for the interaction of PV 3A/3AB proteins with cell membranes (14), 3A can adopt a transmembrane topology when expressed alone, while its precursor 3AB behaves as a non-transmembrane protein. FMVD 3A differs from the rest of picornaviruses in the length of its C-terminus (66 amino acids longer than PV), which could enable 3A to acquire a cytosolic topology, without the contribution of 3B as required in PV. The non-transmembrane association with intracellular membranes and the display of both protein termini to the cytosol are novel evidences of the differences existing among FMDV 3A and those of other picornaviruses.

Acknowledgments

We thank B. Wölk for IgLdR1kdel plasmid, R. Bartenschlager for pMT-DV4A plasmids, E. Martinez-Salas for vaccinia virus VTF7–3, E. Brocchi for MoAb 2C2, M.A. Alonso and I. Sandoval for fruitful advise, as well as the technical assistance received from the Confocal Microscopy Service from the Centro de Biología Molecular Severo Ochoa.

Author Contributions

Conceived and designed the experiments: MGM MAMA LK FS. Performed the experiments: MGM. Analyzed the data: MGM FS LK MAMA. Contributed reagents/materials/analysis tools: MAMA LK. Wrote the paper: MGM MAMA LK FS.

References

1. Pereira HG (1981) Virus disease of food animals; Gibbs EPJ, editor. London: Academic Press. 333–363 p.
2. Domingo E, Baranowski E, Escarmis C, Sobrino F (2002) Foot-and-mouth disease virus. Comp Immunol Microbiol Infect Dis 25: 297–308.
3. Sobrino F, Saiz M, Jimenez-Clavero MA, Nunez JI, Rosas MF, et al. (2001) Foot-and-mouth disease virus: a long known virus, but a current threat. Vet Res 32: 1–30.
4. Belsham GJ (2005) Translation and replication of FMDV RNA. Curr Top Microbiol Immunol 288: 43–70.

5. Sobrino F, Domingo E (2004) Foot and Mouth disease Current Perspectives. Madrid: Horizon Bioscience.
6. Carrillo C, Tulman ER, Delhon G, Lu Z, Carreno A, et al. (2005) Comparative genomics of foot-and-mouth disease virus. J Virol 79: 6487–6504.
7. Forss S, Strebel K, Beck E, Schaller H (1984) Nucleotide sequence and genome organization of foot-and-mouth disease virus. Nucleic Acids Res 12: 6587–6601.
8. Moffat K, Howell G, Knox C, Belsham GJ, Monaghan P, et al. (2005) Effects of foot-and-mouth disease virus nonstructural proteins on the structure and

function of the early secretory pathway: 2BC but not 3A blocks endoplasmic reticulum-to-Golgi transport. J Virol 79: 4382–4395.

9. Gonzalez-Magaldi M, Postigo R, de la Torre BG, Vieira YA, Rodriguez-Pulido M, et al. (2012) Mutations that hamper dimerization of foot-and-mouth disease virus 3A protein are detrimental for infectivity. J Virol.

10. Choe SS, Kirkegaard K (2004) Intracellular topology and epitope shielding of poliovirus 3A protein. J Virol 78: 5973–5982.

11. Liu J, Wei T, Kwang J (2004) Membrane-association properties of avian encephalomyelitis virus protein 3A. Virology 321: 297–306.

12. Datta U, Dasgupta A (1994) Expression and subcellular localization of poliovirus VPg-precursor protein 3AB in eukaryotic cells: evidence for glycosylation in vitro. J Virol 68: 4468–4477.

13. Doedens JR, Giddings TH Jr, Kirkegaard K (1997) Inhibition of endoplasmic reticulum-to-Golgi traffic by poliovirus protein 3A: genetic and ultrastructural analysis. J Virol 71: 9054–9064.

14. Fujita K, Krishnakumar SS, Franco D, Paul AV, London E, et al. (2007) Membrane topography of the hydrophobic anchor sequence of poliovirus 3A and 3AB proteins and the functional effect of 3A/3AB membrane association upon RNA replication. Biochemistry 46: 5185–5199.

15. Towner JS, Ho TV, Semler BL (1996) Determinants of membrane association for poliovirus protein 3AB. J Biol Chem 271: 26810–26818.

16. Knox C, Moffat K, Ali S, Ryan M, Wileman T (2005) Foot-and-mouth disease virus replication sites form next to the nucleus and close to the Golgi apparatus, but exclude marker proteins associated with host membrane compartments. J Gen Virol 86: 687–696.

17. Forss S, Schaller H (1982) A tandem repeat gene in a picornavirus. Nucleic Acids Res 10: 6441–6450.

18. Wimmer E (1982) Genome-linked proteins of viruses. Cell 28: 199–201.

19. Falk MM, Sobrino F, Beck E (1992) VPg gene amplification correlates with infective particle formation in foot-and-mouth disease virus. J Virol 66: 2251–2260.

20. Pacheco JM, Piccone ME, Rieder E, Pauszek SJ, Borca MV, et al. (2010) Domain disruptions of individual 3B proteins of foot-and-mouth disease virus do not alter growth in cell culture or virulence in cattle. Virology 405: 149–156.

21. Garcia-Briones M, Rosas MF, Gonzalez-Magaldi M, Martin-Acebes MA, Sobrino F, et al. (2006) Differential distribution of non-structural proteins of foot-and-mouth disease virus in BHK-21 cells. Virology 349: 409–421.

22. O'Donnell V, Pacheco JM, Henry TM, Mason PW (2001) Subcellular distribution of the foot-and-mouth disease virus 3A protein in cells infected with viruses encoding wild-type and bovine-attenuated forms of 3A. Virology 287: 151–162.

23. Midgley R, Moffat K, Berryman S, Hawes P, Simpson J, et al. (2013) A role for endoplasmic reticulum exit sites in foot-and-mouth disease virus infection. J Gen Virol 94: 2636–2646.

24. Nunez JI, Baranowski E, Molina N, Ruiz-Jarabo CM, Sanchez C, et al. (2001) A single amino acid substitution in nonstructural protein 3A can mediate adaptation of foot-and-mouth disease virus to the guinea pig. J Virol 75: 3977–3983.

25. Beard CW, Mason PW (2000) Genetic determinants of altered virulence of Taiwanese foot-and-mouth disease virus. J Virol 74: 987–991.

26. Pacheco JM, Henry TM, O'Donnell VK, Gregory JB, Mason PW (2003) Role of nonstructural proteins 3A and 3B in host range and pathogenicity of foot-and-mouth disease virus. J Virol 77: 13017–13027.

27. Strauss DM, Glustrom LW, Wuttke DS (2003) Towards an understanding of the poliovirus replication complex: the solution structure of the soluble domain of the poliovirus 3A protein. J Mol Biol 330: 225–234.

28. Lippincott-Schwartz J, Snapp E, Kenworthy A (2001) Studying protein dynamics in living cells. Nat Rev Mol Cell Biol 2: 444–456.

29. White J, Stelzer E (1999) Photobleaching GFP reveals protein dynamics inside live cells. Trends Cell Biol 9: 61–65.

30. Dundr M, Misteli T (2003) Measuring dynamics of nuclear proteins by photobleaching. Curr Protoc Cell Biol Chapter 13: Unit 13 15.

31. Nehls S, Snapp EL, Cole NB, Zaal KJ, Kenworthy AK, et al. (2000) Dynamics and retention of misfolded proteins in native ER membranes. Nat Cell Biol 2: 288–295.

32. Wolk B, Buchele B, Moradpour D, Rice CM (2008) A dynamic view of hepatitis C virus replication complexes. J Virol 82: 10519–10531.

33. De Castro MP (1964) Behaviour of the foot-and-mouth disease virus in cell cultures: susceptibility of the IB-RS-2 cell line. Arq Inst Biol Sao Paulo 31: 63–78.

34. Sobrino F, Davila M, Ortin J, Domingo E (1983) Multiple genetic variants arise in the course of replication of foot-and-mouth disease virus in cell culture. Virology 128: 310–318.

35. Toja M (1997) Caracterización molecular de un virus de la fiebre aftosa y de sus derivados persistentes. Construcción de un clon infeccioso. PhD thesis, Universidad Autónoma de Madrid.

36. Martin-Acebes MA, Herrera M, Armas-Portela R, Domingo E, Sobrino F (2010) Cell density-dependent expression of viral antigens during persistence of foot-and-mouth disease virus in cell culture. Virology 403: 47–55.

37. Laemmli UK (1970) Cleavage of structural proteins during the assembly of the head of bacteriophage T4. Nature 227: 680–685.

38. Miller S, Kastner S, Krijnse-Locker J, Buhler S, Bartenschlager R (2007) The non-structural protein 4A of dengue virus is an integral membrane protein inducing membrane alterations in a 2K-regulated manner. J Biol Chem 282: 8873–8882.

39. Martinez-Salas E, Saiz JC, Davila M, Belsham GJ, Domingo E (1993) A single nucleotide substitution in the internal ribosome entry site of foot-and-mouth disease virus leads to enhanced cap-independent translation in vivo. J Virol 67: 3748–3755.

40. Brown DA, Rose JK (1992) Sorting of GPI-anchored proteins to glycolipid-enriched membrane subdomains during transport to the apical cell surface. Cell 68: 533–544.

41. Virtanen SS, Sandholm J, Yegutkin G, Väänänen KH, Härkönen PL (2010) Inhibition of GGTase-I and FTase disrupts cytoskeletal organization of human PC-3 prostate cancer cells. 815–826 p.

42. Wouters FS, Verveer PJ, Bastiaens PIH (2001) Imaging biochemistry inside cells. Trends in Cell Biology 11: 203–211.

43. Krogh A, Larsson B, von Heijne G, Sonnhammer EL (2001) Predicting transmembrane protein topology with a hidden Markov model: application to complete genomes. J Mol Biol 305: 567–580.

44. Bohm KJ, Stracke R, Unger E (2003) Motor proteins and kinesin-based nanoacutaoric devices. Tsitol Genet 37: 11–21.

45. Snapp EL, Altan N, Lippincott-Schwartz J (2003) Measuring protein mobility by photobleaching GFP chimeras in living cells. Curr Protoc Cell Biol Chapter 21: Unit 21 21.

46. Verkman AS (2002) Solute and macromolecule diffusion in cellular aqueous compartments. Trends Biochem Sci 27: 27–33.

47. Cole NB, Smith CL, Sciaky N, Terasaki M, Edidin M, et al. (1996) Diffusional mobility of Golgi proteins in membranes of living cells. Science 273: 797–801.

48. Snapp EL (2009) Fluorescent proteins: a cell biologist's user guide. Trends Cell Biol 19: 649–655.

49. Gilmore R, Blobel G (1985) Translocation of secretory proteins across the microsomal membrane occurs through an environment accessible to aqueous perturbants. Cell 42: 497–505.

50. Fujiki Y, Hubbard AL, Fowler S, Lazarow PB (1982) Isolation of intracellular membranes by means of sodium carbonate treatment: application to endoplasmic reticulum. J Cell Biol 93: 97–102.

51. Tershak DR (1984) Association of poliovirus proteins with the endoplasmic reticulum. pp. 777–783.

52. Bhattacharya B, Roy P (2010) Role of lipids on entry and exit of bluetongue virus, a complex non-enveloped virus. Viruses 2: 1218–1235.

53. Hogue IB, Grover JR, Soheilian F, Nagashima K, Ono A (2011) Gag induces the coalescence of clustered lipid rafts and tetraspanin-enriched microdomains at HIV-1 assembly sites on the plasma membrane. J Virol 85: 9749–9766.

54. Lu Y, Liu DX, Tam JP (2008) Lipid rafts are involved in SARS-CoV entry into Vero E6 cells. pp. 344–349.

55. Matto M, Rice CM, Aroeti B, Glenn JS (2004) Hepatitis C virus core protein associates with detergent-resistant membranes distinct from classical plasma membrane rafts. J Virol 78: 12047–12053.

56. Rossman JS, Lamb RA (2011) Influenza virus assembly and budding. Virology 411: 229–236.

57. van Geest M, Lolkema JS (2000) Membrane topology and insertion of membrane proteins: search for topogenic signals. Microbiol Mol Biol Rev 64: 13–33.

58. Lorenz H, Hailey DW, Lippincott-Schwartz J (2006) Fluorescence protease protection of GFP chimeras to reveal protein topology and subcellular localization. Nat Methods 3: 205–210.

59. Belov GA, Nair V, Hansen BT, Hoyt FH, Fischer ER, et al. (2012) Complex dynamic development of poliovirus membranous replication complexes. J Virol 86: 302–312.

60. Nunez JI, Molina N, Baranowski E, Domingo E, Clark S, et al. (2007) Guinea pig-adapted foot-and-mouth disease virus with altered receptor recognition can productively infect a natural host. J Virol 81: 8497–8506.

61. Bienz K, Egger D, Pfister T, Troxler M (1992) Structural and functional characterization of the poliovirus replication complex. J Virol 66: 2740–2747.

62. Suhy DA, Giddings TH Jr, Kirkegaard K (2000) Remodeling the endoplasmic reticulum by poliovirus infection and by individual viral proteins: an autophagy-like origin for virus-induced vesicles. J Virol 74: 8953–8965.

63. Bienz K, Egger D, Pfister T (1994) Characteristics of the poliovirus replication complex. Arch Virol Suppl 9: 147–157.

64. Bienz K, Egger D, Rasser Y, Bossart W (1983) Intracellular distribution of poliovirus proteins and the induction of virus-specific cytoplasmic structures. Virology 131: 39–48.

65. Schlegel A, Giddings TH Jr, Ladinsky MS, Kirkegaard K (1996) Cellular origin and ultrastructure of membranes induced during poliovirus infection. J Virol 70: 6576–6588.

66. Teterina NL, Pinto Y, Weaver JD, Jensen KS, Ehrenfeld E (2011) Analysis of poliovirus protein 3A interactions with viral and cellular proteins in infected cells. J Virol 85: 4284–4296.

67. Armer H, Moffat K, Wileman T, Belsham GJ, Jackson T, et al. (2008) Foot-and-mouth disease virus, but not bovine enterovirus, targets the host cell cytoskeleton via the nonstructural protein 3Cpro. J Virol 82: 10556–10566.

68. Gladue DP, O'Donnell V, Baker-Bransetter R, Pacheco JM, Holinka LG, et al. (2014) Interaction of foot-and-mouth disease virus nonstructural protein 3A with host protein DCTN3 is important for viral virulence in cattle. J Virol 88: 2737–2747.

2

Characterization of the *Neurospora crassa* Cell Fusion Proteins, HAM-6, HAM-7, HAM-8, HAM-9, HAM-10, AMPH-1 and WHI-2

Ci Fu[1], Jie Ao[1], Anne Dettmann[2], Stephan Seiler[2,3], Stephen J. Free[1]*

1 Department of Biological Sciences, SUNY University at Buffalo, Buffalo, New York, United States of America, **2** Institute for Biology II, Albert-Ludwigs University Freiburg, Freiburg, Germany, **3** Freiburg Institute for Advanced Studies (FRIAS), Albert-Ludwigs University Freiburg, Freiburg, Germany

Abstract

Intercellular communication of vegetative cells and their subsequent cell fusion is vital for different aspects of growth, fitness, and differentiation of filamentous fungi. Cell fusion between germinating spores is important for early colony establishment, while hyphal fusion in the mature colony facilitates the movement of resources and organelles throughout an established colony. Approximately 50 proteins have been shown to be important for somatic cell-cell communication and fusion in the model filamentous fungus *Neurospora crassa*. Genetic, biochemical, and microscopic techniques were used to characterize the functions of seven previously poorly characterized cell fusion proteins. HAM-6, HAM-7 and HAM-8 share functional characteristics and are proposed to function in the same signaling network. Our data suggest that these proteins may form a sensor complex at the cell wall/plasma membrane for the MAK-1 cell wall integrity mitogen-activated protein kinase (MAPK) pathway. We also demonstrate that HAM-9, HAM-10, AMPH-1 and WHI-2 have more general functions and are required for normal growth and development. The activation status of the MAK-1 and MAK-2 MAPK pathways are altered in mutants lacking these proteins. We propose that these proteins may function to coordinate the activities of the two MAPK modules with other signaling pathways during cell fusion.

Editor: Michael Freitag, Oregon State University, United States of America

Funding: Funding for this study was provided by grants R01GM078589 and 3R01GM078589-04S1 from National Institutes of Health to SF, by grants SE1054/4-2 and SE1054/6-1 from the Deutsche Forschungsgemeinschaft to SS, funds from UB Foundation, and grant SU-12-08 from UB Graduate Student Association Mark Dimond Research Fund to CF. Funding for the confocal microscope was by grant DBI0923133 from National Science Foundation to SUNY University at Buffalo. Funding for the creation of the single gene deletion library was provided by the grant PO1 GM068087. The funders had no role in study design, data collection and analysis, decision to publish, or preparation of the manuscript.

Competing Interests: The authors have declared that no competing interests exist.

* Email: free@buffalo.edu

Introduction

Cell-to-cell fusion between vegetative cells plays a critical role in the life cycles of the filamentous fungi. The fusion between germinating conidia allows the cells to share resources and helps them to establish a colony [1–4]. As the fungal colony matures, cell fusion is important for the movement of resources throughout the colony, a prerequisite for asexual and sexual development. In the model filamentous fungus, *Neurospora crassa*, cell-to-cell fusion plays an important role during colony establishment, as well as during conidiation (asexual development) and protoperithecium formation (sexual development) [5–7]. During colony establishment, fusion between germinating conidia occurs between specialized cells called conidial anastomosis tubes (CATs), which are morphologically and physiologically distinct from germ tubes [7,8]. Germ tubes are wider and exhibit negative chemotrophic interactions, while CATs are thinner and exhibit chemotrophic attraction towards each other [8]. Mutants that are defective in cell fusion can't form an interconnected hyphal network to support nutrient transport within the colony [1,9]. During the *N. crassa* asexual life cycle, wild type colonies transport nutrients from a vegetative hyphal network into the growing aerial hyphae, which generate conidia (asexual spores). Cell fusion mutants are defective in producing the long aerial hyphae typical of wild type cells. They produce short aerial hyphae, which give a "flat" carpet-like conidiation phenotype. [10]. Cell fusion is also important for the *N. crassa* sexual life cycle. Cell fusion mutants are female sterile, and this may be because the efficient transport of amino acids and other nutrients from a vegetative hyphal network into the developing protoperithecia is needed to support sexual development.

Various groups have defined approximately 50 genes required for cell-cell communication and fusion in *N. crassa* [9–17]. Many of these cell fusion genes encode components of the MAK-1 and MAK-2 mitogen-activated protein kinase (MAPK) signal transduction pathways [1,18–24], which are homologous to the yeast cell wall integrity (CWI) and pheromone response cascades, respectively [25–27]. MAK-2 and HAM-1/SO, a protein of unknown molecular function, display oscillatory recruitment to opposing cell tips during CAT communication, suggesting that the chemotrophic interactions between two CATs are coordinated by the MAK-2/SO Ping-Pong signaling behavior [23,28]. NRC-1

Table 1. Strains used in this study.

Strain	Genotype	Strain source
Wild-type 74	OR23-1 V Mat A	FGSC#2489
Wild-type ORS	SL6 Mat a	FGSC#4200
his-3 A	his-3 Mat A	FGSC#6103
his-3 a	his-3 Mat a	This study
histone-1-gfp	Pccg-1::hH1+-sgfp+::his-3+ Mat A	FGSC#9518
lifeact-rfp	Pccg-1::lifeact-rfp::bar+	FGSC#10592
β-tubulin-gfp	Pccg-1::Bml+-sgfp+::his-3+ Mat A	FGSC#9520
Sgfp	Pccg-1::sgfp::his-3+ Mat A	This study
Rfp	Pccg-1::rfp::his-3+ Mat A	This study
grp-sgfp	Pccg-1::grp-sgfp::his-3+ Mat A	This study
rfp-vps-52	Pccg-1::rfp-vps-52::his-3+ Mat A	This study
arg-4-gfp	Pccg-1::arg-4-sgfp::his-3+ Mat A	This study
rfp-vam-3	Pccg-1::rfp-vam-3::his-3+ Mat A	This study
Δham-6 A/a	Δham-6::hygR Mat A/Δham-6::hygR Mat a	FGSC#16993/16903
Δham-6; his-3	Δham-6::hygR, his-3-	This study
Δham-7 A/a	Δham-7::hygR Mat A/Δham-7::hygR Mat a	FGSC#13776/13775
Δham-7; his-3	Δham-7::hygR, his-3-	This study
Δham-8 A/a	Δham-8::hygR Mat A/Δham-8::hygR Mat a	This study/FGSC#17225
Δham-8; his-3	Δham-8::hygR, his-3-	This study
Δham-9 A/a	Δham-9::hygR Mat A/Δham-9::hygR Mat a	This study/FGSC#19549
Δham-9; his-3	Δham-9::hygR, his-3-	This study
Δham-10 A/a	Δham-10::hygR Mat A/Δham-10::hygR Mat a	FGSC#21396/21395
Δham-10; his-3	Δham-10::hygR, his-3-	This study
Δamph-1 A/a	Δamph-1::hygR Mat A/Δamph-1::hygR Mat a	FGSC#12550/12549
Δamph-1; his-3	Δamph-1::hygR, his-3-	This study
Δwhi-2 A/a	Δwhi-2::hygR Mat A/Δwhi-2::hygR Mat a	FGSC#21578/This study
Δwhi-2; his-3	Δwhi-2::hygR, his-3-	This study
mak-1-sgfp	Pccg-1::mak-1-sgfp::bar+	FGSC10299
mak-1-sgfp; Δham-6	Pccg-1::mak-1-sgfp::bar+; Δham-6::hygR	This study
mak-1-sgfp; Δham-7	Pccg-1::mak-1-sgfp::bar+; Δham-7::hygR	This study
mak-1-sgfp; Δham-8	Pccg-1::mak-1-sgfp::bar+; Δham-8::hygR	This study
mak-1-sgfp; Δham-9	Pccg-1::mak-1-sgfp::bar+; Δham-9::hygR	This study
mak-1-sgfp; Δham-10	Pccg-1::mak-1-sgfp::bar+; Δham-10::hygR	This study
mak-1-sgfp; Δamph-1	Pccg-1::mak-1-sgfp::bar+; Δamph-1::hygR	This study
mak-1-sgfp; Δwhi-2	Pccg-1::mak-1-sgfp::bar+; Δwhi-2::hygR	This study
mak-2-sgfp	Pccg-1::mak-2-sgfp::his-3+	This study
mak-2-sgfp; Δham-6	Pccg-1::mak-2-sgfp::his-3+; Δham-6::hygR	This study
mak-2-sgfp; Δham-7	Pccg-1::mak-2-sgfp::his-3+; Δham-7::hygR	This study
mak-2-sgfp; Δham-8	Pccg-1::mak-2-sgfp::his-3+; Δham-8::hygR	This study
mak-2-sgfp; Δham-9	Pccg-1::mak-2-sgfp::his-3+; Δham-9::hygR	This study
mak-2-sgfp; Δham-10	Pccg-1::mak-2-sgfp::his-3+; Δham-10::hygR	This study
mak-2-sgfp; Δamph-1	Pccg-1::mak-2-sgfp::his-3+; Δamph-1::hygR	This study
mak-2-sgfp; Δwhi-2	Pccg-1::mak-2-sgfp::his-3+; Δwhi-2::hygR	This study
so-sgfp	Pccg-1::so-sgfp::his-3+	This study
so-sgfp; Δham-6	Pccg-1::so-sgfp::his-3+; Δham-6::hygR	This study
so-sgfp; Δham-7	Pccg-1::so-sgfp::his-3+; Δham-7::hygR	This study
so-sgfp; Δham-8	Pccg-1::so-sgfp::his-3+; Δham-8::hygR	This study
so-sgfp; Δham-9	Pccg-1::so-sgfp::his-3+; Δham-9::hygR	This study
so-sgfp; Δham-10	Pccg-1::so-sgfp::his-3+; Δham-10::hygR	This study
so-sgfp; Δamph-1	Pccg-1::so-sgfp::his-3+; Δamph-1::hygR	This study

Table 1. Cont.

Strain	Genotype	Strain source
so-sgfp; Δwhi-2	Pccg-1::so-sgfp::his-3+; Δwhi-2::hygR	This study
HA-ham-6; Δham-6	Pham-6::HA-ham-6::his-3+; Δham-6::hygR	This study
HA-ham-7; Δham-7	Pham-7::HA-ham-7::his-3+; Δham-7::hygR	This study
HA-ham-8; Δham-8	Pham-8::HA-ham-8::his-3+; Δham-8::hygR	This study
HA-ham-9; Δham-9	Pham-9::HA-ham-9::his-3+; Δham-9::hygR	This study
HA-amph-1; Δamph-1	Pamph-1::HA-amph-1::his-3+; Δamph-1::hygR	This study
HA-whi-2; Δwhi-2	Pwhi-2::HA-whi-2::his-3+; Δwhi-2::hygR	This study
ham-8-sgfp; Δham-8	Pccg-1::ham-8-sgfp::his-3+; Δham-8::hygR	This study
rfp-ham-8; Δham-8	Pccg-1::rfp-ham-8::his-3+; Δham-8::hygR	This study
rfp-ham-10; Δham-10	Pccg-1::rfp-ham-10::his-3+; Δham-10::hygR	This study
rfp-amph-1; Δamph-1	Pccg-1::rfp-amph-1::his-3+; Δamph-1::hygR	This study

and MEK-2, the upstream MAPKKK and MAPKK in MAK-2 pathway, were also found to have the oscillatory signaling behavior during cell fusion [24]. The *N. crassa* MAK-1 CWI pathway initiates through a set of transmembrane sensors. Signals are integrated by the small GTPase RHO1, which activates a conserved mitogen-activated protein kinase (MAPK) cascade through its interaction with protein kinase C [29–32]. In addition to its function in cell wall stress integration, the CWI pathway is also a central component of the cell-cell communication machinery. The functional relationship between the two signaling pathways during intercellular communication is poorly understood, but evidence for cross-talk between MAK-1 and MAK-2 is provided by work on the striatin interacting phosphatase and kinase (STRIPAK) complex [33,34]. Its subunits, HAM-2, HAM-3, HAM-4, MOB-3, PP2A and PPG-1, are all required for cell-cell communication [9,13,35,36]. Phosphorylation of MOB-3 by MAK-2 is required for nuclear localization of MAK-1 in vegetative hyphae, suggesting that MAK-1-dependent expression of cell fusion genes may be required for establishing cell-cell communication competence.

Among the identified cell fusion genes, there were six genes whose functions were largely uncharacterized. To better understand how cell-to-cell fusion is regulated, these six genes, *ham-6*, *ham-8*, *ham-9*, *ham-10*, *amph-1*, and *whi-2* have been further characterized. The expression patterns, intracellular locations, and how the loss of these genes affects the activation status of the MAK-1 and MAK-2 pathway were examined. A seventh gene, *ham-7*, which has been shown to function as a sensor for the MAK-1 pathway [31], was included in the analysis to examine the cell type expression pattern and cellular location of the HAM-7 sensor. In further characterizing these genes, we demonstrate here that HAM-6, HAM-7 and HAM-8 function as upstream elements in the pathway regulating MAK-1 kinase activity during cell fusion. We further demonstrate that HAM-9, HAM-10, AMPH-1 and WHI-2 are proteins with general functions in regulating *N. crassa* growth, and we suggest that HAM-9, HAM-10, and WHI-2 may provide cross-talk between the two MAP kinase pathways and other signaling cascades during cell fusion.

Materials and Methods

Strains, media and growth conditions

The strains used in this study are listed in Table 1. Wild type *A* (FGSC#2489), wild type *a* (FGSC#4200), *his-3 A* (FGSC#6103),

histone-1-gfp (FGSC#9518) [37], *lifeact-rfp* (FGSC#10592) [38], *β-tubulin-gfp* (FGSC#9520) [37], and *mak-1-gfp* (FGSC#10299) [39] were obtained from Fungal Genetics Stock Center (Kansas City, MO). The *Δham-6* (NCU02767), *Δham-7* (NCU00881), *Δham-8* (NCU02811), *Δham-9* (NCU07389), *Δham-10* (NCU02833), *Δamph-1* (NCU01069) and *Δwhi-2*(NCU10518) strains were obtained from the FGSC *N. crassa* single gene deletion mutant library [40]. All other strains used in this study were either obtained through transformation experiments or mating. The presence of the gene deletion in all of the deletion mutant strains was verified by PCR. The growth media and growth condition for regular strain maintenance, mating, and for conidial anastomosis tube (CAT) formation are available through the FGSC website (www.fgsc.net). The screening procedure to identify cell fusion mutants was performed as previously described [9]. Deletion mutant isolates were further characterized by co-segregation experiments to assess whether the gene deletion (marked by the presence of the hygromycin resistance gene cassette) was responsible for the mutant phenotype [9,40]. Deletion mutants showing co-segregation of hygromycin resistance with the cell fusion mutant phenotypes were verified by complementation experiments.

Plasmid construction and expression of tagged proteins

Plasmids used in this study are listed in Table S1. The primers used to construct GFP-tagged, RFP-tagged and HA-tagged protein constructs are listed in Table S2. GFP and dsRed RFP fusion proteins were generated using the pMF272 and pMF334 vectors [37,41] and were expressed under the control of *ccg-1* promoter at the *his-3* locus [42].

HA-tagged proteins (HA-HAM-6, HA-HAM-8, HA-HAM-9, HA-AMPH-1 and HA-WHI-2) were cloned using the vectors pBM60 and pBM61 [43–45] and were expressed under the control of their own promoters at the *his-3* locus. To identify sites for HA tagging, the protein sequences for HAM-6, HAM-8, HAM-9, AMPH-1 and WHI-2 were used to generate multiple sequence alignments with homologous proteins from other fungi [46]. Protein sequences were also analyzed by the online program Globplot to predict protein structural information [47]. Non-conserved protein sequence regions, which were predicted to be exposed, were chosen for HA tagging. For *ham-6*, *ham-8* and *ham-9*, the HA tag coding sequence was inserted immediately before the stop codon. For *amph-1*, the HA tag coding sequence was inserted between the third and the fourth amino acid codons. For *whi-2*, the peptide sequence VPVDPASGA (amino acid 118–126) was replaced with the HA tag. The cloning of these HA-tagged protein constructs involved two PCR amplification steps. PCR primers were used to amplify approximately 1,500 bp of 5′ UTR sequence, the coding sequence upstream of the HA tag insertion site, and the HA tag coding sequence. A second set of primers were used to amplify the HA tag coding sequence, the rest of the coding sequence of the gene, and the approximately 500 bp 3′ UTR sequence. The two PCR products were then mixed together and used as templates to amplify the entire HA-tagged gene. The primers designed for the two ends of the genes contained an added restriction enzyme site to allow the insertion of the amplified DNA into the multicloning sites of pBM60 or pBM61. An HA-tagged version of HAM-7 with its endogenous promoter was generated by Retrogen Inc. (San Diego, CA). The DNA sequence encoding amino acids 190–198 (YTINILESG), which are located immediately in front of the GPI anchor addition site, were replaced by the HA tag sequence (YPYDVPDYA) in the HA-tagged HAM-7.

Plasmids pgrp-GFP, pRFP-vps-52, parg-4-GFP and pRFP-vam-3 were obtained from the FGSC as fluorescent protein markers for ER, Golgi, mitochondria, and vacuoles respectively [48], and used in co-localization experiments. Plasmids pso-GFP and pmak-2-GFP were kind gifts from Dr. Louise Glass's lab, and were used to study the MAK-2 signal transduction pathway [23].

Analysis of HA-tagged protein expression

To examine the expression of HA-tagged proteins in vegetative hyphae, cultures were grown in 100 ml liquid Vogel's sucrose medium in shaking Erlenmeyer flasks at room temperature for 36 to 48 hours. To examine the expression of HA-tagged proteins in germ tubes and CATs, conidia were used to inoculate 16 ml of Vogel's sucrose medium at a titer of 10^6 per ml and grown in 100×15 mm Petri dishes at 34°C without agitation for 4 hours to allow the formation of germ tubes and CATs [8]. Cells were harvested by filtration using a Büchner funnel and ground in liquid nitrogen. Protein extraction buffer [100 mM Tris/HCL pH 7.4, 1% (w/v) SDS; supplemented with 1X protease cocktail (P-8340 Sigma-Aldrich, St. Louis, MO)] was added, and the cell extracts were collected after centrifugation.

For Western blot analysis of HA-tagged proteins, the protein concentration of the cell extracts was determined by using the DC protein assay kit (BioRad, Hercules, CA). Samples containing 60 μg of protein were subjected to SDS-PAGE and transferred to nitrocellulose membrane. The nitrocellulose membranes were then subjected to Ponceau S red (Sigma-Aldrich, St. Louis, MO) staining to verify equal loading of different protein samples. Western blot experiments with mouse monoclonal anti-HA (Covance, Princeton, NJ) and rabbit anti-mouse IgG-HRP (Sigma-Aldrich, St. Louis, MO) were used to assess the level of protein expression. Chemiluminescent signal was detected by using a ChemiDoc XRS+ System and images were analyzed with Image Lab Software (BioRad, Hercules, CA).

Evaluation of MAK-1 and MAK-2 phosphorylation status

Polyclonal antibody directed against the phosphorylated activation site in common on MAK-1 and MAK-2 was used to evaluate their activation status. To examine the status of the pathways in vegetative hyphae, liquid *N. crassa* cultures were grown at room temperature and harvested by filtration using a Büchner funnel [31]. In some experiments, vegetative cultures were subjected to 8 mM H_2O_2 for 10 minutes just prior to being harvested to determine if the pathways could be activated by oxidative stress. To study the MAK-1 and MAK-2 phosphoryla-

tion status in germ tubes and CATs, conidia were grown for four hours under the conditions described above to allow germ tube and CAT formation. Germlings were harvested by gently scraping them off the culture dishes and collected on a Büchner funnel. The harvested vegetative hyphae samples and germ tubes/CATs samples were ground to a fine powder in a mortar and pestle in liquid nitrogen. Protein extraction for the analysis of the MAK-1 and MAK-2 phosphorylation status was performed as described by Maddi et al. [31].

Preparation of cells expressing HA-tagged proteins for immunolocalization

For each HA-tagged strain, six Petri dishes containing 16 ml of conidial suspension (10^6 conidia/ml) were grown at 34°C for 4 hours to allow germ tube and CAT formation. Cells were harvested by gently scraping them off Petri dishes, and were collected by centrifugation at 6000×g for 5 minutes. Cells were transferred into a microcentrifuge tube and fixed in PBS (phosphate buffered saline) with 3.7% formaldehyde for 15 minutes. After washing with PBS, the samples were incubated for 30 minutes in PBS containing 1 mg/ml of Novozyme 234 (InterSpex Products Inc., Foster City, CA) and 1% bovine serum albumin to digest the cell wall [49,50]. Following the cell wall digestion step, the cells were collected by centrifugation and washed with PBS. Cell samples were then incubated in permeabilization buffer (PBS with 1% BSA and 0.5% Triton-X 100) for 5 minutes to permeabilize the membrane. After the membrane permeabilization step, the cells were collected and washed with PBS. Cell samples were then incubated overnight at 4°C in mouse monoclonal anti-HA primary antibody (Covance, Princeton, NJ) used at a 1:200 dilution in PBS with 1% BSA. After the primary antibody incubation, samples were washed in PBS three times and then incubated for 2 hours in Alexa Fluor 488-conjugated goat anti-mouse secondary antibody (Life Technologies, Carlsbad, CA) used at a 1:100 dilution in PBS with 1% BSA. After three washes, cell samples were used for microscopic observation. All of the incubation buffers and washing buffers used in this assay were supplemented with 1X protease inhibitor cocktail and 1 mM PMSF (Sigma Aldrich, St. Louis, MO).

Live cell imaging for the quantification of CATs

Live-cell imaging was performed using an inverted microscope Diaphot-TMD inverted microscope (Nikon, Japan) to quantify CAT fusion activity for wild type and mutant strains as previously described [8,51]. In brief, 1 ml of fresh conidia at a density of 10^6 per ml were grown in Petri dishes (35×10 mm) at 34°C. Benomyl was added to distinguish CATs from germ tubes. This is because benomyl inhibits the formation of long germ tubes but not CAT formation or fusion [52,53]. After 4 hours of incubation, the cells were examined under the microscope for the presence of CATs and CAT fusion. For each sample, five random images were captured for later quantification of CAT fusion activity. The frequency of cell fusion observed in the mutant samples was compared to the wild type cell fusion frequency to obtain a relative CAT fusion activity for each of the mutants.

Sample preparation and confocal microscopy

For live-cell imaging of CAT formation, an inverted agar block method was adapted for this study [54]. A 1 ml aliquot of conidia expressing GFP-tagged protein or RFP-tagged protein (10^6 conidia per ml) was placed in a fluorodish (World Precision Instruments, FL). A thick agar block was then placed in the middle of the fluorodish to facilitate attachment of the conidia to the

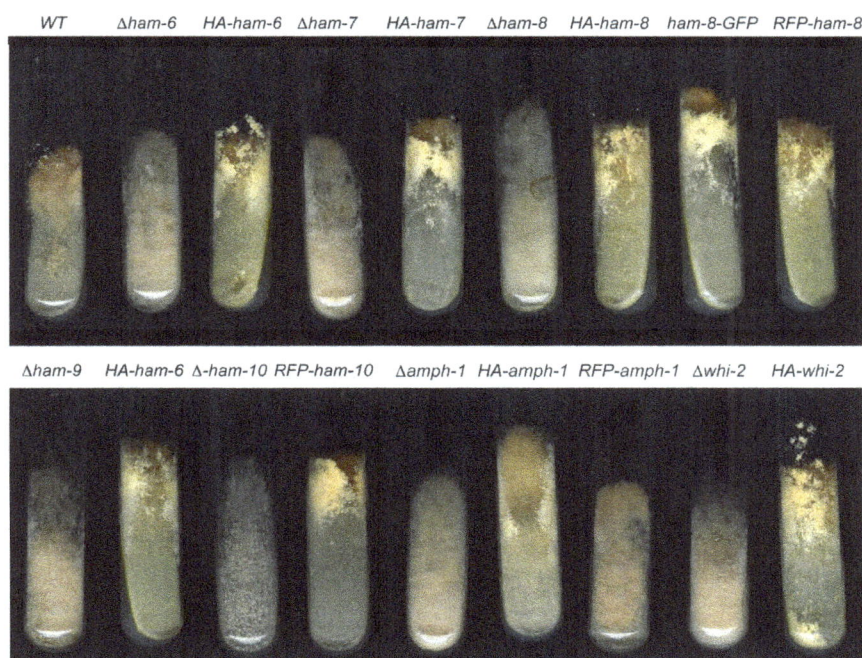

Figure 1. Strains used in this study. Slants containing Vogel's sucrose medium were inoculated with different strain isolates and grown for 4 days. Strains shown in the top panel from left to right include wild type (WT), *Δham-6*, *Δham-6* transformed with *HA-ham-6*, *Δham-7*, *Δham-7* transformed with *HA-ham-7*, *Δham-8*, *Δham-8* transformed with *HA-ham-8*, *Δham-8* transformed with *ham-8-GFP*, and *Δham-8* transformed with *RFP-ham-8*. The bottom panel shows *Δham-9*, *Δham-9* transformed with *HA-ham-9*, *Δham-10*, *Δham-10* transformed with *RFP-ham-10*, *Δamph-1*, *Δamph-1* transformed with *HA-amph-1*, *Δamph-1* transformed with *RFP-amph-1*, *Δwhi-2*, and *Δwhi-2* transformed with *HA-whi-2*.

surface of the cover glass bottom. Benomyl was added in both the liquid medium and the agar block to inhibit germ tube formation. The cells were observed between 3 and 6 hours to follow the formation of CATs.

For imaging of germ tubes and hyphal cells expressing fluorescent protein constructs, conidia (10^5 conidia per ml) were grown in Vogel's sucrose liquid medium for 6 to 8 hours. Cell samples were placed on slides for immediate microscopic observation.

To examine the nuclear localization of MAK-1-GFP and MAK-2-GFP in germ tubes/CATs, conidia (10^6 per ml) were grown in Vogel's sucrose liquid medium at 34°C for 4 hours to allow conidia germination and CAT formation. Cell samples were then collected, fixed, digested with Novozyme, and permeabilized following the procedure described above. After membrane permeabilization, cells were treated with 1 mg/ml RNAase for 30 minutes at room temperature to digest RNA. Cells were then washed once with PBS and stained with propidium iodide (5 µg/ml propidium iodide in PBS) for 20 minutes [55]. After three washing steps, cells were placed on slides for microscopic observation.

Confocal laser scanning microscopy was performed using a Zeiss LSM 710 Confocal Microscope (Carl Zeiss, Inc., Thornwood, NY). Plan-Apochromat 63×/1.40 Oil DIC M27 objective or Plan-Apochromat 40×/1.3 Oil DIC M27 objective lenses were used for imaging. Excitation wavelength and detection wavelength were set up either according to references [37,41,56] or by using the smart setup program of the ZEN2012 image capture software. GFP images were collected at 493 to 598 nm with excitation at 488 nm. RFP images were collected at >570 nm with excitation at 558 nm. When imaging samples with both GFP and RFP signals, the images were collected at 493 to 539 nm and at 554 to

703 nm with excitation at 490 nm and 514 nm. When examining GFP protein samples that have been stained with propidium iodide, the images were collected at 499 to 560 nm and at 572 to 719 nm with excitation at 488 nm and 535 nm. The images were collected sequentially using either a line sequential scanning mode or plane scanning mode. Bright-field images were captured with a transmitted light detector. Time-lapse imaging was performed to evaluate Ping-Pong signaling at time intervals of 10 s to 60 s for periods up to 10 minutes. Images were analyzed with image processing software ZEN lite and Image J.

For the detection of HA-tagged proteins, anti-HA primary antibody was used in conjunction with Alexa Fluor 488 conjugated secondary antibody. Immunofluorescent images were collected at 497 to 622 nm with excitation at 488 nm.

Results

Characterization of cell fusion mutants reveals two functional mutant groups

A screening of plates 110 to 120 from the single gene deletion library identified a new cell fusion gene, *whi-2* (NCU10518), which is a predicted homolog of yeast stress response factor protein Whi2p [57,58]. We confirmed that the cell-fusion defect and hygromycin resistance co-segregated. Moreover, ectopic expression of HA-WHI-2 at the *his-3* locus complemented the *Δwhi-2* mutant phenotypes (Figure 1 and 2), demonstrating that the loss of *whi-2* was responsible for the mutant defects. An examination of the conidial morphology of *Δwhi-2* revealed a conidial phenotype similar to that of the *Δham-10* and *Δamph-1* mutants (Figure S1). Instead of making mature macroconidia, *Δham-10*, *Δamph-1* and *Δwhi-2* produced chains of macroconidia that stopped development at the major constriction stage (Figure S1) [59]. *Δham-10*

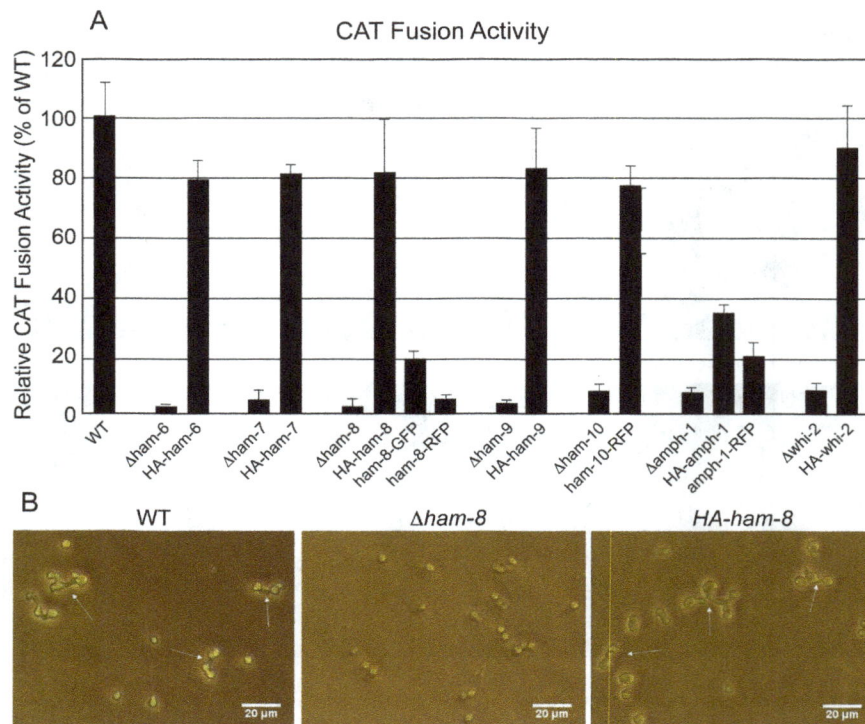

Figure 2. Complementation of CAT fusion activities by different HA, GFP and RFP tagged proteins. A) The levels of CAT fusion activity for the gene deletion mutants and for transformants expressing a tagged version of the deleted gene are shown as a percentile of the cell fusion activity for wild type CATs. B) Photograph of CAT fusion activities in wild type (WT), Δham-8, and Δham-8 transformed with HA-ham-8. Arrows point to examples of CAT fusion in the wild-type and Δham-8 transformed with HA-ham-8 panels.

produced fewer conidia than Δamph-1 and Δwhi-2. In contrast, Δham-6, Δham-7, Δham-8 and Δham-9 generated normal macroconidia (Figure S1).

All of the cell fusion mutants had a flat conidiation phenotype, which is due to a defect in the generation of long aerial hyphae (Figure 1). The phenotypic differences between the Δham-6, Δham-7, Δham-8 and Δham-9 group of mutants, and the Δham-10, Δamph-1 and Δwhi-2 mutants suggest that there are functional differences between the proteins encoded by these two groups of cell fusion genes.

HAM-6, HAM-7 and HAM-8 are specifically expressed in germ tubes/CATs

To examine the cell type expression pattern for these cell fusion proteins, we generated HA-tagged versions that were expressed at the his-3 locus under the control of their own promoters. The HA-tagged proteins fully rescued the mutant developmental and CAT fusion defects of Δham-6, Δham-7, Δham-8, Δham-9 and Δwhi-2 (Figures 1 and 2) [9]. The HA-tagged version of AMPH-1 provided only a partial rescue of Δamph-1 (32.3% of the wild type cell fusion level) (Figures 1 and 2). Western blot analysis was performed to examine the size and expression patterns of the HA-tagged proteins (Figure 3). The predicted MW (molecular weight) for HAM-6, HAM-7, HAM-8, HAM-9, AMPH-1 and WHI-2 are 15.9 KD, 24.4 KD, 57.7 KD, 96.7 KD, 29.9 KD and 32.2 KD respectively. HAM-9, AMPH-1 and WHI-2 are predicted to be cytosolic proteins, and their HA-tagged proteins gave MWs very close to the predicted MWs. HAM-6 and HAM-8 contain three and four predicted TM (transmembrane) domains respectively, and the measured MWs of their HA-tagged proteins were also

very close to their predicted MWs. HAM-7 has been shown to be a GPI-anchored cell wall protein [31]. The measured MW for HA-HAM-7 was 42 KD, 18 KD larger than the predicted MW, which suggests the GPI-anchored cell wall protein is heavily glycosylated.

HA-HAM-6 and HA-HAM-8 displayed a germ tubes/CATs-specific expression pattern, with only a trace amount of expression in vegetative hyphae (Figure 3). HA-HAM-7 was expressed at a very high level in germ tubes and CATs, and at a 5-fold reduced level in vegetative hyphae. In contrast, we determined that HA-HAM-9 and HA-WHI-2 were expressed at about equal level in the germ tubes/CATs and hyphae samples, while HA-AMPH-1 was expressed at higher level in hyphae than in germ tubes/CATs (Figure 3). In summary, these expression experiments support our phenotypic classification of the mutants, and suggest that HAM-6, HAM-7, and HAM-8 form a group of proteins that primarily functions in germlings and during CAT fusion, while HAM-9, WHI-2 and AMPH-1 have general functions during growth and differentiation.

HAM-7 and HAM-8 are found in a punctate pattern near the tips of germ tubes and CATs

In order to determine the location of the cell fusion proteins, we expressed them as GFP- and dsRed RFP-tagged constructs under the control of the ccg-1 promoter in their respective mutant backgrounds (Figure 1). The HAM-9-GFP, RFP-HAM-9, HAM-10-GFP and AMPH-1-GFP fusion proteins failed to rescue the mutant phenotypes. We were unable to detect the GFP and RFP signals from these tagged proteins, suggesting that the tagged proteins were rapidly degraded. HAM-8-GFP and RFP-HAM-8 rescued the Δham-8 conidiation defects, but failed to restore CAT

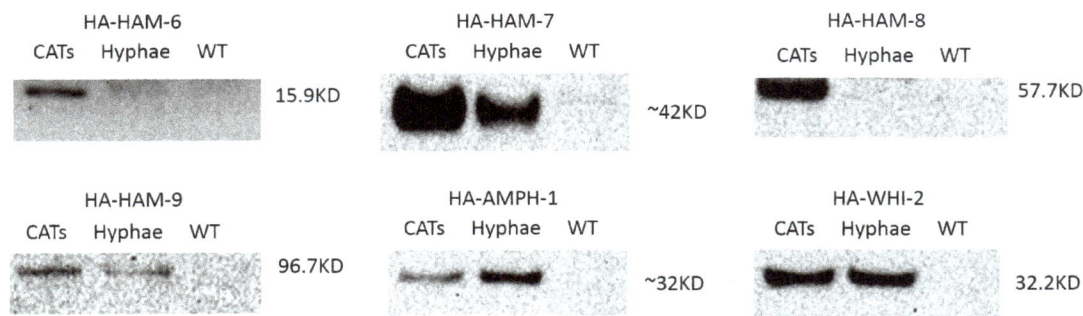

Figure 3. Western blot analyses of HA-tagged proteins' expression patterns. Western blot analyses using anti-HA antibody were performed to detect HA-HAM-6, HA-HAM-7, HA-HAM-8, HA-HAM-9, HA-AMPH-1 and HA-WHI-2 protein levels in four hour germlings (CATs lane) and vegetative hyphae (Hyphae lane). The HA-tagged cell fusion proteins were regulated by their own promoters. Protein samples from wild type germ tubes/CATs were loaded as negative control (WT lane) for each Western blot analysis.

fusion activity, indicating that the tagged HAM-8 proteins were only partially functional (Figures 1 and 2). Moreover, some of the GFP and RFP signals were detected at large vacuolar-like structures, suggesting that the tagged HAM-8 proteins might have been targeted to the vacuole for degradation, and may not reflect the normal localization for HAM-8 (Figure S2A). Co-localization experiments with marker proteins showed that HAM-8-GFP and the vacuolar marker RFP-VAM-3 showed co-localization (Figure S2B). RFP-AMPH-1 gave a partial rescue on both conidiation phenotype and CAT fusion activity (18.5% of wild type cell fusion level) (Figures 1 and 2), and localized in a punctate pattern, suggestive of being associated with small vesicles (Figure 4). We also detected RFP signal in larger vacuolar-like structures, which may represent RFP entering into vacuoles and being degraded (see below). RFP-HAM-10 was the only fluorescent fusion protein that fully rescued both mutant conidiation phenotype and CAT fusion activity (Figures 1 and 2). RFP-HAM-10 was found in vesicular or vacuolar-like structure in germ tubes and CATs (Figures 4 and S3). Because MAK-2 and SO have been detected in association with small vesicles near the tips of CATs, we asked whether the RFP-HAM-10 and RFP-AMPH-1 co-localized with MAK-2-GFP or SO-GFP (Figures S3 and S4). We did not see evidence for the co-localization of RFP-HAM-10 or RFP-AMPH-1 with either

MAK-2-GFP or SO-GFP near the tips of CATs during cell fusion (Figures S3 and S4).

The HA-tagged versions of HAM-6, HAM-7, HAM-8, HAM-9, AMPH-1, and WHI-2, expressed under the control of their endogenous promoters, provided an alternative opportunity to examine the location of these proteins in fixed cells. Except for HA-AMPH-1, which was only partially functional, the HA-tagged version of these proteins fully rescued the mutant defects (Figures 1 and 2). We were unable to get immunolocalization data for HA-HAM-6 and HA-HAM-9, which was not surprising because these two proteins were expressed at very low levels (Figure 3). HA-HAM-7 and HA-HAM-8 were localized in a punctate pattern, suggestive of being found in small vesicles or vacuoles (Figures 5A and 5B). The intensity of the fluorescent signal for HA-HAM-8 near the tip region of the germ tubes/CATs was consistently found to be significantly higher than the signal in the rest of the cell.

HA-AMPH-1 was localized in a punctate pattern, suggestive of being associated with small vesicle, and in the cytosol (Figure 5C), consistent with the localization pattern of RFP-AMPH-1 (Figure 4). Significantly, we did not detect HA-AMPH-1 in large vacuolar-like structures, underscoring our suggestion that the GFP and RFP-tagged fusion proteins detected in large vacuolar-like structures have been targeted to the vacuoles for degradation. HA-WHI-2, which is predicted to be a cytosolic protein, also showed

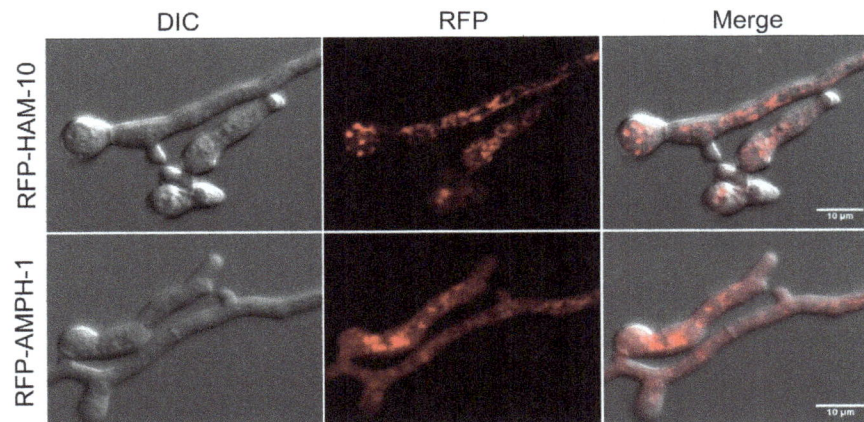

Figure 4. Localization of RFP-HAM-10 and RFP-AMPH-1 in germ tubes/CATs. Confocal microscopic images were taken for CATs expressing RFP-HAM-10 (top row of panels) and RFP-AMPH-1 (bottom row of panels). Images shown from left to right are DIC images, RFP fluorescent images, and merged images.

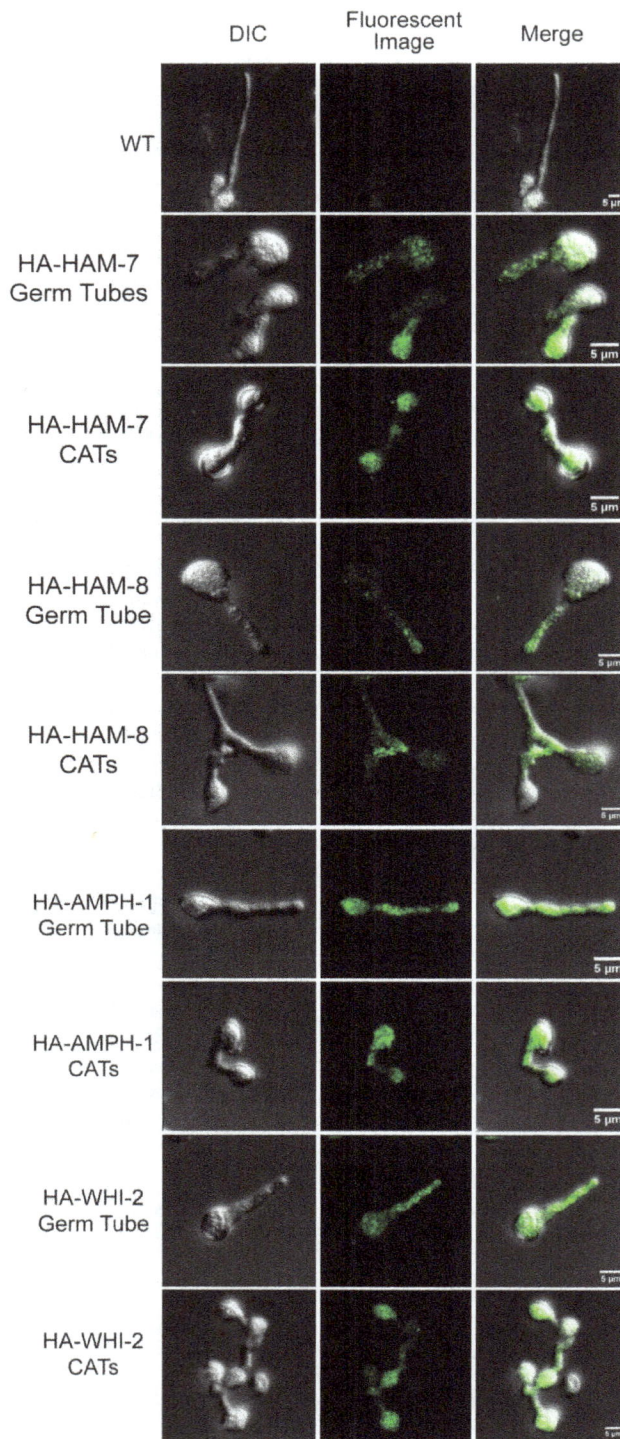

Figure 5. Immunofluorescent localization images for HA-tagged proteins. Anti-HA primary antibody and Alexa Fluor 488-conjugated secondary antibody were used to label HA-tagged protein in fixed germ tubes/CATs. Typical DIC images (left), fluorescent images (middle), and merged images (right) are shown. Images are shown for Wild type (WT) control (top row), HA-ham-7 transformant germ tubes (row 2), and CATs (row 3), HA-ham-8 transformant germ tube (row 4) and CATs (row 5), HA-amph-1 transformant germ tube (row 6) and CATs (row 7), HA-whi-2 transformant germ tube (row 8) and CATs (bottom row).

localization to what appear to be small vesicles or vacuoles as well as to the cytosol. (Figure 5D).

Oscillatory recruitment of MAK-2 and SO to cell tips is abolished in cell fusion mutants

MAK-2 and SO have been shown to be recruited to the cell tips in an oscillatory fashion during CAT fusion [23]. In an effort to identify whether "Ping-Pong" signaling is disrupted in our cell fusion mutants, we generated MAK-2-GFP-expressing and SO-GFP-expressing Δham-6, Δham-7, Δham-8, Δham-9, Δham-10, Δamph-1 and Δwhi-2 isolates by mating cell fusion mutants with MAK-2-GFP-expressing and SO-GFP-expressing wild type strains of the opposite mating type. Germinating conidia of these mutant strains made CAT-like structures in non-cell fusion contexts at very low frequency. Germ tube germination was not affected in cell fusion mutants (Figure 6). Microscopic examination of these strains showed that MAK-2-GFP and SO-GFP never localized at the tip of germ tubes or CAT-like structures (Figure 6).

In separate experiments, mutant conidia expressing MAK-2-GFP and SO-GFP were mixed with an equal number of wild type conidia expressing cytosolic RFP to determine whether the mutant conidia could respond to wild type signals, participate in MAK-2/SO Ping-Pong signaling, and fuse with wild type conidia. We observed that Δamph-1 and Δwhi-2 conidia were able to fuse with wild type at a very low frequency while the remaining mutants never fused with wild type. Interestingly, the few Δamph-1 and Δwhi-2 conidia that participated in cell fusion with wild type displayed normal macroconidial morphologies (Figure 7), while the typical, abnormally-shaped mutant conidia did not (see Figure S1). Because fusions between wild type and Δamph-1 or Δwhi-2 conidia were very rare, we were unable to determine whether the oscillatory recruitment of MAK-2-GFP and SO-GFP to the CAT tip occurred in these germling pairs.

MAK-1 and MAK-2 phosphorylation status is affected in mutant germ tubes/CATs and vegetative hyphae

The phosphorylation of MAK-1 and MAK-2 activates the MAP kinases and is required for cell-cell communication and cell fusion. In order to determine whether HAM-6, HAM-7, HAM-8, HAM-9, HAM-10, AMPH-1 and WHI-2 influence the activity of MAK-1 and MAK-2, we looked at the phosphorylation status of both MAPKs in mutant germlings during CAT-inducing conditions (Figure 8). MAK-1 phosphorylation was dramatically reduced in Δham-6, Δham-7, Δham-8, and Δwhi-2, and slightly reduced in Δham-10. MAK-1 phosphorylation was not reduced in Δham-9 and Δamph-1. These results demonstrate that HAM-6, HAM-7, HAM-8, WHI-2, and perhaps HAM-10 are required for activation of the MAK-1 pathway (Figures 8 and S5). In contrast, MAK-2 phosphorylation was reduced in all cell fusion mutants, which may contribute to the MAK-2/SO signaling defect observed in all of the mutants.

We were also interested in assessing the ability of the mutants to activate the MAPK pathways in response to stress-inducing conditions during vegetative growth. Peroxidase treatment has been used as one way to activate MAK-1 and MAK-2 in vegetative cells, and we used peroxidase treatment to evaluate the activation of the MAP kinase pathways in our mutants. Before peroxidase treatment, the MAK-1 and MAK-2 phosphorylation status in mutant vegetative hyphae was similar to the MAK-1 and MAK-2 phosphorylation status in mutant germ tubes/CATs (Figures 8, 9A and 9B) for Δham-6, Δham-7, Δamph-1 and Δwhi-2. In Δham-8 and Δham-9, the MAK-1 and MAK-2 phosphorylation levels in vegetative hyphae were below the threshold for

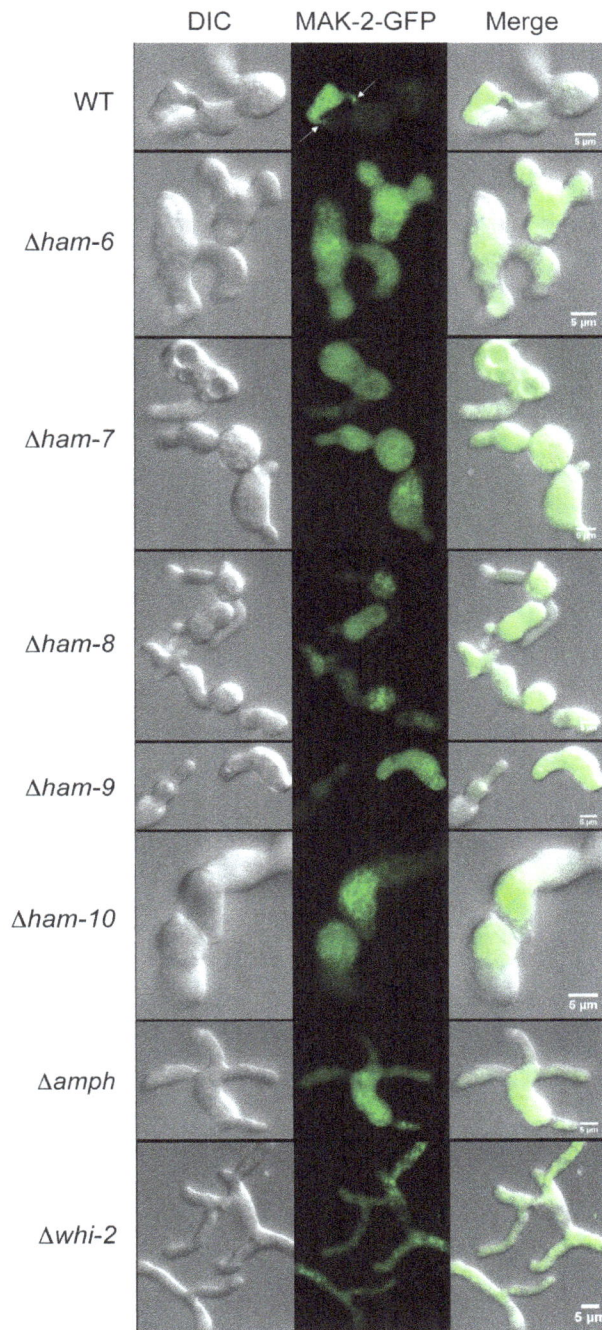

Figure 6. MAK-2-GFP localization in wild type and mutant germ tubes/CATs. MAK-2-GFP expressing wild type (WT) and mutant conidia were grown under CAT induction conditions for 4 hours. DIC images (left column), GFP fluorescent images (middle column), and merged images (right column) are shown. The images show germ tubes/CATs for wild type (WT) (row 1), Δham-6 (row 2), Δham-7 (row 3), Δham-8 (row 4), Δham-9 (row 5), Δham-10 (row 6), Δamph-1 (row 7), and Δwhi-2 (row 8). The arrows in the WT GFP fluorescent image point to the localization of MAK-2-GFP at the sites of cell fusion.

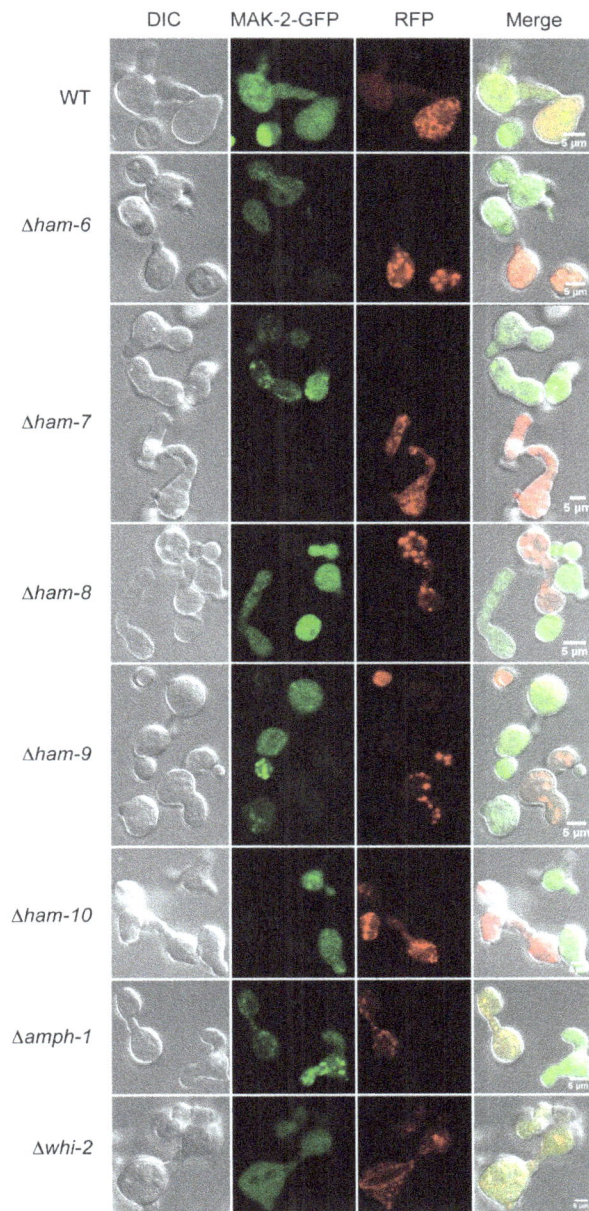

Figure 7. Fusion between MAK-2-GFP-expressing cell fusion mutants and RFP-expressing wild type cells. Conidia samples containing equal number of RFP-expressing wild type conidia and MAK-2-GFP-expressing wild type or mutant conidia were grown under CAT induction conditions for 4 hours. DIC images, GFP fluorescent images, RFP fluorescent images, and merged images for each combination of conidia types are shown in the columns from left to right respectively. Each row shows the images for RFP-expressing wild type conidia mixed with MAK-2-GFP-expressing wild type (WT) (row 1), Δham-6 (row 2), Δham-7 (row 3), Δham-8 (row 4), Δham-9 (row 5), Δham-10 (row 6), Δamph-1 (row 7), and Δwhi-2 (row 8) conidia. Wild type conidia frequently engaged in cell fusion, while Δamph-1 and Δwhi-2 conidia engaged in cell fusion with w conidia at a low frequency.

detection. Stress-induction dramatically increased both MAK-1 and MAK-2 phosphorylation levels in wild type hyphae (Figures 9A and 9B). MAK-1 activation in Δham-6, Δham-7, Δham-9, and Δwhi-2 was strongly reduced. Δham-8 showed an interme-

diate level of MAK-1 activation, while the MAK-1 activation in Δham-10 and Δamph-1 was not significantly affected (Figure 9A).

In these experiments with vegetative hyphae, MAK-2 was activated to much lower levels in Δham-6 and Δham-9 than in wild type hyphae (Figure 9B). Δham-8, Δham-10, Δamph-1 and Δwhi-

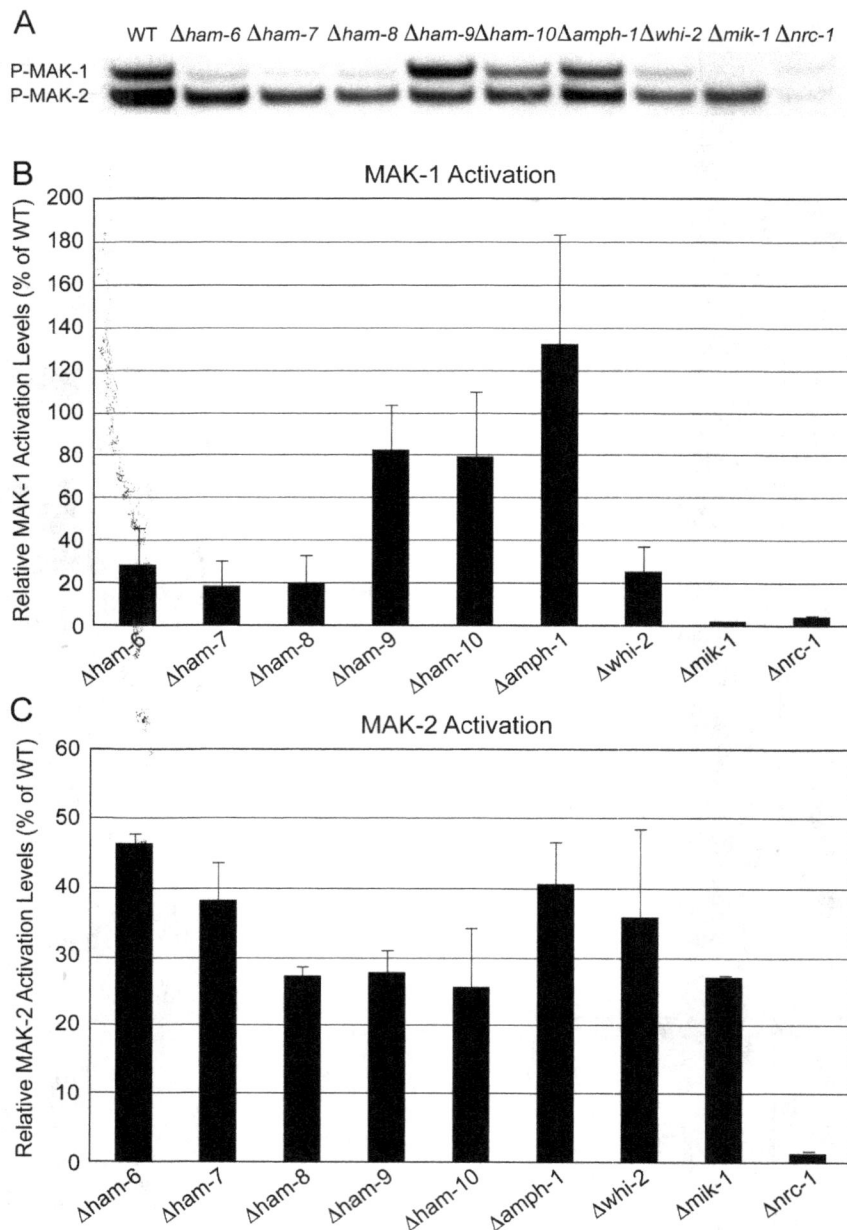

Figure 8. MAK-1 and MAK-2 phosphorylation status in germ tubes/CATs. Western blot analysis using Phospho-p44/42 MAPK antibody was performed to determine MAK-1 and MAK-2 phosphorylation status in wild type (WT) and mutant (Δham-6, Δham-7, Δham-8, Δham-9, Δham-10, Δamph-1, Δwhi-2, Δmik-1, and Δnrc-1) germ tubes/CATS. The positions of the phosphorylated MAK-1 (p-MAK-1) and phosphorylated MAK-2 (p-MAK-2) in the Western blot are noted in the left margin of the figure. B) The relative MAK-1 phosphorylation status in mutant germ tubes/CATs relative to the MAK-1 phosphorylation status in wild type germ tubes/CATs (WT value is set at 100%). C) The relative MAK-2 phosphorylation status in mutant germ tubes/CATs compared to the MAK-2 phosphorylation status in wild type germ tubes/CATs (WT value is set at 100%).

2 showed an intermediate level of MAK-2 activation, while MAK-2 activation was normal in Δham-7 (Figure 9B). The differences in MAK-1 and MAK-2 phosphorylation status between germ tubes/CATs and vegetative hyphae for some of the mutants suggest that there may be differences in how the two MAP kinase pathways are being regulated during the various stages of the N. crassa life cycle.

MAK-1 and MAK-2 nuclear accumulation is normal in mutant germ tubes/CATs

Mutants of the STRIPAK complex have been shown to have a defect in MAK-1 nuclear accumulation, which is regulated by MAK-2 phosphorylation of MOB-3 [34]. MAK-2 nuclear localization is also required during cell fusion [23]. In order to examine if MAK-1 or MAK-2 nuclear accumulation is compromised in the cell fusion mutants, propidium iodide was used to label nuclei in MAK-1-GFP and MAK-2-GFP-expressing mutant strains (Figures S6 and S7). We found that nuclear accumulation

A

MAK-1 Activation

B

MAK-2 Activation

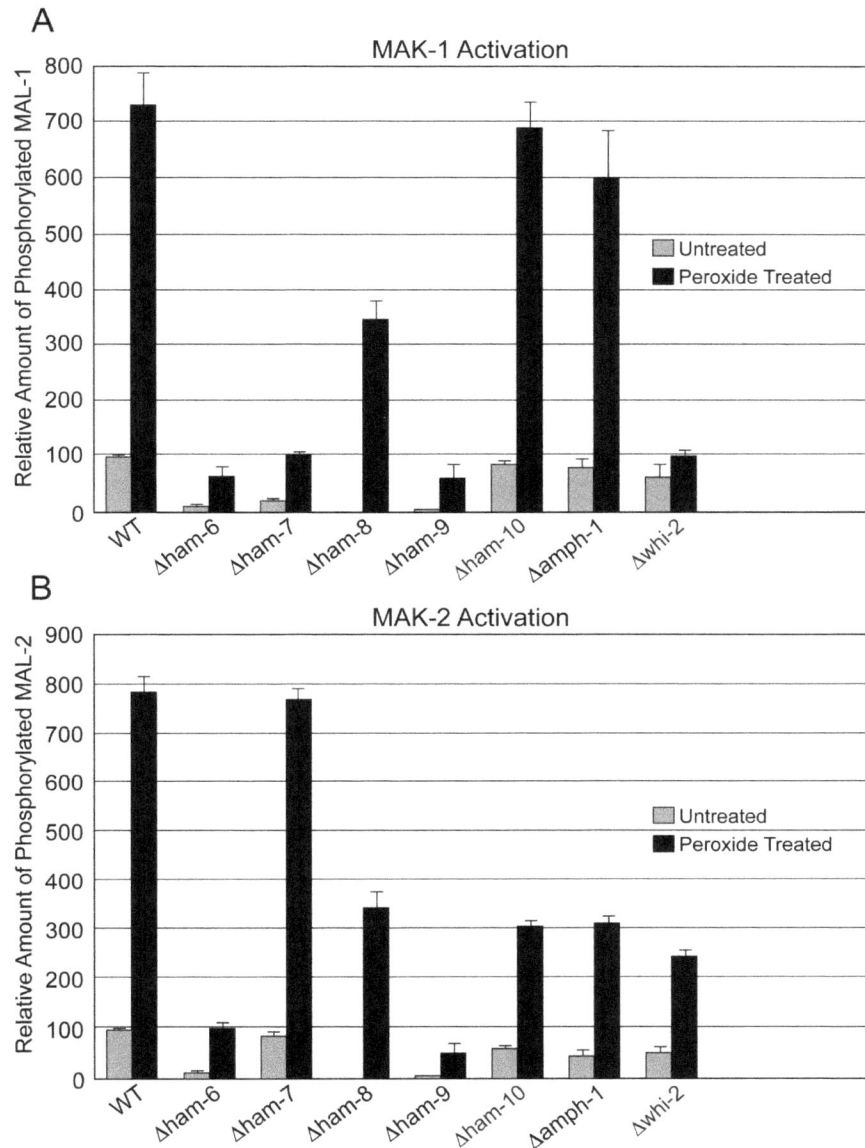

Figure 9. Peroxidase activation of MAK-1 and MAK-2 pathways in cell fusion mutant vegetative hyphal cells. Western blot analyses using Phospho-p44/42 MAPK antibody were performed to evaluate MAK-1 and MAK-2 activation in wild type (WT) and mutant vegetative hyphal cells in response to peroxidase treatments. Quantitative analyses of the Western blots were performed to determine the levels of phosphorylated MAK-1 and MAK-2 in non-stressed and oxidative-stressed samples (wild type, Δham-6, Δham-7, Δham-8, Δham-9, Δham-10, Δamph-1, and Δwhi-2). A) MAK-1 activation in wild type and mutants in response to peroxidase treatment. B) MAK-2 activation in wild type and mutants in response to peroxidase treatment. The levels of MAK-1 and MAK-2 in the non-stressed wild type sample were set as 100% for the quantitative analysis.

of MAK-1 and MAK-2 was normal in all of the cell fusion mutants.

Discussion

Our screen of approximately 11,000 single gene deletion strains from the first 120 plates of the *N. crassa* single gene deletion library identified 25 genes required for cell-to-cell fusion. This screen identified the MAK-1 and MAK-2 MAPK pathways as well as the STRIPAK complex as three major signaling modules regulating cell fusion [1,6,7,24,34,60]. In this report, we focused our research on seven cell fusion genes whose functions were less well-characterized.

HAM-6, HAM-7 and HAM-8 are highly conserved proteins present in all filamentous ascomycetes. The corresponding mutants share the same protoperithecium-deficient, flat conidiation phenotype and are morphologically indistinguishable from each other (Figure 1). They produce abundant macroconidia, but the germinating macroconidia rarely produce CAT-like structures under CAT induction conditions. The *ham-6* gene encodes a 145-amino-acid protein, the *ham-7 gene* encodes a 230-amino-acid protein, and the *ham-8* gene encodes a 597-amino-acid protein. HAM-6 and HAM-8 were predicted to be membrane proteins with 3 and 4 transmembrane domains respectively. HAM-7 is a GPI-anchored cell wall protein, and we have previously shown that it functions as a MAK-1 pathway sensor during hyphal fusion

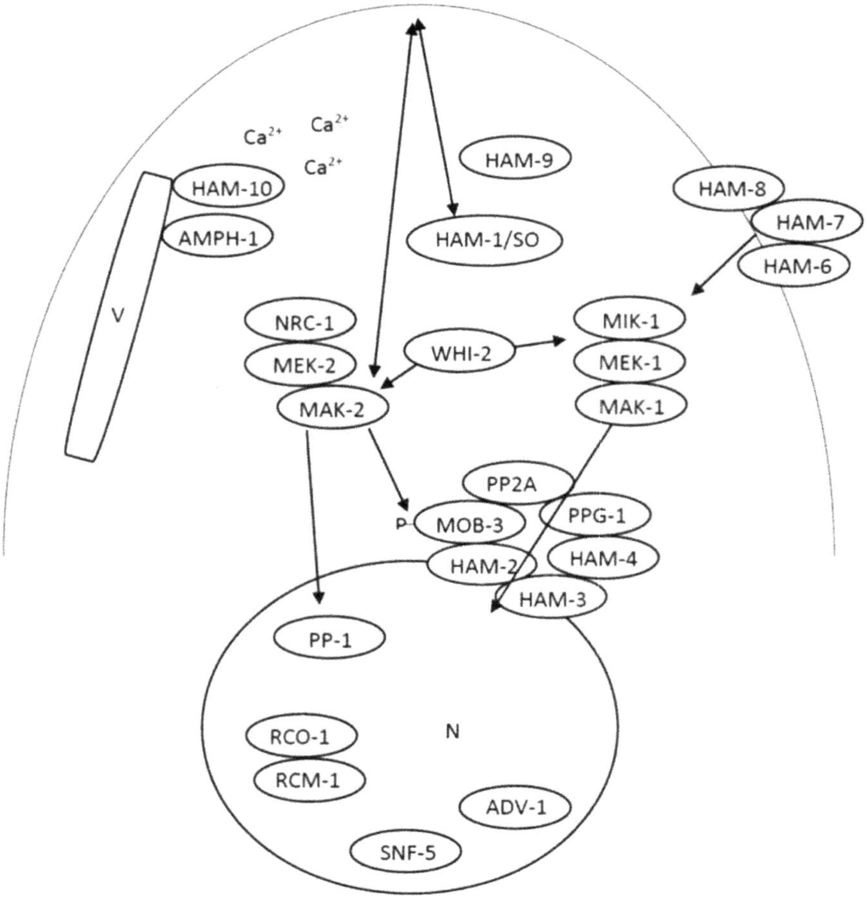

Figure 10. Schematic model for the regulatory network involved in CAT fusion. PP-1, ADV-1, SNF-5, and RCO-1/RCM-1 are transcription factors required for CAT fusion. MIK-1/MEK-1/MAK-1 and NRC-1/MEK-2/MAK-2 are two MAP kinase pathways required for CAT fusion. HAM-2/HAM-3/HAM-4/MOB-3/PP2A/PPG-1 form the STRIPAK complex that regulates MAK-1 nuclear accumulation. HAM-1/SO and MAK-2 engage in Ping-Pong signaling behavior during CAT fusion. HAM-6/HAM-7/HAM-8 are required at the plasma membrane/cell wall for MAK-1 pathway activation. HAM-10 may regulate vesicular trafficking and could potentially respond to calcium signaling during cell fusion. AMPH-1 regulates vesicular trafficking and endocytosis during cell fusion. WHI-2 may regulate the MAP kinase pathways through a general stress response pathway. The role of HAM-9 during CAT fusion remains to be determined.

[31]. We found that the three proteins were expressed at much higher levels in germ tubes/CATs than in vegetative hyphae (Figure 3). HAM-7 and HAM-8 were found to be localized in punctate pattern, suggestive of small vesicular or vacuolar structures (Figures 5A and 5B). The HAM-8 containing structures were found to be concentrated near the tip of the germlings and CATs (Figure 5B). Although HAM-7 has been shown to be a GPI-anchored cell wall protein, we did not see immunolocalization of HAM-7 at the plasma membrane/cell wall boundary. We attribute this to the heavily glycosylated status of HAM-7, which could block to interaction between the glycosylated HAM-7 and the antibody used for immunolocalization. The HAM-7 observed in our localization studies (Figure 5A) may well represent newly synthesized HAM-7 in transit through the secretory pathway that hasn't been fully glycosylated. Tip localization of MAK-2-GFP and SO-GFP was missing in the few CAT-like structures formed by these mutants and the mutants failed to fuse with wild type conidia (Figures 6 and 7). We found that HAM-6, HAM-7 and HAM-8 are required for MAK-1 kinase activation during conidial germination and CAT formation (Figure 8). Despite the lower levels of expression for HAM-6, HAM-7 and HAM-8 in vegetative

cells, MAK-1 phosphorylation was dramatically reduced in Δham-6, Δham-7, and Δham-8 during vegetative hyphal growth. Leeder et al. [1] determined that the expression of the three genes is co-regulated and controlled by the MAK-2 pathway-dependent transcription factor PP-1. In summary, we propose that the GPI-anchored cell wall HAM-7 and the two transmembrane proteins, HAM-6 and HAM-8, function together to regulate the MAK-1 pathway. Given the cell wall/plasma membrane location for the GPI-anchored protein HAM-7, we suggest that the three proteins might participate in a signaling complex at the cell wall/plasma membrane boundary, but our data would also be consistent with a signaling complex localized to intracellular membranes.

The *ham-9* gene encodes an 869-amino-acid protein containing a SAM domain and two PH domains. The SAM domain has been identified in yeast Ste11p (*S. cerevisae* homolog of *N. crassa* NRC-1) [61], and the PH domains have been suggested to play a role in targeting signal transduction proteins to intracellular membrane in signaling events [62]. The C-terminal GFP-tagged and N-terminal RFP-tagged HAM-9 fusion proteins were not functional, precluding any live-imaging analysis. HA-HAM-9 was expressed in both vegetative hyphae and germ tubes/CATs (Figure 3), but its expression level was too low for immunolocalization. The requirement of HAM-9 for both MAK-1 and MAK-2 activation in vegetative hyphae (Figures 9A and 9B), may suggest that HAM-9 regulates cross-communication of the two MAPK pathways during vegetative growth.

The amph-1 gene encodes a 262-amino-acid protein containing a bar domain, a domain frequently involved in protein-protein interaction and regulation of membrane curvature [63]. N. crassa AMPH-1 is a homolog of the yeast Rvs161p and Rvs167p proteins. Rvs161p and Rvs167p are required for endocytosis and cell fusion during yeast mating [63,64]. HA-AMPH-1 and RFP-AMPH-1 localized to small vesicles in the germ tubes/CATs, and some of the vesicles appeared to be associated with the plasma membrane (Figures 4 and 5C), suggesting that N. crassa AMPH-1 plays a role in vesicular trafficking and endocytosis. MAK-1 and MAK-2 activity in Δamph-1 germlings were similar to the wild type control, indicating that AMPH-1 does not affect cell fusion by regulating these pathways. This is consistent with our observation that a few Δamph-1 conidia having wild type morphology were able to participate in cell fusion with wild type conidia (Figure 7). HA-AMPH-1 was expressed in both vegetative hyphae and germlings, suggesting it is a general factor required for all stages of N. crassa life cycle. In summary, we suggest that AMPH-1 functions during vesicular trafficking and endocytosis.

The ham-10 gene encodes a 1,422-amino-acid protein containing a C2 domain near the C terminus. C2 domains function as calcium-dependent lipid-binding domains and are thought to be involved in vesicular trafficking, exocytosis, and signal transduction [65]. HAM-10 tagged with RFP at its N terminus fully rescued Δham-10, but HAM-10 tagged with GFP at the C terminus did not (Figures 1 and 2), suggesting that modification near the C terminal C2 domain may affect the function and stability of HAM-10. RFP-HAM-10 localized in the cytosol and in a punctate pattern, suggestive of a vesicular or vacuolar network location (Figure 4). However, our proposed localization of HAM-10 should be considered as a tentative assignment. We did not demonstrate that the RFP-tag remained attached to HAM-10, nor have we carried out extensive co-localization studies with known vesicle and vacuolar marker proteins to definitively demonstrate co-localization of the RFP-HAM-10 with organelle-specific markers. Our results demonstrate that HAM-10 was not enriched at CAT tips during cell fusion (Figures 4 and S3). The requirement of HAM-10 in both MAPK pathways during different development stages suggests HAM-10 could be a general factor in regulating cell growth.

The whi-2 gene encodes a 298-amino-acid protein with homology to the yeast general stress response protein Whi2p, which has been shown to activate autophagy and mitophagy under nutrient starvation conditions [66,67]. N. crassa Δwhi-2 displayed a conidial development defect (Figure S1). HA-WHI-2 was expressed in both vegetative hyphae and germ tubes/CATs and was localized in cytosol and in a punctate pattern suggestive of small vesicles or vacuoles (Figures 3 and 5D). MAK-2/SO signaling was abolished in Δwhi-2 (Figure 6), but we found that a few mutant macroconidia with wild type morphology were able to participate in cell fusion with wild type conidia (Figure 7). Interestingly, the phosphorylation levels of both MAK-1 and MAK-2 were reduced in germlings and during vegetative hyphal growth (Figures 8 and 9), suggesting WHI-2 functions as a general stress response factor regulating both MAPK pathways.

In summary, our studies on the seven cell fusion genes confirmed that the MAK-1 and the MAK-2 pathways play critical roles during conidia germination and CAT fusion. Figure 10 shows a diagrammatic representation of a CAT tip with many of the proteins we have discussed. The phenotypic characteristics, cell type-specific expression patterns, cellular locations, and MAP kinase activity status of HAM-6, HAM-7 and HAM-8 suggest that the three proteins may form a multimeric sensor complex at the cell wall/plasma membrane or on intracellular vesicles and regulate MAK-1 activation during CAT fusion. Our studies on HAM-9, HAM-10, AMPH-1 and WHI-2 suggest that cell fusion is also affected in mutants lacking proteins with general functions in growth and development. HAM-10, AMPH-1 and WHI-2 clearly play a role in conidial development as well as during CAT fusion. The importance of HAM-9, HAM-10 and WHI-2 for both MAK-1 and MAK-2 signaling may provide opportunities to study cross-talk regulation between MAP kinase pathways and other signaling modules.

Supporting Information

Figure S1   CAT fusion in wild type and mutants. Wild type (WT) and mutant conidia cells were grown under CAT induction conditions for 4 hours. Images for Δham-6, Δham-7, Δham-8, Δham-9, Δham-10, Δamph-1, Δwhi-2, and wild type are shown. The images show that conidia from the mutant isolates are unable to generate CATs. The wild type conidia participate in CAT formation and fusion. The arrows in the Δham-10, Δamph-1, and Δwhi-2 panels point to chains of abnormal conidia. The arrows in the wild type panel point to a site of CAT fusion.

Figure S2   Localization of HAM-8-GFP, RFP-HAM-8, and RFP-VAM-3. Confocal microscopic images were taken of cells expression GFP- and RFP-tagged proteins. A) Images for germ tubes/CATs expressing HAM-8-GFP (top row of panels) and germ tubes/CATs expressing RFP-HAM-8 (bottom row of panels). Fluorescent images (left column), DIC images (middle column), and merged images (right column) are shown. B) Confocal microscopic images were taken for cells expressing both HAM-8-GFP and RFP-VAM-3. GFP fluorescent image (HAM-8-GFP localization in top left panel), DIC image (top right panel), RFP fluorescent image (RFP-VAM-3 localization in bottom left panel), and a merged image (bottom right panel) are shown. Yellow fluorescent signal in the merged image shows co-localization of HAM-8-GFP and RFP-VAM-3.

Figure S3   Localization of RFP-HAM-10 with SO-GFP. Heterokaryotic conidia expressing RFP-HAM-10 and SO-GFP were grown under CAT induction conditions for 4 hours. Confocal microscopic images were taken for CATs engaging in cell fusion. GFP fluorescent image (SO-GFP localization in top left panel), DIC image (top right panel), RFP fluorescent image (RFP-HAM-10 localization in bottom left panel), and a merged image (bottom right panel) are shown. The arrows in the fluorescent images point to a site of cell fusion. Note the presence of SO-GFP and the absence of RFP-HAM-10 at the fusion site.

Figure S4   Localization of RFP-AMPH-1 with MAK-2-GFP. Heterokaryotic conidia expressing RFP-AMPH-1 and MAK-2-GFP were grown under CAT induction conditions for 4 hours. Confocal microscopic images were taken for CATs engaging in cell fusion. GFP fluorescent image (MAK-2-GFP localization in top left panel), DIC image (top right panel), RFP fluorescent image (RFP-AMPH-1 localization in bottom left panel), and a merged image (bottom right panel) are shown. The arrows in the fluorescent images point to a site where cell fusion will occur. Note the presence of MAK-2-GFP and the absence of RFP-AMPH-1 at the tip of CATs.

Figure S5   Ponceau stain for MAK-1 and MAK-2 phosphorylation status in mutant germ tubes/CATs. Ponceau stain image is shown below the Western blot image for the MAK-1 and MAK-2 phosphorylation status in wild type (WT) and mutant germ tubes/CATs. The Western blot image is found as Figure 8 in the manuscript and the Ponceau stain image is given here to demonstrate equal loading of the samples used in the Western blot.

Figure S6   Nuclear localization of MAK-1-GFP in mutant germ tubes. Propidium iodide was used to stain nuclei in MAK-1-GFP-expressing wild type (WT) and mutant germ tubes. The figure shows DIC images, GFP fluorescent images, propidium iodide red fluorescent images, and merged images (from left to right respectively). Images are shown for wild type (WT) (row 1), *Δham-6* (row 2), *Δham-7* (row 3), *Δham-8* (row 4), *Δham-9* (row 5), *Δham-10* (row 6), *Δamph-1* (row 7), and *Δwhi-2* (row 8) germ tubes.

Figure S7   Nuclear localization of MAK-2-GFP in mutant germ tubes. Propidium iodide was used to stain nuclei in MAK-2-GFP-expressing wild type (WT) and mutant germ tubes. The

figure shows DIC images, GFP fluorescent images, propidium iodide red fluorescent images, and merged images (from left to right respectively). Images are shown for wild type (WT) (row 1), *Δham-6* (row 2), *Δham-7* (row 3), *Δham-8* (row 4), *Δham-9* (row 5), *Δham-10* (row 6), *Δamph-1* (row 7), and *Δwhi-2* (row 8) germ tubes.

Acknowledgments

We appreciate the kind gift of pmak-2-GFP and pso-GFP from Dr. Louise Glass. We thank Alan Siegel and James Stamos for the help with microscopy and photography.

Author Contributions

Conceived and designed the experiments: CF AD SS SF. Performed the experiments: CF JA AD SS SF. Analyzed the data: CF JA AD SS SF. Contributed reagents/materials/analysis tools: CF AD SS SF. Contributed to the writing of the manuscript: CF SS SF.

References

1. Leeder AC, Jonkers W, Li J, Glass NL (2013) Germination and Early Colony Establishment in *Neurospora crassa* Requires a MAP Kinase Regulatory Network. Genetics 195: 883–898.
2. Roca MG, Read ND, Wheals AE (2005) Conidial anastomosis tubes in filamentous fungi. FEMS Microbiol Lett 249: 191–198.
3. Richard F, Glass NL, Pringle A (2012) Cooperation among germinating spores facilitates the growth of the fungus, Neurospora crassa. Biology Letters 8: 419–422.
4. Simonin A, Palma-Guerrero J, Fricker M, Glass NL (2012) Physiological significance of network organization in fungi. Eukaryot Cell 11: 1345–1352.
5. Glass NL, Rasmussen C, Roca MG, Read ND (2004) Hyphal homing, fusion and mycelial interconnectedness. Trends in microbiology 12: 135–141.
6. Fleissner A, Simonin AR, Glass NL (2008) Cell fusion in the filamentous fungus, Neurospora crassa. Methods Mol Biol 475: 21–38.
7. Read ND, Fleissner A, Roca MG, Glass NL (2010) Hyphal fusion. In: Borkovich KA, Ebbole DJ, editors. Cellular and molecular biology of filamentous fungi. Washington, DC.: ASM Press. 260–273.
8. Roca MG, Arlt J, Jeffree CE, Read ND (2005) Cell biology of conidial anastomosis tubes in Neurospora crassa. Eukaryot Cell 4: 911–919.
9. Fu C, Iyer P, Herkal A, Abdullah J, Stout A, et al. (2011) Identification and characterization of genes required for cell-to-cell fusion in Neurospora crassa. Eukaryot Cell 10: 1100–1109.
10. Aldabbous MS, Roca MG, Stout A, Huang IC, Read ND, et al. (2010) The ham-5, rcm-1 and rco-1 genes regulate hyphal fusion in Neurospora crassa. Microbiology 156: 2621–2629.
11. Fleissner A, Sarkar S, Jacobson DJ, Roca MG, Read ND, et al. (2005) The so locus is required for vegetative cell fusion and postfertilization events in Neurospora crassa. Eukaryot Cell 4: 920–930.
12. Cano-Dominguez N, Alvarez-Delfin K, Hansberg W, Aguirre J (2008) NADPH oxidases NOX-1 and NOX-2 require the regulatory subunit NOR-1 to control cell differentiation and growth in Neurospora crassa. Eukaryot Cell 7: 1352–1361.
13. Maerz S, Dettmann A, Ziv C, Liu Y, Valerius O, et al. (2009) Two NDR kinase-MOB complexes function as distinct modules during septum formation and tip extension in Neurospora crassa. Mol Microbiol 74: 707–723.
14. Mahs A, Ischebeck T, Heilig Y, Stenzel I, Hempel F, et al. (2012) The essential phosphoinositide kinase MSS-4 is required for polar hyphal morphogenesis, localizing to sites of growth and cell fusion in Neurospora crassa. PLoS One 7: e51454.
15. Schurg T, Brandt U, Adis C, Fleissner A (2012) The Saccharomyces cerevisiae BEM1 homologue in Neurospora crassa promotes co-ordinated cell behaviour resulting in cell fusion. Mol Microbiol 86: 349–366.
16. Palma-Guerrero J, Glass NL (2013) LFD-1 is a component of the membrane merger machinery during cell-cell fusion in *Neurospora crassa*. 27th Fungal Genetics Conference. Asilomar, CA.
17. Read ND, Goryachev AB, Lichius A (2012) The mechanistic basis of self-fusion between conidial anastomosis tubes during fungal colony initiation. Fungal Biology Reviews 26: 1–11.
18. Kothe GO, Free SJ (1998) The isolation and characterization of nrc-1 and nrc-2, two genes encoding protein kinases that control growth and development in Neurospora crassa. Genetics 149: 117–130.
19. Pandey A, Roca MG, Read ND, Glass NL (2004) Role of a mitogen-activated protein kinase pathway during conidial germination and hyphal fusion in Neurospora crassa. Eukaryot Cell 3: 348–358.
20. Li D, Bobrowicz P, Wilkinson HH, Ebbole DJ (2005) A mitogen-activated protein kinase pathway essential for mating and contributing to vegetative growth in Neurospora crassa. Genetics 170: 1091–1104.
21. Park G, Pan S, Borkovich KA (2008) Mitogen-activated protein kinase cascade required for regulation of development and secondary metabolism in Neurospora crassa. Eukaryot Cell 7: 2113–2122.
22. Maerz S, Ziv C, Vogt N, Helmstaedt K, Cohen N, et al. (2008) The nuclear Dbf2-related kinase COT1 and the mitogen-activated protein kinases MAK1 and MAK2 genetically interact to regulate filamentous growth, hyphal fusion and sexual development in Neurospora crassa. Genetics 179: 1313–1325.
23. Fleissner A, Leeder AC, Roca MG, Read ND, Glass NL (2009) Oscillatory recruitment of signaling proteins to cell tips promotes coordinated behavior during cell fusion. Proc Natl Acad Sci U S A 106: 19387–19392.
24. Dettmann A, Illgen J, Marz S, Schurg T, Fleissner A, et al. (2012) The NDR kinase scaffold HYM1/MO25 is essential for MAK2 map kinase signaling in Neurospora crassa. PLoS Genet 8: e1002950.
25. Borkovich KA, Alex LA, Yarden O, Freitag M, Turner GE, et al. (2004) Lessons from the genome sequence of Neurospora crassa: tracing the path from genomic blueprint to multicellular organism. Microbiol Mol Biol Rev 68: 1–108.
26. Rispail N, Soanes DM, Ant C, Czajkowski R, Grunler A, et al. (2009) Comparative genomics of MAP kinase and calcium-calcineurin signalling components in plant and human pathogenic fungi. Fungal Genet Biol 46: 287–298.
27. Saito H (2010) Regulation of cross-talk in yeast MAPK signaling pathways. Curr Opin Microbiol 13: 677–683.
28. Goryachev AB, Lichius A, Wright GD, Read ND (2012) Excitable behavior can explain the "ping-pong" mode of communication between cells using the same chemoattractant. Bioessays 34: 259–266.
29. Vogt N, Seiler S (2008) The RHO1-specific GTPase-activating protein LRG1 regulates polar tip growth in parallel to Ndr kinase signaling in Neurospora. Mol Biol Cell 19: 4554–4569.
30. Khatun R, Lakin-Thomas P (2010) Activation and localization of protein kinase C in *Neurospora crassa*. Fungal Genet Biol 48: 465–473.
31. Maddi A, Dettmann A, Fu C, Seiler S, Free SJ (2012) WSC-1 and HAM-7 are MAK-1 MAP kinase pathway sensors required for cell wall integrity and hyphal fusion in Neurospora crassa. PLoS One 7: e42374.
32. Richthammer C, Enseleit M, Sanchez-Leon E, Marz S, Heilig Y, et al. (2012) RHO1 and RHO2 share partially overlapping functions in the regulation of cell

wall integrity and hyphal polarity in Neurospora crassa. Mol Microbiol 85: 716–733.

33. Bloemendal S, Bernhards Y, Bartho K, Dettmann A, Voigt O, et al. (2012) A homologue of the human STRIPAK complex controls sexual development in fungi. Mol Microbiol 84: 310–323.

34. Dettmann A, Heilig Y, Ludwig S, Schmitt K, Illgen J, et al. (2013) HAM-2 and HAM-3 are central for the assembly of the Neurospora STRIPAK complex at the nuclear envelope and regulate nuclear accumulation of the MAP kinase MAK-1 in a MAK-2-dependent manner. Mol Microbiol. 90: 796–812.

35. Xiang Q, Rasmussen C, Glass NL (2002) The ham-2 locus, encoding a putative transmembrane protein, is required for hyphal fusion in Neurospora crassa. Genetics 160: 169–180.

36. Simonin AR, Rasmussen CG, Yang M, Glass NL (2010) Genes encoding a striatin-like protein (ham-3) and a forkhead associated protein (ham-4) are required for hyphal fusion in Neurospora crassa. Fungal Genet Biol 47: 855–868.

37. Freitag M, Hickey PC, Raju NB, Selker EU, Read ND (2004) GFP as a tool to analyze the organization, dynamics and function of nuclei and microtubules in Neurospora crassa. Fungal Genet Biol 41: 897–910.

38. Berepiki A, Lichius A, Shoji JY, Tilsner J, Read ND (2010) F-actin dynamics in Neurospora crassa. Eukaryot Cell 9: 547–557.

39. Lichius A, Lord KM, Jeffree CE, Oborny R, Boonyarungsrit P, et al. (2012) Importance of MAP kinases during protoperithecial morphogenesis in Neurospora crassa. PLoS One 7: e42565.

40. Colot HV, Park G, Turner GE, Ringelberg C, Crew CM, et al. (2006) A high-throughput gene knockout procedure for Neurospora reveals functions for multiple transcription factors. Proc Natl Acad Sci USA 103: 10352–10357.

41. Freitag M, Selker EU (2005) Expression and visualization of red fluorescent protein (RFP) in Neurospora crassa. Fungal Genet Newsl 52: 14–17.

42. McNally MT, Free SJ (1988) Isolation and characterization of a Neurospora glucose-repressible gene. Curr Genet 14: 545–551.

43. Margolin BS, Freitag M, Selker EU (1997) Improved plasmids for gene targeting at the his-3 locus of Neurospora crassa by electroporation. Fungal Genet Newsl 44: 34–36.

44. Kawabata T, Inoue H (2007) Detection of physical interactions by immuno-precipitation of FLAG- and HA tagged proteins expressed at the his-3 locus in Neurospora crassa. Fungal Genet Newsl 54: 5–8.

45. Honda S, Selker EU (2009) Tools for fungal proteomics: multifunctional neurospora vectors for gene replacement, protein expression and protein purification. Genetics 182: 11–23.

46. Chenna R, Sugawara H, Koike T, Lopez R, Gibson TJ, et al. (2003) Multiple sequence alignment with the Clustal series of programs. Nucleic Acids Res 31: 3497–3500.

47. Linding R, Russell RB, Neduva V, Gibson TJ (2003) GlobPlot: Exploring protein sequences for globularity and disorder. Nucleic Acids Res 31: 3701–3708.

48. Bowman BJ, Draskovic M, Freitag M, Bowman EJ (2009) Structure and distribution of organelles and cellular location of calcium transporters in Neurospora crassa. Eukaryot Cell 8: 1845–1855.

49. Tinsley JH, Minke PF, Bruno KS, Plamann M (1996) p150Glued, the largest subunit of the dynactin complex, is nonessential in Neurospora but required for nuclear distribution. Mol Biol Cell 7: 731–742.

50. Seiler S, Kirchner J, Horn C, Kallipolitou A, Woehlke G, et al. (2000) Cargo binding and regulatory sites in the tail of fungal conventional kinesin. Nat Cell Biol 2: 333–338.

51. Roca MG, Lichius A, Read ND (2010) How to analyze and quantify conidial anastomosis tube (CAT)-mediated cell fusion. The Neurospora protocol guide.

52. Lichius A, Roca MG, Read ND (2010) How to distinguish conidial anastomosis tubes (CATs) from germ tubes, and to discriminate between cell fusion mutants blocked in CAT formation and CAT homing. The Neurospora protocol guide 1–6.

53. Roca MG, Kuo HC, Lichius A, Freitag M, Read ND (2010) Nuclear dynamics, mitosis, and the cytoskeleton during the early stages of colony initiation in Neurospora crassa. Eukaryot Cell 9: 1171–1183.

54. Hickey PC, Jacobson D, Read ND, Glass NL (2002) Live-cell imaging of vegetative hyphal fusion in Neurospora crassa. Fungal Genet Biol 37: 109–119.

55. Maniatis T, Fritsch EF, Sambrook J (1982) Molecular cloning: a laboratory manual. Cold Spring Harbor Laboratory, Cold Spring Harbor, NY.

56. Hickey PC, Swift SR, Roca MG, Read ND (2004) Live-cell Imaging of Filamentous Fungi Using Vital Fluorescent Dyes and Confocal Microscopy. In: Savidge T, Charalabos P, editors. Methods in Microbiology: Academic Press. 63–87.

57. Radcliffe PA, Binley KM, Trevethick J, Hall M, Sudbery PE (1997) Filamentous growth of the budding yeast Saccharomyces cerevisiae induced by overexpression of the WH2 gene. Microbiology 143: 1867–1876.

58. Kaida D, Yashiroda H, Toh-e A, Kikuchi Y (2002) Yeast Whi2 and Psr1-phosphatase form a complex and regulate STRE-mediated gene expression. Genes Cells 7: 543–552.

59. Greenwald CJ, Kasuga T, Glass NL, Shaw BD, Ebbole DJ, et al. (2010) Temporal and spatial regulation of gene expression during asexual development of Neurospora crassa. Genetics 186: 1217–1230.

60. Glass NL, Jacobson DJ, Shiu PK (2000) The genetics of hyphal fusion and vegetative incompatibility in filamentous ascomycete fungi. Annu Rev Genet 34: 165–186.

61. Grimshaw SJ, Mott HR, Stott KM, Nielsen PR, Evetts KA, et al. (2004) Structure of the sterile alpha motif (SAM) domain of the Saccharomyces cerevisiae mitogen-activated protein kinase pathway-modulating protein STE50 and analysis of its interaction with the STE11 SAM. J Biol Chem 279: 2192–2201.

62. Rebecchi MJ, Scarlata S (1998) Pleckstrin homology domains: a common fold with diverse functions. Annu Rev Biophys Biomol Struct 27: 503–528.

63. Ren G, Vajjhala P, Lee JS, Winsor B, Munn AL (2006) The BAR domain proteins: molding membranes in fission, fusion, and phagy. Microbiol Mol Biol Rev 70: 37–120.

64. Youn JY, Friesen H, Kishimoto T, Henne WM, Kurat CF, et al. (2010) Dissecting BAR domain function in the yeast Amphiphysins Rvs161 and Rvs167 during endocytosis. Mol Biol Cell 21: 3054–3069.

65. Sutton RB, Davletov BA, Berghuis AM, Sudhof TC, Sprang SR (1995) Structure of the first C2 domain of synaptotagmin I: a novel Ca2+/phospholipid-binding fold. Cell 80: 929–938.

66. Leadsham JE, Miller K, Ayscough KR, Colombo S, Martegani E, et al. (2009) Whi2p links nutritional sensing to actin-dependent Ras-cAMP-PKA regulation and apoptosis in yeast. J Cell Sci 122: 706–715.

67. Mendl N, Occhipinti A, Muller M, Wild P, Dikic I, et al. (2011) Mitophagy in yeast is independent of mitochondrial fission and requires the stress response gene WHI2. J Cell Sci 124: 1339–1350.

3

Transport and Metabolism Behavior of Brazilein during Its Entrance into Neural Cells

Shuang Zhao[1], Xin-Pei Wang[1], Jing-Fei Jiang[1], Yu-Shuang Chai[1], Yu Tian[2], Tian-Shi Feng[1], Yi Ding[2], Jing Huang[3], Fan Lei[1], Dong-Ming Xing[1], Li-Jun Du[1]*

1 MOE Key Laboratory of Protein Sciences, Laboratory of Molecular Pharmacology and Pharmaceutical Sciences, School of Life Sciences and School of Medicine, Tsinghua University, Beijing, China, 2 Drug Discovery Facility, School of Life Sciences, Tsinghua University, Beijing, China, 3 Department of Chemistry, Virginia Polytechnic Institute and State University, Blacksburg, Virginia, United States of America

Abstract

Brazilein, a natural small molecule, shows a variety of pharmacological activities, especially on nervous system and immune system. As a potential multifunctional drug, we studied the distribution and the transport behavior and metabolic behavior of brazilein in vivo and in vitro. Brazilein was found to be able to distribute in the mouse brain and transport into neural cells. A metabolite was found in the brain and in the cells. Positive and negative mode-MS/MS and Q-TOF were used to identify the metabolite. MS/MS fragmentation mechanisms showed the methylation occurred at the10-hydroxyl of brazilein (10-O-methylbrazilein). Further, catechol-O- methyltransferase (COMT) was confirmed as a crucial enzyme correlated with the methylated metabolite generation by molecular docking and pharmacological experiment.

Editor: Allan Siegel, University of Medicine & Dentistry of NJ - New Jersey Medical School, United States of America

Funding: This work was supported by the National S&T Major Special Project for New Drug R&D Program of China (2012ZX09102-201-008, 2011ZX09101-002-11 and 2012ZX09103-201-041) and the National Natural Science Foundation of China (81073092 and 81374006). The funders had no role in study design, data collection and analysis, decision to publish, or preparation of the manuscript.

Competing Interests: The authors have declared that no competing interests exist.

* Email: lijundu@mail.tsinghua.edu.cn

Introduction

Brazilein (6a,7-dihydro-3,6a,10-trihydroxy-benz[b]indeno[1,2-d]pyran-9(6H)-one, Fig. 1A) is a natural small molecule isolated from dried heartwood of *Caesalpinia sappan* L [1–3]. It has been reported that brazilein exhibits multi-pharmacological activities, such as cardioactive effect [4], immunosuppression [5,6], protection of central and peripheral nerves system [7–9], smooth muscle contraction promotion [10], melanin synthesis suppression [11], anti-oxidant [12,13] and anti-influenza viral activities in vitro [14]. The neural protection effect of brazilein after ischemia/reperfusion injury was studied systematically. It has been reported that this function is correlative with the inflammation suppression effect of brazilein. In the previous research, brazilein was observed that it can inhibit pro-inflammatory cytokine activation, suppress NO production, and inhibit iNOS and cytokine expression [8,9]. Brazilein was also reported to work on 5-HT receptors as an antagonist [15]. Recent studies of brazilein mostly focus on its anticancer activity [16–19]. All of these remind us that brazilein is a potentially valuable drug to be developed.

The pharmacokinetic study in plasma showed that the distribution of $t_{1/2\alpha}$ was 16.89 min, and clearance of $t_{1/2\beta}$ was 280.07 min after intravenous administration of brazilein [20]. Based on previous results, it is necessary to study brazilein's cerebral distribution and transportation through neural cells, in order to well understand the mechanism of brazilein on cerebral protection.

In this research, mice and PC12 cells were used as the models in vivo and in vitro, respectively. HPLC system and confocal microscopic photographs were used to detect the distribution and transportation of brazilein. The distribution of brazilein in the mouse brain was observed and the transport behavior of brazilein through neural cells in vitro was described. A new single substituted metabolite in the mouse brain and in the cell was identified with LC-MS/MS. The related enzyme of this metabolite was confirmed by molecular docking and pharmacological experiment.

Materials and Methods

Animals

Male ICR mice (25–28 g), purchased from Vital River Laboratories (Beijing, China), were kept in the Animal Center of Tsinghua University. Mice were maintained in an environmentally controlled breeding room (temperature 25°C, relative humidity 45–55%, 12 h light/dark cycle). They were fed with standard food pellets and tap water ad libitum. The laboratory animal facility has been accredited by AAALAC (Association for Assessment and Accreditation of Laboratory Animal Care International) and the IACUC (Institutional Animal Care and Use Committee) of Tsinghua University approved all animal protocols used in this study (Approval ID: 2013-DuLJ001).

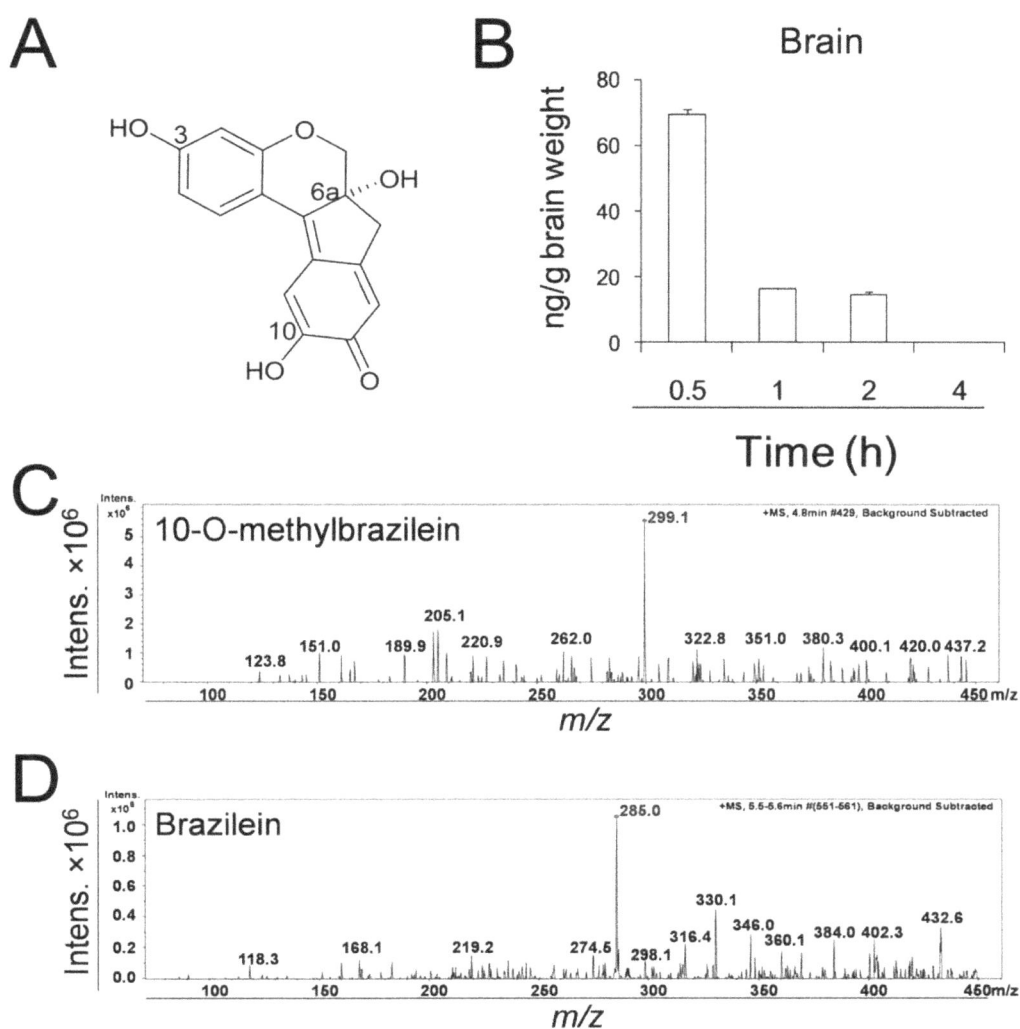

Figure 1. Distribution of brazilein in mice brain and mass spectrum of brazilein and its motabolite. (A) Chemical structure of brazilein. (B) Time course of brazilein contents in mice brain. (C) Mass spectrum of brazilein (*m/z* 285) in mice brains. (D) Mass spectrum of the motabolite (*m/z* 299) in mice brains. The dose of brazilein was 5 mg/kg using introvenouse injection. Data were presented as mean ±S.D. from six independent mice (n = 6).

Cells and reagents

PC12 cells were commercially obtained from the Institute of Basic Medical Sciences, Chinese Academy of Medical Sciences. Methanol and acetonitrile (chromatographic grade), were purchased from Fisher (Germany). All other reagents with analytical grades were purchased from Beijing Chemical Plant.

Experiment procedures

Experiment 1: Brazilein distribution in mice brain. Experimental mice were intravenously administered with brazilein, which was solved in DMSO and then diluted with sterile normal saline solution at a dose of 10 mg/kg (the final concentration of DMSO was controlled less than 1/1000 of concentration). Normal saline with DMSO (same as brazilein groups) served as vehicle control. Then blood and brain samples were taken at 30, 60,120, and 240 min after brazilein administration. The blood samples mixed with heparin were then centrifuged at 4000 rpm for 10 min to get plasma. The brain samples were grinded with saline (pH = 8) to reach homogenate.

Both plasma and brain homogenate were extracted three times with three times the amount of ethyl acetate (ethyl acetate: sample = 3: 1). Then combine and evaporate the supernatant at room temperature. The resulting residue was re-dissolved in 100 μL methanol to determine brazilein and its metabolite with HPLC/MS/MS.

Experiment 2: Drug administration and sample preparing. Brazilein was dissolved in DMSO and diluted with serum-free medium in experiments (DMSO was controlled less than 1/1000 of concentration). The cells used in experiments were cultured to reach a density of 70% and then incubated with serum-free culture medium containing different concentrations of brazilein. After the incubation finished, the drug-containing medium was evaporated, and the cells were washed three times with cold PBS (4°C). After 1.2 mL methanol added, the cells were collected into EP tubes by cell scraper and broken with an ultrasonic cell disruptor. The lysates were centrifuged at 10000 rpm for 5 min and the supernatant was collected. The precipitates were extracted with methanol containing 10%

triethylamine three times. The liquid phase was combined with the supernatant and evaporated with centrifugal concentrator system (Labconco, U.S.) at room temperature. The residue was redissolved in 100 μL of methanol to determine the content of brazilein with HPLC. The precipitates in each tube were redispersed in PBS to quantify the protein content by the method of Bradford using bovine serum albumin (BSA) as a standard [21].

Brazilein transport behavior analysis. Different conditions were applied in order to study brazilein transportation behavior. The cells were administrated with brazilein of different administration doses (0 to 15 μg/mL), at different temperatures (28, 37 and 40°C), with metabolism inhibitor (KCN, Sigma-Aldrich, U.S.) and with catechol-O-methyltransferase inhibitor (entacapone, Selluck Chemicals, U.S.), respectively. Intracellular and extracellular concentrations of brazilein and its metabolite were detected. The intracellular concentration was depicted by the ratios of values determined using HPLC to the protein content determined via Bradford method [21].

Cell culture

PC12 cells were cultured in RPMI1640 at 37°C. The medium, without phenol red, included 10% fetal bovine serum and 5% horse serum.

MTT assay for cytotoxicity

The cytotoxicity of brazilein in PC12 cells was determined using an MTT assay. The MTT assay was operated according to the reference 8.

HPLC/MS System and Conditions

The HPLC system (Waters, U.S.) consisted of a 515 HPLC pump, a 996 Photodiode Array Detector, a Rheodyne 7725i manual injector, and the Empower2 Working Station. Separation was carried out with an XTerra RP$_{18}$ column (5 μm; 3.9×150 mm, Waters). The mobile phase was acetonitrile - water (containing 0.1% formic acid, 10 mM ammonium acetate; 30: 70 v/v). The flow rate was 0.4 mL/min. Detection was performed under a constant temperature (25°C) at the wavelength of 445 nm.

The LC-MS/MS detection was performed with Agilent 1200/ 6340 linear ion-trap LC/MS system. The HPLC condition was the same with that described previously. MS/MS analysis was operated in both positive ion mode and negative ion mode. The parent ions m/z and fragmentation patterns were analyzed to determine compounds.

Q-TOF detection was carried out with Waters Q-TOF LC/ MS, Xevo G2 system. The HPLC condition was described as above and MS detection was conducted in positive ion mode.

Confocal microscopy

PC12 cells were used in experiments when the density reached 70%. Brazilein - treated (5 μg/mL) cells and the control group, no-brazilein-treated cells, were then fixed with 4% paraformaldehyde. Propidium iodide (PI) was used to stain the nucleus of fixed cells. Images were taken with Zeiss LSM 710 Confocal Microscope (Carl Zeiss, Germany) and analyzed using Zen 2009 Light Edition Software. Brazilein and PI were excited at 490 nm and 536 nm, respectively.

Molecular docking

Molecular docking was performed with Autodock 4.2 and presented by PyMOL Molecular Graphics System. The 3D structures of protein COMT with cofactors and brazilein were

downloaded from RCSB Protein Data Bank (PDB code: 1H1D) and Pubchem-NCBI, respectively.

Statistical analysis

Data are expressed as mean ±S.D. Data were statistically analyzed using one-way analysis of variance (ANOVA) with F value determination. The F test was carried out using Excel software for Office 2007 (Microsoft, U.S.). The student's t-test between two groups was performed after the F test. P values below 0.05 were considered statistically significant.

Results and Discussion

Distribution of brazilein

30 min after intravenous administration, brazilein can be found in the mouse brain. And then, it decreased quickly. 4 h after the administration, brazilein was unable to be detected in the mouse brain (Fig. 1B). By using LC-MS, the fragments of brazilein (m/z 285) in the mouse brain were determined (Fig. 1C). Meanwhile, a small amount of metabolite (m/z 299) was detected in the mouse brain using LC-MS (Fig. 1D).

Brazilein can be detected by HPLC-UV

The MTT assay showed that 10 μg/mL brazilein produced significant cytotoxicity to PC12 cells in 24 h (Fig. 2A). Therefore, the safe dose of 5 μg/mL was used in experiments. Under the HPLC conditions described in methods, the chromatographs of blank cells without brazilein, brazilein standards and the test sample in which the cells incubated with 5 μg/mL brazilein for 4 h, were shown in Fig. 2B. The chromatograph of the test sample indicated that an unknown compound was produced after the administration of brazilein (Fig. 2B(III) peak1). The UV-VIS absorption spectrum of this unknown compound was similar to that of brazilein (Figs. 2C–D), which implied that this compound might be a structural analogue of brazilein (Fig. 2B(III) peak2). It is named brazilein-X temporarily. The retention time of brazilein-X and brazilein were 6.01 min and 7.63 min, respectively. This indicates that the brazilein-X is more polarity than brazilein. Besides these two peaks, there is no other peak on HPLC graph (up to 60 min).

The calibration curve was plot linear of HPLC peak areas over the concentration range of 0.1, 0.2, 0.5, 1, 2 and 5 μg/mL of brazilein. The correlation coefficient (r^2) was 0.9994, indicating a good linear relationship between peak areas and brazilein concentrations. The relationship was quantified by the equation: $y = 4.88 \times 10^6 x - 3845.23$, where x represented the concentration of brazilein and y represented the peak areas. The intra-day and inter-day precision were evaluated using three different concentrations (Table 1 and Table 2). Maximal CV value was 2.59% for intra-day and 5.29% for inter-day precision, indicating HPLC is quite a precise detection method for brazilein. The recovery of brazilein was determined by comparing the data obtained by standard with the same concentration brazilein after the whole extraction procedure described in methods. The average recovery was 46.77%.

Brazilein in PC12 cells

We used confocal assay to detect the entrance of brazilein into PC12 cells. Brazilein emits green fluorescence at an excitation wavelength of 449 nm. In the experiments, the green fluorescence can be observed after 1 h of brazilein (5 μg/mL) into the medium, and it's getting more obviously at 4 h. (Figs. 3A–B).

The concentration of intracellular and extracellular brazilein was quantified by HPLC. The concentration versus time curve of

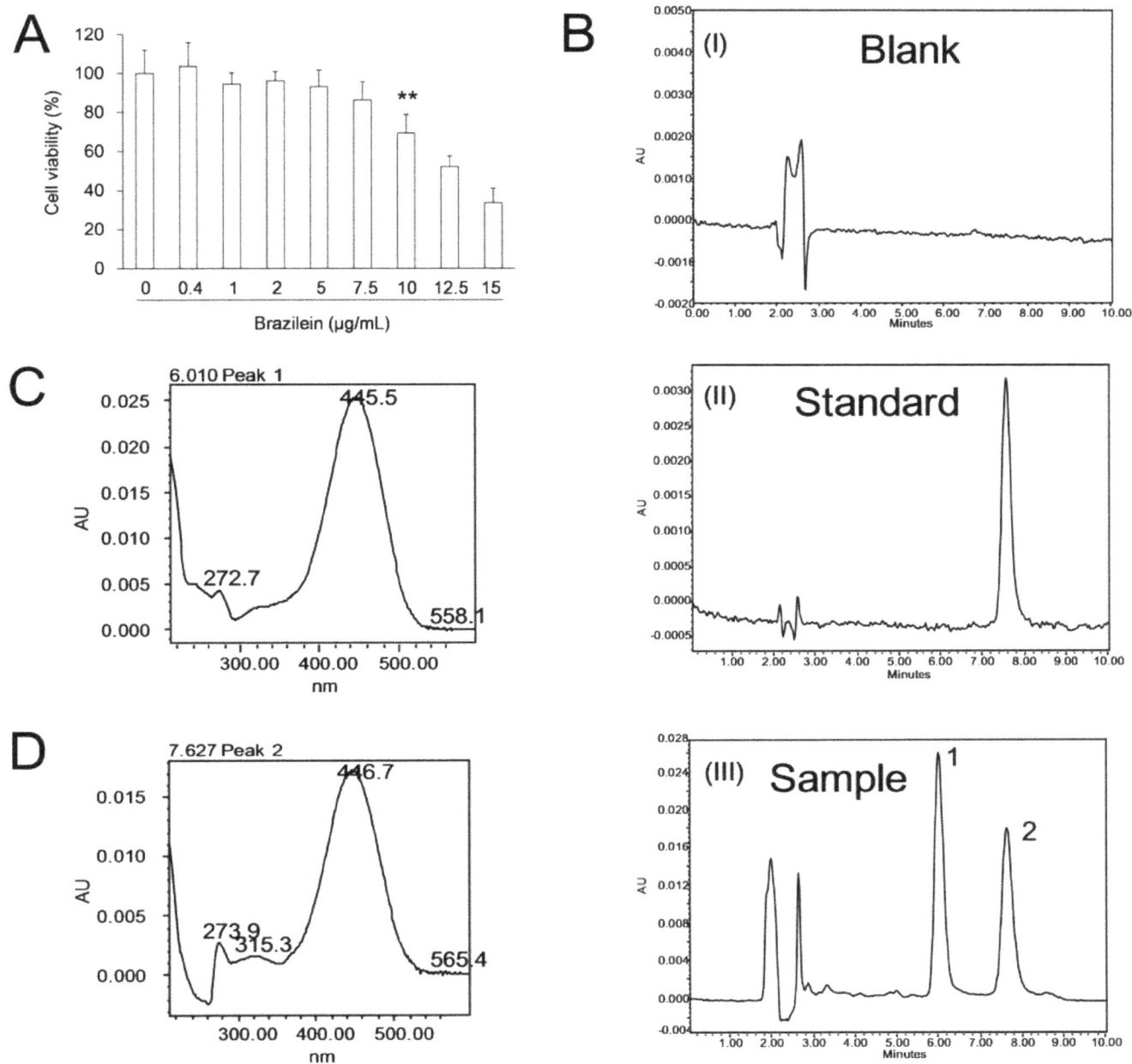

Figure 2. Cytotoxicity assay and HPLC detection of brazilein. (A) The chemical structure formula of brazilein. (B) Cytotoxicity of brazilein in the MTT assay. In the assay, the group with 0 µg/ml brazilein was considered as the control. Data were presented as mean ±S.D. from six independent experiments (n = 6). ** $p<0.01$ *v.s.* the control. (B) The HPLC chromatograms of brazilein: B (I) is the chromatograms of cell lysate as the blank control; B (II) represents the brazilein standard; B (III) represents the cells administrated with 5 µg/mL brazilein for 4 h. Peak 2 in B (III) shares the same retention time with brazilein standard while peak 1 is ahead of the standard. (C) UV spectrogram of peak 1. (D) UV spectrogram of peak 2 (brazilein standard).

brazilein in cells showed that the intracellular brazilein reached the maximum at 4 h, and then gradually decreased (Fig. 3C). Meanwhile, the concentration of extracellular brazilein decreased (Fig. 3D). The negative control (medium containing brazilein but without cells) exhibited that concentration of brazilein did not change over time and no new metabolites were produced. This suggested that brazilein indeed entered the cell, therefore decreased the extracellular concentration and increased the intracellular concentration. While brazilein was detected in the

cells, a metabolite (named brazilein-X temporarily) can be found. Along the time, the intracellular metabolite increased, as well as the metabolite in the medium increased (Figs. 3E–F). It is suggested that the metabolite was generated in the cells and discharged into the medium.

Brazilein enters cells by passive transportation

In our prior MTT assay experiment, 20 µg/mL brazilein would produce significant cytotoxicity to PC12 cells in 4 h, and 15 µg/

Table 1. Validation of the intra- and inter-day precision of brazilein.

Spiked concentration (μg/ml)	Measured concentration (μg/ml)[a]	Accuracy (%)	CV (%)
intra-day			
0.2	0.189±0.005	94.58	2.59
1	0.953±0.018	95.25	1.82
5	4.877±0.067	97.53	1.33
inter-day			
0.2	0.184±0.010	91.81	5.29
1	0.968±0.034	96.79	3.42
5	4.861±0.130	97.21	2.60

[a]Each value represents the mean ± S.D. (n = 3).

mL brazilein was the safe dosage without any cytotoxicity to the cells (Fig. S1 in File S1). Therefore, we choose 15 μg/mL brazilein as a safety dosage in the 4 hour-experiment. In the experiment, the intracellular concentration of brazilein increased in a concentration-dependent manner (brazilein range of 0 to 15 μg/mL), the correlation coefficient was 0.9851 (Fig. 4A). This indicated that brazilein did not appear transport saturation phenomenon within the dose range, while the saturation might occur in the process of active transportation or endocytosis because of the restrictions of the transporter numbers.

When KCN was introduced to cells as an energy generation inhibitor, it was showed that either low concentration (0.25 mM) or high concentration (2.5 mM) of KCN did not significantly change the content of brazilein in cells or in medium (Figs. 4C–D). This suggested that the entrance of brazilein is an energy-independent process, implying that the transportation of brazilein was not an active transport or endocytosis process but a passive transportation. Under the high temperature (40°C), the intracellular brazilein was found to significantly increase, while extracellular brazilein decreased (Figs. 4E–F). This indicated that brazilein transport process was temperature-dependent.

When the administrated brazilein increased, the generation of brazilein-X increased before reaching at a plateau region (Fig. 4B). The correlation coefficient of the logarithmic fitting is 0.9831. KCN also significantly inhibited brazilein-X generation (Fig. 4G). In addition, under 37°C and 40°C, the content of intracellular brazilein-X has significantly exceeded that under 28°C (Fig. 4H). These results indicated that brazilein-X was generated in dose-dependent, energy-dependent and temperature-dependent manners.

Identification of the metabolite with Ion trap MS and Q-TOF MS under positive and negative mode

LC/MS analysis showed that the molecular weight of the metabolite brazilein-X was 299, which was 14 more than the molecular weight of brazilein (Fig. 5). The MS/MS results showed that brazilein-X shared the same fragmentation pattern with brazilein in positive ion mode, and the m/z of each fragment of brazilein-X was 14 more than the corresponding fragment of brazilein. In the negative mode of MS/MS, the quasi-molecular ion peaks of brazilein-X lost a fragment with molecular weight of 15. It can be speculated that this new metabolite was a methylation of brazilein (methyl-brazilein). We used high resolution MS (Waters Q-TOF LC/MS, XevoG2) to identify the methylation of brazilein. The mass spectra were matched with our previous results (Figs. S2 and S3 in File S1). We used the Elemental Composition Analysis Function of the mass spectrometry to characterize the metabolite. This elemental composition analysis is based on isotope ratio. It showed that compared with brazilein, the metabolite had one more carbon and two more hydrogen, which supported our inference of methylation of brazilein (Tables. S1 & S2).

There are three hydroxyl groups that are the potential methylation sites in brazilein. To determine the location of methylation, we proposed the fragmentation mechanisms via positive ion mode of MS/MS (Fig. 6A) in the work by Hulme *et al.* in 2005 [22]. They rationalized the fragmentation mechanism of brazilein and some analogues, but without the fragmentation of methyl-brazilein under positive ESI. The m/z of 189 in methyl-brazilein is proposed to follow the same fragmentation mechanism of m/z 175 of brazilein. The m/z of 137 in methyl-brazilein is proposed to follow the same fragmentation mechanism of m/z 123 of brazilein (Fig. 6B). The methylation site and the chemical structure were therefore determined and the new metabolite might be determined to be 10-O-methylbrazilein (Fig. 6C).

Table 2. Recovery of brazilein.

Spiked concentration (μg/ml)	Measured concentration (μg/ml)[a]	Recovery (%)	CV (%)
0.2	0.097±0.007	48.53	6.95
1	0.462±0.017	46.22	3.73
5	2.278±0.124	45.55	5.44

[a]Each value represents the mean ±S.D. (n = 3).

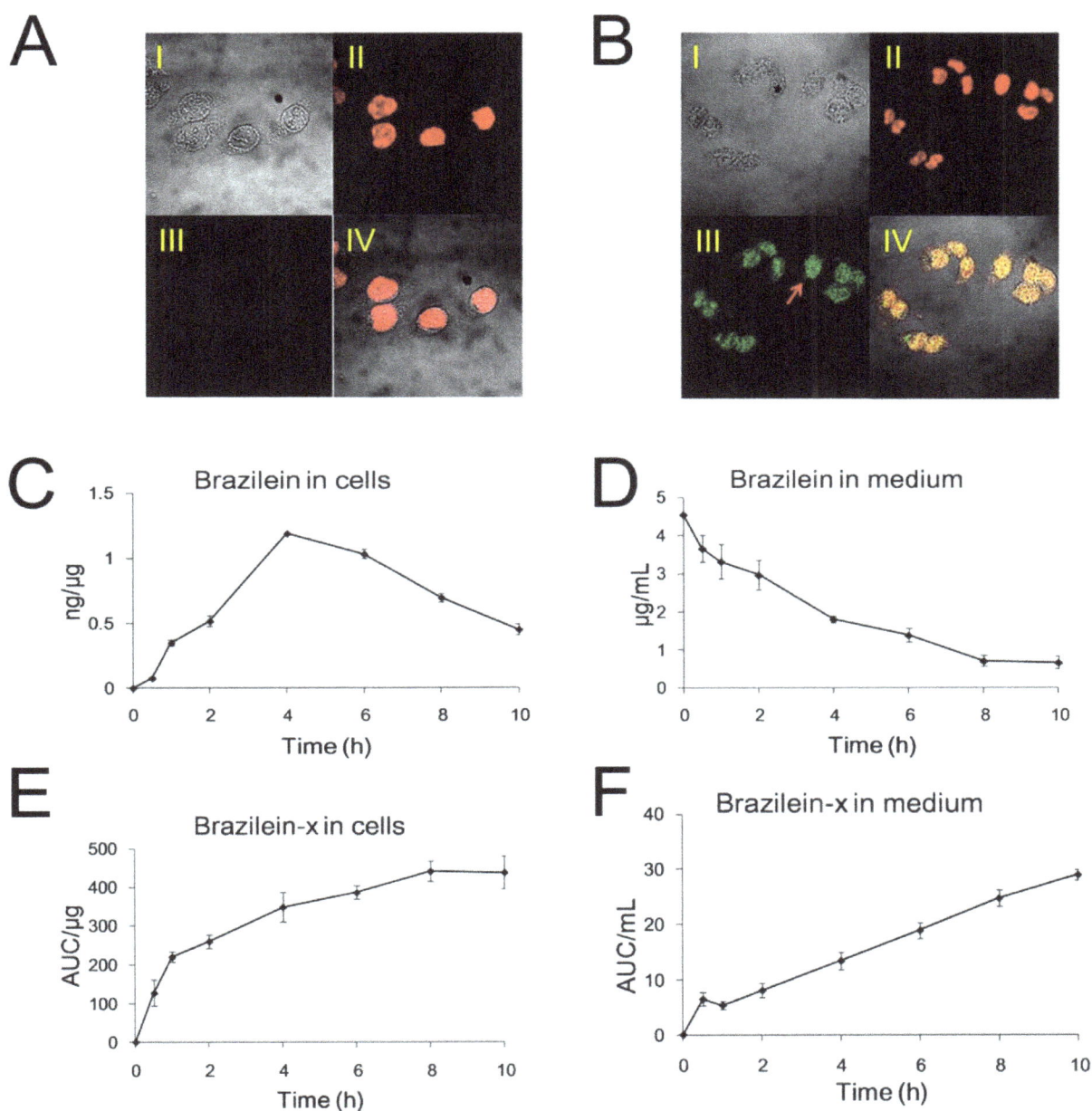

Figure 3. Entrance of brazilein into cells. (A) Confocal microscopy image of cells without brazilein. (B) Image of cells administrated with 5 µg/mL brazilein for 4 h. A (I) and B (I) represent PC12 cells in bright field; A (II) and B(II) represent the nucleus stained by PI; B (III) represents the fluorescence of brazilein; (IV) is the merged image of (I), (II) and (III) in A and B, respectively. (C–D) Concentration-time curve of brazilein in cells and in medium. (E–F) Concentration-time curve of brazilein-X in cells and in medium. Brazilein was added to the culture with concentration of 5 µg/mL. Data were presented as mean ±S.D. from three independent experiments (n = 3).

The phenomenon that the peak of 10-O-methylbrazilein appeared earlier than the peak of brazilein in the HPLC chromatogram could support the inference of 10-O-methylbrazilein (Fig. 6C). As a general rule, methylation of a hydroxyl group decreases the molecular polarity. Then the retention time in the RP-HPLC will increase, which is contradicted with the observed phenomenon in our research. These could be explained that the intra-molecular hydrogen bond formed between 10-hydroxyl group and the adjacent carbonyl group in brazilein, causing the molecular polarity reduced. Methylation process was able to break

this hydrogen bonding, increasing the polarity and shortening the retention time.

In previous chemical research, some of its analogues have been isolated from *Caesalpinia sappan* L, including brazilin (the hydrogenation form of brazilein), 3′-O-methylbrazilin and neo-protosappanin [23]. In 2009, Yen, C. T. *et al*. reported the total synthesis of brazilein and its derivatives of replacing the hydroxyl groups in brazilein and brazilin [24]. Disubstituted and trisubstituted derivatives, such as trimethyl brazilin [25], have been synthesized [26]. But region selective single substituted compounds have not been reported yet. Previous structure-activity research of

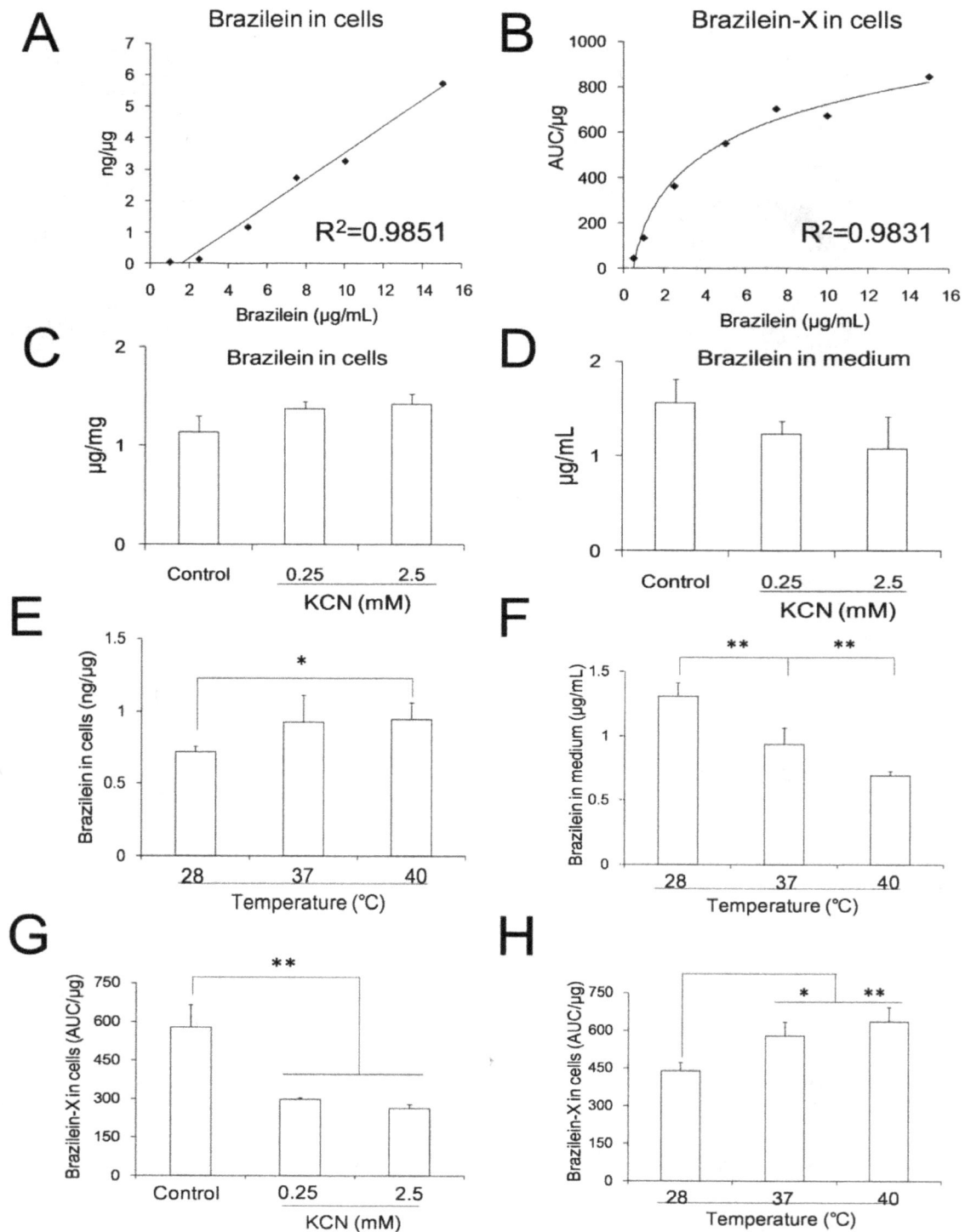

Figure 4. Effects of administrated dose, inhibitor of cytochrome oxidase and temperature on transportation and metabolism of brazilein. (A) – (B) Concentrations of brazilein and brazilein-X in cells when cells were administrated with different doses of brazilein for 4 h. Linear fitting and logarithm fitting are used to analyze data in (A) and (B). (C) – (D) Effect of cytochrome oxidase inhibitor (KCN) on concentration of brazilein in cells and in medium. (E) – (F) Effect of temperature on concentration of brazilein in cells (E) and in medium (F). (G) Effect of cytochrome oxidase inhibitor (KCN) on concentration of brazilein-X in cells. (H) Effect of temperature on concentration of brazilein-X in cells. In Figures (C) – (H), cells were administrated with 5 μg/mL brazilein for 4 h. Data were presented as mean ±S.D. from three independent experiments (n = 3). * $p < 0.05$, ** $p < 0.01$ *v.s.* controls.

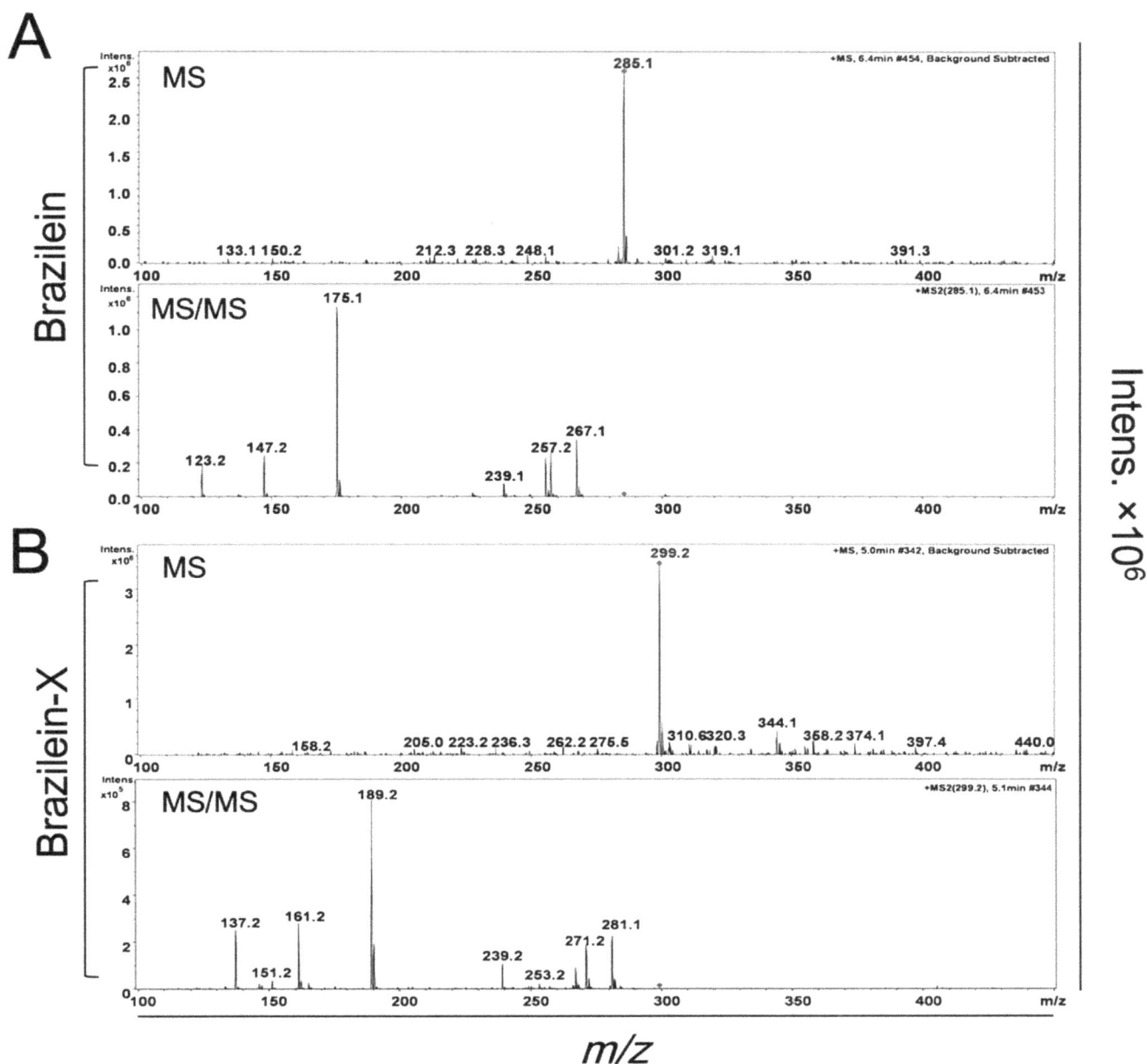

Figure 5. Mass spectrum (MS, MS/MS) of brazilein and brazilein-X in positive ion mode. (A) Mass spectrum of brazilein (*m/z* 285). (B) Mass spectrum of brazilein-X (*m/z* 299).

brazilein in our laboratory showed that it was difficult to selective replace a single hydroxyl group by chemical methods because of the similarity between 3-hydroxyl and 10-hydroxyl in brazilein.

10-O-methylbrazilein, as a single-substituted derivative, was not reported in the previous studies. Research groups rarely reported single-substituted derivatives of brazilein by chemical reaction [27,28]. Thus, the chemical properties and biological activities of this new analogue were yet unknown. The issue merits further study.

COMT contributes to the methylation of brazilein

Computer-based molecular docking was conducted to further demonstrate the inferences above and to understand the enzymatic reaction of brazilein methylation. The results showed that brazilein was able to combine with the active sites of catechol-O-methyltransferase (COMT) and the S-adenosyl methionine (SAM), which is a common co-substrate involved in methyl group transfers and served as methyl donor (Fig. 7A, PDB code: 1H1D). The hydrogen bonds between brazilein and COMT predicted by docking were showed in Fig. 7B. It is showed that Lys144 and Mg^{2+} ion linked with the 10-hydroxyl which was methylated in brazilein, and the carbonyl group adjacent to the 10-hydroxyl had hydrogen bonds with carboxylate of Glu199 and amidogen of Asn170. Brazilein is able to well fit into the electron density map of COMT (Fig. 7C). The detailed analysis with docking predicted that brazilein could combine with the COMT similarly with the reported substrates and inhibitors of COMT [29,30].

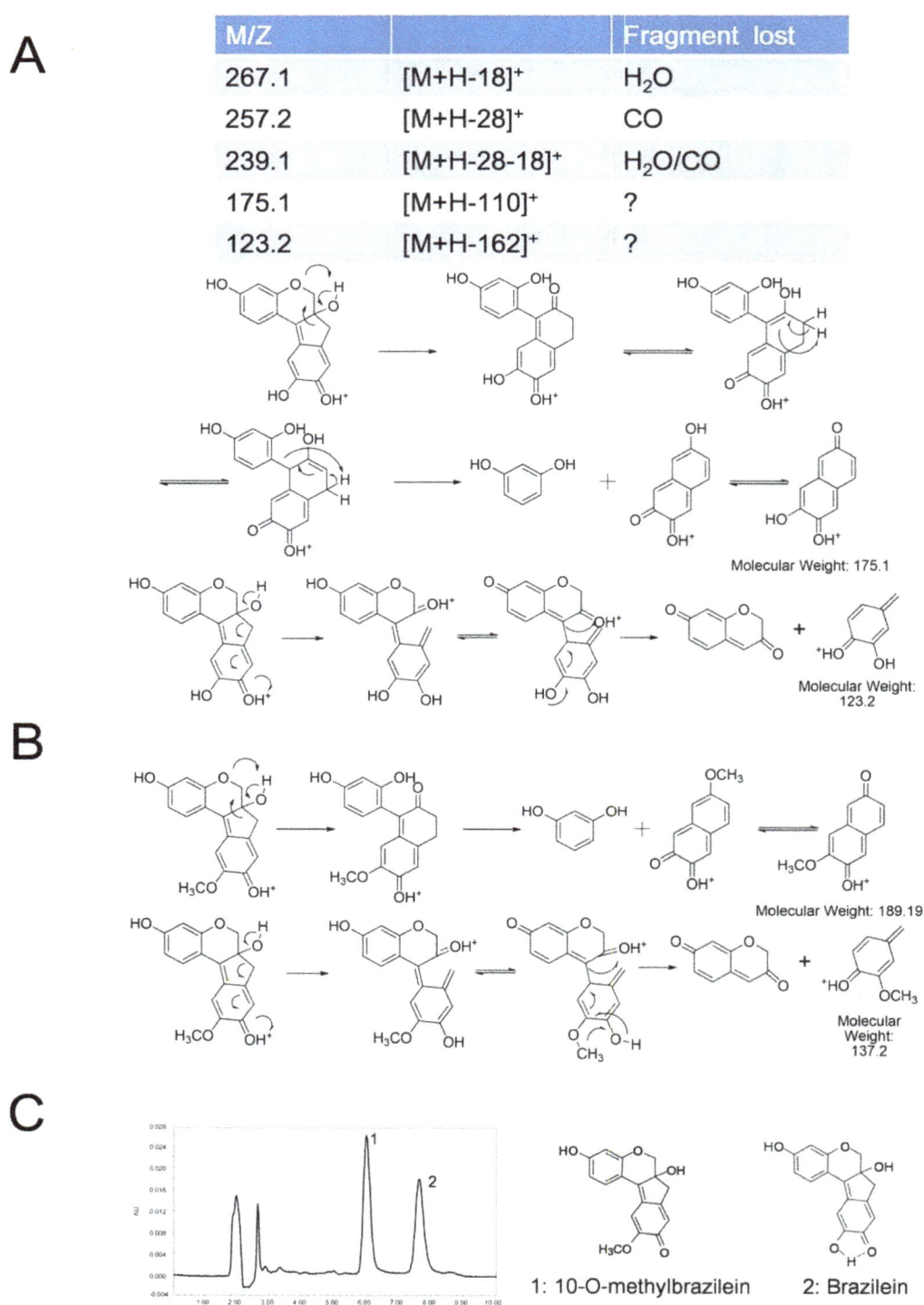

Figure 6. MS analysis of brazilein and its metabolitebrazilein-X (10-O-methylbrazilein). (A) Positive ion ESI mass spectra and fragmentation of brazilein. (B) Fragmentation of methyl-brazilein. (C) HPLC chromatogram of 10-O-methylbrazilein and brazilein.

In the experiments, different concentrations (15, 150, 1500 nM) of entacapone, an inhibitor of COMT, were administrated to PC12 cells with 5 μg/mL brazilein, in order to confirm the relationship between COMT and brazilein metabolite. Results showed that entacapone significantly inhibited the transition of brazilein to metabolite and rendered in a dose-dependent manner

(Fig. 7D). These results supported the computer's prediction, indicating that COMT could methylate brazilein into 10-O-methylbrazilein.

Common substrates of catechol-O-methyltransferase are catechol derivatives, including endogenous molecules such as dopamine, epinephrine and norepinephrine [31] and varied exogenous

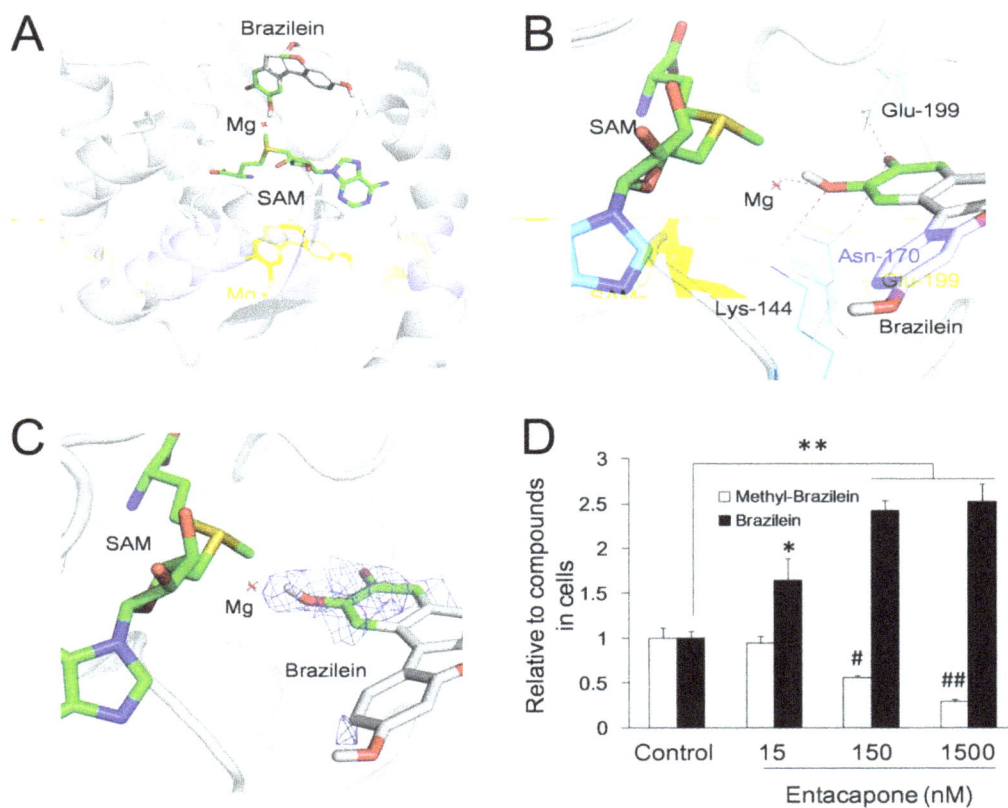

Figure 7. Methylation of brazilein by catechol-O-methyltransferase (COMT). (A) Molecular Docking model of brazilein and COMT with cofactors S-adenosyl methionine (SAM, served as methyl donor) and Mg^{2+} (PDB code: 1H1D). The white cartoon represents COMT. Red dot represents Mg^{2+}. SAM and brazilein are showed as stick structure (colored by atom type: greencarbon, redoxygen, bluenitrogen, yellowsulfur). (B) Hydrogen bonds between brazilein and COMT with Mg^{2+}. Residues linking to brazilein by hydrogen bonds are labeled and showed of blue lines. Red line of dashes represents the hydrogen bonds. (C) The electron density map of COMT around brazilein. Molecular docking was performed using Auto-Dock Vina.Figures were drawn using PyMOL Molecular Graphics System. (E) Effect of COMT inhibitor (entacapone) on transformation of brazilein to methyl-brazilein (10-O-methylbrazilein). Cells were administrated with 5 μg/mL brazilein for 4 h. Data were presented as mean ±S.D. from three independent experiments (n = 3). * $p<0.05$, ** $p<0.01$ v.s. controls. # $p<0.05$, ## $p<0.01$ v.s. controls.

compounds like 3,5-dinitrocatechol and catechol containing adenine replacement [32]. Tolcapone and entacapone which was used in this research have been used as inhibitors of COMT in the therapy of Parkinson's disease [33]. Previous researchers have extensively studied the mechanism of COMT methylation process from structural perspective [30] and functional perspective [34]. Docking model in this research displayed the combination between brazilein and COMT complex with SAM and Mg^{2+} (PDB code: 1H1D), which was corresponding to the proposed catalytic mechanism [27,35]. From molecular level, brazilein bond to the enzyme active site which is near the surface of the enzyme and was close to the methyl donor SAM and cofactor Mg^{2+}. From view of bond level, Lys144, in which NH_2 acted as the catalytic core to deprotonate hydroxyl in catechol, displayed a connection to the 10-hydroxyl of brazilein through hydrogen bonding. Mg^{2+}, Asn170 and Glu199 also had hydrogen bonds with hydroxyl and carbonyl groups for "anchoring" effect.

Though brazilein does not have catechol structure, the computer-based molecular docking and the pharmacological experiment implied that brazilein was a substrate of COMT. This may ascribe to the carbonyl group adjacent to the 10-hydroxyl, which makes the similar spatial structure and bond connection as catechol. It suggested that other compounds with similar structure might also be the substrates of COMT. Because

the known COMT competitive inhibitors are almost catechol structure, this discovery may provide a new thought to search and design COMT inhibitors, especially served as drugs for Parkinson's disease. However, this kind of inhibitors may still have similar side effect with existed Parkinson's disease drugs, such as constipation. Tolcapone and entacapone also have this side effect. In addition, biosynthesis with COMT is probably a new method to synthesize the regioselective single-methylated brazilein and may extend to other similar compounds. All these assumptions are remained to be studied.

Conclusions

Taken together, brazilein is able to distribute in the mouse brain and enter PC12 cells via a passive transportation. The transportation of brazilein was a dose-dependent, non-saturated, energy-independent and temperature-dependent process. During this process, brazilein could be transformed into 10-O-methylbrazilein in the brain and neural cells. COMT contributes to the transformation. These results are of benefit to understand the neural protect effects and the metabolism of brazilein.

Supporting Information

File S1   Figure S1, MTT assay of brazilein for 4 hours. In the assay, the group with 0 µg/mL brazilein was considered as the control. Data were presented as mean ±S.D. from six independent experiments (n = 6). ****** $p < 0.01$ *v.s.* the control. **Figure S2,** Mass spectrum of brazilein and the metabolite in high resolution MS (Waters Q-TOF LC/MS, XevoG2). **Figure S3,** MS/MS of brazilein and the metabolite in high resolution MS (Waters Q-TOF LC/MS, XevoG2).

Author Contributions

Conceived and designed the experiments: SZ LJD. Performed the experiments: SZ XPW JFJ YT. Analyzed the data: SZ YSC TSF. Contributed reagents/materials/analysis tools: TSF YD FL DMX LJD. Wrote the paper: SZ JH LJD.

References

1. Kim DS, Baek NI, Oh SR, Jung KY, Lee IS, et al. (1997) NMR assignment of Brazilein. Phytochemistry 46: 177–178.
2. De Oliveira LFC, Edwards HGM, Velozo ES, Nesbitt M (2002) Vibrational spectroscopic study of brazilin and brazilein, the main constituents of brazilwood from Brazil. Vib Spectrosc 28: 243–249.
3. Wang ZX, Ding Y, Du LJ, Wang W (2008) Quality standard of Caesalpinia-sappan L. Cent South Pharm 6: 460–465.
4. Zhao YN, Pan Y, Tao JL, Xing DM, Du LJ (2006) Study on Cardioactive effects of brazilein. Pharmacology 76: 76–83.
5. Oh SR, Kim DS, Lee IS, Jung KY, Lee JJ, et al. (1998) Anticomplementary activity of constituents from the heartwood of Caesalpiniasappan. Planta Med 64: 456–458.
6. Ye M, Xie WD, Lei F, Meng Z, Du LJ, et al. (2006) Brazilein, an Important Immunosuppressive Component from CaesalpiniaSappan L. Int Immunopharmacol 6: 426–432.
7. Cao J, Li LS, Liu B, Liu LY, Yin WT, et al. (2011) Activation of growth-associated protein by intragastric brazilein in motor neuron of spinal cord connected with injured sciatic nerve in mice. Chem Res Chin Univ 27: 254–257.
8. Shen J, Zhang HY, Lin H, Su H, Du LJ, et al. (2007) Brazilein Protects the Brain against Focal Cerebral Ischemia Reperfusion Injury Correlating to Inflammatory Response Suppression. Eur J Pharmacol 558: 88–95.
9. Zhang Z, Cao J, Yin W, Li L, Jiang R, et al. (2011) Brazilein intervention for sciatic nerve injury in BALB/c mice. Neural Regen Res 6: 908–913.
10. Shen J, Yip S, Wang ZX, Wang W, Xing DM, et al. (2008) Brazilein-induced Contraction of Rat Arterial Smooth Muscle Involves Activation of Ca^{2+} Entry and ROK, ERK Pathways. Eur J Pharmacol 580: 366–371.
11. Mitani K, Takano F, Kawabata T, Allam AE, Ota M, et al. (2013) Suppression of melanin synthesis by the phenolic constituents of sappanwood (caesalpinia-sappan). Planta Med. 79: 37–44.
12. Hu J, Yan XL, Wang W, Wu H, Hua L, et al. (2008) Antioxidant activity in vitro of three constituents from Caesalpiniasappan L. Tsinghua Sci Technol 13: 474–479.
13. Kabbash A, Yagi A, Ishizu T, Haraguchi H, Fujioka T, et al. (2008) Isolation and identification of a new homoisoflavan with potent antioxidant activity from Commelinaelegans. Saudi Pharm J 16: 25–32.
14. Liu AL, Shu SH, Qin HL, Lee SMY, Wang YT, et al. (2009) In vitro anti-Influenza viral activities of constituents from caesalpiniasappan. Planta Med 75: 337–339.
15. Wang XK, Lei F, Chai YS, Wang YG, Zhan HL, et al. (2011) Brazilein, a selective 5-HT receptors antagonist in mouse uterus. Chin J Nat Med 9: 338–344.
16. Hsieh CY, Tsai PC, Chu CL, Chang FR, Chang LS, et al. (2013) Brazilein suppresses migration and invasion of MDA-MB-231 breast cancer cells. Chem Biol Interact 204: 105–115.
17. Liang CH, Chan LP, Chou TH, Chiang FY, Yen CM, et al. (2013) Brazilein from Caesalpiniasappan L. antioxidant inhibits adipocyte differentiation and induces apoptosis through caspase-3 activity and anthelmintic activities against hymenolepis nana and anisakis simplex. Evid Based Complement Alternat Med No.864892. DOI: 10.1155/2013/864892.
18. Tao L, Li J, Zhang J (2011) Brazilein overcame ABCB1-mediated multidrug resistance in human leukaemiaK562/AO2 cells. Afr J Pharm Pharmacol 5: 1937–1944.
19. Zhong X, Wu B, Pan YJ, Zheng S (2009) Brazilein inhibits survivin protein and mRNA expression and induces apoptosis in hepatocellular carcinoma HepG2 cells. Neoplasma 56: 387–392.
20. Lan JQ, Hu J, Li BN, Xing DM, Liu CC, et al. (2008) Determination of brazilein in rat plasma after intravenous administration by HPLC. Biomed Chromatogr 22: 1201–1205.
21. Bradford MM (1976) A rapid sensitive method for the quantitation of microgram quantities of protein utilizing the principle of proteindye binding. Anal Biochem 72: 248–254.
22. Hulme AN, McNab H, Peggie DA, Quye A (2005) Negative ion electrospray mass spectrometry of neoflavonoids. Phytochemistry 66: 2766–2770.
23. Namikoshi M, Nakata H, Yamada H, Nagai M, Saitoh T (1987) Homoiso-flavonoids and Related Compounds. II. Isolation and Absolute Configurations of 3, 4-Dihydroxylated Homoisoflavans and Brazilins from Caesalpiniasappan L. Chem Pharm Bull 35: 2761–2773.
24. Yen CT, Nakagawa-Goto K, Hwang TL, Wu PC, Morris-Natschke SL, et al. (2010) Antitumor Agents.271: Total Synthesis and Evaluation of Brazilein and Analogs as Anti-inflammatory and Cytotoxic Agents. Bioorg Med Chem Lett 20: 1037–1039.
25. Davis FA, Chen BC (1993) Enantioselective Synthesis of (+)-O-Trimethylsappa-none B and (+)-O-Trimethylbrazilin. J Org Chem 58: 1751–1753.
26. Huang Y, Zhang J, Pettus TR (2005) Synthesis of (+/−)-brazilin using IBX. Org Lett 7: 5841–5844.
27. Pan CX, Guan YF, Zhang HB (2012) Synthesis of Aza-brazilin/diarylindan-Based Hybrid. Chin J Org Chem 32: 1116–1120.
28. Wang XQ, Zhang HB, Yang XD, Zhao JF, Pan CX (2013) Enantioselective total synthesis of (+)-brazilin, (−)-brazilein and (+)-brazilide A. Chem Commun 49: 5405–5407.
29. Vidgren J, Svensson LA, Liljas A (1994) Crystal structure of catechol O-methyltransferase. Nature 368: 354–358.
30. Bonifácio MJ, Archer M, Rodrigues ML, Matias PM, Learmonth DA, et al. (2002) Kinetics and crystal structure of catechol-o-methyltransferase complex with co-substrate and a novel inhibitor with potential therapeutic application. Mol Pharmacol 62: 795–805.
31. Mannisto PT, Kaakkola S (1999) Catechol-O-methyltransferase (COMT): biochemistry, molecular biology, pharmacology, and clinical efficacy of the new selective COMT inhibitors. Pharmacol Rev 51: 593–628.
32. Ellermann M, Paulini R, Jakob-Roetne R, Lerner C, Borroni E, et al. (2011) Molecular recognition at the active site of catechol-O-methyltransferase (COMT): adenine replacements in bisubstrate inhibitors. Chemistry 17: 6369–6381.
33. Haasio K (2010) Toxicology and Safety of COMT Inhibitors. Int Rev Neurobiol 95: 163–189.
34. Lotta T, Vidgren J, Tilgmann C, Ulmanen I, Melén K, et al. (1995) Kinetics of human soluble and membrane-bound catechol O-methyltransferase: a revised mechanism and description of the thermolabile variant of the enzyme. Biochemistry 34: 4202–4210.
35. Zheng YJ, Bruice TC (1997) A theoretical examination of the factors controlling the catalytic efficiency of a transmethylation enzyme-catechol O-methyltransferase. J Am Chem Soc 119: 8137–8145.

4

Sperm-Associated Antigen 6 (SPAG6) Deficiency and Defects in Ciliogenesis and Cilia Function: Polarity, Density, and Beat

Maria E. Teves[1], Patrick R. Sears[3], Wei Li[1], Zhengang Zhang[1,4], Waixing Tang[5], Lauren van Reesema[1], Richard M. Costanzo[6], C. William Davis[7], Michael R. Knowles[7], Jerome F. Strauss III[1,2], Zhibing Zhang[1,2]*

1 Department of Obstetrics and Gynecology, Virginia Commonwealth University, Richmond, Virginia, United States of America, 2 Department of Biochemistry and Molecular Biology, Virginia Commonwealth University, Richmond, Virginia, United States of America, 3 Cystic Fibrosis Center, University of North Carolina, Chapel Hill, North Carolina, United States of America, 4 Department of Infectious Diseases, Tongji Medical College, Huazhong University of Science and Technology, Wuhan, Hubei, China, 5 Department of Otorhinolaryngology, University of Pennsylvania, Philadelphia, Pennsylvania, United States of America, 6 Department of Physiology and Biophysics, Virginia Commonwealth University, Richmond, Virginia, United States of America, 7 Department of Cell & Molecular Physiology of Medicine, University of North Carolina, Chapel Hill, North Carolina, United States of America

Abstract

SPAG6, an axoneme central apparatus protein, is essential for function of ependymal cell cilia and sperm flagella. A significant number of Spag6-deficient mice die with hydrocephalus, and surviving males are sterile because of sperm motility defects. In further exploring the ciliary dysfunction in Spag6-null mice, we discovered that cilia beat frequency was significantly reduced in tracheal epithelial cells, and that the beat was not synchronized. There was also a significant reduction in cilia density in both brain ependymal and trachea epithelial cells, and cilia arrays were disorganized. The orientation of basal feet, which determines the direction of axoneme orientation, was apparently random in Spag6-deficient mice, and there were reduced numbers of basal feet, consistent with reduced cilia density. The polarized epithelial cell morphology and distribution of intracellular mucin, α-tubulin, and the planar cell polarity protein, Vangl2, were lost in Spag6-deficient tracheal epithelial cells. Polarized epithelial cell morphology and polarized distribution of α-tubulin in tracheal epithelial cells was observed in one-week old wild-type mice, but not in the Spag6-deficient mice of the same age. Thus, the cilia and polarity defects appear prior to 7 days post-partum. These findings suggest that SPAG6 not only regulates cilia/flagellar motility, but that in its absence, ciliogenesis, axoneme orientation, and tracheal epithelial cell polarity are altered.

Editor: Yulia Komarova, University of Illinois at Chicago, United States of America

Funding: This research was supported by NIH grant HD076257, Virginia Commonwealth University Presidential Research Incentive Program (PRIP) and Massey Cancer Award (to ZZ), NIH grants HD37416 (JFS), and HL071798 (MRK). The authors also thank Pamela J. Gigliotti for technical support with the tissue processing at the VCU Biological Macromolecule Core Facility, supported, in part, with the funding from NIH-NCI Cancer Center Core Grant 5P30CA016059. The authors declare no competing financial interests. The funders had no role in study design, data collection and analysis, decision to publish, or preparation of the manuscript.

Competing Interests: The authors have declared that no competing interests exist.

* Email: zzhang4@vcu.edu

Introduction

Mammalian SPAG6 is the orthologue of PF16, a component of the central apparatus of the "9+2" axoneme of the green algae model organism, *Chlamydomonas reinhardtii* [1]. In *Chlamydomonas*, PF16 protein is present along the length of the flagella, and immunogold labeling localizes the PF16 protein to a single microtubule of the central pair. Mutations in the *Chlamydomonas* PF16 gene cause flagellar paralysis, and PF16 is believed to be involved in C1 central microtubule stability and flagellar motility [2]. In addition to *Chlamydomonas reinhardtii*, SPAG6/PF16 has been shown to regulate flagellar motility in other models, including trypanosomes, Plasmodium, and Giardia [3], [4,5].

Gene targeting has been used to create mice lacking SPAG6 [6]. Approximately 50% of *Spag6*-deficient animals died from hydrocephalus before adulthood, and males surviving to maturity were infertile. Even though an abnormal axoneme ultrastructure

was discovered in the *Spag6*-deficient sperm [6], cilia of brain ependymal cells and trachea epithelial cells from the mutant mice contained "9+2" axonemes that appeared to be grossly intact [7]. However, brain ependymal cells of *Spag6*-deficient mice are functionally defective since hydrocephalus develops.

In further characterizing the cilia abnormalities of *Spag6*-deficient mice, we discovered that ciliary beat frequency was significantly reduced. The mutant mice also had fewer trachea and ependymal cilia, and these cilia were arrayed in a random fashion on the cell surface. The central pair orientation differed significantly between cilia of the *Spag6*-deficient mice, reflecting the random orientation of basal feet. The polarized epithelial cell morphology and distribution of α-tubulin and planar cell polarity protein, Vangl2, were lost in *Spag6*-deficient tracheal epithelial cells. These findings suggest that mouse SPAG6 has multiple functions: it regulates cilia/flagellar motility through the central pair apparatus; but also plays a role in ciliogenesis, axoneme orientation, and cell polarity.

Materials and Methods

Spag6 and Spag16L mutant mice

Spag6 and *Spag16L* mutant mice were generated previously in our laboratory [6,8]. All animal work was approved by Virginia Commonwealth University's Institutional Animal Care & Use Committee (protocol #AM10297 and AD10000167) in accordance with Federal and local regulations regarding the use of non-primate vertebrates in scientific research.

High-speed video analysis of ciliary beat frequency

Ciliary beat frequency was assessed with the Sisson–Ammons video analysis (SAVA) system (Ammons Engineering, Mt. Morris, MI) [9]. Tracheas from wild type and *Spag6*-deficient mice (3 weeks old) were removed and video movies were taken within five minutes with an Nikon Eclipse TE-2000 inverted microscope (×40 phase-contrast objective) equipped with an ES-310 Turbo monochrome high-speed video camera (Redlake, San Diego, CA) set at 125 frames per second. The ciliary beat pattern was evaluated on slow-motion playbacks.

Transmission electron microscopy

For transmission electron microscopy (TEM), the samples (from 3 week old mice) were cut into small sections (2×2 mm) and fixed in 2.5% glutaraldehyde (PH = 7. 4) for 6–8 hours at 4°C. They were washed and post fixed in 2% OsO4 for 1 hour, at 4°C. The tissue was dehydrated through ascending series of ethanol concentrations and embedded in araldite CY212. Semi thin sections (1 μm) were cut and stained with toluidine blue. Ultra-thin sections (60–70 nm) were cut and stained with uranyl acetate and alkaline lead citrate.

Scanning electron microscopy

Specimens (from 3 week old mice) were fixed with 1.5% glutaraldehyde and 1.5% paraformaldehyde in 0.1 M sodium phosphate buffer, pH 7.3 for 3 hours at room temperature and postfixed for two hours in 2% osmium tetroxide in 0.1 sodium phosphate buffer. After dehydration in graded ethanol, samples for scanning electron microscopy (SEM) were dried in a critical-point dryer (Polaron, Watford, UK), mounted on stubs, and coated with gold-palladium in a cool sputter coater (Fisons Instruments Uckfield, UK). The specimens were examined using a scanning electron microscope DSM 960 (Zeiss Oberkochen, Germany).

Histology

H&E and Periodic acid-Schiff (PAS) staining on mouse trachea (from 1 and 3 week old mice) were carried out using standard procedures. 5 μm sections were cut for experiments.

Immunofluorescence staining of brain and trachea

Brain and trachea from wild-type and *Spag6*-mutant mice (3 week old) were fixed with 4% paraformaldehyde in 0.1 M PBS (pH 7.4), and 5 μm paraffin sections were made. For the immunofluorescence, the method described by Tsuneoka was used [10]. The sections were incubated with an anti-Vangl2 or anti-acetylated tubulin primary antibody at 4°C for overnight. Slides were washed with PBS and incubated for 1 hour at room temperature with Alexa 488-conjugated anti-mouse IgG secondary antibody (1:500; Jackson ImmunoResearch Laboratories) or Cy3-conjugated anti-rabbit IgG secondary antibody (1:5000; Jackson ImmunoResearch Laboratories). Following secondary antibody incubation, the slides were washed again three times in PBS, mounted using VectaMount with 4′, 6-diamidino-2-pheny-lindole (DAPI) (Vector Laboratories, Burlingame, CA), and sealed with a cover slip. Images were captured by confocal laser-scanning microscopy (Leica TCS-SP2 AOBS).

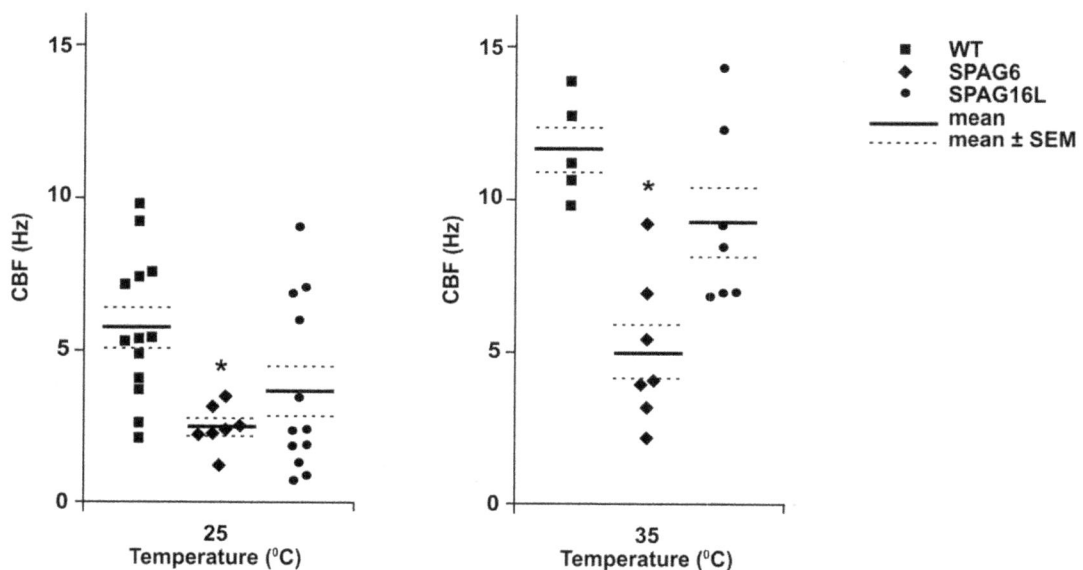

Figure 1. Trachea ciliary beat frequency (CBF) is dramatically decreased in *Spag6*-mutant mice. Graph showing ciliary beat frequency for wild-type, *Spag6*, and *Spag16L* mutant mice. The mean CBF in *Spag6*-deficient mice was significantly lower than that in the wild type and *Spag16L* mice at both 25°C (n = 13, 7 and 12 for wild-type, *Spag6* mutant, and *Spag16L* mutant mice respectively) and 35°C (n = 5, 7 and 7 for wild-type, *Spag6* mutant, and *Spag16L* mutant mice respectively). *p<0.05. ANOVA was conducted to determine significant difference.

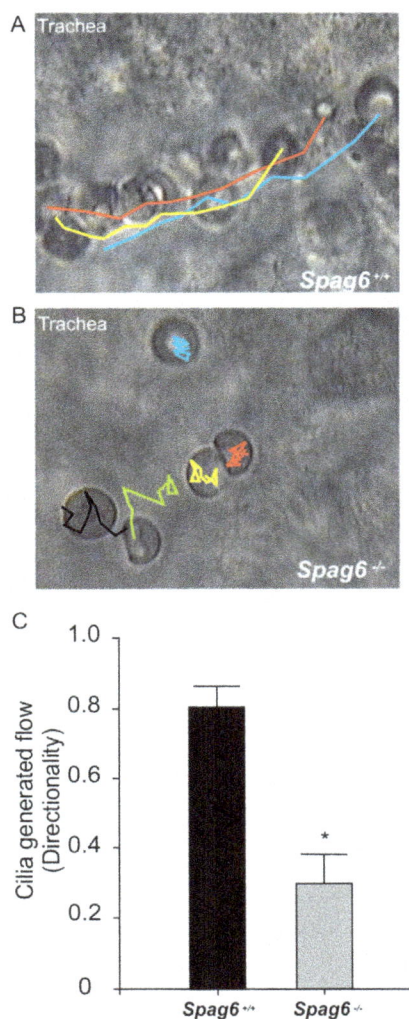

Figure 2. Cilia-generated flow is significantly reduced in *Spag6*-deficient tracheal epithelium. A) Longitudinal view of tracheal epithelia from wild-type mouse showing the tracking of movement of blood cells. B) Longitudinal view of tracheal epithelia from *Spag6* knockout mouse showing the tracking of movement of blood cells. C) Cilia generated flow was quantified by analyzing the directionality of movement of blood cells. * Significant differences (p<0.05) vs. wild-type. Data are presented as mean ± SEM. The colors indicates movement track of individual blood cells.

Video-microscopy

Tracheas were collected from three week old wild-type and *Spag6* knockout mice. Trachea sections were placed luminal side down on a coverslip containing some blood drops in 37°C PBS. Cilia movement and blood cells flows were observed with differential interference contrast microscopy using an inverted microscope (Nikon) equipped with a 100 X oil immersion objective. Images were recorded at 30 frames per second with SANYO color CCD, Hi-resolution camera (VCC-3972, Sanyo Electric Co, Japan) and Pinnacle Studio HD (Ver. 14.0, Pinnacle Systems, Inc., Mountain View, CA, USA) software. Several randomly selected areas were imaged for each sample. Quantification of blood cells directionality was performed with ImageJ software and plugin MTrackJ (NIH). 200 blood cells were tracked for each sample. Directionality was defined as the net displace-

ment achieved divided by the total distance traveled. A directionality of 1 indicated the blood cell moved in a straight line, while a directionality of 0 represents a random movement approach. Data represent mean ± SEM of three mice for each genotype.

Western blot analysis

Equal amounts of protein (50 μg/lane) were heated to 95°C for 10 minutes in sample buffer, loaded onto 10% sodium dodecyl sulfate-polyacrylamide gels, electrophoretically separated, and transferred to polyvinylidene difluoride membranes (Millipore, Billerica, MA). Membranes were blocked (Tris-buffered saline solution containing 5% nonfat dry milk and 0.05% Tween 20 (TBST)) and then incubated overnight with indicated antibodies at 4°C. After washing in TBST, the blots were incubated with second antibodies for 1 hour at room temperature. After washing, the proteins were detected with Super Signal chemiluminescent substrate (Pierce, Rockford, IL).

Quantitative analysis of basal foot orientation

A reference line was drawn for each image. For each basal foot, a vector connecting the center of the basal body and the protrusion of the basal foot was drawn. The angle between this vector and the reference line was measured manually using ImageJ software (NIH). Five images were analyzed from each mouse, and three wild-type and three mutant mice were used for the analysis. The mean angle was calculated for each cell using Oriana 4.0 software (Kovach Computing Services). The mean angle was defined as mean ciliary direction (shown as 0° in each circular plot graph). Deviation from the mean angle was calculated for all of the basal feet analyzed. Deviation angles of the basal feet were pooled and plotted on a circular graph using Oriana 4.0 software.

Statistical methods

Significant difference of axoneme number and basal feet number between wild-type and *Spag6*-deficient mice was calculated using student *t* test. Significant difference of CBF among wild-type, *Spag6*, and *Spag16L*-deficient mice was calculated using ANOVA. * = significant at 0.05. Statistical analysis of tracheal epithelial cell basal body rootlet orientation was carried out with Oriana 4.0 (Kovach Computing Services) circular statistics software.

Results and Discussion

A recent study reported that the mouse has two copies of the *Spag6* gene, the one previously studied on chromosome 16, which is proposed to have evolved from the parental isoform, *Spag6-BC061194*, which is located on chromosome 2 [11]. Even though the amino acid sequences of the two SPAG6 proteins are 97% identical, the nucleotide sequences of the two *Spag6* genes are significantly different. We confirmed that the *Spag6*-deficient mouse we created and studied retains the *Spag6*-BC061194 gene, so that the phenotypes we have described represent the solely the impact of the loss of the "evolved" *Spag6* gene (data not shown).

Ciliary beat frequency (CBF) of tracheal cilia was measured in *Spag6*-deficient and littermate wild-type mice. Compared to the wild-type mice, baseline CBF was significantly reduced in the *Spag6*-deficient mice at both room temperature (25°C) and 35°C (Fig. 1). Of note, mutation of the *Spag16L* gene, which encodes a central apparatus protein, SPAG16L, that interacts with SPAG6, does not cause CBF abnormalities or uncoordinated cilia beat [12], despite the fact that *Spag16L* mutant mice are infertile due

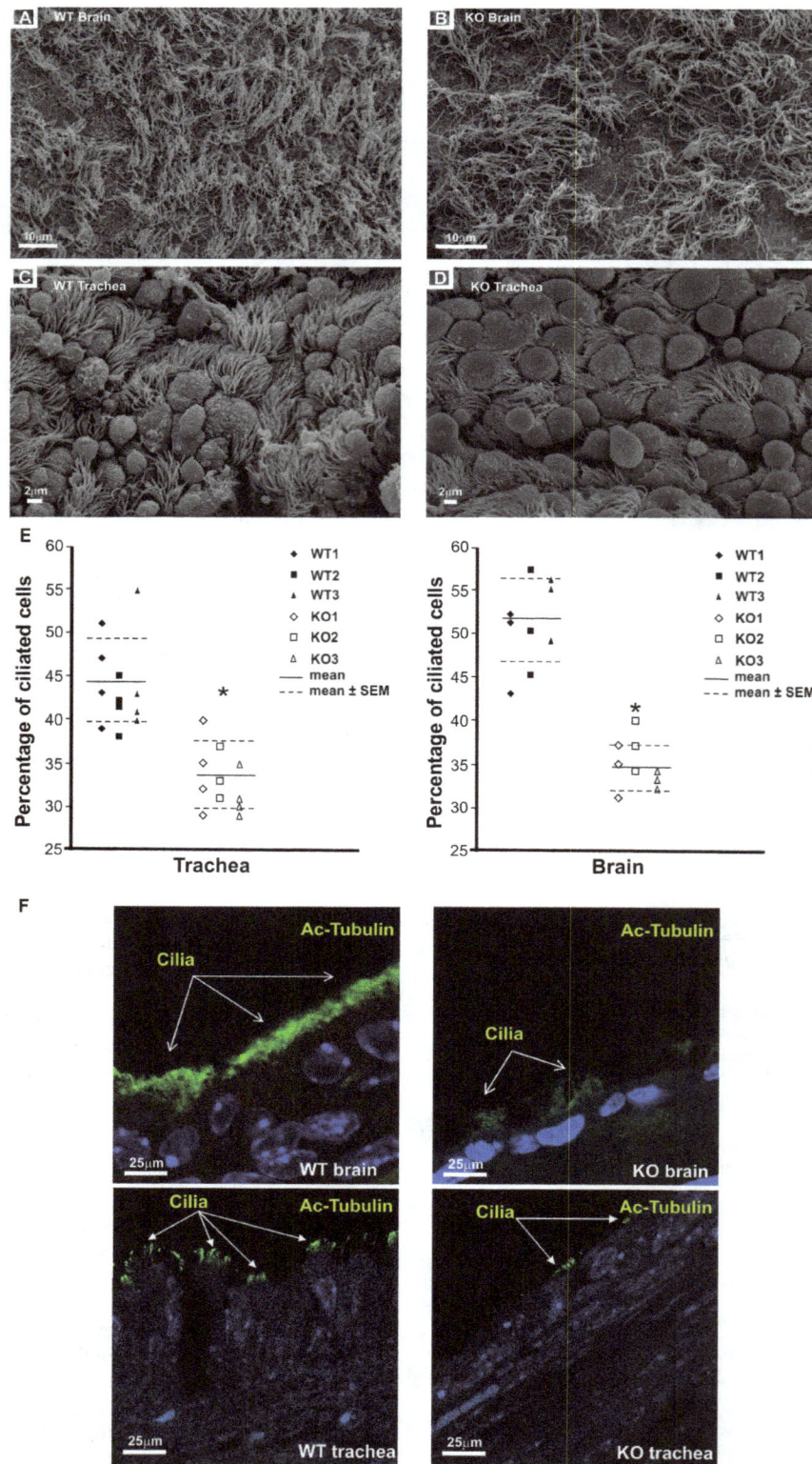

Figure 3. Analyses of cilia in the trachea epithelial cells and brain ependymal cells by scanning electronic microscopy and immunofluorescence staining. Representative images from SEM analyses from wild-type and *Spag6*-deficient mice. Cilia of the wild type mice were well-organized in both brain ependymal cells (A) and trachea epithelial cells (C). In contrast, there was a dramatic reduction in cilia density in both brain ependymal cells (B) and trachea epithelial cells (D) of the *Spag6*-deficient mice, and the cilia were disorganized. To calculate the percentage of ciliated cells, cells with cilia and total cells were counted from three (brain) or four (trachea) SEM images from each mouse, and ratio was calculated (E). Three wild-type and three *Spag6*-deficient mice were analyzed. * Significant differences vs. wild-type (p<0.05). Brain and trachea

sections from wild type and *Spag6*-deficient mice were examined by immunofluorescence staining using an antibody targeting acetylated tubulin. In the wild type mice, the cilia-containing signal was continuously observed along the surface of the epithelial cells. However, in the *Spag6*-deficient mice, the signal was discontinuous (F).

to a severe sperm motility defect [8]. Thus, SPAG6 has functional roles in tracheal cilia that are distinct from those of SPAG16L.

Tracheal ciliary beating was also observed by video microscopy. Consistent with the CBF results, cilia from wild-type mice beat at a faster rate, and the beat was coordinated, with all the cilia beating in the same direction at a specific time point (Video S1). The metachronal beating resulted in a directional flow as shown by the movement of particles/blood cells (Video S2 and Fig. 2A). However, cilia from *Spag6*-deficient mice beat at a much slower rate, and the beating was largely uncoordinated. At specific time points, some cilia beat in one direction, but others in an opposite direction (Video S3). Significantly reduced directed flow was observed as the blood cells collected at the beginning of the tracheal tubes (Video S4, Fig. 2B and Fig. 2C).

Scanning electron microscopy (SEM) was carried out on three (3 week old) wild-type and three *Spag6*-deficient mice of the same age to examine cilia orientation. Tracheal and ependymal cilia in the wild-type animals are anchored to the cell surface in organized arrays (Fig. 3A, 3C). In contrast, cilia arrays of *Spag6*-deficient mice were disorganized (Fig. 3B, 3D, Figure S1C and Figure S1D). In addition, there was a dramatic reduction in cilia density in both brains and tracheas of the *Spag6*-deficient mice, and the percentage of ciliated cells was significantly lower in the mutant mice (Fig. 3E). Cilia in these tissues were further examined by immunofluorescence staining using an antibody to acetylated tubulin. In wild-type mice, the cilia signal was continuous along the surface of the epithelial cells, and extended away from the cell surface into the lumen (Fig. 3F, left panel). However, in the *Spag6*-deficient mice, the signal was discontinuous, and the signal extended to a lesser extent from cell surface than in wild-type tissues (Fig. 3F, right panel).

Transmission electron microscopy (TEM) was conducted to examine the orientation of the central pair microtubules in four wild-type and four *Spag6*-deficient mice. In wild-type animals, orientation of the two central microtubules of all the cilia in the ependymal cells (Fig. 4A) and tracheal epithelial cells (Fig. 4C) was consistent, as shown by the similar orientation of lines connecting the two microtubules in all the axonemes. In the *Spag6*-deficient mice, the axoneme structure appeared normal, but the orientation of the two central microtubules was random; lines connecting central microtubules pointed to one direction in some axonemes, while the lines pointed to a different direction in others (Fig. 4B, and Fig. 4D). To compare the cilia number in the brain ependymal and trachea epithelial cells of wild-type and *Spag6*-deficient mice, the axoneme number was counted from ten TEM images randomly selected from each group. The *Spag6*-deficient mice had significantly lower axoneme numbers than that in the wild-type mice (p<0.05, Fig. 4E).

The ciliary beat orientation is determined by the orientation of the basal feet [13]. The basal feet were examined in the brain ependymal and trachea epithelial cells of wild-type and the *Spag6*-deficient mice by TEM. In the wild-type mice, the basal feet were present in both brain ependymal cells (Fig. 5A) and trachea epithelial cells (Fig. 5C), and they were organized in a similar orientation. However, in the *Spag6*-deficient mice, even though basal feet were morphologically intact, the orientation was random. Like the orientation of the central microtubules in the *Spag6*-deficient mice, some basal feet pointed to one direction, others pointed in different direction (Fig. 5B, and Fig. 5D). The

basal feet number was also counted from the TEM images, and the number was significantly reduced in both brain ependymal (p<0.05) and trachea epithelial cells (p<0.05) of the *Spag6*-deficient mice (Fig. 5E). To determine if the organization of basal feet in *Spag6*-mutant mice is significantly different compared to wild-type animals, basal foot orientation of tracheal epithelial cells was analyzed in three *Spag6* mutant mice (Figure S2A) and three wild-type mice (Figure S2B). Statistical analysis demonstrated that there was significant difference in the r_{cell} metric between the mutant and wild-type mice (Fig. 5F), suggesting that intracellular planar polarity was lost in the mutant mice.

To investigate tissue-level cell polarity, histological sections of tracheas from three (3 week old) wild type and three *Spag6*-deficient mice were examined by light microscopy. H&E staining revealed that in the wild-type mice, two or three rows of nuclei were present in the pseudostratified columnar epithelium lining of the trachea, and the nuclei were oval in shape and oriented in a basal/apical distribution (Fig. 6A). However, the *Spag6*-deficient epithelial cells did not form the pseudostratified columnar morphology. Only one row of nuclei was present, and the cells lay relatively flat along the basement membrane, with most cells having round nuclei (Fig. 6B). The *Spag6*-deficient tracheal epithelial cells not only lose the polarized morphology, the polarized distribution of mucin was also absent. In the wild-type mice, PAS staining demonstrated that mucin was localized in apical region of epithelial cells (Fig. 6C). However, this pattern was never seen in the epithelial cells of *Spag6*-deficient mice. In contrast, mucin was present throughout the cytoplasm (Fig. 6D). The polarized morphology of wild-type tracheal epithelial cells was also observed in low magnification TEM images, and mucin was observed on the trachea surface (Fig. 6E). However, the polarized pattern was lost in the mutant mice, and mucin was not detected on the surface of the tracheal epithelial cells (Fig. 6F).

The localization of the planar cell polarity protein, Vangl2, was examined in the tracheas of three wild-type and three *Spag6*-deficient mice by immunofluorescence staining. Even though there was no difference in total expression level of the protein in trachea/lung between wild-type and the *Spag6*-deficient mice by Western blot analysis (Figure S3), it appears that in wild-type trachea epithelial cells from three-week old mice, Vangl2 signal was more intense in the apical region (Figure S4A). This polarized distribution was not evident in the *Spag6*-deficient epithelial cells (Figure S4B).

Immunofluorescence staining was also conducted on tracheas from three week old wild-type and age-matched *Spag6*-deficient mice using an anti-α-tubulin antibody. In the trachea of wild-type mice, cilia were intensely stained. Inside the epithelial cells, a strong signal was visualized in the apical regions (Figure S4C). However, in the *Spag6*-deficient mice, the signal was evenly distributed throughout the whole cell body (Figure S4D). The microtubule distribution pattern is consistent with that of the planar cell polarity (PCP) proteins, suggesting that polarized microtubules might contribute to the localization of PCP proteins.

One-week mice were analyzed, when the mutant mice did not show obvious abnormalities related to reduced ciliary motility, such as hydrocephalus. As in the three-week old wild-type mice, epithelial cells in the one-week old wild-type mice are polarized as shown by H&E staining (Fig. 7A), and immunofluorescence staining using an anti-acetylated tubulin antibody revealed a

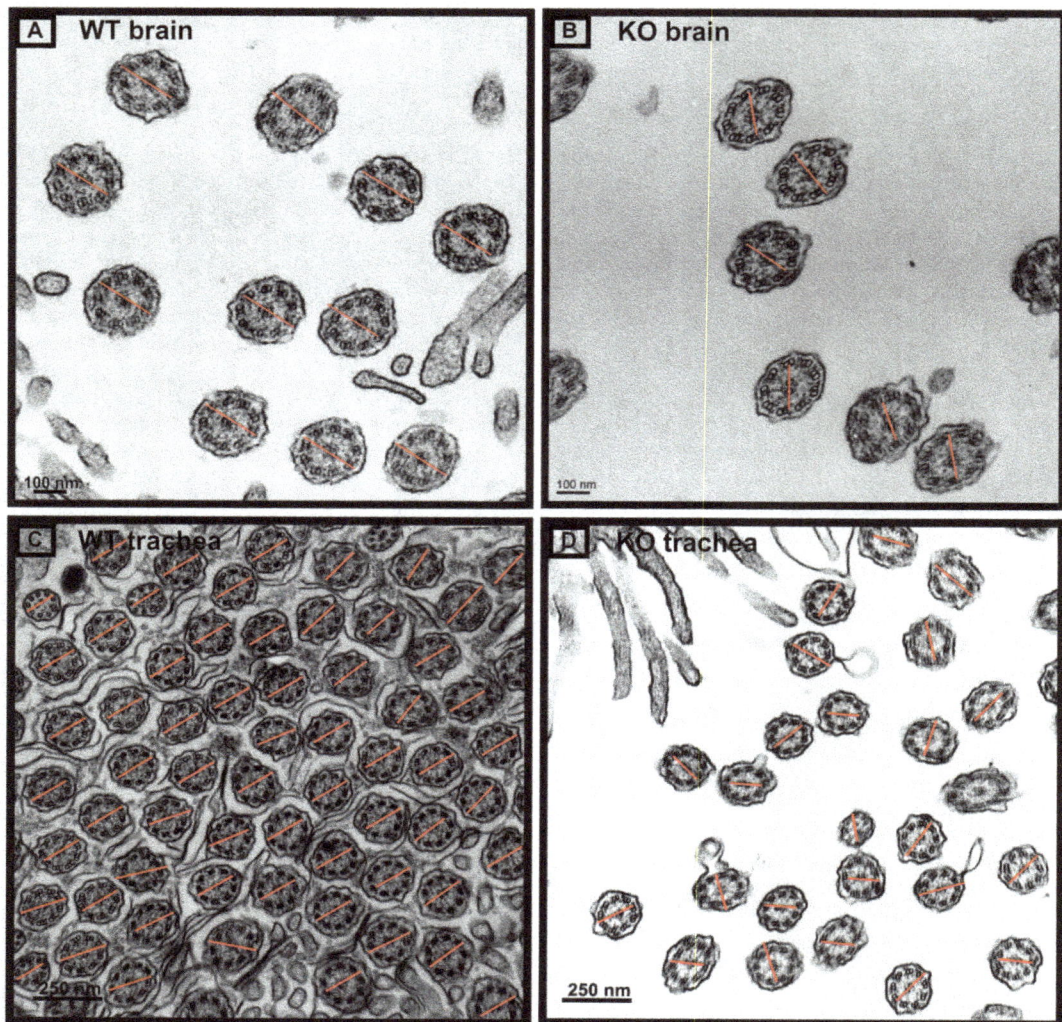

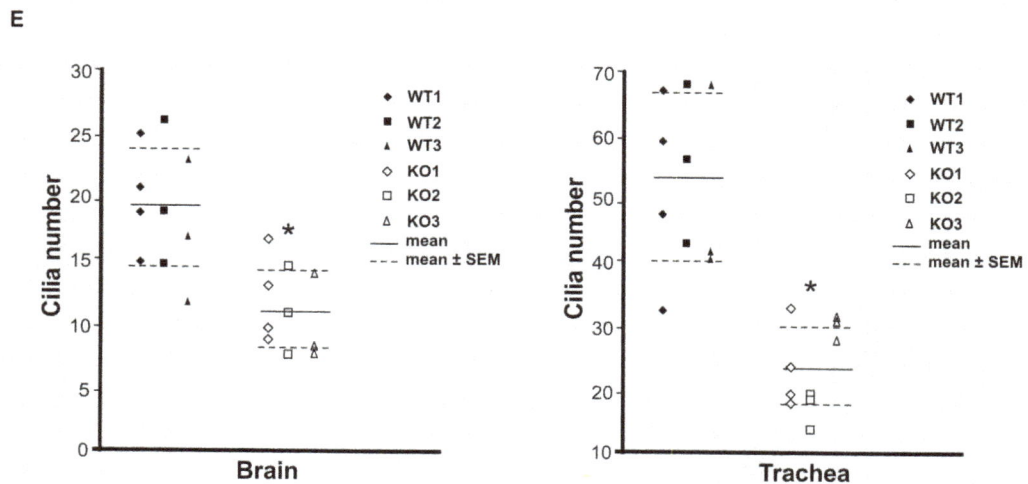

Figure 4. Examination of rotational polarity of ciliary axoneme of brain ependymal cells and trachea epithelial cells by transmission electronic miscroscopy. Axoneme cross-sectional images were taken with a transmission electron microscope. The rotational polarity of each axoneme was evaluated by the angle of the line connecting the central pair. Notice that the orientation of the lines in wild type mice is similar (A: brain; C: trachea). while the orientation of the lines in the *Spag6*-deficient mice varies among axonemes (B: brain; D: trachea). E. Average axoneme number counted from ten images randomly selected from each group. Three or four images were counted from each mouse, and three wild-type and three mutant mice were examined. Horizontal lines represent the means and SEMs. * p<0.05.

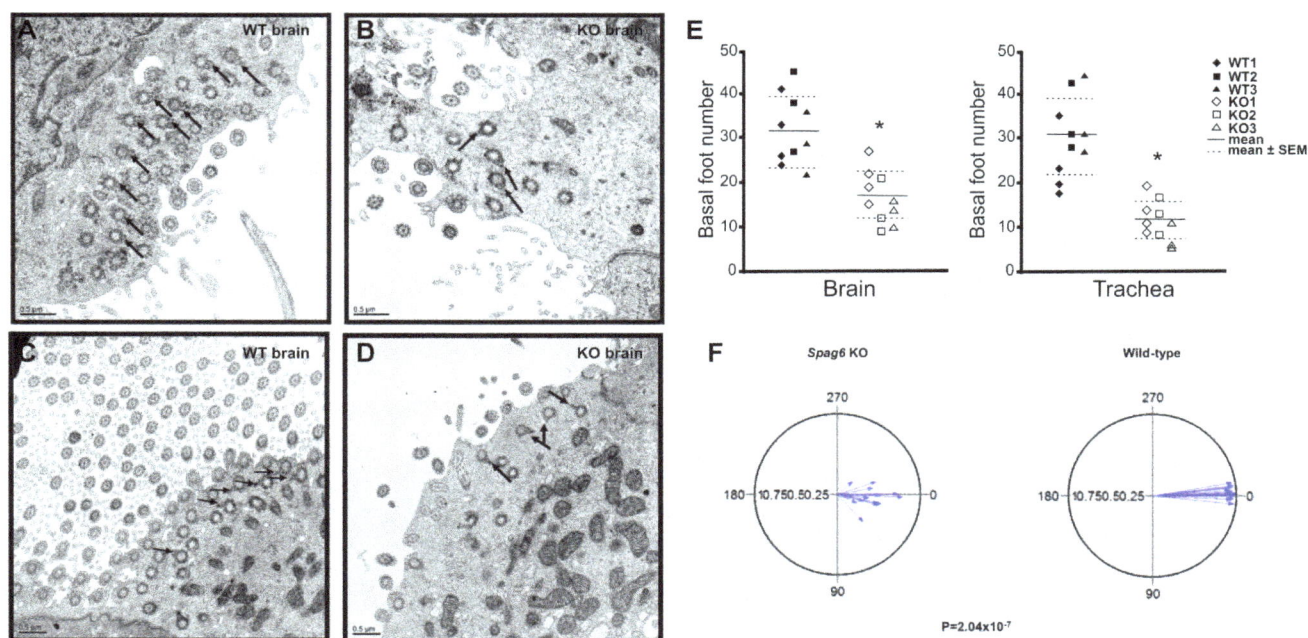

Figure 5. Basal feet polarity of brain ependymal cells and trachea epithelial cells was lost in the *Spag6*-deficient mice. Basal body images were taken with a transmission electron microscope. Notice that the basal feet point to the same orientation in the wild type animals (A: brain; C: trachea). However, they point to different orientation in the *Spag6*-deficient mice (B: brain; D: trachea). The number of basal body in the *Spag6*-deficient mice was significantly reduced in both brain and trachea. The arrows point to the basal feet. E. Average basal feet number counted from ten images randomly selected from each group. Horizontal lines represent means and SEMs. Three or four images were counted from each mouse, and three wild-type and three mutant mice were examined. * $p<0.05$. F. Circular plots of tracheal epithelial cell basal feet orientation in *Spag6*-deficient (left) and wild-type mice (right). For each mouse, basal foot orientation from five images was analyzed. For each image, the angel for one basal foot orientation was set as $0°$ (or $360°$), angels of the rest basal feet were measured. Each plot represents the combined data from three mice as shown in Figure S2 (* $p<0.001$ between the two groups).

strong cilia signal (Fig. 7C). However, the epithelial cells of the *Spag6*-deficient mice did not show the polarized morphology. Compared to the wild-type mice, cilia number is dramatically reduced (Fig. 7B, 7D). The polarized distribution of α-tubulin is obvious in the wild-type mice (Figure S5A), but not in the *Spag6*-deficient mice (Figure S5B). These findings indicate that the cilia number and orientation defects are present earlier than 7 days post-partum.

It has been reported that a feedback loop generated by fluid flow contributes to cilia polarization [14,15]. Indeed, our findings of disorganization and reduced number of cilia in ependymal and tracheal epithelial cells in *Spag6*-mutants, and alterations in basal body alignment are similar to those previously described in other mutant mice as a consequence of reduced ciliary motility. Studies involving Jhy$^{lacZ/lacZ}$ mice showed disorganization and altered axonemal structure of the ependymal cilia. However, the hydrocephalus appeared to be unrelated to abnormal brain development or patterning [16]. Observations in *ktu*-mutant mice revealed that the PCP protein, Vangl1, localized asymmetrically in ependymal and tracheal epithelial cells, while the alignment of basal bodies only differed from wild-type mice in brain ependymal cells, suggesting that ciliary motility was required in the alignment of brain ependymal cells, but not for airway cilia [17].

There are other possible mechanisms, in addition to ciliary motility defects, as causal factors of the phenotypes in SPAG6-mutant mice. In multi-ciliated cells, basal bodies are replicated deep within the cytoplasm, and their apical movement and docking are thought to involve regulated actin assembly [18,19] and vesicle trafficking [20]. Indeed, actin is enriched at the apical

surface of ciliated epithelial cells [21,22]. Disruption of the actin cytoskeleton blocks basal body migration and ciliogenesis [19]. In this case, the ciliogenesis defect is associated with the failure of basal body docking at the apical plasma membrane. *Spag6*-deficient cells may have a disrupted actin cytoskeleton affecting basal body docking, which traps some basal bodies inside the cytoplasm where they are degraded, with the result that fewer cilia develop in the brain ependymal and tracheal epithelial cells.

A recent study demonstrated that silencing of the *Spag6* gene in Xenopus larvae gives rise to disruption of orientation of basal bodies, suggesting that the planar cell polarity mechanism might be involved [14]. PCP refers to the polarization of a field of cells within the plane of a cell sheet [23]. It is a downstream branch of Wnt signaling [24]. This form of polarization is required for diverse cellular processes in vertebrates, including convergent extension (CE) [25] and the establishment of PCP in epithelial tissues and ciliogenesis [26,27]. In multi-ciliated cells, planar polarity is present in two distinct models, termed rotational polarity and tissue-level polarity [28,29]. The former refers to the alignment of the basal bodies within each multi-ciliated cell, and the latter to the coordination of many multi-ciliated cells across the tissue. SEM and high magnification TEM studies clearly demonstrated that *Spag6*-deficient epithelial cells in the brain and trachea lost rotational polarity. Alternatively, SPAG6 may cause ciliogenesis defects through a role in basal bodies. Pearson et al. reported that SPAG6 was present in newly assembled basal bodies [30], and SPAG6 localizes to the center of the transition zone at the site of central pair assembly [31].

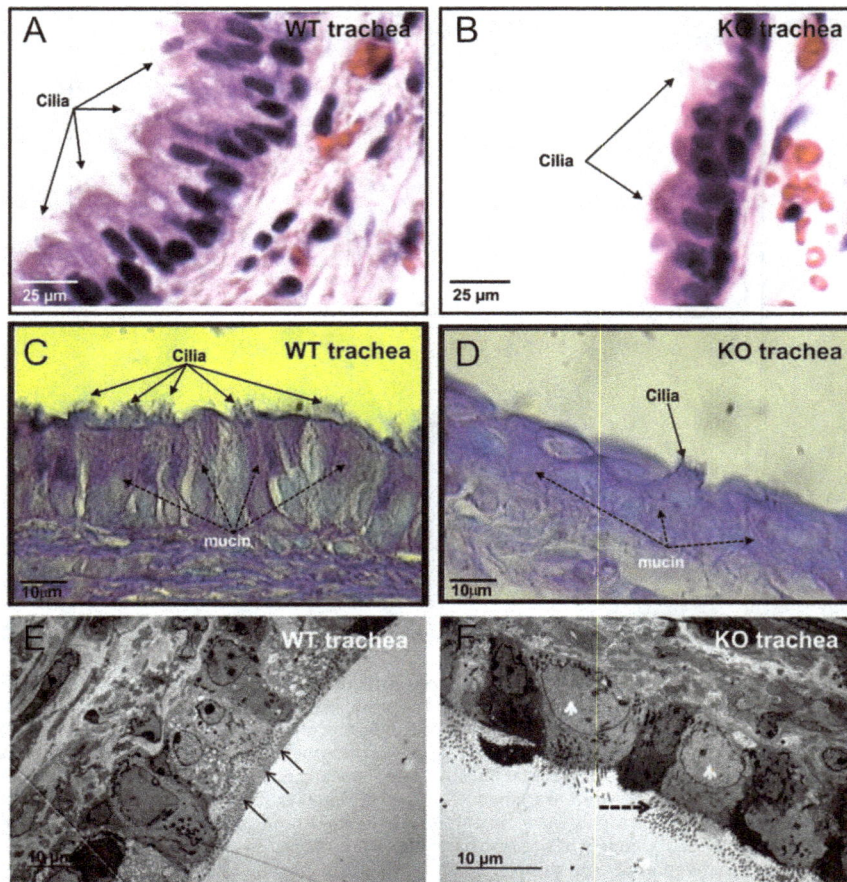

Figure 6. Examination of trachea epithelial cell polarity in three-week old mice. H&E stained tissue demonstrating that two or three rows of nuclei were seen in the pseudostratified columnar epithelium lining the trachea of the wild-type mice, and the nuclei were oval shape and the two poles were at basal/apical distribution (Fig. 6A). In contrast, in the *Spag6*-deficient mice, the cells lie flat along the basement membrane, most cells had round nuclei (Fig. 6B). PAS staining demonstrated that mucin was localized in apical region of epithelial cells (Dashed arrows in Fig. 6C). However, this pattern was never seen in the epithelial cells of *Spag6*-deficient mice, mucin was present through the whole cytoplasm (Fig. 6D). The above-mentioned differences between wild-type and *Spag6*-deficient mice were confirmed by TEM with low magnification. The wild-type epithelial cells show polarized pattern, and mucin was found along the surface in all the three mice analyzed (arrows in Fig. 6E), where the cilia axonemes were located. However, in the mutant mice, the epithelial cells lost this pattern, and the cells look larger than those in the wild-type mice (arrow heads), and no mucin was found in any of the three mice analyzed (Fig. 6F). Three wild-type and three mutant mice were analyzed and the results were similar.

Several recent studies revealed that the PCP signaling cascade is a central regulator of the orientation of cilia-mediated fluid flow. Disruption of core PCP genes, including the Dishevelled (Dvl1), Celsr2 and Celsr3 resulted in a randomization of rotational polarity [32,33,15]. PCP signaling also controls the tissue-level polarity of multi-ciliated cells, PCP proteins, van Gogh-like 2 (Vangl2) and Frizzled are in this case [29]. A more recent study indicated that PCP proteins, including Vangl1, Vangl2, Prickle2 (PK2), Dishevelled1 (Dvl1) and Dvl2 localize asymmetrically to the tracheal epithelial cell cortex [34].

Little is known about the regularity of orientation of basal bodies in multi-ciliated cells. Previous studies of multi-ciliated cells suggested that microtubules attached to the basal feet link basal bodies to one another, and also to the apical junctions [35]. The classic planar polarity in the Drosophila wing epithelial cells is also associated with sub-apical microtubules [36,37]. These microtubules are planar polarized, with their plus ends enriched at the distal face of cells, where Dvl1 and Frizzled localize. It appears that the apical microtubule network is an upstream regulator of

PCP signaling [38,39]. It is suggested that a similar planar polarized web of microtubules may also influence planar polarity of basal bodies. In addition, basal bodies in multi-ciliated cells make complex connections to both actin and cytokeratin networks, and these may be involved in polarization [40]. We have previously shown that SPAG6 is a microtubule binding protein [1]. It may play a role in stabilizing the microtubule system. In the absence of SPAG6, sub-apical microtubule stability might be affected, which could result in disruption of polarized PCP distribution, causing the basal bodies to lose their polarized localization.

Defects in mammalian cilia or flagella motility/function caused by mutations in central apparatus genes appear to depend upon the genetic background and cellular context. For instance, mutation of the *Pcdp1* gene results in several phenotypes commonly associated with primary ciliary dyskinesia. Homozygous mutants on a C57BL/6J background develop severe hydrocephalus and mainly die within the first week of life. However, on other genetic backgrounds (129S6/SvEvTac), mice

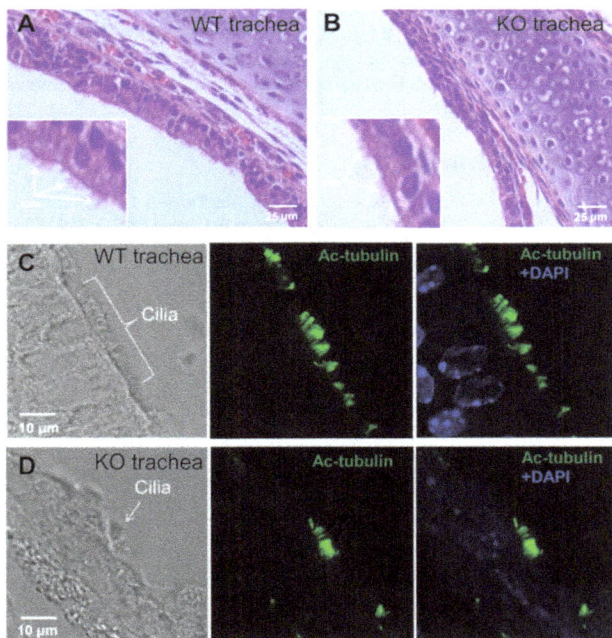

Figure 7. Examination of trachea epithelial cell polarity in one-week old mice. H&E stained tissue demonstrating the polarized pattern of epithelial cells in the trachea of wild type mice (Fig. 7A), but not in the *Spag6*-deficient mice (Fig. 7B). Acetylated tubulin signal is abundant along the tracheal epithelial cells in the wild-type mice (Fig. 7C), the signal is dramatically reduced in the *Spag6*-deficient mice (Fig. 7D). Three wild-type and three mutant mice were analyzed and similar results were observed.

develop either mild or no hydrocephalus with survival to adulthood. The respiratory epithelial cilia have a normal ultrastructure, but beat with reduced frequency. Interestingly, the male mice are infertile, producing sperm with no visible flagella, suggesting that the mechanisms regulating the biogenesis of cilia and flagella are likely to be different [41]. Tracheal epithelial cilia from *Spef2*-deficient mice beat at lower frequency and have a normal 9+2 axonemal structure without apparent defects in the dynein arms, but epididymal sperm lack recognizable axonemal structures [42]. SPAG16L-null mice show no evidence of cilia dysfunction, such as hydrocephalus, sinusitis, and bronchial infection [8], and have tracheal epithelial cells with motile cilia [12]. However, males are infertile due to severe sperm motility defects, even though the sperm have a normal axoneme ultrastructure [8]. In the *Spag17*-mutant mouse, the rapid neonatal demise is associated with a profound respiratory phenotype characterized by immotile cilia and defects in the 9+2 axonemal structure [12]. This is not observed in cilia from knockouts of the *Spag6* and *Spag16* genes. Although the genetic background of mutant mice may significantly influence the phenotypes, the observations summarized above suggest that specific central pair genes may have unique roles in different cell types.

In conclusion, our studies demonstrate that SPAG6 deficiency causes multiple abnormalities in the function of cilia and ciliated cells, including defects in sperm flagellar motility; ciliogenesis; ciliary beat; axoneme orientation; cell morphology and polarity.

Supporting Information

Figure S1 Analysis of cilia in the trachea epithelial cells and brain ependymal cells by scanning electronic microscopy. Tracheas and brains from wild type and Spag6-deficient mice were processed for SEM. Notice that cilia in the brains (A) and trachea (C) of the wild-type animals sit on the cell surface in a highly ordered state. However, cilia in the ependymal cells (B) and trachea (D) of Spag6-deificent mice appeared to be disordered on the cell surface. Fig. S1 shows cilia in the trachea epithelial cells and brain ependymal cells by scanning electronic microscopy with high magnification.

Figure S2 Circular plots of tracheal epithelial cell basal foot orientation in three individual *Spag6*-deficient (upper) and three individual wild-type mice (lower). Five TEM images were randomly selected from each mouse and basal foot orientations were measured. Arrow direction represents the mean vector of cilium orientation per cell; arrow length is the length of the mean vector, with longer arrows indicating stronger coordination of orientation. r_{cell} is the length of mean vector and describes rotational orientation.

Figure S3 Analysis of Vangl2 protein expression level in the lung/trachea by Western blotting. Lungs/tracheas from three wild-type and three *Spag6*-deficient mice were homogenized and Western blotting was performed with anti-Vangl2 antibody, the membrane was striped and re-probed with an anti-actin antibody as a loading control. There was no difference in Vangl2 protein expression level between the wild-type and *Spag6*-deficient mice.

Figure S4 Examination of Vangl2 and α–tubulin localization in trachea epithelial cells of three-week old mice. The distribution of the PCP protein, Vangl2, and, α-tubulin was examined by immunofluorescence staining. More intense signal was detected in the apical regions in wild-type trachea epithelial cells (arrows in A for Vangl2 and arrowheads in C for α-tubulin). These proteins appeared to be distributed evenly throughout the cytoplasm in cells from *Spag6* mutant mice (dashed arrows in B). In the trachea of wild-type mice, cilia were also intensively stained by an anti-α-tubulin antibody (arrows in C and D).

Figure S5 Examination of α-tubulin localization in trachea epithelial cells in one-week old mice. Distribution of α-tubulin is polarized in the wild-type mice (arrowheads in upper panel). However, the polarized pattern is not seen in the *Spag6*-deficient mice (lower panel), where α-tubulin is evenly distributed throughout the cytoplasm.

Video S1 Trachea ciliary beat observed by video microscopy. Cilia from wild-type mice (*Spag6* knockout littermate) beat at a fast rate, and the beat was coordinated, with all the cilia beating in the same direction at a specific time point.

Video S2 Airway clearance in wild-type mice. Real time video showing the efficiency of ciliated epithelium in moving particles (blood cells) in the trachea. Arrows indicate the direction flow.

Video S3 Cilia from *Spag6*-deficient mice beat at much slower rate, and the beating is largely uncoordinated. At specific time points, some cilia beat in one direction, while others beat in the opposite direction.

Video S4 *Spag6*-deficient mice fail to clear particles from the airway. Uncoordinated cilia from tracheal epithelium failed to generate blood cell flow. Arrows indicate the presence of blood cells stacked at the beginning of tracheal tube. Video is shown in real time.

Acknowledgments

We thank Dr. Eszter K. Vladar in Dr. Jeffrey Axelrod's laboratory at Department of Pathology, Stanford University for assistance with circular statistics. We thank Dr. Bruce. K. Rubin, Professor and Chair of Department of Pediatric, Virginia Commonwealth University for his comments and edits. We thank Ling Zhang, associate professor of Wuhan University of Science and Technology for assistance with statistics on cilia and basal foot number.

Author Contributions

Conceived and designed the experiments: ZZ. Performed the experiments: MET WL PRS ZGZ WT LVR ZZ. Analyzed the data: CWD MRK JFS ZZ. Contributed reagents/materials/analysis tools: RMC MRK JFS ZZ. Wrote the paper: ZZ JFS MET.

References

1. Sapiro R, Tarantino LM, Velazquez F, Kiriakidou M, Hecht NB, et al. (2000) Sperm antigen 6 is the murine homologue of the Chlamydomonas reinhardtii central apparatus protein encoded by the PF16 locus. Biol Reprod 62: 511–518.
2. Smith EF, Lefebvre PA (1996) PF16 encodes a protein with armadillo repeats and localizes to a single microtubule of the central apparatus in Chlamydomonas flagella. J Cell Biol 132: 359–370.
3. Branche C, Kohl L, Toutirais G, Buisson J, Cosson J, et al. (2006) Conserved and specific functions of axoneme components in trypanosome motility. J Cell Sci 119(Pt 16): 3443–3455.
4. Straschil U, Talman AM, Ferguson DJ, Bunting KA, Xu Z, et al. (2010) The Armadillo repeat protein PF16 is essential for flagellar structure and function in Plasmodium male gametes. PLoS One 5: e12901.
5. House SA, Richter DJ, Pham JK, Dawson SC (2011) Giardia flagellar motility is not directly required to maintain attachment to surfaces. PLoS Pathog 7: e1002167.
6. Sapiro R, Kostetskii I, Olds-Clarke P, Gerton GL, Radice GL, et al. (2002) Male infertility, impaired sperm motility, and hydrocephalus in mice deficient in sperm-associated antigen 6. Mol Cell Biol 22: 6298–6305.
7. Zhang Z, Tang W, Zhou R, Shen X, Wei Z, et al. (2007) Accelerated mortality from hydrocephalus and pneumonia in mice with a combined deficiency of SPAG6 and SPAG16L reveals a functional interrelationship between the two central apparatus proteins. Cell Motil Cytoskeleton 64(5): 360–376.
8. Zhang Z, Kostetskii I, Tang W, Haig-Ladewig L, Sapiro R, et al. (2006) Deficiency of SPAG16L causes male infertility associated with impaired sperm motility. Biol Reprod 74(4): 751–759.
9. Sisson JH, Stoner JA, Ammons BA, Wyatt TA (2003) All-digital image capture and whole field analysis of ciliary beat frequency. J Microsc 211: 103–111.
10. Tsuneoka M, Nishimune Y, Ohta K, Teye K, Tanaka H, et al. (2006) Expression of Mina53, a product of a Myc target gene in mouse testis. Int J Androl 29: 323–330.
11. Qiu H, Gołas A, Grzmil P, Wojnowski L (2013) Lineage-specific duplications of Muroidea Faim and Spag6 genes and atypical accelerated evolution of the parental Spag6 gene. J Mol Evol 77(3): 119–29.
12. Teves ME, Zhang Z, Costanzo RM, Henderson SC, Corwin FD, et al. (2013) Spag17 is Essential for Motile Cilia Function and Neonatal Survival. Am J Respir Cell Mol Biol 48(6): 765–772.
13. Kunimoto K, Yamazaki Y, Nishida T, Shinohara K, Ishikawa H, et al. (2012) Coordinated ciliary beating requires Odf2-mediated polarization of basal bodies via basal feet. Cell 148: 189–200.
14. Mitchell B, Jacobs R, Li J, Chien S, Kintner C (2007) A positive feedback mechanism governs the polarity and motion of motile cilia. Nature 447: 97–101.
15. Guirao B, Meunier A, Mortaud S, Aguilar A, Corsi JM, et al. (2010) Coupling between hydrodynamic forces and planar cell polarity orients mammalian motile cilia. Nat Cell Biol 12: 341–350.
16. Appelbe OK, Bollman B, Attarwala A, Triebes LA, Muniz-Talavera H, et al. (2013) Disruption of the mouse Jhy gene causes abnormal ciliary microtubule patterning and juvenile hydrocephalus. Dev Biol 382(1): 172–85.
17. Matsuo M, Shimada A, Koshida S, Saga Y, Takeda H (2013) The establishment of rotational polarity in the airway and ependymal cilia: analysis with a novel cilium motility mutant mouse. Am J Physiol Lung Cell Mol Physiol 304(11): L736–45.
18. Dawe HR, Farr H, Gull K (2007) Centriole/basal body morphogenesis and migration during ciliogenesis in animal cells. J Cell Sci 120: 7–15.
19. Boisvieux-Ulrich E, Lainé MC, Sandoz D (1990) Cytochalasin D inhibits basal body migration and ciliary elongation in quail oviduct epithelium. Cell Tissue Res 259: 443–454.
20. Sorokin SP (1968) Reconstructions of centriole formation and ciliogenesis in mammalian lungs. J Cell Sci 3: 207–230.
21. Park TJ, Haigo SL, Wallingford JB (2006) Ciliogenesis defects in embryos lacking inturned or fuzzy function are associated with failure of planar cell polarity and Hedgehog signaling. Nat Genet 38: 303–311.
22. Pan J, You Y, Huang T, Brody SL (2007) RhoA-mediated apical actin enrichment is required for ciliogenesis and promoted by Foxj1. J Cell Sci 120: 1868–1876.
23. Goodrich LV, Strutt D (2011) Principles of planar polarity in animal development. Development 138: 1877–1892.
24. Wallingford JB, Mitchell B (2011). Strange as it may seem: the many links between Wnt signaling, planar cell polarity, and cilia. Gene Dev 25: 201–213.
25. Ybot-Gonzalez P, Savery D, Gerrelli D, Signore M, Mitchell CE, et al. (2007) Convergent extension, planar-cell-polarity signalling and initiation of mouse neural tube closure. Development 134: 789–799.
26. Dworkin S, Jane SM, Darido C (2011) The planar cell polarity pathway in vertebrate epidermal development, homeostasis and repair. Organogenesis 7: 202–208.
27. Wallingford JB (2010) Planar cell polarity signaling, cilia and polarized ciliary beating. Curr Opin Cell Biol 22: 597–604.
28. Mirzadeh Z, Han YG, Soriano-Navarro M, García-Verdugo JM, Alvarez-Buylla A (2010) Cilia organize ependymal planar polarity. J Neurosci 30: 2600–2610.
29. Mitchell B, Stubbs JL, Huisman F, Taborek P, Yu C, et al. (2009) The PCP pathway instructs the planar orientation of ciliated cells in the Xenopus larval skin. Curr Biol 19: 924–929.
30. Pearson CG, Giddings TH Jr, Winey M (2009) Basal body components exhibit differential protein dynamics during nascent basal body assembly. Mol Biol Cell 20: 904–914.
31. Kilburn CL, Pearson CG, Romijn EP, Janet B, Meehl JB, et al. (2007) New Tetrahymena basal body protein components identify basal body domain structure. J Cell Biol 178: 905–912.
32. Park TJ, Mitchell BJ, Abitua PB, Kintner C, Wallingford JB (2008) Dishevelled controls apical docking and planar polarization of basal bodies in ciliated epithelial cells. Nat Genet 40: 871–879.
33. Tissir F, Qu Y, Montcouquiol M, Zhou L, Komatsu K, et al. (2010) Lack of cadherins Celsr2 and Celsr3 impairs ependymal ciliogenesis, leading to fatal hydrocephalus. Nat Neurosci 13(6):700–707.
34. Vladar EK, Bayly RD, Sangoram AM, Scott MP, Axelrod JD (2012) Microtubules enable the planar cell polarity of airway cilia. Curr Biol 22(23): 2203–2212.
35. Sandoz D, Chailley B, Boisvieux-Ulrich E, Lemullois M, Laine MC, et al. (1988) Organization and functions of cytoskeleton in metazoan ciliated cells. Biol Cell 63: 183–193.
36. Eaton S, Wepf R, Simons K (1996) Roles for Rac1 and Cdc42 in planar polarization and hair outgrowth in the wing of Drosophila. J Cell Biol 135:1277–1289.
37. Turner CM, Adler PN (1998) Distinct roles for the actin and microtubule cytoskeletons in the morphogenesis of epidermal hairs during wing development in Drosophila. Mech Dev 70: 181–192.
38. Shimada Y, Yonemura S, Ohkura H, Strutt D, Uemura T (2006) Polarizedtransport of Frizzled along the planar microtubule arrays in Drosophila wing epithelium. Dev Cell 10: 209–222.
39. Hannus M, Feiguin F, Heisenberg CP, Eaton S (2002) Planar cell polarization requires Widerborst, a B' regulatory subunit of protein phosphatase 2A. Development 129: 3493–3503.
40. Chailley B, Nicolas G, Lainé MC (1989) Organization of actin microfilaments in the apical border of oviduct ciliated cells. Biol Cell 67: 81–90.
41. Lee L, Campagna DR, Pinkus JL, Mulhern H, Wyatt TA, et al. (2008) Primary ciliary dyskinesia in mice lacking the novel ciliary protein Pcdp1. Mol Cell Biol 28(3): 949–957.
42. Sironen A, Kotaja N, Mulhern H, Wyatt TA, Sisson JH, et al. (2011) Loss of SPEF2 function in mice results in spermatogenesis defects and primary ciliary dyskinesia. Biol Reprod 85(4): 690–701.

5

Neonatal NMDA Receptor Blockade Disrupts Spike Timing and Glutamatergic Synapses in Fast Spiking Interneurons in a NMDA Receptor Hypofunction Model of Schizophrenia

Kevin S. Jones[1,2], Joshua G. Corbin[2], Molly M. Huntsman[3]*

1 Biology Department, Howard University, Washington, DC, United States of America, 2 Center for Neuroscience Research, Children's National Medical Center, Washington, DC, United States of America, 3 Department of Pharmaceutical Sciences, Skaggs School of Pharmacy and Pharmaceutical Sciences, and Department of Pediatrics, School of Medicine, University of Colorado, Anschutz Medical Campus, Aurora, CO, United States of America

Abstract

The dysfunction of parvalbumin-positive, fast-spiking interneurons (FSI) is considered a primary contributor to the pathophysiology of schizophrenia (SZ), but deficits in FSI physiology have not been explicitly characterized. We show for the first time, that a widely-employed model of schizophrenia minimizes first spike latency and increases GluN2B-mediated current in neocortical FSIs. The reduction in FSI first-spike latency coincides with reduced expression of the Kv1.1 potassium channel subunit which provides a biophysical explanation for the abnormal spiking behavior. Similarly, the increase in NMDA current coincides with enhanced expression of the GluN2B NMDA receptor subunit, specifically in FSIs. In this study mice were treated with the NMDA receptor antagonist, MK-801, during the first week of life. During adolescence, we detected reduced spike latency and increased GluN2B-mediated NMDA current in FSIs, which suggests transient disruption of NMDA signaling during neonatal development exerts lasting changes in the cellular and synaptic physiology of neocortical FSIs. Overall, we propose these physiological disturbances represent a general impairment to the physiological maturation of FSIs which may contribute to schizophrenia-like behaviors produced by this model.

Editor: Kenji Hashimoto, Chiba University Center for Forensic Mental Health, Japan

Funding: This study was supported by NIH NINDS R01 NS053719 (Molly M Huntsman); NIH NICHD R01DA020140 (Joshua G Corbin); Children's Research Institute: EUNICE KENNEDY SHRIVER NICHD Training Grant 5P30HD040677 (Kevin S Jones). The funders had no role in study design, data collection and analysis, decision to publish, or preparation of the manuscript.

Competing Interests: The authors have declared that no competing interests exist.

* Email: Molly.Huntsman@UCDenver.edu

Introduction

Deficits to inhibitory neurotransmission are highly implicated in the etiology of schizophrenia (SZ) [1], as immunohistochemical analyses of post-mortem brain tissue often reveal decreases in the expression of one or more biochemical markers for gamma-aminobutyric acid (GABA) signaling [1]. Expression of the calcium binding protein, parvalbumin (PV), is particularly diminished in the neocortex of many SZ patients [2], which implies dysfunction of PV-expressing interneurons [3]. PV-expressing interneurons are physiologically distinguished by their capacity to discharge action potentials ("spikes") at very high frequency and are thus termed "fast spiking" interneurons [4]. FSIs are interconnected via chemical and electrical synapses [5–7] which helps synchronize their own firing patterns [8,9], and pace the firing patterns of large networks of pyramidal cells [10]. FSIs are thus integral for generating neural oscillations [11,12], which incidentally, are often compromised in SZ patients. Although FSI dysfunction is highly inferred in the pathophysiology of SZ [13], physiological support for this hypothesis is lacking, particularly at the single cell level.

The NMDA receptor hypofunction model of SZ is founded on the discovery that acute administration of non-competitive NMDA antagonist (e.g. PCP, ketamine, and MK-801) evokes behaviors in healthy humans that are highly reminiscent of psychosis in SZ patients [14] [15]. Moreover, these drugs elicit behavioral deficits in animal models that closely model aspects of SZ [16]; [17] and also replicate disruptions in GABAergic biochemical markers. Administration of non-competitive NMDA receptor antagonist during early development is a particularly robust approach to model SZ-like biochemical deficits to GABA signaling [18–23]. Electrophysiological characterization of the NMDA hypofunction model of SZ has recently begun [24], but remains incomplete. Since direct physiological evaluation of FSIs in SZ patients is unfeasible, electrophysiological characterization of FSIs in animal models of SZ may be an expedient approach to identify specific impairments in FSI function.

In this study neonatal mice were treated with the NMDA receptor antagonist, MK-801, on postnatal day (PND) 6–8. The impact of neonatal MK-801 treatment on FSI physiology was then assessed during adolescence. This approach allowed us to directly test the hypothesis that transient disruption of NMDA receptor activity during early development causes persistent impairments to the function of neocortical FSIs. Whole-cell patch-clamp electrophysiology revealed that neonatal MK-801 treatment dramatically altered the spiking kinetics and action potential dynamics of FSIs.

Pharmacological analysis revealed an increase in GluN2B-mediated NMDA current at excitatory synapses of FSIs from MK-801-treated mice. Immunohistochemical analyses identified congruent changes in the expression of key ion channel subunits that corroborate both sets of physiological data.

Methods and Materials

Experimental Animals

Ethics Statement. All animal use procedures were carried out in strict accordance with National Institutes of Health Guide for the Care and Use of Laboratory Animals and were approved by the Institutional Animal Care and Use Committee at Children's National Medical Center.

To aid visualization of FSIs, we utilized transgenic mice that expressed the fluorescent reporter, Enhanced Yellow Fluorescent Protein (EYFP), exclusively in PV^+ interneurons. These $PV\text{-}Cre^{+/-}$; $Rosa26EYFP^{loxP+}$ mice were obtained by crossing a transgenic strain expressing cre recombinase under the control of the endogenous parvalbumin promoter ($PV\text{-}cre$); and a strain expressing a loxP-flanked $Rosa26\text{-}EYFP$ marker (Jackson Laboratories, Maine). Only male mice were used in this study as sexually dimorphic responses to MK-801 have been reported [25,26]. Male mouse pups were randomly assigned to the control or experimental group on PND6 and administered a subcutaneous injection of 0.75 mg/kg MK-801 (Tocris, USA) or an equal volume of saline for three consecutive days.

Preparation of Brain Slices for Electrophysiology

Three to six week-old mice were anesthetized by carbon dioxide exposure and decapitated ($N = 9$, from 4 litters (vehicle-treated); $N = 13$, from 5 litters (MK-801-treated). Brains were rapidly removed to ice-cold, oxygenated (95% O_2/5% CO_2) sucrose-based slicing solution (in mM): 234 sucrose, 11 glucose, 24 $NaHCO_3$, 2.5 KCl, 1.25 $NaH_2PO_4*H_2O$, 10 $MgSO_4$ and 0.5 $CaCl_2$. Brains were trimmed, glued to a stage, and immersed in cold slicing solution. The somatosensory cortex (S1) was cut into 300 uM thick coronal slices with a vibrating microtome (Leica 1200S, Germany) and transferred to an incubation chamber containing oxygen-saturated artificial cerebral spinal fluid (aCSF) comprised of (in mM): 126 NaCl, 26 $NaHCO_3$, 10 glucose, 2.5 KCl, 1.25 $NaH_2PO_4*H_2O$, 2 $MgCl_2*6H_2O$, and 2 $CaCl_2*2H_2O$; pH 7.4. Modified coronal slices were used in a subset of experiments to retain a complete thalamocortical circuit [27,28]. Slices were incubated at 32°C for one hour, then placed at ambient temperature (21–23°C) for 30 minutes prior to recording. Brain slices were placed into a recording chamber and visualized with a fixed staged, upright microscope (Nikon, FN1) equipped with 4x and 60x objectives, fluorescent filters, infrared (IR) illumination, Nomarski optics, and an IR-sensitive video camera (Cool Snap EZ, Photometrics).

Electrophysiological recordings

Whole-cell patch-clamp recordings of FSIs were obtained under continuous perfusion of oxygenated aCSF (~2 ml/min). Non-

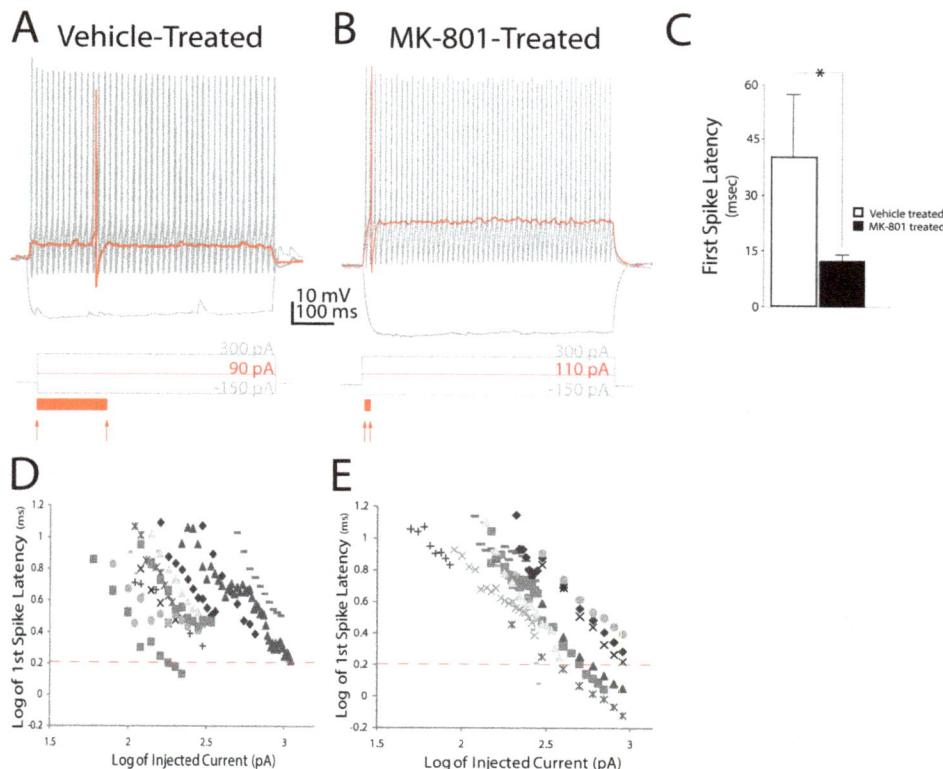

Figure 1. Postnatal NMDA receptor blockade decreases first-spike latency in neocortical layer IV FSIs. (A, B) Prototypical discharge patterns of FSIs from a vehicle- or MK-801-treated mouse. Traces show voltage response to 600 ms injections of current at saturating (black trace), or threshold (red trace) amplitude. Red arrows below trace indicate onset of current injection and spike threshold (defined as dVm/dt>10). Solid red bar below trace illustrates delay to first spike (C) Quantification of delay to first spike at rheobase response. Values are means ± S.E.M. *$P<0.05$, vs. control. (D, E). A log-log plot of first spike delay versus amplitude of injected current in layer IV FSIs from vehicle- (D) or MK-801–treated mice (E). Dashed lines on the y-axes illustrate spike latency of log 0.3 ms (or 2.0 ms $\log_{10}$).

Table 1. Electrical properties of layer IV FS neurons in the Somatosensory Cortex.

	Vehicle-Treated	MK-801-Treated
Passive Intrinisic Properties		
R_{in} (MΩ)	**103.0±8.1 (9)**	**154.8±13.4** (13)
V_{rest} (mV)	−65.5±1.8 (9)	−66.9±1.1 (13)
τ_m (ms)	6.8±0.8 (9)	8.5±0.8 (11)
I_h (mV)	4.00±0.5 (7)	3.75±0.9 (7)
Active Intrinsic Properties		
I_{th} (pA)	190±45 (9)	140±20 (13)
Max. Spike Freq. (Hz)	161.8±9.9 (9)	138.7±16.9 (9)
AP half-width (ms)	**0.41±0.02 (9)**	**0.47±0.01** (13)
AP accommodation ratio	**0.94±0.10 (9)**	**0.73±0.02*** (13)
AP threshold (mV)	−41.9±3.2 (9)	−34.0±6.5 (13)
AHP Amplitude (mV)	17.1±2.5 (9)	15.5±1.3 (13)
Ratio 1st/2nd AHP Amp.	1.02±0.01 (9)	1.01±0.003 (13)
$t_{AHP\ peak}$ (ms)	3.41±0.08 (9)	3.63±0.34 (12)

*$P<0.05$;
**$P<0.01$.
Vrest, resting membrane potential; R_{in}, input resistance; τ_m, membrane time constant. Data are mean ± SEM. Numbers in parantheses are N.Selected values shown in bold are significantly different from the same parameter measured in control cells.
*Value is significantly ($P<0.05$) different from control.
**Value is significantly ($P<0.01$) different from control.
Resting membrane potential (Vrest) was recorded within the first two minutes after break in. Input resistance (Rin) was calculated from the voltage response measured at the end of hyperpolarizing current steps (600 ms). Membrane time constant (τ) was obtained by fitting a single exponential to the relaxation of membrane potential (Vm) to a −300 pA current injection. Ih current was calculated by the formula Vm2−Vm1, where Vm2 was measured at the end of a −300 pA current step and Vm1 was measured at the maximal hyperpolarization. Depolarizing currents were injected to elicit action potentials (AP) and the AP firing patterns were used to characterize accommodation ratio, APl duration at half-width, AP amplitude and rise time. Single AP properties were measured from the first AP elicited by a minimally supra-threshold current step which elicited a continuous train of APs. AP firing threshold was determined by differentiating the voltage trace (dV/dt) evoked by the current step in current-clamp mode. The voltage and time at which the first AP discharged were taken as the value at which dV/dt≥10. In all recordings, this value exceeded baseline values (an average of the preceding 10–100 ms) by≥20-fold. AP half-width was defined as the period between the half-amplitude point in the rising and decaying phase of the AP. The ratio of the 1st AP/2nd AP was calculated using the peak amplitudes as one measure of accommodation. The ratio of the interval between the first APs and the average interval of the last three APs during a train was calculated as a second measure of AP accommodation. The after hyperpolarization (AHP) amplitude was calculated as the difference from AP threshold to the maximum trough of hyperpolarization.

filamented glass pipettes (King Precision Glass, USA) were pulled on a vertical puller (PC-10, Narishige) to obtain electrodes with resistances of 3–5 MΩ when filled with an intracellular solution (in mM): 70 KCl, 70 potassium gluconate, 10 HEPES, 10 EGTA, 2 MgCl$_2$, 4 Mg-ATP, 0.3 Na-GTP. Spontaneous excitatory and inhibitory events were acquired using a cesium-based intracellular solution comprised of (in mM): 20 KCl, 100 cessium methanesulfonate, 10 HEPES, 4 Mg-ATP, 0.3 Na-GTP, 3 QX-314 [29].

A GΩ seal was formed between the cell and glass pipette and a solenoid-controlled vacuum transducer was used to apply brief suction pulses (120 psi at 20–50 ms) to break into the cell. Recordings were acquired with pClamp 10 and a MultiClamp 700B amplifier (Molecular Devices, Sunnyvale, CA). Capacitance and series resistance compensation were used (typically 80%), and series resistance was continuously monitored throughout experiments. Putative FSIs were visually identified by EYFP fluorescence and then biophysical responses to brief hyperpolarizing and depolarizing current injections were characterized to confirm identity. The rheobase current (Ith) was determined as the minimum current amplitude capable of discharging a single spike for three consecutive trials. To accurately measure Ith, the amplitude of injected current was incrementally increased by 1–5 pA per sweep. Most cells discharged a single spike at Ith, (9/10 cells recorded in 10 slices from 9 vehicle-treated animals; 13/17 cells recorded in 17 slices from 13 MK-801-treated animals). Any cell which did not discharge a single spike was omitted from

the study. Phase plots were constructed by calculating the first derivative (dV/dt) of I$_{th}$ and plotting the value versus membrane potential (Vm).

For evoked recordings, depolarizing current pulses were delivered once every 15 seconds (Isoflex, A.M.P.I.; CPI, Carl Pisaturo, Stanford University). A minimal stimulation protocol was used to evoke EPSCs with a failure rate of approximately 50%. The amplitude of the current pulse ranged from 8.8–150 µA, with a fixed duration of 0.2 ms. EPSCs were evoked at the thalamocortical synapses of layer IV FSIs by placing a 25 µm concentric bipolar microelectrode (FHC, Bowdoin, ME) directly in contact with fibers projecting from the ventrobasal complex of the thalamus. EPSC amplitude was measured from the baseline to the peak of the initial response. The averaged EPSC decay time was fit to the double exponential function: $f(t) = A_1 exp^{-t/\tau^1} + A_2 exp^{-t/\tau^2}$. These fits were used to determine a weighted time constant: $\tau_W = (\tau_1\alpha_1 + \tau_2\alpha_2)/(\alpha_1 + \alpha_2)$ where α and τ is the amplitude and time constant, respectively. NMDA currents were pharmacologically isolated by adding 20 uM glycine (Sigma), 20 uM 6,7-Dinitroquinoxaline-2,3-dione (DNQX, Tocris) and 10 uM SR 95531 hydrobromide (Gabazine, Tocris) to the perfusate.

Analysis of Electrophysiology Data

Electrophysiological analyses were performed offline using Clampfit v10.2 (Molecular Devices, Sunnyvale CA) and Mini Analysis Program (Syanptosoft, Decatur, GA).

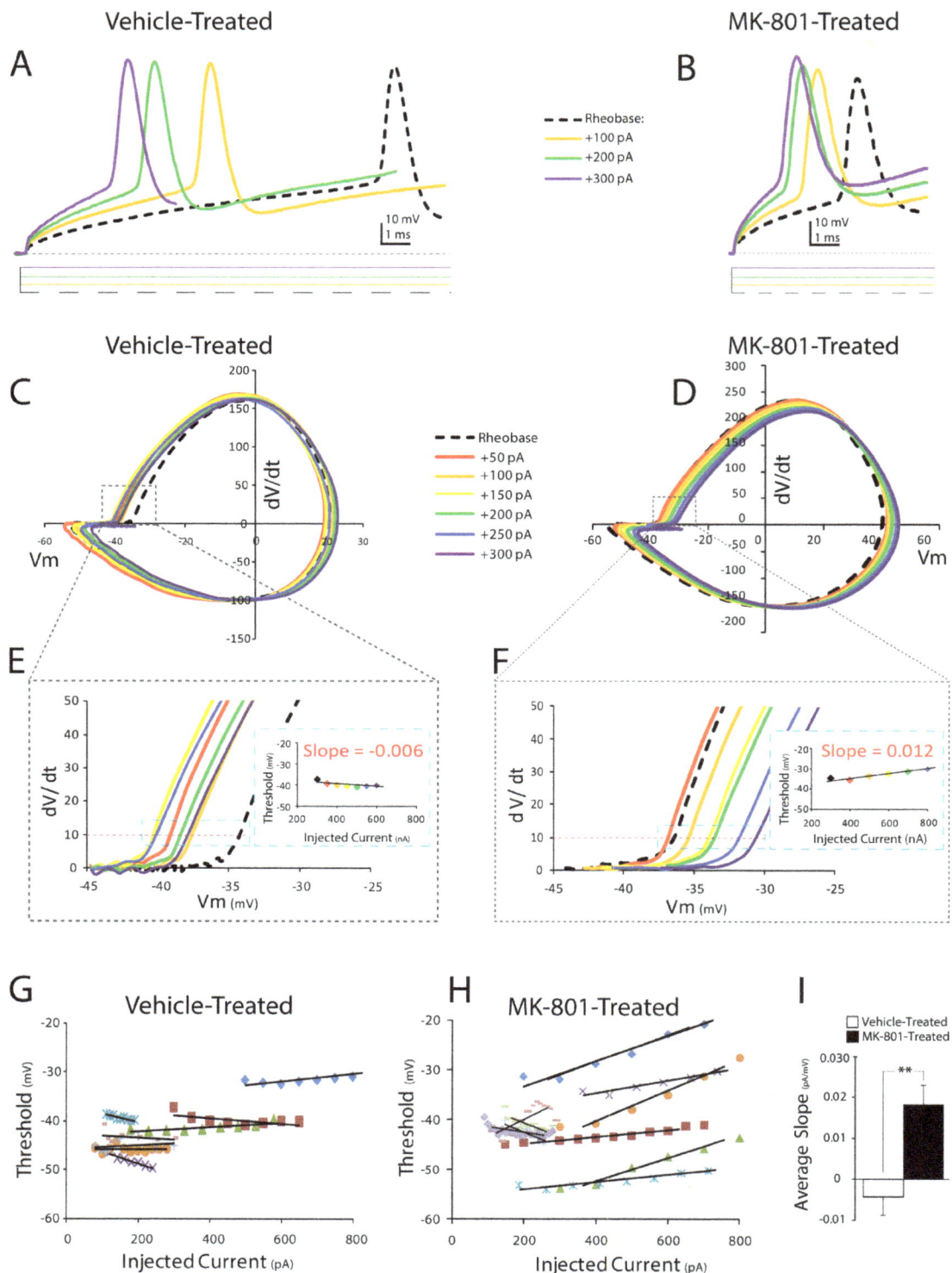

Figure 2. Postnatal NMDA receptor blockade destabilizes spike threshold in neocortical layer IV FSIs. (A, B) Representative voltage responses of first spikes elicited from FSIs at rheobase (I_{th}); I_{th}+100 pA; I_{th}+200; and I_{th}+300 pA current steps. For clarity, only the initial spike in each trace is displayed, and the intervening, +50 pA, current steps are omitted. (C, D) Phase plots of the traces shown in A, B. Phase plots are shown for the first spike discharged during injection at I_{th} and I_{th} +: 50, 100, 150, 200, 250 and 300 pA. (E, F) Regions of interest indicated by the dashed line are displayed at higher magnification. Red dashed line indicates spike threshold (dV/dt = 10). Insets are plots of injected current amplitude versus first spike threshold of displayed cell. (G, H) Plots of injected current amplitude versus first spike threshold. (I) Mean slope of injected current amplitude versus first spike threshold in (G, H). Values are means ± S.E.M. **P<0.005, vs. control by ANOVA.

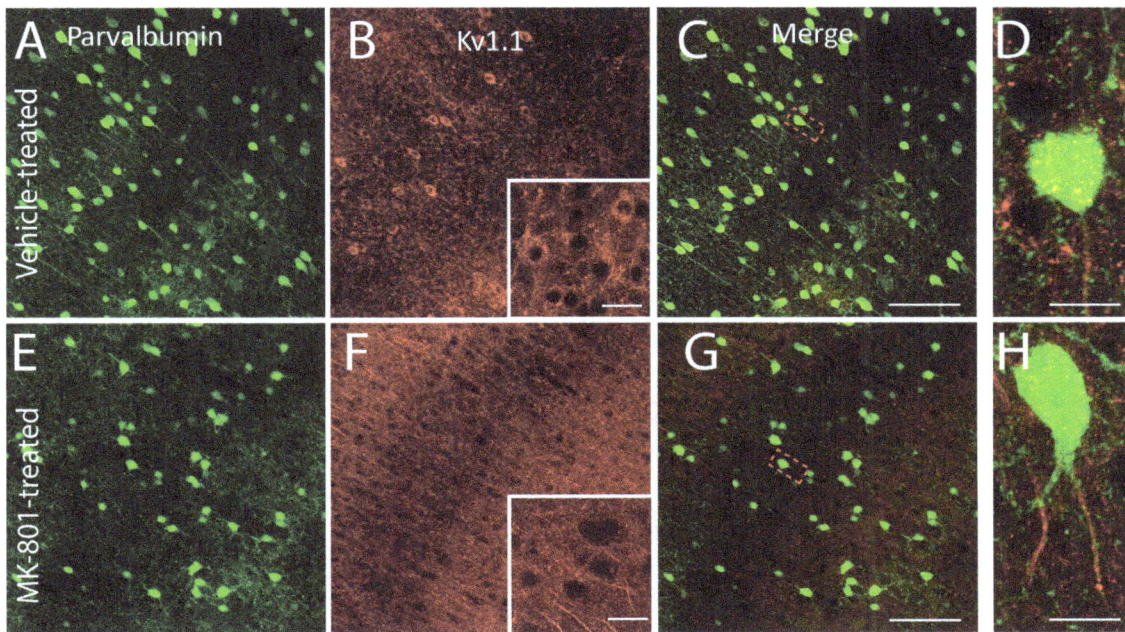

Figure 3. Postnatal NMDA receptor blockade reduces somatic expression of Kv1.1. (A–H) Representative confocal micrographs (maximal intensity, flattened Z-stack, z separation of 1.03 um) of somatosensory cortex from adolescent mice (PND42) neonatally injected with vehicle (A–D), or MK-801 (E–H). (D, H) Staining for the Kv1.1 potassium channel subunit is dramatically reduced in FSIs from MK-801-injected mice. Insets in (B) and (F) show higher magnification of layer IV regions of interest. Green = Parvalbumin; Red = Kv1.1. Scale bars = 100 um (A–C; E–G) or 10 um (D, H and insets in B, F).

Immunohistochemistry

Mice were anesthetized with isoflurane. An incision was made into the thoracic cavity, the rib cage was removed and mice were transcardially perfused with 20 mL of ice cold phosphate buffered saline (PBS) followed by 20 mL of an ice cold 4% paraformaldehyde (PFA) solution. The brain was removed and fixed overnight in 4% PFA. Brains were immersed in agar and cut into coronal sections (50 uM). Primary antibody staining was performed by incubating free-floating sections in a solution of PBST (PBS+0.2% Tween) +10% normal goat serum with the one or more of the following antibodies: mouse, anti-parvalbumin (Sigma, St. Louis, MO); goat, anti-NR2B (Abcam, Cambridge, MA); rabbit anti-parvalbumin (Savant, Switzerland); and mouse Kv1.1 (UC Davis/NIH NeuroMab Facility). Sections were shaken overnight at 4°C, then washed five times in PBST and incubated overnight at 4°C in a solution of PBST+10% normal donkey serum and the following secondary antibodies: donkey, anti-rabbit Alexa 488 (Invitrogen); donkey, anti-mouse Alexa 555 or donkey anti-goat Alexa 555. Sections were washed, mounted, and covered with a glass cover slip. Confocal micrographs were captured on a Leica LSM 510 confocal microscope equipped with 10x, 20x, 40x oil, 63x oil, and 100x oil objectives. Post-hoc immunohistochemical analysis confirmed all fluorescently labeled cells were positive for parvalbumin.

Statistical Analysis

Data are expressed as mean +/− SEM. Differences among experimental groups were considered statistically significant at $p <$ .05. ANOVA and two-tailed, Student's T-Test was used to compare changes between each experimental group.

Results

Neonatal NMDA Receptor Blockade Minimizes First-Spike Latency in Neocortical FSIs

The electrophysiological development of FSIs in the mouse somatosensory cortex (S1) has been comprehensively characterized [30], and FSIs from S1 may also be highly vulnerable to perinatal NMDA receptor blockade [18]. We therefore chose FSIs from S1 as the subject of this study to maximize sensitivity to MK-801 treatment.

The onset of action potential discharge (or "spiking") varies widely in neocortical FSIs and is inversely proportional to stimulus strength [31]. Strong stimuli elicit spikes from FSIs rapidly, whereas near-threshold level stimuli elicit spikes with a temporal delay, termed first-spike latency [31]. We found that first-spike latency was significantly shorter in FSIs from MK-801-treated mice compared to vehicle-treated mice [MK-801-treated: 12.5 ± 1.7 ms (SEM), $n = 11$ cells; *versus* Vehicle-treated: 43.8 ± 17.0 ms, $n = 9$ cells; $P = 0.02$] (Figures 1A, B). Since first-spike latency is inversely proportional to stimulus intensity, we examined the relation of stimulus intensity to first-spike latency to more closely assess the impact of neonatal MK-801 treatment on the input-output function of FSIs. A log-log plot confirmed first-spike latency is, indeed, inversely proportional to stimulus intensity when spikes are discharged by low-to-moderate strength stimuli (e.g. current amplitude of $<2x$ I_{th}) (Figure 1D). However, when spikes were discharged by high intensity stimuli (e.g. $>2x$ of I_{th}), the relation of stimulus intensity to first-spike latency became non-linear and a stable, stimulus-insensitive function appeared in FSIs from saline-treated mice (Figure 1D). By contrast, first-spike latency in FSIs from MK-801 treated mice remained inversely proportional at all stimulus intensities (Figure 1E). This permitted spikes with substantially shorter latencies to be discharged by high

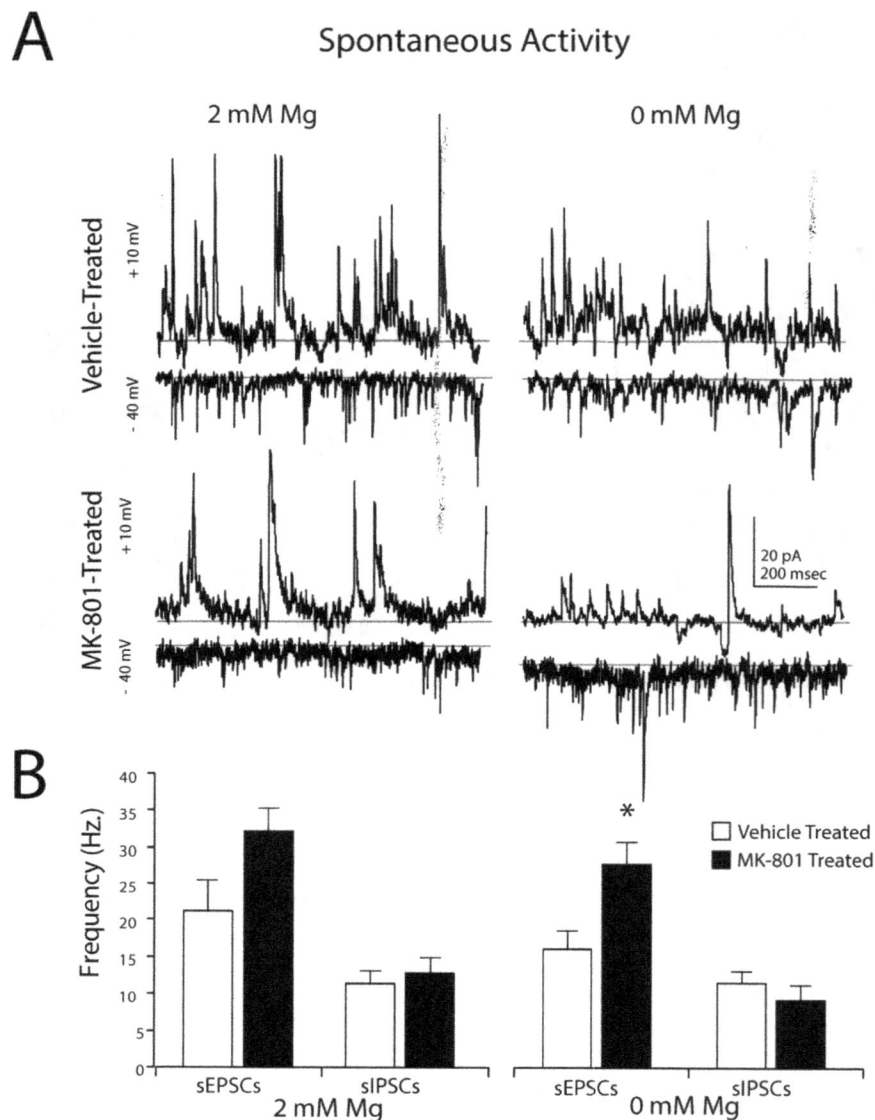

Figure 4. Postnatal NMDA receptor blockade increases glutamatergic input to neocortical FSIs. (A) Representative whole-cell patch clamp recordings from FSIs in acute brain slices from the somatosensory cortex of adolescent mice (PND21-P25) neonatally-treated with vehicle or MK-801. Spontaneous activity was recorded in 2 mM (left traces) or 0 mM (right traces) of added Mg^{2+}. Recordings were obtained in voltage clamp mode. sIPSCs (top traces) were recorded at the sEPSC reversal potential ($\sim+10$ mV) and sEPSCs (bottom traces) were recorded at the Cl- reversal potential (~-40 mV). (B) Bar graphs of mean sEPSCs and sIPSCs frequency obtained during 5 min recordings. Values are means $\pm$ S.E.M. *$P<0.05$, vs. control by ANOVA.

intensity stimulation. These data suggest first-spike latency is comprised of at least two components in FSIs which exhibit distinct relationships to stimulus intensity. Moreover, neonatal NMDA receptor blockade appears to shorten the stimulus-insensitive component of first-spike latency.

Neonatal NMDA Receptor Blockade Alters Several Intrinsic Spiking Properties of Neocortical FSIs

Many intrinsic properties of FSIs are optimized to deliver precisely timed spikes [32] that exhibit very little frequency accommodation during spike trains [4]. For instance, FSIs exhibit a low input resistance, fast membrane time constant, and narrow spike width [33]. All of these attributes help FSIs discharge spikes with high temporal precision and at a sustained rate. We

characterized several active and passive membrane properties of FSIs to determine the impact of neonatal MK-801 treatment. The input resistance of FSIs from MK-801 treated mice was 50.3% higher than vehicle-treated mice [Rin; MK-801-treated: 154.8±13.4 MΩ (SEM), $n = 13$; Vehicle-treated: 103.0±8.1 MΩ. $n = 9$; $P<0.01$, Table 1]. Spikes discharged from the FSIs of MK-801-treated mice were 14.6% broader than vehicle-treated mice [AP ½ width: MK-801-treated: 0.47±0.01 ms (SEM), $n = 13$; Vehicle-treated: 0.41±0.02 ms, $n = 9$; $P<0.01$, Table 1] and displayed significantly more spike frequency accommodation [spike frequency accommodation ratio: MK-801-treated: 0.73±0.02 ms (SEM), $n = 13$; Vehicle-treated: 0.94±0.10 ms, $n = 9$; $P<0.05$, Table 1]. The maximum firing rate of FSIs from MK-801-treated mice was 17% lower than FSIs

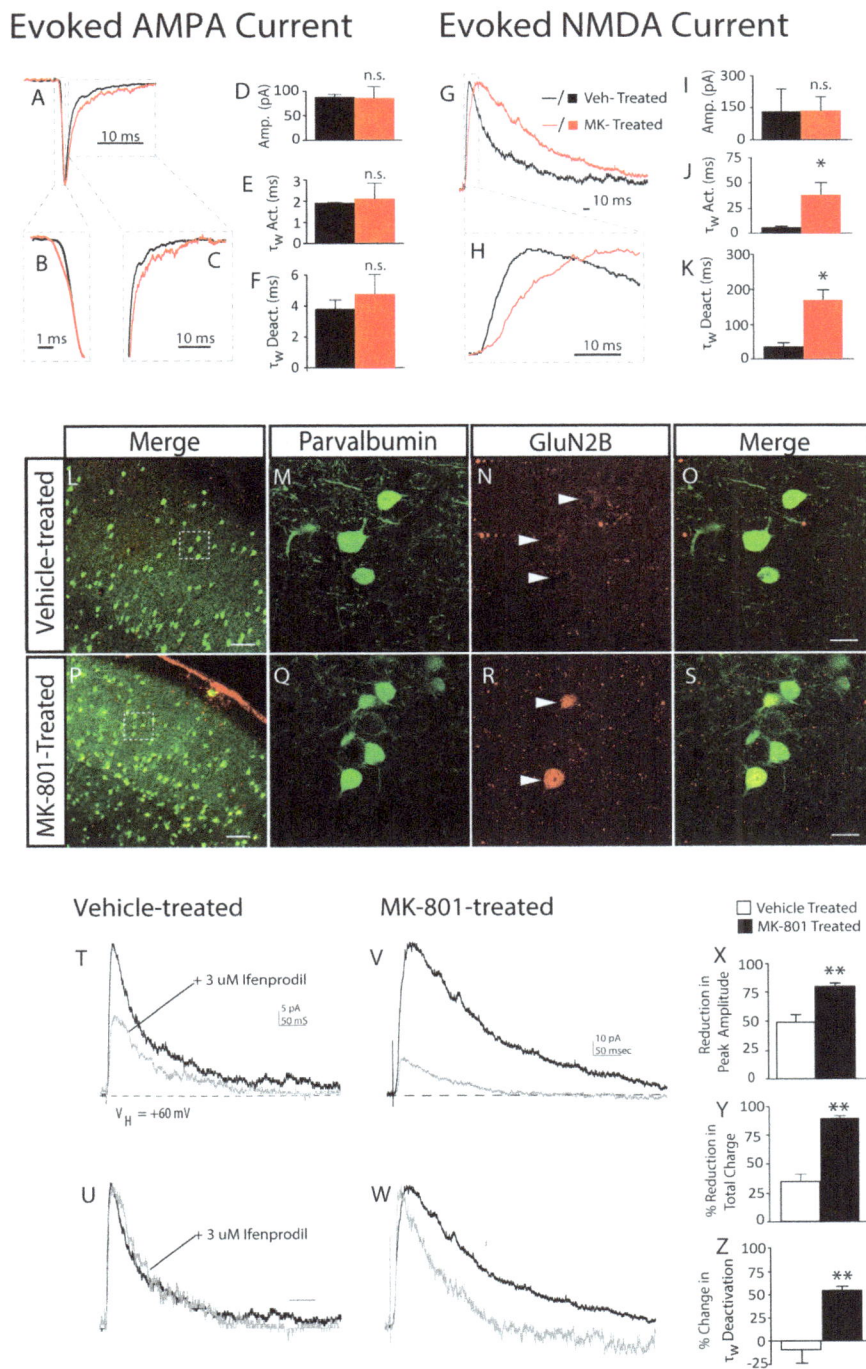

Figure 5. Postnatal NMDA receptor blockade increases expression of functional NR2B receptors in neocortical FSIs. Prototypical excitatory post-synaptic currents were evoked onto layer IV FSIs in a thalamocortical slice preparation. (A–C) AMPA-mediated responses were similar in vehicle-treated mice (black traces) and MK-801-treated mice (red traces). ($V_{Hold-AMPA} = -60$ mV). Traces shown are the average of 5–10 responses obtained at the same stimulus intensity from vehicle-treated mouse (black lines); or MK-801-treated mouse (red lines). Dashed boxes designate regions of interest in activation and decay kinetics. (D–F) Bar graphs of mean current amplitude and mean weighted activation tau (τ_wAct.) and deactivation tau (τ_wDeact.) of evoked AMPA current evoked from vehicle-treated mice (black bars) and MK-801-treated mice (red bars). (G, H) The kinetics of NMDA-mediated responses are slower in MK-801-treated mice (red traces). ($V_{Hold-NMDA} = +60$ mV). Traces shown are the average of 5–10 responses obtained at the same stimulus intensity. Dashed boxes designate regions of interest in activation and decay kinetics. (I–K) Bar graphs of mean current amplitude and mean weighted activation tau (τ_w) and deactivation tau (τ_w) of evoked NMDA current from vehicle and MK-801-treated mice. Values are means ± S.E.M. *$P<0.05$, vs. control by ANOVA. (L–S) Representative confocal micrographs of layer IV somatosensory cortex from mice neonatally treated with vehicle (L–O), or MK-801 (P–S). (L, P) Dashed boxes designate regions of interest in layer IV selected for higher magnification as shown in (M–O), and (Q–S). Green = Parvalbumin; Red = GluN2B. Scale bars = 100 um (L, P) or 10 um (M–O; Q–S). Images are shown as a maximum intensity projection of single images at a Z-spacing of 1.0 uM. (N, R) Arrowheads denote GluN2B staining as puncta or broad staining in brain slices from a vehicle-treated, or MK-801-treated mouse, respectively. (T–W) Representative traces of ifenprodil blockade on monosynaptic

NMDA current evoked onto a FSI from an adolescent mouse neonatally treated with vehicle (T) or MK-801 (V) Traces shown are averages of 5–10 responses to the same stimulus intensity before (black lines) and after (grey lines) 10 minute wash in of 3 uM Ifenprodil. (U, W) Same traces as shown in (T) and (V) scaled on y-axis to facilitate visual comparison of the actions of ifenprodil on activation and decay kinetics. Average traces were fit, as previously described. (X, Y, Z) Quantitative analysis of impact of ifenprodil on mean peak amplitude (X); mean reduction in total charge (Y); and mean change in τ_w decay (Z). ($N = 4$ cells from 4 animals (vehicle-treated); and 3 cells from 3 animals (MK-801-treated).

from vehicle-treated mice, although this difference was not statistically significant ($P = 0.11$). Together these data suggest neonatal NMDA receptor blockade disrupts both active and passive membrane properties of neocortical FSIs which could subsequently impair spiking behavior.

Neonatal NMDA Receptor Blockade Destabilizes Spike Threshold Dynamics

The spike threshold of layer IV neocortical neurons can be dynamic [34]. We characterized the spike threshold dynamics of neocortical FSIs to assess the impact of neonatal NMDA receptor blockade. We first examined spike threshold at I_{th}. I_{th} was determined, as previously described, and phase plots were constructed for each cell to allow spike threshold to be determined graphically (Figure 2). Spike threshold at I_{th} was about 8 mv more depolarized in FSIs from MK-801-treated mice, although this difference was not statistically significant [Vehicle-treated: -41.9 ± 3.2, $n = 9$ versus MK-801-treated: -34.0 ± 6.5 mV, $n = 13$; $P = 0.34$, Table 1]. The amplitude of the I_{th} current was incrementally increased in successive sweeps to examine the dynamics of spike threshold (Figure 2C–F). The spike threshold of FSIs from vehicle-treated mice was examined across a broad range of stimulus intensities and found to be stable and independent of current amplitude (Figure 2G). By contrast, FSIs from MK-801-treated mice exhibited a positive correlation to current amplitude [slope in MK-801-treated FSIs: 0.018 ± 0.004 pA mV^{-1}, $n = 8$; versus slope of vehicle-treated FSIs: -0.004 ± 0.004 pA mV^{-1}, $n = 8$; $P = 0.003$] (Figure 2I). These data suggest stimulus intensity has a larger influence on the spike threshold of FSIs from MK-801-treated mice, and that the dynamics of spike threshold are also less stable than in FSIs from vehicle-treated mice.

Neonatal NMDA Receptor Blockade Reduces Kv1.1 Expression in Neocortical FSIs

Both first-spike latency and spike threshold are strongly modulated by K$^+$ channels comprised from Kv1 potassium

channel subunits [34]. In neocortical FSIs, first-spike latency is specifically mediated by the Kv1.1 subunit [31]. We therefore hypothesized the alterations in first-spike latency and spike threshold dynamics observed in FSIs from MK-801-treated mice resulted from disturbances in the expression of the Kv1.1 subunit. Immunohistochemical analysis revealed that neonatal MK-801 treatment dramatically reduced staining for the Kv1.1 subunit in neocortical FSIs (Figures 3A–H). These findings are consistent with our physiological characterizations and demonstrate that transient, NMDA receptor blockade during early development can reduce expression of the Kv1.1 subunit in the FSIs of adolescent mice.

Neonatal NMDA Receptor Blockade Increases Mg^{2+}-Sensitive sEPSCs onto FSIs

We next evaluated the impact of neonatal NMDA receptor blockade on synaptic input to FSIs. Whole-cell patch clamp recordings from neocortical FSIs revealed the average frequency of sEPSCs recorded from FSIs of MK-801-treated mice was 50% higher than vehicle-treated mice, although statistical significance was not reached [Vehicle-treated: 21.2 ± 4.4 Hz, $n = 6$; versus MK-801-treated: 31.4 ± 2.7, $n = 8$; $P = 0.06$] (Figure 4A). NMDA-mediated EPSCs may be highly sensitive to Mg^{2+} block [35,36], so we repeated these experiments in superfusate with 0 mM added Mg^{2+} (Figures 4A, B). In nominal Mg$^+$, the frequency of sEPSCs recorded from the FSIs of MK-801-treated mice was 60% higher than sEPSCs recorded from vehicle-treated mice and statistically significant [Vehicle-treated: 16.1 ± 2.51 Hz, $n = 4$; versus MK-Treated: 26.1 ± 2.21 Hz, $n = 8$; $P = 0.03$] (Figure 4B). There was no difference in sIPSC frequency at either concentration of Mg^{2+} (Figure 4B), thus inhibitory currents were not examined further.

Neonatal NMDA Receptor Blockade Increases NMDA Current at the Thalamocortical Synapses of FSIs

The increase in sEPSC frequency that occurs in 0 mM Mg^{2+} superfusate could result from increased NMDA-mediated events

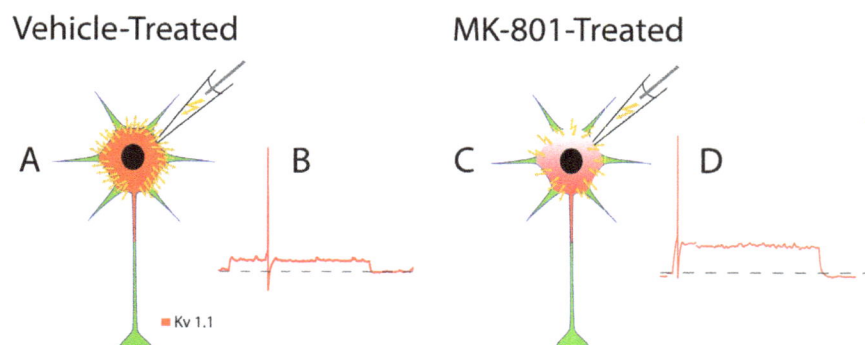

Figure 6. Schematic depiction of how neonatal MK-801 treatment impacts the spike timing of an FSI from an adolescent mouse. (A) A significant proportion of the current injected during whole-cell patch-clamp (yellow lightning bolts) dissipates through the high density of Kv1.1-containing K$^+$ channels expressed in FSIs from vehicle-treated mice. (B) The time required to depolarize the cell membrane to threshold is increased and first-spike latency is temporally expanded. (C) By contrast, a lower proportion of injected current dissipates FSIs from MK-801-treated mice due to reduced expression of Kv1.1-containing K$^+$ channels. (D) The time required to depolarize to threshold is reduced and first-spike latency is therefore minimized.

or increased α-amino-3-hydroxy-5-methyl-4-isoxazolepropionic acid (AMPA)-mediated events. Moreover, since layer IV FSIs receive excitatory input from both thalamocortical circuits [28,37] and local layer IV spiny stellate cells [38] the origin of the Mg^{2+}-sensitive sEPSCs was not clear. To more clearly distinguish the source of the Mg^{2+}-sensitive sEPSCs we utilized a thalamocortical slice preparation. In this preparation, monosynaptic AMPA currents were evoked onto FSIs from vehicle-treated or MK-801-treated mice ($V_{HOLD} = -60$ mV). The average amplitude of AMPA current was similar between the two groups [Vehicle-treated: 97.2 ± 13.0 pA, $n = 3$; MK-801-treated: 87.3 ± 23.6 pA, $n = 3$; $P = 0.36$] (Figures 5A, D), and no difference was detected in the 10–90% rise time [Vehicle-treated: 1.8 ± 0.1 ms, $n = 3$; MK-801-treated: 2.1 ± 0.7 ms, $n = 3$; $P = 0.32$] or decay time [Vehicle-treated: 3.0 ± 1.2 ms, $n = 3$; MK-801-treated: 4.8 ± 1.3 ms, $n = 3$; $P = 0.18$] of evoked AMPA current (Figures 5B, C, E, F).

Although there was no difference in the peak amplitude of evoked NMDA current [Vehicle-treated: 131.4 ± 105.5 pA, $n = 3$; MK-801-treated: 137.3 ± 65.4 pA, $n = 3$; $P = 0.49$] (Figure 5G, I), both the 10–90% rise time and the decay time of evoked NMDA current was significantly slower in FSIs from MK-801-treated mice: [10–90% rise time: MK-801-treated: 37.9 ± 12.4 ms, $n = 3$; Vehicle-treated: 5.2 ± 1.7 ms, $n = 3$; $P = 0.02$], [Average decay time: MK-801-treated: 170.4 ± 9.7 ms, $n = 3$; Vehicle-treated: 54.3 ± 12.0 ms, $n = 3$, $P = 0.02$] (Figures 5G, H, J, K). Together the increase in Mg^{2+}-sensitive sEPSCs, and slower kinetics of the evoked NMDA current suggests neonatal NMDA receptor blockade alters the composition of NMDA receptors expressed at the thalamocortical synapses of layer IV FSIs.

NMDA receptors are comprised from heteromeric subunits [39] and the kinetic properties of the receptor channel are mostly determined by GluN2 subunits [40]. By adolescence, neocortical NMDA current is predominated by GluN2A and GluN2B-comprised receptors [41]. The GluN2B subunit confers slower channel kinetics to NMDA receptors [42], thus we hypothesized neonatal MK-801 exposure altered expression of GluN2B subunits in FSIs. Immunohistochemical analysis revealed neonatal MK-801 treatment increased GluN2B staining in FSIs (Figures 5R, S). The functionality of the surplus GluN2B subunits was examined by measuring the sensitivity of evoked NMDA current to the GluN2B-selective antagonist, ifenprodil [43]. Bath application of ifenprodil dramatically increased the decay rate of NMDA current evoked onto FSIs from MK-801-treated mice, but not vehicle-treated mice [% change τ_W: Vehicle-treated, $-9.9 \pm 13.7\%$, $n = 3$; % change τ_W: MK-801-treated, $55.6 \pm 4.0\%$, $n = 3$, $P = 0.008$] (Figures 5T, U, Z). Similarly, ifenprodil attenuated the peak amplitude and total charge of NMDA current evoked from MK-801-treated mice more strongly than vehicle-treated mice [% peak amplitude reduction I_{NMDA}: Vehicle-treated, $49.5 \pm 1.7\%$, $n = 3$; MK-801-treated, $80.4 \pm 2.4\%$, $n = 3$, $P = 0.009$] [% reduction Q_{NMDA}: Vehicle-treated, $35.1 \pm 6.7\%$, $n = 3$; MK-801-treated, $90.1 \pm 1.7\%$, $n = 3$; $P = 0.009$] (Figures 5X, Y). These data demonstrate neonatal MK-801 treatment substantially increased the proportion of GluN2B-mediated NMDA current at the thalamocortical synapses of layer IV FSIs.

Discussion

Here we demonstrate that neonatal MK-801 treatment disrupts both intrinsic and synaptic functions of neocortical FSIs in adolescent mice. These findings support our hypothesis that transient neonatal NMDA receptor blockade causes persistent alterations in FSI physiology, and we now propose a mechanism by which these alterations could be mediated.

Neonatal MK-801 Treatment May Minimize First-Spike Latency by Reducing Kv1.1 Expression in FSIs

Neonatal MK-801-treatment reduces expression of the Kv1.1 subunit (Figure 3) which we propose shortens first-spike latency in FSIs (Figure 1). First, consider that near resting membrane potential a substantial fraction of Kv1.1-comprised K^+ channels are partially open [44]. Thus, during low amplitude stimulation a significant portion of the current injected during whole-cell patch clamp of FSIs leaks out as K^+ ions (Figure 6A). This efflux of K^+ ions slows the rate of membrane depolarization [45] [44] and delays spike discharge [44] (Figure 6B). Since neonatal MK-801 treatment reduces Kv1.1 expression (Figure 6C), the fraction of current that leaks during whole-cell patch clamp is minimized, increasing the rate of membrane depolarization, and minimizing first-spike latency (Figure 6D). This interpretation is further supported by the finding that the resting input resistance (R_{in}) of FSIs from MK-801-treated mice is 50% higher than FSIs from vehicle-treated mice (Table 1) ($P < 0.01$). Moreover, the amplitude of I_{th} in FSIs from MK-801-treated mice was smaller than vehicle-treated mice, although the difference did not reach statistical significance (Table 1). Together, these data support our interpretation that first-spike latency is shortened in FSIs from MK-801-treated mice because of reduced expression of the Kv1.1 subunit.

Neonatal MK-801-treatment May Increase the Spike Threshold Dynamics of FSIs

Compared to FSIs from vehicle-treated mice, the spike threshold of FSIs from MK-801-treated mice is more dynamic and inversely correlated to stimulus intensity (Figure 2G, H). During periods of sustained or intense stimulation the spike threshold of FSIs from MK-801-treated mice could adapt to more depolarized values [46] and thereby reduce excitability. Interestingly, FSIs from MK-801-treated mice exhibit a reduced maximum spike rate and an increased spike accommodation ratio (Table 1). Increased spike-threshold dynamics may help explain these findings. It is not clear how neonatal MK-801 treatment might increase the spike-threshold dynamics of FSIs. However, K+ channels comprised from the Kv1.1 subunit have been shown to regulate spike-threshold variability [31,34,44]. Thus it is conceivable that the changes in FSI spike-threshold dynamics could result from changes in the expression and/or function of K+ channels comprised from the Kv1.1 subunit.

Neonatal MK-801 Treatment May Impair Physiological Maturation of Neocortical FSIs

Our mechanistic model of how neonatal MK-801 treatment shortens first-spike latency was derived from a relatively straight-forward interpretation of our results. However, a more expansive consideration of our findings supports the notion that neonatal MK-801 treatment may broadly impair physiological maturation of FSIs. Below we consider this idea within the context of our most salient findings–disrupted membrane properties, shortened first-spike latency, and increased GluN2B NMDA current.

The membrane properties and spiking characteristics of mature FSIs are highly specialized to facilitate the discharge of brief, precisely timed spikes. Specifically, spikes from mature FSIs exhibit a narrow spike width, low input resistance, fast membrane time constant, and low frequency accommodation ratio [47] [32] [33]. Curiously, these properties are not innately optimized, but rather undergo extensive refinement during the postnatal development of FSIs. For example, FSIs from immature mice have a high input resistance, slow membrane time constant, broad spike width and exhibit considerable spike frequency accommodation

[33]. These immature features oppose the rapid and precise spiking activity that characterizes mature FSIs [47]. However, FSIs undergo rapid postnatal development, and by P18 they attain mature membrane properties and spiking characteristics that remain stable through adulthood [30].

In this study the membrane properties and spiking characteristics of FSIs were evaluated in adolescent mice aged adolescent mice aged PND 26 ± 2.3. Notably, several of the membrane properties and spiking characteristics of FSIs from MK-801 treated mice were physiologically immature compared to FSIs from age-matched, vehicle-treated mice. For instance, the average membrane resistance and spike accommodation ratio of FSIs from MK-801-treated mice were significantly less mature than FSIs from vehicle-treated mice (Table 1). Similarly, the membrane time constant (τ_m), steady-state firing rate, and instantaneous firing rate of FSIs from MK-801-treated mice were also less mature than FSIs from vehicle-treated mice (Table 1), though these differences did not reach statistical significance. Overall, the membrane properties and spiking characteristics of FSIs from MK-801-treated mice exhibited a level of physiological maturity comparable to FSIs from healthy mice aged ~P13–15 [30]. We therefore conclude that transient, neonatal MK-801 treatment impairs biophysical maturation of neocortical FSIs. More broadly, these data suggest a role for NMDA signaling in the normal development of FSIs.

Neonatal MK-801 Treatment May Impair Developmental Regulation of First-Spike Latency

FSIs can be distinguished by first-spike latency [48], but the development of this pattern has not been fully characterized. However, since first-spike latency is highly dependent on the Kv1.1 subunit (Goldberg et al 2008), it is likely to correlate with the developmental expression of the Kv1.1 subunit. In neonatal mice the expression of Kv1.1 mRNA is very low in neocortical FSIs, but the expression increases more than 40-fold by P21 [33]. If expression of the Kv1.1 subunit is monotonically regulated, first-spike latency in FSIs may progress with a similar developmental trajectory. If so, our finding that transient, neonatal MK-801 treatment reduces expression of the Kv1.1 subunit and shortens first-spike latency in the FISIs of adolescent mice further supports the notion that NMDA signaling contributes to the physiological development of neocortical FSIs.

Neonatal MK-801 Treatment May Impair Maturation of Excitatory Synapses in Neocortical FSIs

The subunit composition of NMDA receptors is developmentally regulated. During early development, neocortical NMDA current is comprised mostly from GluN2B-containing receptors

[49,50]. During adolescence a stereotyped developmental switch occurs and the proportion of NMDA current mediated by GluN2B-containing receptors in FSIs diminishes [24]. Thus, the proportion of GluN2B-mediated NMDA current is approximately inversely proportional to FSI development. In this study NMDA current was characterized from the FSIs of adolescent mice (PND 42–45). Sensitivity to the GluN2B-specific antagonist, ifenprodil, was used to determine the contribution of GluN2B-mediated NMDA current. As expected, NMDA current evoked from the FSIs of vehicle-treated mice was comprised of only 35% GluN2B-mediated current, whereas NMDA current evoked from the FSIs of MK-801-treated mice was comprised of about 90% GluN2B-mediated current (Figures 5T–Z). While this study was in progress two reports emerged which corroborate our findings. In the first report, neonatal administration of a GluN2A-specific antagonist increased the proportion of GluN2B-mediated NMDA current evoked from FS cells in the somatosensory cortex of adolescent mice [24]. In the second report, perinatal administration of phencyclidine increased the expression of the GluN2B subunit in the prefrontal cortex of adolescent rats [51]. Together, these data suggest transient disruption of NMDA signaling may indelibly alter the ontogeny of excitatory synapses throughout the neocortex.

In summary, neonatal NMDA receptor blockade remains a robust approach for modeling many schizophrenia-like behavioral deficits in animals [17]. However, the pathophysiological mechanisms which give rise to the behavioral deficits are unclear. In this study we discovered that neonatal NMDA receptor blockade simultaneously *minimizes* first-spike latency and *increases* the proportion of GluN2B-mediated NMDA current at thalamocortical synapses onto neocortical FSIs. Both of these results are consistent with our general notion that neonatal MK-801 treatment impairs the physiological maturation of neocortical FSIs, and since layer IV FSIs process the majority of thalamic sensory input, it is conceivable that the disturbances in FSI physiology we report contribute to the pathophysiology underlying the schizophrenia-like behaviors.

Acknowledgments

K.S.J. wishes to expresses deep gratitude to G.H.J. for many insightful discussions and unflagging support.

Author Contributions

Conceived and designed the experiments: MMH JGC KSJ. Performed the experiments: KSJ. Analyzed the data: MMH KSJ. Contributed reagents/materials/analysis tools: MMH. Wrote the paper: MMH JGC KSJ.

References

1. Lewis DA, Hashimoto T, Volk DW (2005) Cortical inhibitory neurons and schizophrenia. Nat Rev Neurosci 6: 312–324.
2. Inan M, Petros TJ, Anderson SA (2012) Losing your inhibition: Linking cortical GABAergic interneurons to schizophrenia. Neurobiol Dis.
3. Eyles DW, McGrath JJ, Reynolds GP (2002) Neuronal calcium-binding proteins and schizophrenia. Schizophr Res 57: 27–34.
4. Kawaguchi Y (1995) Physiological subgroups of nonpyramidal cells with specific morphological characteristics in layer II/III of rat frontal cortex. J Neurosci 15: 2638–2655.
5. Hestrin S, Galarreta M (2005) Electrical synapses define networks of neocortical GABAergic neurons. Trends Neurosci 28: 304–309.
6. Monyer H, Markram H (2004) Interneuron Diversity series: Molecular and genetic tools to study GABAergic interneuron diversity and function. Trends Neurosci 27: 90–97.
7. Gibson JR, Beierlein M, Connors BW (1999) Two networks of electrically coupled inhibitory neurons in neocortex. Nature 402: 75–79.
8. Pangratz-Fuehrer S, Hestrin S (2011) Synaptogenesis of electrical and GABAergic synapses of fast-spiking inhibitory neurons in the neocortex. J Neurosci 31: 10767–10775.
9. Bacci A, Huguenard JR, Prince DA (2003) Functional autaptic neurotransmission in fast-spiking interneurons: a novel form of feedback inhibition in the neocortex. J Neurosci 23: 859–866.
10. Fries P, Nikolic D, Singer W (2007) The gamma cycle. Trends Neurosci 30: 309–316.
11. Sohal VS, Zhang F, Yizhar O, Deisseroth K (2009) Parvalbumin neurons and gamma rhythms enhance cortical circuit performance. Nature 459: 698–702.
12. Cardin JA, Carlen M, Meletis K, Knoblich U, Zhang F, et al. (2009) Driving fast-spiking cells induces gamma rhythm and controls sensory responses. Nature 459: 663–667.
13. Lewis DA, Curley AA, Glausier JR, Volk DW (2012) Cortical parvalbumin interneurons and cognitive dysfunction in schizophrenia. Trends Neurosci 35: 57–67.

14. Luby ED, Cohen BD, Rosenbaum G, Gottlieb JS, Kelley R (1959) Study of a new schizophrenomimetic drug; sernyl. AMA Arch Neurol Psychiatry 81: 363–369.

15. Krystal JH, Karper LP, Seibyl JP, Freeman GK, Delaney R, et al. (1994) Subanesthetic effects of the noncompetitive NMDA antagonist, ketamine, in humans. Psychotomimetic, perceptual, cognitive, and neuroendocrine responses. Arch Gen Psychiatry 51: 199–214.

16. Jentsch JD, Roth RH (1999) The neuropsychopharmacology of phencyclidine: from NMDA receptor hypofunction to the dopamine hypothesis of schizophrenia. Neuropsychopharmacology 20: 201–225.

17. Lim AL, Taylor DA, Malone DT (2012) Consequences of early life MK-801 administration: long-term behavioural effects and relevance to schizophrenia research. Behav Brain Res 227: 276–286.

18. Wang CZ, Yang SF, Xia Y, Johnson KM (2008) Postnatal phencyclidine administration selectively reduces adult cortical parvalbumin-containing interneurons. Neuropsychopharmacology 33: 2442–2455.

19. Rujescu D, Bender A, Keck M, Hartmann AM, Ohl F, et al. (2006) A pharmacological model for psychosis based on N-methyl-D-aspartate receptor hypofunction: molecular, cellular, functional and behavioral abnormalities. Biol Psychiatry 59: 721–729.

20. Abekawa T, Ito K, Nakagawa S, Koyama T (2007) Prenatal exposure to an NMDA receptor antagonist, MK-801 reduces density of parvalbumin-immunoreactive GABAergic neurons in the medial prefrontal cortex and enhances phencyclidine-induced hyperlocomotion but not behavioral sensitization to methamphetamine in postpubertal rats. Psychopharmacology (Berl) 192: 303–316.

21. Coleman LG Jr, Jarskog LF, Moy SS, Crews FT (2009) Deficits in adult prefrontal cortex neurons and behavior following early post-natal NMDA antagonist treatment. Pharmacol Biochem Behav 93: 322–330.

22. Rotaru DC, Lewis DA, Gonzalez-Burgos G (2012) The role of glutamatergic inputs onto parvalbumin-positive interneurons: relevance for schizophrenia. Rev Neurosci 23: 97–109.

23. Powell SB, Sejnowski TJ, Behrens MM (2012) Behavioral and neurochemical consequences of cortical oxidative stress on parvalbumin-interneuron maturation in rodent models of schizophrenia. Neuropharmacology 62: 1322–1331.

24. Zhang Z, Sun QQ (2011) Development of NMDA NR2 subunits and their roles in critical period maturation of neocortical GABAergic interneurons. Dev Neurobiol 71: 221–245.

25. Wintrip N, Nance DM, Wilkinson M (1998) Sexually dimorphic MK801-induced c-fos in the rat hypothalamic paraventricular nucleus. Neurosci Lett 242: 151–154.

26. Wessinger WD (1995) Sexual dimorphic effects of chronic phencyclidine in rats. Eur J Pharmacol 277: 107–112.

27. Fleidervish IA, Binshtok AM, Gutnick MJ (1998) Functionally distinct NMDA receptors mediate horizontal connectivity within layer 4 of mouse barrel cortex. Neuron 21: 1055–1065.

28. Agmon A, Connors BW (1991) Thalamocortical responses of mouse somatosensory (barrel) cortex in vitro. Neuroscience 41: 365–379.

29. Turrigiano GG, Nelson SB (2004) Homeostatic plasticity in the developing nervous system. Nat Rev Neurosci 5: 97–107.

30. Goldberg EM, Jeong HY, Kruglikov I, Tremblay R, Lazarenko RM, et al. (2011) Rapid developmental maturation of neocortical FS cell intrinsic excitability. Cereb Cortex 21: 666–682.

31. Goldberg EM, Clark BD, Zagha E, Nahmani M, Erisir A, et al. (2008) K+ channels at the axon initial segment dampen near-threshold excitability of neocortical fast-spiking GABAergic interneurons. Neuron 58: 387–400.

32. Gupta A, Wang Y, Markram H (2000) Organizing principles for a diversity of GABAergic interneurons and synapses in the neocortex. Science 287: 273–278.

33. Okaty BW, Miller MN, Sugino K, Hempel CM, Nelson SB (2009) Transcriptional and electrophysiological maturation of neocortical fast-spiking GABAergic interneurons. J Neurosci 29: 7040–7052.

34. Higgs MH, Spain WJ (2011) Kv1 channels control spike threshold dynamics and spike timing in cortical pyramidal neurones. J Physiol 589: 5125–5142.

35. Monyer H, Burnashev N, Laurie DJ, Sakmann B, Seeburg PH (1994) Developmental and regional expression in the rat brain and functional properties of four NMDA receptors. Neuron 12: 529–540.

36. Qian A, Buller AL, Johnson JW (2005) NR2 subunit-dependence of NMDA receptor channel block by external Mg2+. J Physiol 562: 319–331.

37. Swadlow HA (1995) Influence of VPM afferents on putative inhibitory interneurons in S1 of the awake rabbit: evidence from cross-correlation, microstimulation, and latencies to peripheral sensory stimulation. J Neurophysiol 73: 1584–1599.

38. Sun QQ, Huguenard JR, Prince DA (2006) Barrel cortex microcircuits: thalamocortical feedforward inhibition in spiny stellate cells is mediated by a small number of fast-spiking interneurons. J Neurosci 26: 1219–1230.

39. Laube B, Kuhse J, Betz H (1998) Evidence for a tetrameric structure of recombinant NMDA receptors. J Neurosci 18: 2954–2961.

40. Paoletti P (2011) Molecular basis of NMDA receptor functional diversity. Eur J Neurosci 33: 1351–1365.

41. Cull-Candy S, Brickley S, Farrant M (2001) NMDA receptor subunits: diversity, development and disease. Curr Opin Neurobiol 11: 327–335.

42. Erreger K, Dravid SM, Banke TG, Wyllie DJ, Traynelis SF (2005) Subunit-specific gating controls rat NR1/NR2A and NR1/NR2B NMDA channel kinetics and synaptic signalling profiles. J Physiol 563: 345–358.

43. Williams K (1993) Ifenprodil discriminates subtypes of the N-methyl-D-aspartate receptor: selectivity and mechanisms at recombinant heteromeric receptors. Mol Pharmacol 44: 851–859.

44. Gittelman JX, Tempel BL (2006) Kv1.1-containing channels are critical for temporal precision during spike initiation. J Neurophysiol 96: 1203–1214.

45. Dodson PD, Barker MC, Forsythe ID (2002) Two heteromeric Kv1 potassium channels differentially regulate action potential firing. J Neurosci 22: 6953–6961.

46. Fontaine B, Pena JL, Brette R (2014) Spike-threshold adaptation predicted by membrane potential dynamics in vivo. PLoS Comput Biol 10: e1003560.

47. Doischer D, Hosp JA, Yanagawa Y, Obata K, Jonas P, et al. (2008) Postnatal differentiation of basket cells from slow to fast signaling devices. J Neurosci 28: 12956–12968.

48. Li P, Huntsman MM (2014) Two functional inhibitory circuits are comprised of a heterogeneous population of fast-spiking cortical interneurons. Neuroscience 265: 60–71.

49. Barth AL, Malenka RC (2001) NMDAR EPSC kinetics do not regulate the critical period for LTP at thalamocortical synapses. Nat Neurosci 4: 235–236.

50. Liu XB, Murray KD, Jones EG (2004) Switching of NMDA receptor 2A and 2B subunits at thalamic and cortical synapses during early postnatal development. J Neurosci 24: 8885–8895.

51. Owczarek S, Hou J, Secher T, Kristiansen LV (2011) Phencyclidine treatment increases NR2A and NR2B N-methyl-D-aspartate receptor subunit expression in rats. Neuroreport 22: 935–938.

6

21-Benzylidene Digoxin: A Proapoptotic Cardenolide of Cancer Cells That Up-Regulates Na,K-ATPase and Epithelial Tight Junctions

Sayonarah C. Rocha[1], Marco T. C. Pessoa[1], Luiza D. R. Neves[1], Silmara L. G. Alves[2], Luciana M. Silva[6], Herica L. Santos[1], Soraya M. F. Oliveira[3], Alex G. Taranto[3], Moacyr Comar[3], Isabella V. Gomes[4], Fabio V. Santos[4], Natasha Paixão[7], Luis E. M. Quintas[7], François Noël[7], Antonio F. Pereira[8], Ana C. S. C. Tessis[8,9], Natalia L. S. Gomes[10], Otacilio C. Moreira[10], Ruth Rincon-Heredia[11], Fernando P. Varotti[5], Gustavo Blanco[12], Jose A. F. P. Villar[2], Rubén G. Contreras[11]*, Leandro A. Barbosa[1]*

1 Laboratório de Bioquímica Celular, Universidade Federal de São João del Rei, Campus Centro-Oeste Dona Lindú, Divinópolis, MG, Brazil, 2 Laboratório de Síntese Orgânica, Universidade Federal de São João del Rei, Campus Centro-Oeste Dona Lindú, Divinópolis, MG, Brazil, 3 Laboratório de Bioinformática, Universidade Federal de São João del Rei, Campus Centro-Oeste Dona Lindú, Divinópolis, MG, Brazil, 4 Laboratório de Biologia Celular e Mutagenicidade, Universidade Federal de São João del Rei, Campus Centro-Oeste Dona Lindú, Divinópolis, MG, Brazil, 5 Laboratório de Bioquímica de Parasitos, Universidade Federal de São João del Rei, Campus Centro-Oeste Dona Lindú, Divinópolis, MG, Brazil, 6 Laboratório de Biologia Celular e Inovação Biotecnológica, Fundação Ezequiel Dias, Belo Horizonte, MG, Brazil, 7 Laboratório de Farmacologia Bioquímica e Molecular, Instituto de Ciências Biomédicas, Universidade Federal do Rio de Janeiro, Rio de Janeiro, RJ, Brazil, 8 Laboratório de Bioquímica Microbiana, Instituto de Microbiologia Paulo Góes, Universidade Federal do Rio de Janeiro, Rio de Janeiro, RJ, Brazil, 9 Instituto Federal de Educação, Ciência e Tecnologia do Rio de Janeiro (IFRJ), Rio de Janeiro, RJ, Brazil, 10 Laboratório de Biologia Molecular e Doenças Endêmicas, Instituto Oswaldo Cruz/Fiocruz, Rio de Janeiro, RJ, Brazil, 11 Department of Physiology, Biophysics and Neurosciences, Center for Research and Advanced Studies (Cinvestav), Mexico City, Mexico, 12 Department of Molecular and Integrative Physiology, University of Kansas Medical Center, Kansas City, Kansas, United States of America

Abstract

Cardiotonic steroids are used to treat heart failure and arrhythmia and have promising anticancer effects. The prototypic cardiotonic steroid ouabain may also be a hormone that modulates epithelial cell adhesion. Cardiotonic steroids consist of a steroid nucleus and a lactone ring, and their biological effects depend on the binding to their receptor, Na,K-ATPase, through which, they inhibit Na^+ and K^+ ion transport and activate of several intracellular signaling pathways. In this study, we added a styrene group to the lactone ring of the cardiotonic steroid digoxin, to obtain 21-benzylidene digoxin (21-BD), and investigated the effects of this synthetic cardiotonic steroid in different cell models. Molecular modeling indicates that 21-BD binds to its target Na,K-ATPase with low affinity, adopting a different pharmacophoric conformation when bound to its receptor than digoxin. Accordingly, 21-DB, at relatively high μM amounts inhibits the activity of Na,K-ATPase α_1, but not α_2 and α_3 isoforms. In addition, 21-BD targets other proteins outside the Na,K-ATPase, inhibiting the multidrug exporter Pdr5p. When used on whole cells at low μM concentrations, 21-BD produces several effects, including: 1) up-regulation of Na,K-ATPase expression and activity in HeLa and RKO cancer cells, which is not found for digoxin, 2) cell specific changes in cell viability, reducing it in HeLa and RKO cancer cells, but increasing it in normal epithelial MDCK cells, which is different from the response to digoxin, and 3) changes in cell-cell interaction, altering the molecular composition of tight junctions and elevating transepithelial electrical resistance of MDCK monolayers, an effect previously found for ouabain. These results indicate that modification of the lactone ring of digoxin provides new properties to the compound, and shows that the structural change introduced could be used for the design of cardiotonic steroid with novel functions.

Editor: Shree Ram Singh, National Cancer Institute, United States of America

Funding: This work was supported by FAPEMIG (Fundação de Amparo a Pesquisa do Estado de Minas Gerais) APQ-01114-12, APQ-02596-12, CNPq (Conselho Nacional de Desenvolvimento Científico e Tecnológico) 472394/2012-6 and NIH DK081431. The authors are grateful for the financial and structural support offered by the University of São Paulo through the NAP-CatSinQ (Research Core in Catalysis and Chemical Synthesis). The funders had no role in study design, data collection and analysis, decision to publish, or preparation of the manuscript.

Competing Interests: The authors have declared that no competing interests exist.

* Email: leaugust@yahoo.com.br (LAB); rcontrer@fisio.cinvestav.mx (RGC)

Introduction

Cardiotonic steroids are common substances in the plant kingdom that confer a competitive advantage against depredation, either to the plant that synthesizes them, or to the animals that accumulates them from the diet [1]. Some experimental evidence shows that animals can endogenously synthesize cardiac steroids and that these substances play an important role in the regulation of blood pressure, cell proliferation and cell death [2,3]. The basic structure of cardiotonic steroids consists of a steroid backbone with a cis/trans/cis configuration and a lactone moiety at position 17β, also known as the aglycone. In addition, these compounds frequently have a sugar attached at position 3β of the steroidal nucleus, for which they are commonly known as cardiac glycosides. The nature of the lactone ring distinguishes the cardenolides, that have an unsaturated butyrolactone ring, from the bufadienolides, which present an α-pyrone ring. The only structural difference between the two main cardenolides that are used therapeutically, digoxin and digitoxin, is a single extra hydroxyl group (–OH) on digoxin, which considerably alters its pharmacokinetics. Digitoxin, is more lipophilic, shows strong binding to plasma proteins, is almost entirely metabolized in the liver and exhibits a very long half-life. In contrast, digoxin displays weak binding to plasma proteins, is not extensively metabolized and is excreted in a primarily unaltered form by the kidneys [2,3].

The main pharmacological effect of therapeutic doses of digoxin and digitoxin is their positive inotropic effect. This is mediated through the inhibition of the myocardial cell Na,K-ATPase, which results in an increase in the cytosolic levels of Na^+ and secondarily Ca^{2+}, due to reduction in transport of the Na^+ dependent Na^+/Ca^{2+} exchanger. The increased Ca^{2+} in the myocardial cell cytosol leads to further filling of the sarcoplasmic reticulum with Ca^{2+}, which will be readily available for its release upon stimulation to produce enhanced muscle contraction [3–5].

In the context of cell biology, there are two areas in which cardiac steroids are of potential interest: cancer and cell adhesion. Both these phenomena are closely related, as shown by the loss of cell adhesion that takes place during cancer proliferation and metastasis. The effect of cardiac steroids on cell adhesion is poorly understood and is the focus of intense research [6–8]. Since the 1960s, several interesting antitumor effects have been observed for digitalis [9,10], as well as for other cardenolides [11–15] and the related cardiac glycosides, the bufadienolides [16–19].

Human cancer predominantly affects epithelia [20], which cells are highly polarized and bound to each other through junctional complexes, including tight junctions (TJs). This last structure confers epithelia their capacity to function as effective barriers in the separation of biological compartments [21]. Loss of TJs is involved in cancer progression and metastasis [6,22], for example, a decrease in the expression of the tight junction protein claudin (Cldn)-7, is involved in the dissemination of breast cancer cells [23]. Furthermore, the importance of Na,K-ATPase for the formation of junctional complexes has been well established [24,25], as is the requirement of cell-cell contacts for the polarized expression of the Na,K-ATPase in lateral plasma membrane of epithelial cells [26]. Interestingly, the β1 subunit of Na,K-ATPase itself functions as a cell adhesion molecule in astrocytes [27] and epithelia [28–31].

A decrease in cell surface expression of the β1 subunit has been associated with epithelial-to-mesenchymal transition, a process involved in tumor invasiveness and metastasis [25,32]. Treatment with several types of cardiac steroids causes various effects on cell junctions. High concentrations of ouabain (≥300 nM) and other cardiac steroids disrupt tight, adherens and communicating junctions as well as desmosomes of epithelial cells [24,33,34]. On the contrary, low concentrations of ouabain (10 nM) increase the sealing of TJs [35], accelerate cell polarity (as determined by the development of primary cilia [36]) and increase cell communication through communicating junctions [30].

Although studies on the biological effects of available cardiac steroids on tumor cells are rapidly developing, this is not the case for newly synthesized cardiotonic steroids. Some synthetic cardiotonic steroids have been tested for their anticancer activity [17,37–44]. Oleandrin and 9-hydroxy-2″oxovoruscharin, a hemi-

synthetic cardenolide, is under clinical trials and has demonstrated anticancer activity both *in vitro* and *in vivo*, in multiresistance cancer cells cultures [15,43,44]. Other studies demonstrated that several modifications of the sugar moiety of digoxin and digitoxin, can increase the anticancer effect of these compounds [45–48].

Several authors have described the synthesis of digoxin derivatives in which the addition of styrene groups to the lactone ring moiety produced some interesting biological activity [49,50]. However, there are yet no reports concerning the anticancer activity or the effects of these digoxin derivatives on Na,K-ATPase activity.

The aim of this work was to study the biological effects of 21-benzylidene digoxin (21-BD), a digoxin derivative with an additional styrene group in the C21 carbon of the lactone ring, in cancer and normal cells.

Materials and Methods

General procedure for the synthesis of 21-benzylidene digoxin

Benzaldehyde (0.18 ml, 1.8 mmol), digoxin (0.469 g, 0.6 mmol), anhydrous K_2CO_3 (0.249 g, 1.8 mmol) and 60 ml of methanol were added into a round bottom flask. After stirring for 6 h at 70°C, the solvent was evaporated in a rotary evaporator. The crude product was diluted with 20 ml of water and extracted with hot ethyl acetate (3×30 ml). The organic layer was washed with brine, dried over anhydrous Na_2SO_4 and concentrated under vacuum. The crude product was then purified via silica column chromatography (CH_2Cl_2/MeOH 11:1). After purification, the pure product was diluted in tetrahidrofurano (THF), precipitated with hexane and concentrated under reduced pressure to give 21-BD (0.325 g, 0.37 mmol, 62%) as a white solid.

Cell culture

HeLa (human cervix carcinoma; ATCC CCL2) and RKO-AS45-1 (colon carcinoma; ATCC CRL-2579) cells were cultured in RPMI or DMEM/HAM F-10 (1:1) medium (Sigma, St. Louis, MO, USA). CHO-K1 (ATCC CCL61) and MDCK-II cells (canine renal; ATCC CCL-34) were grown in DMEM. All media were supplemented with 10% fetal bovine serum (FBS; Hyclone) (CDMEM), 60 mg/ml streptomycin and 100 mg/ml penicillin. Cells were seeded at a density of 5×10^5 cells/cm^2 and incubated in a humidified atmosphere with 5% CO_2. The culture medium was changed every 48 h to avoid nutrient depletion.

Insect cell culture and viral infections

Sf-9 insect cells were grown in Grace's medium with 3.3 g/l lactalbumin hydrolysate, 3.3 g/l yeastolate, and supplemented with 10% (v/v) fetal bovine serum, 100 units/ml penicillin, 100 μg/ml streptomycin and 0.25 μg/ml Fungizone. Cells were grown in suspension cultures and were transferred to 150 mm tissue culture plates before infection. Infections were performed as previously described [51]. After 72 h at 27°C, cells were scraped from the culture plates, centrifuged at 1,500×*g* for 10 min and suspended in 10 mM imidazole hydrochloride (pH 7.5) and 1 mM EGTA. Cells were homogenized on ice using a potter-Elvehjem homogenizer and the lysate was centrifuged for 10 min at 1,000×*g*. The supernatant was removed, centrifuged for an additional 10 min at 12,000×*g*, and the final pellet was suspended in 250 mM sucrose, 0.1 mM EGTA, and 25 mM imidazole HCl, pH 7.4.

Cytotoxicity assay

Compound cytotoxicity effect was assessed using the MTT (3-(4,5-dimethylthiazol-2-yl)-2,5-diphenyltetrazolium bromide) tetrazolium salt colorimetric method (Sigma, St. Louis, MO, USA). Briefly, the cells were plated in 96-well plates (1×10^5 cells/well) and incubated for 24 h at 37°C to confluence. Then, the wells were washed with culture medium and digoxin or 21-BD was added at different concentrations (0.05–500 μM). After incubation for 24 or 48 h, the plates were treated with MTT. Readings were performed in a Spectramax M5e microplate reader (Molecular Devices, Sunnyvale, CA, USA) at 550 nm. Cytotoxicity was scored as the percentage of reduction of absorbance, relative to untreated control cultures [52]. All experiments were performed in triplicate. The results were expressed as the mean of LC_{50} (the drug concentration that reduced cell viability to 50%).

Apoptosis Assay

Comet assay. The single-cell gel electrophoresis assay (comet assay) was performed according to previously published protocols [53,54]. For this, CHO-K1 cells were treated with 20, 35 or 50 μM digoxin or 21-DB, which are concentrations that do not affect cell viability according to the MTT assay. The cells were seeded in 24-well plates. The next day cells were washed twice with PBS, and incubated 3 h with the corresponding compounds in culture media without serum. The negative and positive control groups were treated with PBS, and methyl methanesulfonate (400 μM), respectively. After 24 h, the cells were washed twice with PBS and detached using a trypsin-EDTA solution. Trypsin was inactivated with 3.0 ml of complete medium, followed by centrifugation (5 min, 150×g). The pellet was then resuspended in 500 μl of PBS, and 30 μl aliquots of the cell suspensions were mixed with 70 μl of low melting point agarose (0.5%). These mixtures were placed on slides pre-coated with normal melting point agarose (1.5%) and covered with coverslips. The coverslips were removed after five minutes, and the slides were immersed overnight in lysis solution (NaCl 2.5 M, EDTA 100 mM, Tris 10 mM, pH 10; Triton X-100 1% and DMSO 10%). Next, slides were washed with PBS and maintained for 40 min in a horizontal electrophoresis box filled with cold alkaline buffer (EDTA 1 mM, NaOH 300 mM, pH>13). Electrophoresis was conducted at 0.86 V/cm and 300 mA (20 minutes) and the slides were subsequently neutralized (0.4 M de Tris, pH 7.5), fixed with methanol and stained with ethidium bromide. Visual analyses were performed under fluorescence [55,56].

Phosphatidylserine translocation. HeLa cells were plated in 60 mm diameter Petri dishes at confluence and incubated overnight in CDMEM. They were then treated with 2 μM digoxin or 50 μM 21-BD in CDMEM for 6, 12 or 24 h. After incubation, suspended cells were recovered from the media by centrifugation at 150×g, 20°C for 5 minutes. Attached cells were obtained by a mild protease treatment 1 ml Accutase plus 3 ml PBS), centrifuged as indicated above and added to the cells obtained from the media. Annexin-V bound to the outer leaflet of the plasma membrane was measured with a comercial kit (Annexin-V-Fluos Stainig kit, Roche, Germany) following manufacturer instructions. Briefly, cell suspension was incubated with a mixture of Annexin-F-FITC and propidium iodide in saline buffer for 15 min. Cell suspension was analyzed by flow cytometry (FACSCalibur flow cytometer, FowJo software).

Transepithelial electrical resistance (TER)

MDCK cells were cultured on Transwell permeable supports (3415, Corning Inc., NY, USA) the transepithelial electrical resistance was measured with an EVOM and the EndOhm-6

system (World Precision Instruments, Sarasota, FL, USA). Final values were obtained by subtracting the resistance of the bathing solution and the empty insert. The results are expressed in ohms·cm² ($\Omega \cdot cm^2$) as a percentage of the control.

Immunofluorescence

After TER measurements, MDCK monolayers were washed three times with ice-cold PBS/Ca^{2+}, fixed with 4% paraformaldehyde for 30 min at 4°C, permeabilized with 0.1% Triton X-100 for 5 min, blocked for 30 min with 3% BSA and treated for 1 h at 37°C with a specific primary antibody. The monolayers were then rinsed 3 times with PBS/Ca^{2+}, incubated with an appropriate FITC or TRICT-labeled antibodies for 30 min at room temperature and rinsed as indicated above. Filters with the cells were excised with a scalpel and mounted in Vectashield (Vector Labs, Burlingame, CA, USA). The preparations were examined with a Leica confocal SP5 microscope (Leica Microsystems, Wetzlar, Germany). The captured images were imported into FIJI, version 2.8 (National Institutes of Health, Baltimore, USA), to obtain maximum projections and into the GNU Image Manipulation Program (GIMP) to normalize brightness and contrast in all images and construct figures.

Immunoblotting

Western blot analysis of whole-cell protein extracts was performed as described previously [57]. Briefly, MDCK and HeLa cells were washed with PBS and solubilized with radioimmunoprecipitation assay (RIPA) buffer [10 mM piperazine-N,N'-bis(2-ethanesulfonic acid), pH 7.4, 150 mM NaCl, 2 mM ethylenediamine-tetraacetic acid (EDTA), 1% Triton X-100, 0.5% sodium deoxicholate, and 10% glycerol] containing protease inhibitors (Complete Mini; Roche Diagnostics, Indianapolis, IN). The protein content of the cell lysate was measured (BCA protein assay reagent; Pierce Chemical, Rockford, IL) and prepared for SDS-polyacrylamide gel electrophoresis (PAGE) by boiling in sample buffer. The resolved proteins were electrotransfered to a polyvinylidene difluoride membrane (Hybond-P; GE Healthcare, Little Chalfont, Buckinghamshire, United Kingdom). The proteins of interest were then detected with the specific polyclonal or monoclonal antibodies indicated in each case, followed by species-appropriate peroxidase-conjugated antibodies (Zymed Laboratories, South San Francisco, CA) and chemiluminescent detection (ECL PLUS; GE Healthcare).

Detection of Na,K-ATPase and claudin expression via real-time quantitative RT-PCR

For real-time quantitative RT-PCR (RT-qPCR), total RNA was extracted from the cell samples using TRIzol (Life technologies, USA). The concentration was estimated via spectrophotometry with a Nanodrop ND2000 (Thermo Scientific, USA). All reverse transcriptase reactions were performed using 5 μg of RNA with the Superscript III kit (Invitrogen, USA) according to the manufacturer's instructions. All RNA samples were reverse transcribed simultaneously to minimize the inter assay variation associated with the reverse transcription reaction. Real-time quantitative PCR was performed on an ABI Prism 7500 fast sequence detection system (Applied Biosystems) using Go Taq qPCR master mix (Promega, USA). The following primers and concentrations were employed (Murphy et Al, 2004): Na,K-ATPase α_1 subunit (GenBank accession number NM_000701): Fw (300 nM): 5′-TGTCCAGAATTGCAGGTCTTTG-3, Rv (300 nM): 5′-TGCCCGCTTAAGAATAGGTAGGT-3′; Na,K-ATPase β_1 subunit (GenBank accession number NM_001677): Fw

(300 nM): 5'-ACC AAT CTT ACC ATG GAC ACT GAA-3', Rv (300 nM): 5'-CGG TCT TTC TCA CTG TAC CCA AT-3'; Cldn-2 Fw GGTGGGCATGAGATGCACT, Rv CAC-CACCGCCAGTCTGTCTT; Cldn-4 Fw TGCAC-CAACTGCGTGGAGGATGAG, Rv ACCAC-CAGCGGGTTGTAGAAGTCC. The applied PCR conditions were as follows: 50°C for 2 minutes, followed by 40 cycles at 95°C for 15 sec and 60°C for 1 min. Each of these primer sets generated a unique PCR product, as confirmed by the obtained melting curves. The PCR assays were performed in triplicate, and the data were pooled. Relative quantitative measurement of target gene levels was performed using the $\Delta\Delta Ct$ method of Livak et al. [58]. As endogenous housekeeping control genes, we employed the glyceraldehyde 3-phosphate dehydrogenase (GAPDH, GenBank accession number NM_002046) and ribosomal 18S subunit genes (GenBank accession number NR_003286) [59]. The following primers and concentrations were used: GAPDH Fw (300 nM): 5'-ATGTTCGTCATGGGTGTGAA-3', GAPDH Rv (300 nM): 5'-GGTGCTAAGCAGTTGGTGGT-3', 18S Fw (300 nM): 5'-CAGCCACCCGAGATTGAGCA-3', 18S Rv (300 nM): 5'-TAGTAGCGACGGGCGGTGTG-3'.

Molecular modeling analyses

Initially, digoxin, ouabain, and 21-BD (Figure 1A) were generated and refined via the semi-empirical PM6 method [60] implemented in Gaussian 09 W software [61]. The crystallographic structure of Na,K-ATPase (target protein) complexed with ouabain was obtained from the Protein Data Bank (PDB ID: 4HYT [62] with a 3.40 Å resolution, using the α_1 subunit of Na,K-ATPase). The magnesium ion and three interstitial water molecules were kept in the binding site. Next, a rigid re-dock of ouabain was carried out to validate our system. Subsequently, digoxin and 21-BD were docked against the ATPase. The docking analyses were conducted using AutoDock Vina 1.0.2 [63]. The applied search algorithm was Iterated Local Search Global Optimizer for global optimization. In this process, a succession of steps with mutation and local optimization (the Broyden-Fletcher-Goldfarb-Shanno [BFGS] method) were conducted, and each step followed the Metropolis criterion [64]. In this study, a grid box was constructed exploring all active sites, which was defined as a cube with the geometric center in ouabain, with dimensions of 20×20×24 Å, spaced points of 1 Å and X, Y and Z coordinates of −27.065, 20.469 and −69.469, respectively. All molecular modeling figures were constructed using DS Visualizer 3.1 [65].

Preparation of Pdr5p plasma membranes

Plasma membranes containing the overexpressed Pdr5p protein were prepared from the *Saccharomyces cerevisiae* mutant strain AD124567 as previously reported [66].

Fractionation of cell lysates and preparation of membrane fractions

A total of 2.5×10^5 cells were grown in 75 cm^2 culture bottles and treated with digoxin and 21-BD. After 48 hours of treatment, cells were washed three times with cold PBS and scrapped from the culture bottle with a rubber policeman in a membrane preparation buffer (6 mM Tris [pH 6.8], 20 mM imidazole, 0.25 M Sucrose, 0.01% SDS, 3 mM EDTA and 2 mM PMSF). The cells were homogenized in a potter-Elvehjem homogenizer, using ten stokes in ice. Then, they were sonicated in an ultrasonic cell disruptor on ice for 10 s at 45% power. The sample was subjected to centrifugation at 20,000×g for 90 min at 4°C. The

supernatant was discarded and the pellet resuspended in 250 µl of membrane preparation buffer. Finally, this sample was sonicated for 10 s at 25% power until complete homogenization.

Na,K-ATPase preparation from rat brain hemispheres

Brain hemispheres from adult male Wistar rats were rapidly collected after diethylether anesthesia and decapitation. The protocols used for the use of rats were approved by the Institutional Commission for Ethics in the Use of Animals, process code DFBCICB011 and conformed to the Guide for the Care and Use of Laboratory Animals, published by US National Institute of Health (NIH publication No. 85-23, revised in 1996). The brain tissues, used as a source of ouabain-sensitive (α_2/α_3) Na,K-ATPase isoforms, were homogenized in 250 mM buffered sucrose, 2 mM dithiothreitol, 0.1 mM PMSF and 5 mM Tris/HCl (pH 7.4) with a motor-driven Teflon Potter-Elvehjem homogenizer. Subsequently, chaotropic treatment with 2 M KI was performed for 1 h under constant stirring, followed by centrifugation three times at 100,000×g for 1 h. The final pellet was resuspended in 250 mM sucrose, 0.1% sodium deoxycholate and 20 mM maleate/Tris (pH 7.4), then stored overnight at −20°C and subjected to differential centrifugation after thawing [67]. The pellets were resuspended in the same buffer without PMSF and stored in liquid N$_2$.

Na,K-ATPase preparation from mouse kidney

Kidney tissue was isolated from adult mice and was homogenized in 250 mM sucrose, 0.1 mM EGTA, and 25 mM imidazole HCl, pH 7.4, using a motor-driven Teflon Potter-Elvehjem homogenizer. The sample was then subjected to centrifugation at 4,500×g for 10 min. The resulting supernatant was centrifuged at 70,000×g for 1 h. The final pellet was resuspended in the homogenization solution and used for Na,K-ATPase activity assays. All experimental protocols involving mice were approved by the University of Kansas Medical Center Institutional Animal Care and Use Committee.

NTPase Assays

Enzymatic activities were assayed using ATP as a substrate in standard medium (50 µl final volume) containing 100 mM Tris-HCl pH 7.5, 4 mM MgCl$_2$, 75 mM KNO$_3$, 7.5 mM NaN$_3$, and 0.3 mM ammonium molybdate in the presence of 3 mM ATP. The reaction was initiated by the addition of 13 µg/ml of the plasma membrane preparations, then maintained at 37°C for 60 min and stopped by the addition of 1% SDS, as previously described [68]. The released inorganic phosphate (Pi) was measured as described elsewhere [69]. Using stock solutions in DMSO, 21-BD and digoxin were added up to a 5% v/v final concentration. The difference in ATPase activity in the presence or absence of 3 µM oligomycin corresponded to a Pdr5p-mediated ATPase activity.

Measurement of the effect of 21-BD on Na,K-ATPase activity

The brain hemisphere preparations were incubated at 37°C for 2 h in medium containing 87.6 mM NaCl, 3 mM KCl, 3 mM MgCl$_2$, 3 mM ATPNa$_2$, 1 mM EGTA, 10 mM sodium azide and 20 mM maleic acid/Tris (pH 7.4), and the cell membrane preparations were incubated at 37°C for 1 h in 120 mM NaCl, 20 mM KCl, 2 mM MgCl$_2$, 3 mM ATPNa$_2$ and 50 mM HEPES, (pH 7.5), both in the absence and presence of 1 mM ouabain or increasing concentrations of 21-BD and digoxin. Na,K-ATPase activity was determined by measuring the Pi released according to a colorimetric method described previously [69], and specific

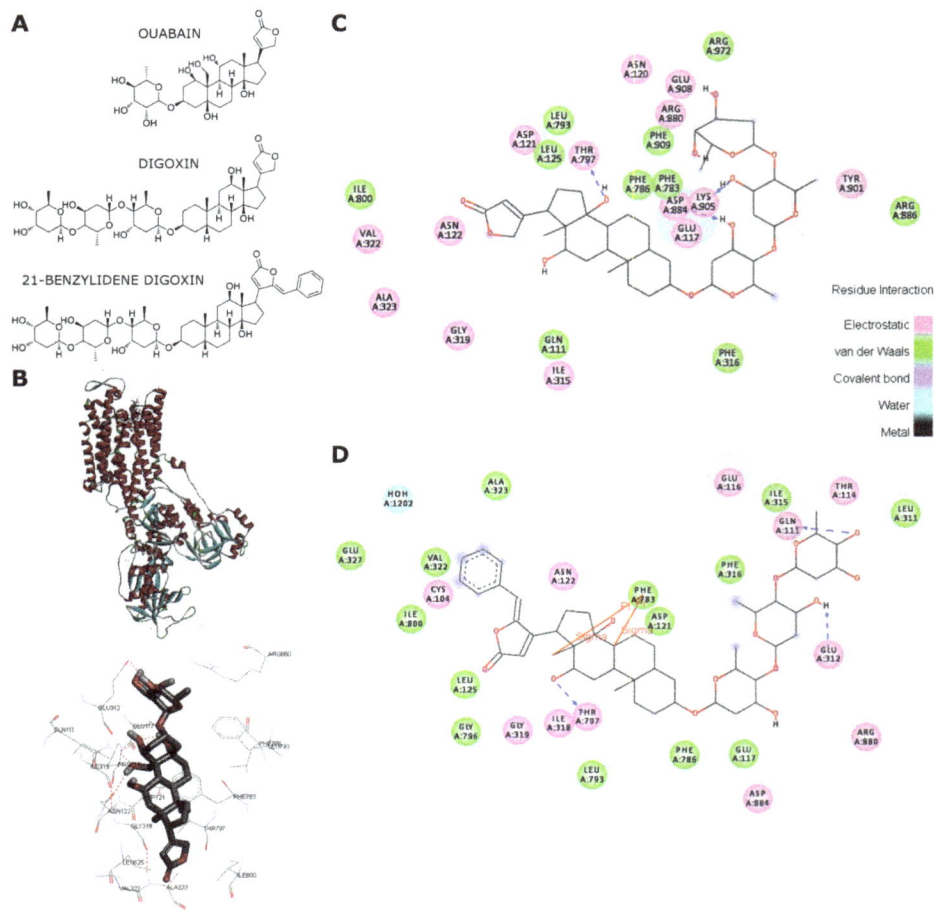

Figure 1. Structure and pharmacophoric conformation of 21-BD. (A) Chemical structure of ouabain, digoxin, and 21-BD. (B) Left: whole structure of Na,K-ATPase (PDB: 4HYT) showed in solid ribbon representation, where the alpha-helix, beta-sheets and turns are in red, blue and gray, respectively; right: highlight of the binding site with ouabain shown in tube representation. Dashed red lines indicate hydrogen bonds. Only polar hydrogen atoms were showed for a better visualization. Pharmacophoric conformation of, (C) digoxin and (D) 21-BD; polar and nonpolar interactions are depicted by magenta and green colors, respectively. Dashed lines indicate hydrogen bonds. Residue interactions are color coded as indicated in the inserted scale.

activity was considered as the difference between the total and ouabain-resistant ATPase activities [70].

Ouabain binding

HeLa cells were plated on 24 multi well plates at confluence and cultured overnight. Monolayers were then rapidly washed three times with potassium free saline buffer (140 mM NaCl, 1.8 mM $CaCl_2$, 5 mM sucrose, 10 mM Tris-HCl pH 7.4 at room temperature) and incubated 30 min with total binding solution (0.1×10^{-6} ^{3}H-ouabain plus 0.9×10^{-6} cold ouabain in potassium free saline buffer) under gentle agitation and at room temperature. Then monolayers were washed four times, 1 min each, with ice cold 0.1 M $MgCl_2$ and dissolved with 400 µl of SDS 1%. ^{3}H activity was measured in samples of 350 µl by scintillation counting. Total ^{3}H-ouabain binding was competed by adding to the total binding solution the necessary amount of 21-BD to reach the concentrations indicated in results. A subset of monolayers was exposed to competed binding solution (total binding solution plus 0.5×10^{-4} M cold ouabain) to measure the unspecific binding and subtract it to the total binding values.

Statistical analyses

Statistical analyses were performed with GraphPad Prism software, version 5. The results were expressed as mean ± error standard of the mean. Statistical significance in a one-way analysis of variance (ANOVA), followed by a Bonferroni's selected pairs comparison test, was set at $P < 0.05$ (*), $P < 0.01$ (**), or $P < 0.001$ (***) vs. the control condition, and "n" represents the number of independent experiments.

Results

We synthesized 21-BD via a simple stereo selective vinylogous aldol reaction, according to Xu et al. [50], and characterized the final product through conventional NMR, HRMS and IR analysis (Figure 1A; Figure S1 and Data S1).

To perform the molecular modeling analyses we started with the geometric optimization of ligands, through a semi-empirical approach, to correct geometric parameters such as bond lengths and refine the structure. The re-dock structure obtained with the software Autodock Vina, shows that the refined and the crystallographic ligand share the same conformation at the active site, with a root mean square deviation value of 2.24 Å for the best solution, which indicated the accuracy of the methodology used (see Figure S2A). Figure 1B shows a simulation of the docking of ouabain to its binding site in the Na,K-ATPase α_1 isoform. As shown, the enzyme preserves the secondary structural elements of the ATPase family (Figure 1B, left) and that the active site is composed of Gln111, Glu117, Pro118, Asn122, Leu125, Glu312, Ile315, Gly319, Val322, Ala323, Phe783, Phe786, Leu793, Ile800, Arg880 (Figure 1B, right), in agreement with known structures [71].

Following these molecular modeling analyses we docked ouabain, digoxin and 21-BD to Na,K-ATPase (Figure S2B and Figure 1C and D). Simulations resulted in binding energies for ouabain, digoxin, and 21-BD of -9.8, -1.9, and -10.0 kcal/mol, respectively. These data suggest that 21-BD may have a similar binding affinity to the Na,K-ATPase than ouabain and digoxin. However, both these last compounds exhibit different pharmacophoric conformations in the binding site, principally at the level of the lactone ring. Figures 1C and D highlight the most important intermolecular interactions between the ligands and the target protein. Digoxin binds to the enzyme via hydrogen bonding with Thr797, Asp884 and Lys905 aminoacids. The electrostatic and hydrophobic interaction occur with Glu117, Asn120, Asp121, Asn122, Ile315, Gly319, Val322, Ala323, Arg880, Asp884, Tyr901, Glu908 and Gln111, Leu125, Phe316, Phe783, Phe786, Leu793, Arg886, Phe909, Arg972, respectively (Figure 1C). In contrast to natural cardiac glycosides, 21-BD complexes to the Na,K-ATPase through Gln111, Glu312, and Thr797 hydrogen bonding. In addition, the aromatic ring reaches a hydrophobic pocket formed principally by Cys104, Val322, Ala323, Glu327, Ile800 (Figure 1D). Moreover, the pharmacophoric conformation is completely different from that of the natural glycosides, whose aromatic moiety binds to water molecules and magnesium ion.

To directly determine binding of 21-BD to the ouabain site of the Na,K-ATPase, we first tested if the synthetic steroid could compete with ^{3}H-ouabain binding in HeLa cells, which express the α_1 isoform of the Na,K-ATPase [72]. As shown in Figure 2A, 21-BD significantly competed with ouabain binding at concentrations in the μM range, indicating that the affinity of the Na,K-ATPase for this ligand is low. We next studied if 21-BD affects Na,K-ATPase activity of a membrane preparation from rat cerebral hemispheres. This tissue is mainly composed of the ouabain sensitive α_2 and α_3 isoforms of the enzyme, which comprise approximately 80% of the total Na,K-ATPase activity, with the remaining 20% corresponding to the α_1 isoform [73]. Digoxin inhibited most of the Na,K-ATPase activity, with an IC_{50} of 219 ± 40 nM, corresponding to the high-affinity sites of the α_2 and α_3 isoforms, and at higher concentrations it further inhibited the α_1 isoform. In contrast, 21-BD had no effect on rat brain Na,K-ATPase, even at the highest concentration tested (100 μM) (Figure 2B).

To further determine if 21-BD had any effect on the α_1 isoform of the Na,K-ATPase, we tested its activity of two different sources of Na,K-ATPase rich in α_1. We used membrane fractions from Sf9 insect cells exogenously expressing the rat Na,K-ATPase α_1 and β_1 subunits and membrane fractions from mouse kidney, which primarily contains the Na,K-ATPase α_1 isoform. The expression in Sf9 cells provides a useful system to specifically investigate effects on the exogenously expressed Na,K-ATPase, since these cells virtually lack expression of an endogenous Na,K-ATPase. As shown in Fig. 2C–D, 21-BD inhibited activity of $\alpha_1\beta_1$ from Sf9 cells, but had little effect on parallel preparations expressing only the β_1 subunit of the enzyme, which by lacking the catalytic α_1 subunit serve as a control. The effect of 21-BD only took place at high concentrations (100 μM) of the compound. Similarly, 21-BD

inhibited the mouse kidney Na,K-ATPase with similar kinetics. These results agree with the ^{3}H-ouabain binding experiments in HeLa cells, suggesting that 21-BD interacts with Na,K-ATPase with low affinity.

Although the results described above suggest that Na,K-ATPase α_1 is a receptor for 21-BD, the possibility exists that 21-BD could be also targeting another protein/s in the membrane of the cells. Steroids have been shown to inhibit the activity of yeast and mammalian ATP-dependent efflux pumps [74,75]. Therefore, we investigated the effect of 21-BD on Pdr5p, a member of the *Saccharomyces cerevisiae* ABC transporters family that shares many substrates and inhibitors with the mammalian P-glycoprotein [76]. Interestingly, after incubation with membrane preparations from *Saccharomyces cerevisiae*, 21-BD showed a concentration-dependent inhibitory effect on NTPase activity, with an IC_{50} of 1.25 ± 0.36 μM. In contrast, digoxin had no significant effect on Pdr5p (Figure 2E). This suggests that, while 21-BD has an effect on its Na,K-ATPase target, it can also affect other proteins, such as the ATP dependent transporter of the plasma membrane.

To further study the effects of 21-BD, its action was tested and compared to that of digoxin in whole cells. Thus, the compounds were directly applied to HeLa cervix cells, which express the Na,K-ATPase α_1 isoform, and RKO colorectal cancer cells that, in addition to α_1, express the α_3 isoform [77]. After incubation with the compounds for 48 h, cells were harvested and Na,K-ATPase activity determined. As expected, treatment with 150 nM digoxin for 48 h inhibited Na,K-ATPase activity (Figure 3A, green columns). Surprisingly, 10 μM 21-BD increased Na,K-ATPase activity in both cell lines (Figure 3A, red columns). To explore whether the increase in Na,K-ATPase activity was caused by an increase in expression of the Na,K-ATPase, the levels of Na,K-ATPase subunits in the cells was determined by RT-PCR. As depicted if Figure 3B, incubation with 10 μM 21-BD for 48 h increased mRNA of the Na,K-ATPase α_1 and β_1 subunits in HeLa cells.

To test the possibility that 21-BD could exhibit anticancer effects, such as those described for digoxin, we used HeLa, RKO cancer cells, and normal epithelial MDCK cells. Both digoxin and 21-BD showed a cytotoxic effect on HeLa cells after treatment for 24 and 48 h. Digoxin produced the classical time- and dose-dependent decrease in viability (LC_{50} of 2.2 ± 0.8 μM), being approximately 25 times more potent that 21-BD (LC_{50} of $56,16\pm8,12$ μM), (Figure 4A). In RKO cells, digoxin decreased viability with higher potency, with an LC_{50} of 0.42 ± 0.1 μM, compared to that of 21-BD, which had a LC_{50} of 55.81 ± 15.15 μM. Thus, while RKO cells showed a higher sensitivity to digoxin than HeLa cells (Figure 4B), they exhibited the same sensitivity to 21-BD than HeLa cells. Surprisingly, different from digoxin, 21-BD increased the viability of MDCK cells (Figure 4C). This unexpected result may depend on the fact that 21-BD is not toxic to MDCK cells and that it may induce a higher mitochondrial activity to metabolize the substrate in the MTT assay.

Digoxin is known to induce apoptosis in several cell types [78,79]. Therefore, the effect of 21-BD in reducing HeLa and RKO cell viability could be due to induction of cell apoptosis. To assess this possibility, we followed two crucial events of the apoptotic process after 21-BD treatment: DNA fragmentation and the translocation of phosphatidylserine from the inner towards the outer leaflet of the plasma membrane of CHO-K1 cells. As shown in Figures 5A and B, 21-BD causes primary DNA fragmentation as determined by comet assay. The amount of phosphatidylserine translocated was measured through the binding of fluoresceinated annexin-V and the amount of necrotic cells was estimated by the

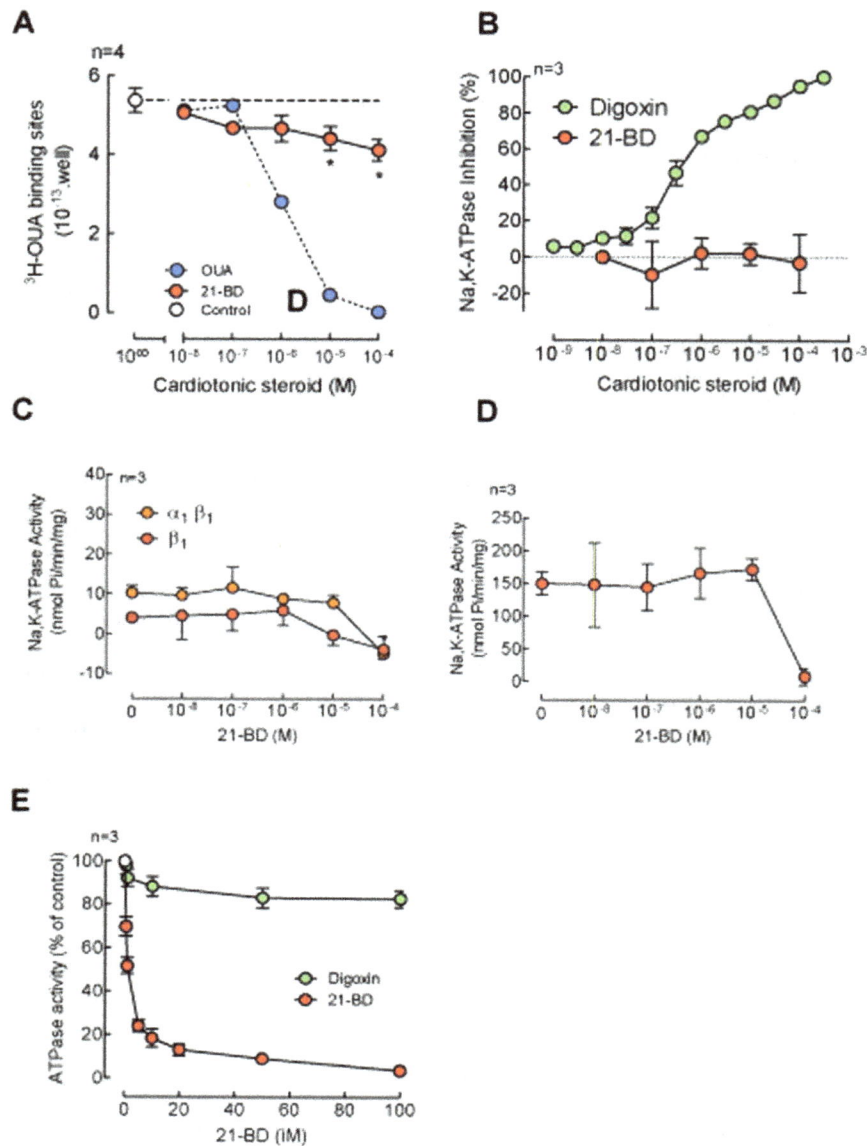

Figure 2. 21-BD effect on Na,K-ATPase and Pdr5p activity. (A) 21-BD competition of ^{3}H-ouabain binding on HeLa cells; the control for maximal binding is represented with a white circle and a long dashed line, competition of ouabain and 21-BD is shown with blue and red circles respectively. (B) Inhibition of rats brain hemisphere Na,K-ATPase after 2 h incubation with digoxin (green circles) or 21-BD (red circles). (C) Effect of 21-BD on the Na,K-ATPase activity on proteins expressed in Sf9 insect cells, Na,K-ATPase activity was measured on Sf9 cells expressing the rat $\alpha_1 \beta_1$ (orange circles) or β_1 (red circles) after 15 min treatment with the indicated concentrations of 21-BD. (D) Dose-response curve for the effects of 21-DB on Na,K-ATPase activity of mouse kidney membrane preparations. E) Effect of 21-BD (red circles) or digoxin (green circles) on the activity of the Pdr5p transporter.

incorporation of propidium iodide into the cell nucleus (Figure 5C). We found that 21-BD increases phosphatidylserine translocation (red bar, Apoptotic) but not necrosis (Necrotic, red bar). Altogether, these results show that 21-BD is able to induce cell apoptosis.

The integrity of cell junctions is a key factor in cancer biology, as several authors have demonstrated that alterations of tight junctions are associated with cell transformation and metastasis [6,23,41,80,81]. We tested whether 21-BD treatment modifies transepithelial electrical resistance (TER) or the distribution of the tight junction proteins Cldn-2 and -4 and ZO-1. For this purpose, we used MDCK cells, which are a well-known model for tight

junction studies [82]. We first examined whether 21-BD altered TER. As shown in Figure 6A, 50 μM 21-BD induces a sustained increase in TER for at least 87 h of treatment and in a dose-dependent manner. This change in TER is associated with an increase in the expression of tight junction proteins. Figure 6B shows that 21-BD increases the cellular content of claudin-4 mRNA at all concentrations tested (Figure 6B, red circles) and claudin-2 mRNA only at the lowest concentration (Figure 6B, red triangles). Conversely, digoxin increases claudin-4 (Figure 6B, green circles) but not claudin-2 mRNA (Figure 6B, green triangles). The increase in claudin-4 mRNA results in a corresponding increment of the protein at high 21-BD concen-

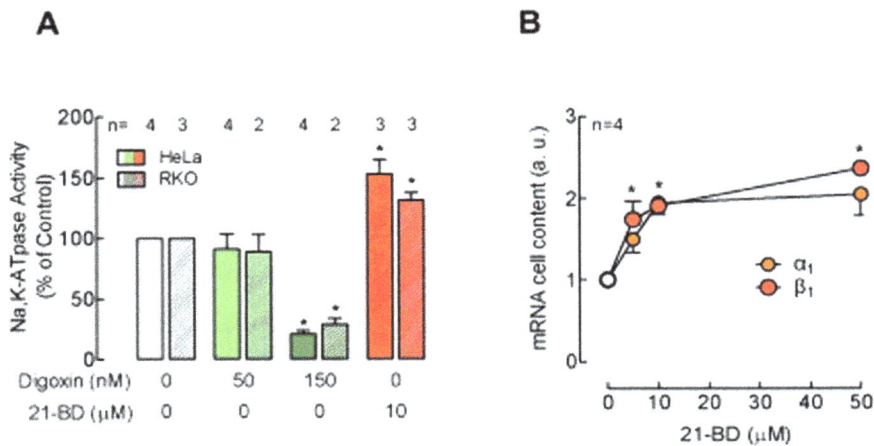

Figure 3. 21-BD increases Na,K-ATPase expression in cancer cells. (A) Na,K-ATPase activity after incubation of HeLa cells with 21-BD or digoxin for 48 h with different concentrations of 21-BD. B) mRNA content of the Na,K-ATPase α_1 and β_1 subunits of HeLa cells after 48 h incubation with various concentrations of 21-BD.

trations (Figure 6C) while claudin-2 protein strongly decreases (Figure 6C, Cldn-2, red bars). 21-BD also induces the increment of ZO-1, an important peripheral membrane protein of tight junctions (Figure 6C).

Figure 7 illustrates that 21-BD increases the expression of claudin-4 and ZO-1 at the tight junction (arrows) and the cytoplasm of MDCK cells (arrow heads D and F vs A and C), displaying their characteristic "chicken fence" like pattern of expression of these junctional proteins. Conversely, 21-BD reduces the expression of claudin-2 (Figure 7E vs B). The coordinated antagonistic variation of claudins -4 and -2 has also been shown in epithelial cells treated with other factors, e.g. EGF [83,84].

21-BD also increases the cellular content of the α_1 Na,K-ATPase at the highest concentration tested (Figure 8A) and its localization in the plasma membrane (Figure 8C, arrow) and in the cytoplasm (Figure 8C, arrow heads) of MDCK cells.

Discussion

We have synthetized a digoxin derivative, 21-BD, that reduces cell viability by inducing apoptosis, and increases the hermeticity of tight junctions through the up-regulation of claudin-4 and ZO-1 and the down-regulation of claudin-2 junctional proteins. These effects take place after 12 to 48 h incubation with 21-BD. These delayed responses most probably result from the activation of

signaling cascades that result in changes in expression of different genes, as has been reported for other cardenolides [85–87]. An important question is whether 21-BD triggers those effects as a result of its binding to the Na,K-ATPase or to another membrane protein. 21-BD is able to displaces ^{3}H-ouabain from the Na,K-ATPase (Figure 2A), demonstrating that 21-BD has the capacity to bind to the Na,K-ATPase with a low affinity and that 21-BD occupancy of the sites in the pump may trigger cellular responses via the classical Na,K-ATPase mediated cascade of intracellular mediators. Molecular modeling also indicates the binding of 21-BD to the α-Na,K-ATPase (Figure 1D), although with a pharmacophoric conformation very different from that of digoxin. Also, supporting the role of the Na,K-ATPase as a 21-BD receptor is the regulation that this cardiotonic steroid exerts on the tight junctions in MDCK cells, which is similar to that caused by ouabain [35]. In addition, 21-BD induces apoptosis like other cardiotonic steroids do. Nevertheless, it is known that adrenocortical bovine cells express high affinity binding sites for ouabain that are distinct from the Na,K-ATPase [88]. Here we show that 21-BD, but not digoxin, is able to inhibit the activity of the yeast ABC ATPase Pdr5p (Figure 2E), suggesting the possibility that 21-BD might, in addition to the Na,K-ATPase, act through a different receptor.

Our docking results show that the aromatic ring of 21-BD may reach a hydrophobic pocket in the binding site (Figure 1),

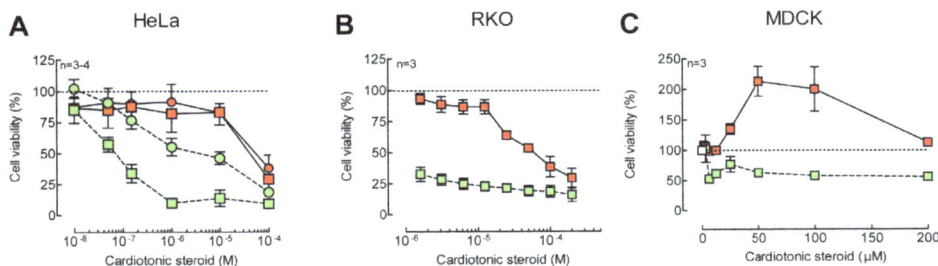

Figure 4. High concentrations of 21-BD reduce cell viability of HeLa and RKO. HeLa (A) or RKO (B) cells were treated with digoxin (green symbols) or 21-BD (red symbols) for 24 (circles) or 48 (squares) h. Viability was measured by MTT reduction assay. 100 and 25 µM 21-BD induced the statistically significant reduction of HeLa and RKO viability, respectively (p<0.0084). Digoxin reduces HeLa viability starting with 150 µM for 24 h and 50 µM for 48 h (p<0.001). RKO cells have a higher sensitivity to digoxin that induces statistically significant differences starting from 1.6 µM for 48 h (p<0.0001).

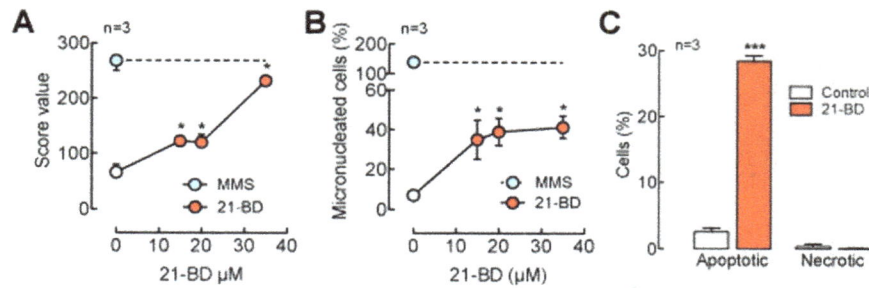

Figure 5. 21-BD induces apoptosis in HeLa and CHO-K1 cells. (A) Score value obtained from the comet assay of CHO-K1 cells incubated 24 h with 21-BD at different concentrations (red circles). (B) Micronucleated cells percentage of CHO-K1 cultures incubated with 21-BD at different concentrations for 24 h. A 24 h incubation with 0.4 mM Methyl methanesulfonate (MMS) was used as a control (A, B, blue circles). (C) Apoptotic and necrotic HeLa cells after 24 h of incubation in control media (white bars), media with 50 μM 21-BD (red bars) or 2 μM digoxin (green bars) for 24 h. Apoptosis and necrosis were detected by flow cytometry ussing an annexin-V translocation assay and the incorporation of propidium iodide in to the nucleus, respectively. $P<0.01$.

suggesting that intermolecular hydrophobic interactions may be optimized by linking alkyl groups to the aromatic ring or by increasing the aliphatic chain between lactone and aromatic rings. In other words, extending the molecule in the direction of the hydrophobic pocket may improve the affinity of new compounds through extra van der Waals interactions with the enzyme.

Additionally, the docking simulations revealed a large change in the pharmacophoric conformation of 21-DB at the lactone ring, which may explain the biological effect of this compound.

Retrospective clinical observations performed 40 years ago suggest that cardiac steroids have anticancer effects [10,89,90]; however, few studies have been performed to elucidate the

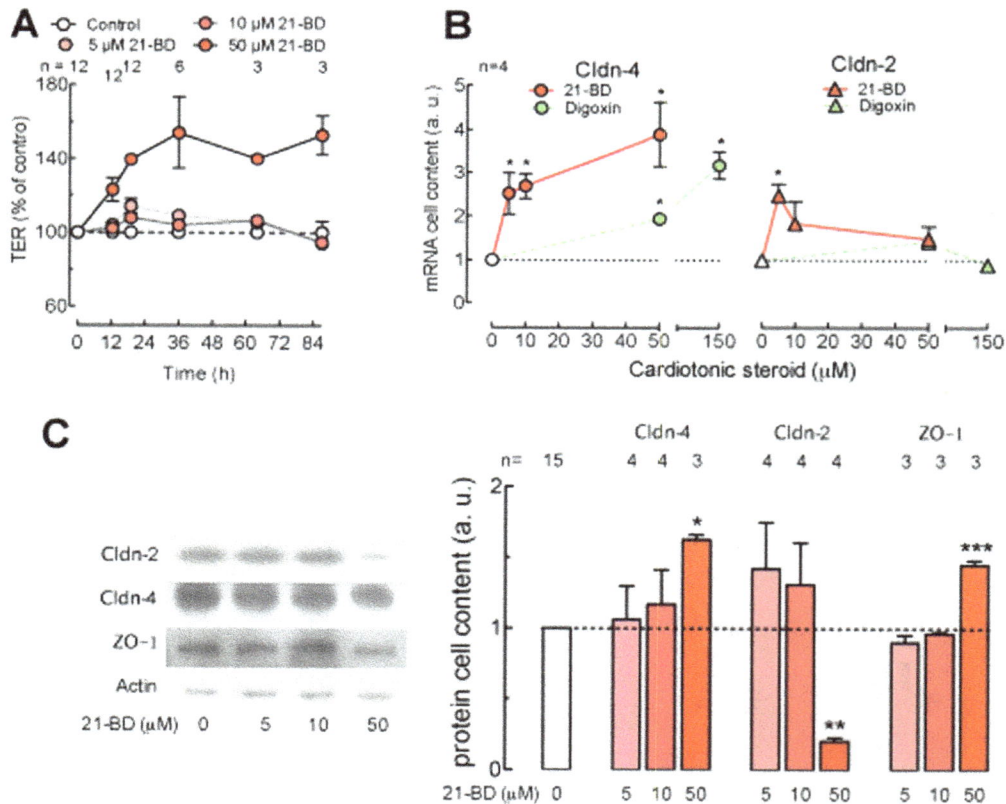

Figure 6. 21-BD regulates tight junctions. MDCK cells were cultured in transwell permeable supports and treated with 5, 10 and 50 μM 21-BD. (A) TER was measured as a function of time. The control TER data (white circles, dotted line) averaged 183 ± 8 $\Omega.cm^2$ (n = 13) and were normalized to 100%. 5 and 10 μM 21-BD provoke transient small increases of TER, while 50 μM 21-BD causes a stronger and a sustained TER increase (red circles). (B) MDCK cells were incubated 48 h with different concentrations of 21-BD (red symbols) or digoxin (green symbols). mRNA cell content of claudins -4 (circles) and -2 (triangles) were measured by quantitative real time PCR. (C) Protein cell content of the tight junction integral membrane proteins claudins -4 and -2 and the membrane-associated protein ZO-1 as a function of 21-BD concentration in the media for 48 h. Images from the left part of the figure C are representative immunoblots and the graph in the right part is the statistical analysis.

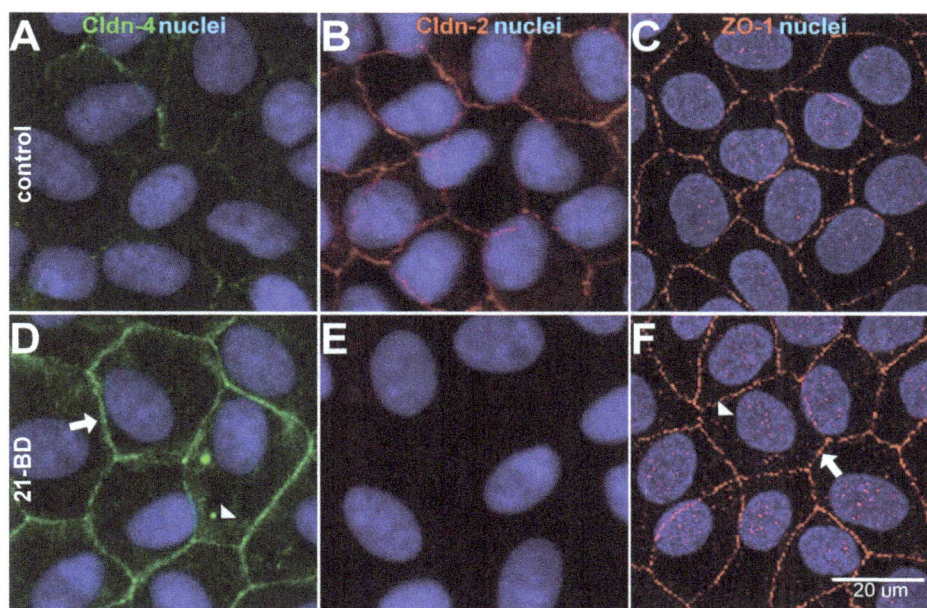

Figure 7. 21-BD regulates tight junctions proteins localization. Confluent monolayers of MDCK cells, grown on filters, were maintained in control medium (A, B, C) or treated with 50 µM 21-BD for 48 h (D, F, G) and processed for immunofluorescence, using antibodies against the TJs proteins: the integral membrane proteins claudin-4 (Cldn-4, A, D green) and claudin-2 (Cldcn-2, B, E red) and the peripheral membrane protein ZO-1 (C, E, red). Nuclei were stained with TOPRO (blue). 21-BD increases claudin-4 and ZO-1 expression at the tight junction (arrows) and in the cytoplasma (arrow heads), while simultaneously reduces the expression of claudin-2.

underlying mechanisms of these effects [3,4]. It has been proposed that cardiotonic steroids induce necrosis through the inhibition of the ion transport activity of the Na,K-ATPase, the resulting sustained increase of the cytosolic calcium concentration and the activation of cellular signaling cascades [16,91–94]. Nevertheless, low concentrations of ouabain (10 nM) do not induce apoptosis in MDCK cells [35], while high concentrations (300 nM - 3 µM) induce necrotic processes such as cell swelling independent of Na,K-ATPase pumping activity [95] and permeabilization of lysosomal membranes [42]. These findings indicate that cardiotonic steroids induce cell death though complex mechanisms named oncosis, a type of cell death that involves apoptotic (caspase activation) as well as necrotic (cell swelling) processes [95]. Our data indicate that 21-BD induces DNA damage, which is a signature characteristic of apoptosis but is also compatible with oncosis. *In vitro* studies show that other mechanisms may be related to cell responses triggered by cardiac glycosides. For instance, bufalin, a hydrophobic cardiotonic steroid, downregulates the expression of Cyclin A, Bcl-2 and Bcl-X_L and upregulates

the expression of p21 and Bax in ovarian endometrial cyst stromal cells, affecting cell cycle progression and inducing apoptosis [96]. Digoxin applied at low concentrations (<10 nM) prevents apoptosis in HeLa cells, whereas at higher concentrations induces the release of cytochrome-c, thereby triggering apoptosis [97]. UNBS1450 induces death in A549 lung cancer cells through the NFκB signaling pathway [42]. The diversity of reported effects indicates the need for further studies aimed to elucidate the possible mechanisms involved in cell death induced by digoxin and other types of cardiac steroids.

21-BD is moderately cytotoxic for HeLa and RKO cells (LC_{50}, of approximately 50 µM). This cytotoxicity is correlated with the induction of DNA damage. Studies addressing the genotoxicity of cardiac glycosides are scarce [98,99]. Given that cardiac glycosides produce reactive oxygen species (ROS) [4,100] and that these substances are well-known genotoxic agents [101], it is possible that 21-BD could induce an increase in ROS. Alternatively, 21-BD may inhibit topoisomerase II, as is known for other cardiac glycosides, including digoxin [91].

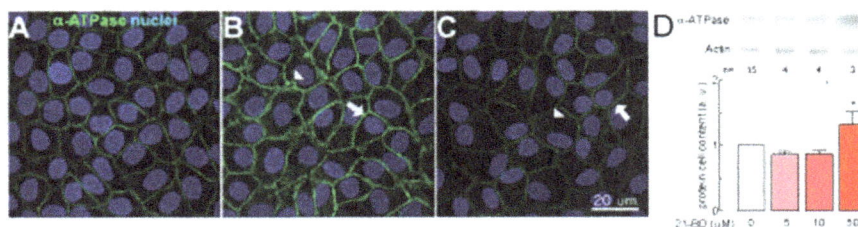

Figure 8. 21-BD increases the expression of Na,K-ATPase in MDCK cells. (A) Protein cell content of the α_1 subunit of the Na,K-ATPase of confluent monolayers of MDCK cells grown on filters in control medium (white bar) or treated with different concentrations of 21-BD (red bars) for 48 h; upper part of the figure A shows representative immunoblots of the α_1 subunit of the Na,K-ATPase and actin, the lower part the densitometric analysis. (B and C) Na,K-ATPase α_1 subunit stained with a fluoresceinated antibody (B, C, white) or Topro (blue) to detect the nuclei.

When directly applied to HeLa and RKO cells and in relatively low concentrations, 21-BD induces the increase of Na,K-ATPase activity through up-regulation of the α_1 and β_1 subunits of the enzyme (Figures 3A, B). In contrast, in relatively high concentrations, 21-BD inhibits the activity of the Na,K-ATPase α_1 isoform of cell homogenates (Fig. 2C, D). This effect appears to be isoform specific, since 21-DB has no effect on the $\alpha2$ and α_3 isoforms of brain that have a high affinity for for cardiac steroids. The requirement of high concentrations of 21-DB to inhibit α_1 shows the low affinity of this Na,K-ATPase isoform for the compound. The effects that we observe with lower concentrations of 21-BD are consistent with the activation of the signaling functions of the Na,K-ATPase that are known to trigger a variety of cellular phenomena including induction of endocytosis of junctional proteins [34,83] and the Na,K-ATPase itself [102–104], the detachment of epithelial cells from the substrate and from themselves [24,105], cell proliferation [106], protection from cell death caused by the addition of serum [107], cell survival [108] and the sealing of tight junction through the upregulation of claudins [35]. The effects produced by 21-BD depend on the cellular context. While this compound reduces cell viability and induces apoptosis in cells derived from tumors (HeLa and RKO), it does not affect the viability of normal kidney epithelial cells (MDCK) as indicated by the increase in the transepithelial electrical resistance. The overexpression of claudin and the strengthening of tight junctions is an important mechanism against the phenotypic characteristics of cancer, which can regulate events such cell migration, transformation, proliferation and invasiveness [109–113].

Our results suggests that the modification of lactone ring of digoxin and maybe other cardiotonic steroids, could be an interesting alternative for chemical modification of this class of compounds. The effects presented by 21-BD appear to be unrelated with the classical effects that the cardiotonic steroids exert on intracellular ion changes through changes in Na,K-ATPase activity and raises the possibility that 21-BD (and maybe other cardiotonic steroids) may be involved with other proteins different from the Na,K-ATPase. Interestingly, digoxin has been shown to bind to other proteins besides Na,K-ATPase, such as the retinoic acid receptor-related orphan nuclear receptor RORγt. Digoxin-21-salicylidene is another digoxin derivative, with similar modification than 21-BD, that does not have high cytotoxic effect and selectively binds to RORγt, reducing progression of autoimmune encephalitis [49]. So, the decrease of cardiotonic steroid cytotoxicity opens the opportunity to use this class of compounds in other diseases.

Surprisingly, Na,K-ATPase is not the only enzyme that is responsive to 21-BD. Although Na,K-ATPase activity was not directly affected by 21-BD, the activity of the yeast multidrug exporter Pdr5p was largely inhibited by 21-BD. Cholesterol-derived steroids can block the pumping function of this transporter, showing an IC_{50} in the micromolar range, and some derivatives are considered interesting leads for use as drugs for the putative treatment of multidrug-resistant (MDR) tumors [74,75]. It is also known that several cardiotonic steroids exhibit anticancer activity in cells that express the chemotherapy resistance phenotype [43,44,114–117]. This activity may be explained if as suggested by our 21-BD results, these cardiotonic steroids, have

also the capacity to inhibit MDR proteins. Alternatively, anti MDR activity may be caused by downregulation of c-Myc or MDR-related genes, the inhibition of the glycosylation of MDR-related proteins or the inhibition of glycolysis [43]. Our results show, for the first time, that a cis-trans-cis steroid is also able to cause direct inhibition of Pdr5p activity and opens the possibility that 21-BD may act as a drug to reverse the MDR phenotype.

Another important action of 21-BD is its interaction with junctional complexes. This increases TER as a result of the upregulation of the tight junction proteins Claudin-4 and ZO-1, which increase the sealing of tight junctions [118–120], and downregulates the levels of Claudin-2, which is expressed in epithelia with low TER [121,122]. Additionally, 21-BD induces an increase in the levels of Claudin-4 mRNA. This effect shows that ouabain is not the sole cardiotonic steroid that alters the molecular composition of tight junctions and TER [24,35], but that 21-BD can induce these effects as well.

Conclusion

The combination of an styrene group with the lactone moiety of digoxin produced a cardiotonic steroid, 21-BD, that enhances Na,K-ATPase catalytic activity in intact cells and stimulates Na,K-ATPase mediated signaling events. The later include reduction of cell viability via apoptosis in cancer, but not normal epithelial cells and increase in the closure of tight junctions of epithelial cells. The chemical modification of cardiotonic steroids with the styrene group has the potential to provide compounds with new functional properties that could results beneficial to treat cancer and perhaps other diseases in which cell proliferation occurs.

Supporting Information

Figure S1 Synthesis of 21-BD.

Figure S2 Structure and pharmacophoric conformation of ouabain. (A) Crystallographic (green) and docked (red) structures of ouabain. (B) Pharmacophoric conformation of ouabain. Ouabain complexes with the receptor via hydrogen bonding between Gln111, Glu117, Asp121, Asn122, Glu312 and Thr797. In addition, eletrostatic and hydrophobic interactions perform around the steroid core composed by Pro118, Leu125, Gly319, Phe783, Phe786, Leu793, Ile800, Arg880 and Asp121, Ile315, Val322, Ala323, respectively.

Author Contributions

Conceived and designed the experiments: FPV LMS AGT MC HLS FVS LEQ FGN AFP OCM GB JAFPV RGC LAB. Performed the experiments: SCR MTCP LDRN SLGA SMFO IVG ACSCT NLSG RRH NP. Analyzed the data: FPV LMS AGT MC FVS LEQ FGN AFP OCM GB JAFPV RGC LAB. Contributed reagents/materials/analysis tools: FPV LMS AGT MC HLS FVS LEQ FGN AFP OCM GB JAFPV RGC LAB. Contributed to the writing of the manuscript: FPV LMS AGT MC FVS LEQ FGN AFP OCM GB JAFPV RGC LAB.

References

1. Brower LP, Moffitt CM (1974) Palatability dynamics of cardenolides in the monarch butterfly. Nature 249: 280–283.

2. Haux J (1999) Digitoxin is a potential anticancer agent for several types of cancer. Med Hypotheses 53: 543–548.

3. Bagrov AY, Shapiro JI, Fedorova OV (2009) Endogenous Cardiotonic Steroids: Physiology, Pharmacology, and Novel Therapeutic Targets. Pharmacol Rev 61: 9–38.

4. Schoner W, Scheiner-Bobis G (2007) Endogenous and exogenous cardiac glycosides and their mechanisms of action. Am J Cardiovasc Drug 7: 173–189.

5. Blaustein MP (1993) Physiological effects of endogenous ouabain: control of intracellular Ca2+ stores and cell responsiveness. Am J Physiol 264: C1367–1387.

6. Chen JQ, Contreras RG, Wang R, Fernandez SV, Shoshani L, et al. (2006) Sodium/potassium ATPase (Na+, K+-ATPase) and ouabain/related cardiac glycosides: A new paradigm for development of anti- breast cancer drugs? Breast Cancer Res Treat 96: 1–15.

7. Martin TA, Jiang WG (2001) Tight junctions and their role in cancer metastasis. Histol Histopathol 16: 1183–1195.

8. Escudero-Esparza A, Jiang WG, Martin TA (2011) The Claudin family and its role in cancer and metastasis. Front Biosci-Landmrk 16: 1069–1083.

9. Shirator O (1967) Growth Inhibitory Effect of Cardiac Glycosides and Aglycones on Neoplastic Cells - in Vitro and in Vivo Studies. Gann 58: 521–528.

10. Stenkvist B, Bengtsson E, Dahlqvist B, Eriksson O, Jarkrans T, et al. (1982) Cardiac-Glycosides and Breast-Cancer, Revisited. New Engl J Med 306: 484–484.

11. Simpson CD, Mawji IA, Anyiwe K, Williams MA, Wang XM, et al. (2009) Inhibition of the Sodium Potassium Adenosine Triphosphatase Pump Sensitizes Cancer Cells to Anoikis and Prevents Distant Tumor Formation. Cancer Res 69: 2739–2747.

12. Ihenetu K, Qazzaz HM, Crespo F, Fernandez-Botran R, Valdes R (2007) Digoxin-like immunoreactive factors induce apoptosis in human acute T-cell lymphoblastic leukemia. Clin Chem 53: 1315–1322.

13. Winnicka K, Bielawski K, Bielawska A, Miltyk W (2007) Apoptosis-mediated cytotoxicity of ouabain, digoxin and proscillaridin A in the estrogen independent MDA-MB-231 breast cancer cells. Arch Pharm Res 30: 1216–1224.

14. Winnicka K, Bielawski K, Bielawska A, Surazynski A (2008) Antiproliferative activity of derivatives of ouabain, digoxin and proscillaridin A in human MCF-7 and MDA-MB-231 breast cancer cells. Biol Pharm Bull 31: 1131–1140.

15. Van Quaquebeke E, Simon G, Andre A, Dewelle J, El Yazidi M, et al. (2005) Identification of a novel cardenolide (2″-oxovoruscharin) from Calotropis procera and the hemisynthesis of novel derivatives displaying potent in vitro antitumor activities and high in vivo tolerance: Structure-activity relationship analyses. J Med Chem 48: 849–856.

16. Sun L, Chen TS, Wang XP, Chen Y, Wei XB (2011) Bufalin Induces Reactive Oxygen Species Dependent Bax Translocation and Apoptosis in ASTC-a-1 Cells. Evid-Based Compl Alt: 1–12.

17. Cunha GA, Resck IS, Cavalcanti BC, Pessoa CO, Moraes MO, et al. (2010) Cytotoxic profile of natural and some modified bufadienolides from toad Rhinella schneideri parotoid gland secretion. Toxicon 56: 339–348.

18. Takai N, Ueda T, Nishida M, Nasu K, Narahara H (2008) Bufalin induces growth inhibition, cell cycle arrest and apoptosis in human endometrial and ovarian cancer cells. Int J Mol Med 21: 637–643.

19. Yu CH, Kan SF, Pu HF, Jea Chien E, Wang PS (2008) Apoptotic signaling in bufalin- and cinobufagin-treated androgen-dependent and -independent human prostate cancer cells. Cancer Sci 99: 2467–2476.

20. Cairns J (1975) Mutation selection and the natural history of cancer. Nature 255: 197–200.

21. Cereijido M, Contreras RG, Shoshani L, Flores-Benitez D, Larre I (2008) Tight junction and polarity interaction in the transporting epithelial phenotype. Biochim Biophys Acta 1778: 770–793.

22. Rajasekaran SA, Hu J, Gopal J, Gallemore R, Ryazantsev S, et al. (2003) Na,K-ATPase inhibition alters tight junction structure and permeability in human retinal pigment epithelial cells. Am J Physiol Cell Physiol 284: C1497–1507.

23. Kominsky SL, Argani P, Korz D, Evron E, Raman V, et al. (2003) Loss of the tight junction protein claudin-7 correlates with histological grade in both ductal carcinoma in situ and invasive ductal carcinoma of the breast. Oncogene 22: 2021–2033.

24. Contreras RG, Shoshani L, Flores-Maldonado C, Lazaro A, Cereijido M (1999) Relationship between Na+,K+-ATPase and cell attachment. J Cell Sci 112: 4223–4232.

25. Rajasekaran SA, Palmer LG, Moon SY, Soler AP, Apodaca GL, et al. (2001) Na,K-ATPase activity is required for formation of tight junctions, desmosomes, and induction of polarity in epithelial cells. Mol Biol Cell 12: 3717–3732.

26. Contreras RG, Avila G, Gutierrez C, Bolivar JJ, Gonzalezmariscal L, et al. (1989) Repolarization of Na+-K+ Pumps during Establishment of Epithelial Monolayers. Am J Physiol 257: C896–C905.

27. Gloor S, Antonicek H, Sweadner KJ, Pagliusi S, Frank R, et al. (1990) The Adhesion Molecule on Glia (Amog) Is a Homolog of the Beta-Subunit of the Na,K-Atpase. J Cell Biol 110: 165–174.

28. Shoshani L, Contreras RG, Roldan ML, Moreno J, Lazaro A, et al. (2005) The polarized expression of Na+,K+-ATPase in epithelia depends on the association between beta-subunits located in neighboring cells. Mol Biol Cell 16: 1071–1081.

29. Padilla-Benavides T, Roldan ML, Larre I, Flores-Benitez D, Villegas-Sepulveda N, et al. (2010) The Polarized Distribution of Na+, K+-ATPase: Role of the Interaction between beta Subunits. Mol Biol Cell 21: 2217–2225.

30. Cereijido M, Contreras RG, Shoshani L, Larre I (2012) The Na+-K+-ATPase as self-adhesion molecule and hormone receptor. Am J Physiol Cell Physiol 302: C473–C481.

31. Vagin O, Dada LA, Tokhtaeva E, Sachs G (2012) The Na-K-ATPase alpha(1)beta(1) heterodimer as a cell adhesion molecule in epithelia. Am J Physiol Cell Physiol 302: C1271–1281.

32. Rajasekaran SA, Huynh TP, Wolle DG, Espineda CE, Inge LJ, et al. (2010) Na,K-ATPase Subunits as Markers for Epithelial-Mesenchymal Transition in Cancer and Fibrosis. Mol Cancer Ther 9: 1515–1524.

33. Watsky MA, Rae JL (1992) Dye Coupling in the Corneal Endothelium - Effects of Ouabain and Extracellular Calcium Removal. Cell Tissue Res 269: 57–63.

34. Rincon-Heredia R, Flores-Benitez D, Flores-Maldonado C, Bonilla-Delgado J, Garcia-Hernandez V, et al. (2014) Ouabain induces endocytosis and degradation of tight junction proteins through ERK1/2-dependent pathways. Exp Cell Res 320: 108–118.

35. Larre I, Lazaro A, Contreras RG, Balda MS, Matter K, et al. (2010) Ouabain modulates epithelial cell tight junction. P Natl Acad Sci USA 107: 11387–11392.

36. Larre I, Castillo A, Flores-Maldonado C, Contreras RG, Galvan I, et al. (2011) Ouabain modulates ciliogenesis in epithelial cells. P Natl Acad Sci USA 108: 20591–20596.

37. Jensen M, Schmidt S, Fedosova NU, Mollenhauer J, Jensen HH (2011) Synthesis and evaluation of cardiac glycoside mimics as potential anticancer drugs. Bioorgan Med Chem 19: 2407–2417.

38. Langenhan JM, Peters NR, Guzei IA, Hoffmann M, Thorson JS (2005) Enhancing the anticancer properties of cardiac glycosides by neoglycorandomization. P Natl Acad Sci USA 102: 12305–12310.

39. Langenhan JM, Endo MM, Engle JM, Fukumoto LL, Rogalsky DR, et al. (2011) Synthesis and biological evaluation of RON-neoglycosides as tumor cytotoxins. Carbohyd Res 346: 2663–2676.

40. Ye M, Qu GQ, Guo HZ, Guo D (2004) Novel cytotoxic bufadienolides derived from bufalin by microbial hydroxylation and their structure-activity relationships. J Steroid Biochem 91: 87–98.

41. Mijatovic T, Lefranc F, Van Quaquebeke E, Van Vynckt F, Darro F, et al. (2007) UNBS1450: A new hemi-synthetic cardenolide with promising anti-cancer activity. Drug Develop Res 68: 164–173.

42. Mijatovic T, Mathieu V, Gaussin JF, De Neve N, Ribaucour F, et al. (2006) Cardenolide-induced lysosomal membrane permeabilization demonstrates therapeutic benefits in experimental human non-small cell lung cancers. Neoplasia 8: 402–412.

43. Mijatovic T, Kiss R (2013) Cardiotonic steroids-mediated Na+/K+-ATPase targeting could circumvent various chemoresistance pathways. Planta medica 79: 189–198.

44. Mijatovic T, Dufrasne F, Kiss R (2012) Cardiotonic steroids-mediated targeting of the Na(+)/K(+)-ATPase to combat chemoresistant cancers. Curr Med Chem 19: 627–646.

45. Elbaz HA, Stueckle TA, Wang HYL, O'Doherty GA, Lowry DT, et al. (2012) Digitoxin and a synthetic monosaccharide analog inhibit cell viability in lung cancer cells. Toxicol Appl Pharm 258: 51–60.

46. Wang HYL, Wu BL, Zhang Q, Kang SW, Rojanasakul Y, et al. (2011) C5′-Alkyl Substitution Effects on Digitoxigenin alpha-L-Glycoside Cancer Cytotoxicity. Acs Med Chem Lett 2: 259–263.

47. Hinds JW, McKenna SB, Sharif EU, Wang HY, Akhmedov NG, et al. (2013) C3′/C4′-Stereochemical Effects of Digitoxigenin alpha-L-/alpha-D-Glycoside in Cancer Cytotoxicity. ChemMedChem 8: 63–69.

48. Wang HYL, Rojanasakul Y, O'Doherty GA (2011) Synthesis and Evaluation of the alpha-D-/alpha-L-Rhamnosyl and Amicetosyl Digitoxigenin Oligomers as Antitumor Agents. Acs Med Chem Lett 2: 264–269.

49. Huh JR, Leung MWL, Huang PX, Ryan DA, Krout MR, et al. (2011) Digoxin and its derivatives suppress T(H)17 cell differentiation by antagonizing ROR gamma t activity. Nature 472: 486–490.

50. Xu HW, Wang JF, Liu GZ, Hong GF, Liu HM (2007) Facile synthesis of gamma-alkylidenebutenolides. Org Biomol Chem 5: 1247–1250.

51. Sanchez G, Timmerberg B, Tash J, Blanco G (2006) The Na,K-atpase alpha4 isoform from humans has distinct enzymatic properties and is important for sperm motility. J Androl: 42–42.

52. Hilario FF, de Paula RC, Silveira MLT, Viana GHR, Alves RB, et al. (2011) Synthesis and Evaluation of Antimalarial Activity of Oxygenated 3-alkylpyridine Marine Alkaloid Analogues. Chem Biol Drug Des 78: 477–482.

53. Singh NP, McCoy MT, Tice RR, Schneider EL (1988) A simple technique for quantitation of low levels of DNA damage in individual cells. Exp Cell Res 175: 184–191.

54. Tice RR, Andrews PW, Hirai O, Singh NP (1991) The single cell gel (SCG) assay: an electrophoretic technique for the detection of DNA damage in individual cells. In: Witmer CR, Snyder RR, Jollow DJ, Kalf GF, and Kocsis IG, editors. Biological Reactive Intermediates IV, Molecular and Cellular Effects and their Impact on Human Health. New York, NY: Plenum Pressley, T. A. 157–164.

55. Tice RR, Agurell E, Anderson D, Burlinson B, Hartmann A, et al. (2000) Single cell gel/comet assay: Guidelines for in vitro and in vivo genetic toxicology testing. Environ Mol Mutagen 35: 206–221.

56. Garcia O, Mandina T, Lamadrid AI, Diaz A, Remigio A, et al. (2004) Sensitivity and variability of visual scoring in the comet assay - Results of an

inter-laboratory scoring exercise with the use of silver staining. Mutat Res 556: 25–34.

57. Shoshani L, Contreras RG, Roldan ML, Hidalgo MR, Fiorentino R, et al. (2005) The polarized expression of Na+,K+-ATPase depends on the association between beta-subunits located in neighboring cells. J Gen Physiol 126: 40a–41a.

58. Livak KJ, Schmittgen TD (2001) Analysis of relative gene expression data using real-time quantitative PCR and the 2(T)(-Delta Delta C) method. Methods 25: 402–408.

59. Chechi K, Gelinas Y, Mathieu P, Deshaies Y, Richard D (2012) Validation of Reference Genes for the Relative Quantification of Gene Expression in Human Epicardial Adipose Tissue. Plos One 7: e32265.

60. Stewart JJP (2007) Optimization of parameters for semiempirical methods V: Modification of NDDO approximations and application to 70 elements. J Mol Model 13: 1173–1213.

61. Frisch MJT, Schlegel GW, Scuseria HB, Robb GE, Cheeseman MA, et al. (2009) Gaussian 09, Revision A.1. Gaussian, Inc, Wallingford CT,.

62. Laursen M, Yatime L, Nissen P, Fedosova NU (2013) Crystal structure of the high-affinity Na+K+-ATPase-ouabain complex with Mg2+ bound in the cation binding site. Proc Natl Acad Sci U S A 110: 10958–10963.

63. Trott O, Olson AJ (2010) Software News and Update AutoDock Vina: Improving the Speed and Accuracy of Docking with a New Scoring Function, Efficient Optimization, and Multithreading. J Comput Chem 31: 455–461.

64. Nocedal JWS (1999) Numerical Optmization. Springer Verlag: Berlin.

65. Discovery Studio 3.1 Help. (2010) San Diego: Accelrys Software Inc.

66. Rangel LP, Fritzen M, Yunes RA, Leal PC, Creczynski-Pasa TB, et al. (2010) Inhibitory effects of gallic acid ester derivatives on Saccharomyces cerevisiae multidrug resistance protein Pdr5p. Fems Yeast Res 10: 244–251.

67. Bettero GM, Salles L, Figueira RRM, von Poser G, Rates SMK, et al. (2011) In Vitro Effect of Valepotriates Isolated from Valeriana glechomifolia on Rat P-Type ATPases. Planta Med 77: 1702–1706.

68. Dulley JR (1975) Determination of Inorganic-Phosphate in Presence of Detergents or Protein. Anal Biochem 67: 91–96.

69. Fiske CH, Subbarow Y (1925) THE COLORIMETRIC DETERMINATION OF PHOSPHORUS. J Biol Chem 66: 375–400.

70. Noel F, Pimenta PHC, dos Santos AR, Tomaz ECL, Quintas LEM, et al. (2011) Delta(2,3)-Ivermectin ethyl secoester, a conjugated ivermectin derivative with leishmanicidal activity but without inhibitory effect on mammalian P-type ATPases. N-S Arch Pharmacol 383: 101–107.

71. Ogawa H, Shinoda T, Cornelius F, Toyoshima C (2009) Crystal structure of the sodium-potassium pump (Na+,K+-ATPase) with bound potassium and ouabain. Proc Natl Acad Sci U S A 106: 13742–13747.

72. Zahler R, Zhang ZT, Manor M, Boron WF (1997) Sodium kinetics of Na,K-ATPase alpha isoforms in intact transfected HeLa cells. J Gen Physiol 110: 201–213.

73. Pocas ESC, Costa PRR, da Silva AJM, Noel F (2003) 2-Methoxy-3,8,9-trihydroxy coumestan: a new synthetic inhibitor of Na+,K+-ATPase with an original mechanism of action. Biochem Pharmacol 66: 2169–2176.

74. Perez-Victoria FJ, Conseil G, Munoz-Martinez F, Perez-Victoria JM, Dayan G, et al. (2003) RU49953: a non-hormonal steroid derivative that potently inhibits P-glycoprotein and reverts cellular multidrug resistance. Cell Mol Life Sci 60: 526–535.

75. Conseil G, Perez-Victoria JM, Renoir JM, Goffeau A, Di Pietro A (2003) Potent competitive inhibition of drug binding to the Saccharomyces cerevisiae ABC exporter Pdr5p by the hydrophobic estradiol-derivative RU49953. Biochim Biophys Acta 1614: 131–134.

76. da Silva FR, Tessis AC, Ferreira PF, Rangel LP, Garcia-Gomes AS, et al. (2011) Oroidin Inhibits the Activity of the Multidrug Resistance Target Pdr5p from Yeast Plasma Membranes. J Nat Prod 74: 279–282.

77. Yang PY, Menter DG, Cartwright C, Chan D, Dixon S, et al. (2009) Oleandrin-mediated inhibition of human tumor cell proliferation: importance of Na,K-ATPase alpha subunits as drug targets. Mol Cancer Ther 8: 2319–2328.

78. Alonso E, Cano-Abad MF, Moreno-Ortega AJ, Novalbos J, Milla J, et al. (2013) Nanomolar ouabain elicits apoptosis through a direct action on HeLa cell Cross Mark mitochondria. Steroids 78: 1110–1118.

79. Winnicka K, Bielawski K, Bielawska A, Miltyk W (2010) Dual effects of ouabain, digoxin and proscillaridin A on the regulation of apoptosis in human fibroblasts. Nat Prod Res 24: 274–285.

80. Lopardo T, Lo Iacono N, Marinari B, Giustizieri ML, Cyr DG, et al. (2008) Claudin-1 Is a p63 Target Gene with a Crucial Role in Epithelial Development. Plos One 3: e2715.

81. Ikari A, Atomi K, Takiguchi A, Yamazaki Y, Hayashi H, et al. (2012) Enhancement of cell-cell contact by claudin-4 in renal epithelial madin-darby canine kidney cells. J Cell Biochem 113: 499–507.

82. Cereijido M, Robbins ES, Dolan WJ, Rotunno CA, Sabatini DD (1978) Polarized Monolayers Formed by Epithelial-Cells on a Permeable and Translucent Support. J Cell Biol 77: 853–880.

83. Garcia-Hernandez V, Flores-Maldonado C, Rincon-Heredia R, Verdejo-Torres O, Bonilla-Delgado J, et al. (2014) EGF regulates claudin-2 and -4 expression through STAT3 and Src in MDCK cells. J Cell Physiol (Epub ahead of print).

84. Flores-Benitez D, Rincon-Heredia R, Razgado LF, Larre I, Cereijido M, et al. (2009) Control of tight junctional sealing: roles of epidermal growth factor and prostaglandin E2. Am J Physiol Cell Physiol 297: C611–620.

85. Liu B, Anderson SL, Qiu J, Rubin BY (2013) Cardiac glycosides correct aberrant splicing of IKBKAP-encoded mRNA in familial dysautonomia derived cells by suppressing expression of SRSF3. FEBS J 280: 3632–3646.

86. Chen ZC, Yu BC, Chen LJ, Cheng JT (2013) Increase of peroxisome proliferator-activated receptor delta (PPARdelta) by digoxin to improve lipid metabolism in the heart of diabetic rats. Horm Metab Res 45: 364–371.

87. Didiot MC, Hewett J, Varin T, Freuler F, Selinger D, et al. (2013) Identification of Cardiac Glycoside Molecules as Inhibitors of c-Myc IRES-Mediated Translation. J Biomol Screen 18: 407–419.

88. Ward SC, Hamilton BP, Hamlyn JM (2002) Novel receptors for ouabain: studies in adrenocortical cells and membranes. Hypertension 39: 536–542.

89. Stenkvist B, Bengtsson E, Eriksson O, Holmquist J, Nordin B, et al. (1979) Cardiac-Glycosides and breast cancer. Lancet 1: 563–563.

90. Stenkvist B, Bengtsson E, Eklund G, Eriksson O, Holmquist J, et al. (1980) Evidence of a Modifying Influence of Heart Glucosides on the Development of breast cancer. Anal Quant Cytol 2: 49–54.

91. Bielawski K, Winnicka K, Bielawska A (2006) Inhibition of DNA topoisomerases I and II, and growth inhibition of breast cancer MCF-7 cells by ouabain, digoxin and proscillaridin A. Biol Pharm Bull 29: 1493–1497.

92. Kometiani P, Liu LJ, Askari A (2005) Digitalis-induced signaling by Na+/K+-ATPase in human breast cancer cells. Mol Pharmacol 67: 929–936.

93. Li ZC, Zhang ZB, Xie JX, Li X, Tian J, et al. (2011) Na/K-ATPase Mimetic pNaKtide Peptide Inhibits the Growth of Human Cancer Cells. J Biol Chem 286: 32394–32403.

94. Liu LJ, Askari A (2005) Digitalis-induced growth arrest in human breast cancer cells: On the importance and mechanism of amplification of digitalis signal through Na/K-ATPase. J Gen Physiol 126: 72a–72a.

95. Platonova A, Koltsova S, Maksimov GV, Grygorczyk R, Orlov SN (2011) The death of ouabain-treated renal epithelial C11-MDCK cells is not mediated by swelling-induced plasma membrane rupture. J Membr Biol 241: 145–154.

96. Nasu K, Nishida M, Ueda T, Takai N, Bing S, et al. (2005) Bufalin induces apoptosis and the G0/G1 cell cycle arrest of endometriotic stromal cells: a promising agent for the treatment of endometriosis. Mol Hum Reprod 11: 817–823.

97. Ramirez-Ortega M, Maldonado-Lagunas V, Melendez-Zajgla J, Carrillo-Hernandez JF, Pastelin-Hernandez G, et al. (2006) Proliferation and apoptosis of HeLa cells induced by in vitro stimulation with digitalis. Eur J Pharmacol 534: 71–76.

98. Snyder RD, Green JW (2001) A review of the genotoxicity of marketed pharmaceuticals. Mutat Res 488: 151–169.

99. Brambilla G, Martelli A (2009) Update on genotoxicity and carcinogenicity testing of 472 marketed pharmaceuticals. Mutat Res 681: 209–229.

100. Huang YT, Chueh SC, Teng CM, Guh JH (2004) Investigation of ouabain-induced anticancer effect in human androgen-independent prostate cancer PC-3 cells. Biochem Pharmacol 67: 727–733.

101. Emerit I (1994) Reactive Oxygen Species, Chromosome Mutation, and Cancer - Possible Role of Clastogenic Factors in Carcinogenesis. Free Radical Bio Med 16: 99–109.

102. Rosen H, Glukhman V, Feldmann T, Fridman E, Lichtstein D (2004) Cardiac steroids induce changes in recycling of the plasma membrane in human NT2 cells. Mol Biol Cell 15: 1044–1054.

103. Gupta S, Yan Y, Malhotra D, Liu J, Xie Z, et al. (2012) Ouabain and insulin induce sodium pump endocytosis in renal epithelium. Hypertension 59: 665–672.

104. Liu J, Kesiry R, Periyasamy SM, Malhotra D, Xie Z, et al. (2004) Ouabain induces endocytosis of plasmalemmal Na/K-ATPase in LLC-PK1 cells by a clathrin-dependent mechanism. Kidney Int 66: 227–241.

105. Rincon-Heredia R, Flores-Benitez D, Flores-Maldonado C, Bonilla-Delgado J, Garcia-Hernandez V, et al. (2013) Ouabain induces endocytosis and degradation of tight junction proteins through ERK1/2-dependent pathways. Exp Cell Res 320: 108–118.

106. Tian J, Li X, Liang M, Liu LJ, Xie JX, et al. (2009) Changes in Sodium Pump Expression Dictate the Effects of Ouabain on Cell Growth. J Biol Chem 284: 14921–14929.

107. Aizman O, Uhlen P, Lal M, Brismar H, Aperia A (2001) Ouabain, a steroid hormone that signals with slow calcium oscillations. P Natl Acad Sci USA 98: 13420–13424.

108. Oselkin M, Tian DZ, Bergold PJ (2010) Low-dose cardiotonic steroids increase sodium-potassium ATPase activity that protects hippocampal slice cultures from experimental ischemia. Neurosci Lett 473: 67–71.

109. Lelievre SA (2010) Tissue polarity-dependent control of mammary epithelial homeostasis and cancer development: an epigenetic perspective. Journal Mammary Gland Biol Neoplasia 15: 49–63.

110. Escudero-Esparza A, Jiang WG, Martin TA (2011) The Claudin family and its role in cancer and metastasis. Front Biosci (Landmark Ed): 16: 1069–1083.

111. Kwon MJ (2013) Emerging roles of claudins in human cancer. Int J Mol Sci 14: 18148–18180.

112. Runkle EA, Mu D (2013) Tight junction proteins: from barrier to tumorigenesis. Cancer Lett 337: 41–48.

113. Ding L, Lu Z, Lu Q, Chen YH (2013) The claudin family of proteins in human malignancy: a clinical perspective. Cancer Manag Res 5: 367–375.

114. Mijatovic T, Jungwirth U, Heffeter P, Hoda MAR, Dornetshuber R, et al. (2009) The Na+/K+-ATPase is the Achilles Heel of multi-drug-resistant cancer cells. Cancer Lett 282: 30–34.

115. Zhao M, Bai L, Wang L, Toki A, Hasegawa T, et al. (2007) Bioactive cardenolides from the stems and twigs of Nerium oleander. J Nat Prod 70: 1098–1103.

116. Piacente S, Masullo M, De Neve N, Dewelle J, Hamed A, et al. (2009) Cardenolides from Pergularia tomentosa Display Cytotoxic Activity Resulting from Their Potent Inhibition of Na+/K+-ATPase. J Nat Prod 72: 1087–1091.

117. Efferth T, Davey M, Olbrich A, Rucker G, Gebhart E, et al. (2002) Activity of drugs from traditional Chinese medicine toward sensitive and MDR1- or MRP1-overexpressing multidrug-resistant human CCRF-CEM leukemia cells. Blood Cells Mol Dis 28: 160–168.

118. Kiuchi-Saishin Y, Gotoh S, Furuse M, Takasuga A, Tano Y, et al. (2002) Differential expression patterns of claudins, tight junction membrane proteins, in mouse nephron segments. J Am Soc Nephrol 13: 875–886.

119. Reyes JL, Lamas M, Martin D, Namorado MD, Islas S, et al. (2002) The renal segmental distribution of claudins changes with development. Kidney Int 62: 476–487.

120. Furuse M, Furuse K, Sasaki H, Tsukita S (2001) Conversion of Zonulae occludentes from tight to leaky strand type by introducing claudin-2 into Madin-Darby canine kidney I cells. J Cell Biol 153: 263–272.

121. Colegio OR, Van Itallie CM, McCrea HJ, Rahner C, Anderson JM (2002) Claudins create charge-selective channels in the paracellular pathway between epithelial cells. Am J Physiol Cell Physiol 283: C142–C147.

122. Morita K, Furuse M, Fujimoto K, Tsukita S (1999) Claudin multigene family encoding four-transmembrane domain protein components of tight junction strands. P Natl Acad Sci USA 96: 511–516.

7

On the Dynamics of the Adenylate Energy System: Homeorhesis vs Homeostasis

Ildefonso M. De la Fuente[1,2,3,4]*, **Jesús M. Cortés**[4,5], **Edelmira Valero**[6], **Mathieu Desroches**[7], **Serafim Rodrigues**[8], **Iker Malaina**[9,4], **Luis Martínez**[2,4]

1 Institute of Parasitology and Biomedicine "López-Neyra", CSIC, Granada, Spain, **2** Department of Mathematics, University of the Basque Country UPV/EHU, Leioa, Spain, **3** Unit of Biophysics (CSIC, UPV/EHU), and Department of Biochemistry and Molecular Biology University of the Basque Country, Bilbao, Spain, **4** Biocruces Health Research Institute, Hospital Universitario de Cruces, Barakaldo, Spain, **5** Ikerbasque: The Basque Foundation for Science, Bilbao, Basque Country, Spain, **6** Department of Physical Chemistry, School of Industrial Engineering, University of Castilla-La Mancha, Albacete, Spain, **7** INRIA Paris-Rocquencourt Centre, Paris, France, **8** School of Computing and Mathematics, University of Plymouth, Plymouth, United Kingdom, **9** Department of Physiology, University of the Basque Country UPV/EHU, Bilbao, Spain

Abstract

Biochemical energy is the fundamental element that maintains both the adequate turnover of the biomolecular structures and the functional metabolic viability of unicellular organisms. The levels of ATP, ADP and AMP reflect roughly the energetic status of the cell, and a precise ratio relating them was proposed by Atkinson as the adenylate energy charge (AEC). Under growth-phase conditions, cells maintain the AEC within narrow physiological values, despite extremely large fluctuations in the adenine nucleotides concentration. Intensive experimental studies have shown that these AEC values are preserved in a wide variety of organisms, both eukaryotes and prokaryotes. Here, to understand some of the functional elements involved in the cellular energy status, we present a computational model conformed by some key essential parts of the adenylate energy system. Specifically, we have considered (I) the main synthesis process of ATP from ADP, (II) the main catalyzed phosphotransfer reaction for interconversion of ATP, ADP and AMP, (III) the enzymatic hydrolysis of ATP yielding ADP, and (IV) the enzymatic hydrolysis of ATP providing AMP. This leads to a dynamic metabolic model (with the form of a delayed differential system) in which the enzymatic rate equations and all the physiological kinetic parameters have been explicitly considered and experimentally tested *in vitro*. Our central hypothesis is that cells are characterized by changing energy dynamics (*homeorhesis*). The results show that the AEC presents stable transitions between steady states and periodic oscillations and, in agreement with experimental data these oscillations range within the narrow AEC window. Furthermore, the model shows sustained oscillations in the Gibbs free energy and in the total nucleotide pool. The present study provides a step forward towards the understanding of the fundamental principles and quantitative laws governing the adenylate energy system, which is a fundamental element for unveiling the dynamics of cellular life.

Editor: Marie-Joelle Virolle, University Paris South, France

Funding: This study was funded by the University of Basque Country (UPV/EHU): University-Society grant US11/13 and Ministerio de Economía y Competitividad (Spain), Project No. BFU2013-44095-P. The funders had no role in study design, data collection and analysis, decision to publish, or preparation of the manuscript.

Competing Interests: The authors have declared that no competing interests exist.

* Email: mtpmadei@ehu.es

Introduction

Living cells are essentially highly evolved dynamic reactive structures, in which the most complex known molecules are synthesized and destroyed by means of a sophisticated metabolic network characterized by hundreds to thousands of biochemical reactions, densely integrated, shaping one of the most complex dynamic systems in nature [1,2].

Energy is the fundamental element for the viability of the cellular metabolic network. All cells demand a large amount of energy to keep the entropy low in order to ensure their self-organized enzymatic functions and to maintain their complex biomolecular structures. For instance, during growth conditions it has been observed that in microbial cells the protein synthesis accounts for 75% of the total energy, and the cost of DNA replication accounts for 2% of the energy [3,4].

Although different nucleosides can bind to three phosphates which may serve to store biochemical energy i.e., GTP, (d)CTP, (d)TTP and (d)UTP [5], there exists a consensus that adenosine 5'-triphosphate (ATP) is the principal molecule for storing and transferring energy in cells. All organisms, from the simplest bacteria to human cells, use ATP (Mg-ATP) as their major energy source for metabolic reactions [6–8], and the levels of ATP, ADP and AMP reflect roughly the energetic status of the cell [7]. ATP is originated from different classes of metabolic reactions, mainly substrate-level phosphorylation, cellular respiration, photophosphorylation and fermentation, and it is used by enzymes and structural proteins in all main cytological processes, i.e., motility, cell division, biosynthetic reactions, cell cycle, allosteric regulations, and fast synaptic modulation [7–9].

In the living cell, practically all bioenergetic processes are coupled with each other via adenosine nucleotides, which are consumed or regenerated by the different enzymatic reactions. In fact, the most important regulatory elements involved in the coupling of catabolic and anabolic reactions are ATP, ADP and AMP [7]. The adenosine nucleotides are not only tied to the metabolic pathways involved in the cell's energetic system but also act as allosteric control of numerous regulatory enzymes allowing that changes in ATP, ADP and AMP levels can practically regulate the functional activity of the overall multienzymatic network of cell [10–13].

A characteristic of the temporal evolution of ATP, ADP and AMP concentrations is their complexity [14]. Extensive experimental studies have shown that metabolism exhibits extremely large and complex fluctuations in the concentrations of individual adenosine nucleotides, which are anything but stationary [14–16]. In fact, under normal conditions inside the cell, the time evolution of the adenosine-5′-triphosphate is subjected to marked variations presenting transitions between quasi-steady states and oscillatory behaviors [15,16]. For instance, complex ATP rhythms were reported to occur in: myxomycetes [17,18], neurons [19], yeast [16], embrionary cells [20,21], myocytes [22], islet β-cells [23,24], keratinocytes [25], hepatocytes [26], red blood cells [27] and L and MEL cells [28]. Many of these oscillations have clearly non-periodic behaviors [19,26], and ADP and AMP also exhibit complex oscillatory patterns [29–31]. In addition to ATP ultradian oscillations, specific circadian rhythms have also been reported, which occur with a period close to 24 hours (the exogenous period of the Earth's rotation) [27,32,33].

Oscillatory behavior is a very common phenomenon in the temporal dynamics of the concentration for practically all cell metabolites. Indeed, during the last four decades, the studies of biochemical dynamical behaviors, both in prokaryotic and eukaryotic organisms, have shown that in cellular conditions spontaneous molecular oscillations emerge in most of the fundamental metabolic processes. For instance, specific biochemical oscillations were reported to occur in: free fatty acids [34], NAD(P)H concentration [35], biosynthesis of phospholipids [36], cyclic AMP concentration [37], actin polymerization [38], ERK/MAPK metabolism [39], mRNA levels [31], intracellular free amino acid pools [40], cytokinins [41], cyclins [42], transcription of cyclins [43], gene expression [44–47], microtubule polymerization [48], membrane receptor activities [49], membrane potential [50,51], intracellular pH [52], respiratory metabolism [53], glycolysis [54], intracellular calcium concentration [55], metabolism of carbohydrates [56], beta-oxidation of fatty acids [57], metabolism of mRNA [58], tRNA [59], proteolysis [60], urea cycle [61], Krebs cycle [62], mitochondrial metabolic processes [63], nuclear translocation of the transcription factor [64], amino acid transports [65], peroxidase-oxidase reactions [66], protein kinase activities [67] and photosynthetic reactions [68]. In addition, experimental observations in *Saccharomyces cerevisiae* during continuous culture have shown that the majority of metabolome also shows oscillatory dynamics [69].

Persistent properties in oscillatory behaviours have also been observed in other studies, e.g., DNA sequences [70–71], NADPH series [72], K$^+$ channel activity [73], biochemical processes [74,75], physiological time series [76,77], and neural electrical activity [78,79].

Likewise, it has been observed that genomic activity shows oscillatory behavior. For instance, under nutrient-limited conditions yeast cells have at least 60% of all gene expressions oscillating with an approximate period of 300 min [80]. Other experimental observations have shown that practically the entire transcriptome exhibits low-amplitude oscillatory behavior [81] and this phenomenon has been described as a genomewide oscillation [47,81–83].

At a global metabolic level, experimental studies have shown that the cellular metabolic system resembles a complex multioscillator system [69,81,83], what allows for interpretation that the cell is a complex metabolic network in which multiple autonomous oscillatory and quasi-stationary activity patterns simultaneously emerge [84–89].

Cells are open dynamic systems [90,91], and when they are exposed to unbalanced conditions, such as metabolic stress, physiological processes produce drastic variations both in the concentration of the adenosine nucleotides [15,16,92,93] and in their molecular turnovers [94]. Tissues such as skeletal and cardiac muscles must sustain very large-scale changes in ATP turnover rate during equally large changes in work. In many skeletal muscles, these changes can exceed 100-fold [95].

The ratio of ATP, ADP and AMP is functionally more important than the absolute concentration of ATP. Different ratios have been used as a way to test the metabolic pathways which produce and consume ATP. In 1967, Atkinson proposed a simple index to measure the energy status of the cell, defined as AEC = ([ATP] +0.5[ADP])/([ATP] + [ADP] + [AMP]) [96].

The AEC is a scalar index ranging between 0 and 1. When all adenine nucleotide pool is in form of AMP the energy charge (AEC) is zero, and the system is completely discharged (zero concentrations of ATP and ADP). With only ADP, the energy charge is 0.5. If all adenine nucleotide pool is in form of ATP the AEC is 1.

The first experimental testing of this equation showed that (despite of extremely large fluctuations in the adenosine nucleotide concentrations), many organisms under optimal growth conditions maintained their AEC within narrow physiological values, between AEC = 0.7 and AEC = 0.95, stabilizing in many cases at a value close to 0.9. Atkinson and coauthors concluded that for these values of AEC, the major ATP-producing reactions are in balance with the major ATP-consuming reactions; for very unfavorable conditions the AEC drops off provoking cells to die [97–101].

During the last four decades, extensive biochemical studies have shown that the narrow margin of the AEC values is preserved in a wide variety of organisms, both eukaryotes and prokaryotes. For instance, AEC values between 0.7 and 0.95 have been reported to occur in cyanobacteria [102,103], mollicutes (mycoplasmas) [104], different bacteria both gram positive and gram negative as *Dinoroseobacter shibae* [105], *Streptococcus lactis* [106], *Bacillus licheniformis* [107], *Thermoactinomyces vulgaris* [108], *Escherichia coli* [109], *Myxococcus xanthus* [110] and *Myxococcus Coralloides* [111], different eukaryotic cells as zooplankton [112], algae [113], yeast [114], neurons [115,116], erythrocytes [117], astrocytes [118], platelets [119], spermatozoa [120], embryonic kidney cells (HEK) [121], skeletal muscle [122], liver tissue [123], fungi [124], different microorganisms of mangrove soils and water (fungi, bacteria and algae) [125] and plants [126–128].

Studies of different species of plants, over long periods of time, have demonstrated a close relationship between AEC and cellular growth, e.g. leaf tissue collected bimonthly from *Spartina patens*, *S. cynosuroides*, *S. alterniflora* and *Distichlis spicata* showed that the adenylate energy charge peaked in spring and summer at 0.78–0.85 and then declined in late summer and early fall [129]. In the case of organisms better adapted to cold, such as winter wheat cells (*Triticum aestivum*) that are cultivated from September to December in the Northern Hemisphere, the ATP levels were shown to decrease gradually when the cells were exposed to various low temperature stresses (ice encasement at −1°C); however, even after 5 weeks of icing when cell viability was severely reduced, AEC values remained high, about 0.8 [130].

There is a long history of quantitative modelling of ATP production and turnover, dating back to Sel'kov's model on glycolytic energy production from 1968 [131], later developed by Goldbeter [132], as well as by Heinrich and Rapoport [133]. In this context, Sel'kov also published a kinetic model of cell energy metabolism with autocatalytic reaction sequences for glycolysis and glycogenolysis in which oscillations of the adenylate energy charge were observed [134].

However, the first adenylate energy system was developed by Reich and Sel'kov in 1974 [135]. This system was modeled with first-order kinetics by using ordinary differential equations.

Here, in order to further understanding of the elements that determine the cellular energy status of cells we present a computational model conformed by some key essential parts of the adenylate energy system. Specifically, the model incorporates (I) the main synthesis process of ATP for cell from ADP (ATP synthase), (II) the catalyzed phosphotransfer reaction for interconversion of adenine nucleotides (ATP, ADP and AMP) (adenylate kinase), (III) the enzymatic hydrolysis of ATP yielding ADP (kinase and ATPase reactions) and (IV) the enzymatic hydrolysis of ATP providing AMP (enzymatic processes of synthetases). The metabolic model has been analyzed by using a system of delay differential equations in which the enzymatic rate equations and all the physiological kinetic parameters have been explicitly considered and experimentally tested *in vitro* by other groups. We have used a system of delay-differential equations fundamentally to model the asynchronous metabolite supplies to the enzymes.

The numerical analysis shows that the AEC can perform transitions between oscillations and steady state patterns in a stabilized way, similar to what happens in the prevailing conditions inside the cell. The max and min values of the oscillations range within a physiological window validated by experimental data.

We finally suggest that rather than a permanent physiological stable state (*homeostasis*), the living systems seem to be characterized by changing energy dynamics (*homeorhesis*).

Methods

Cells require a permanent generation of energy flow to keep the functionality of its complex metabolic structure which integrates a large ensemble of enzymatic processes, interconnected by a network of substrate fluxes and regulatory signals [4].

To understand some elements that determine the energy status of cells we have studied the dynamics of the main biochemical reactions interconverting ATP, ADP and AMP. Specifically, we have developed a model for the basic structure of the adenylate energy system which represents the fundamental biochemical reactions interconverting ATP, ADP and AMP coupled to the main fluxes of adenine nucleotides involved in catabolic and anabolic processes (Figure 1).

The essential metabolic processes incorporated into the adenylate energy model are the following:

I. First, we have assumed the oxidative phosphorylation as the main synthesis source of ATP in the cell.

As is well known, the enzymatic oxidation of nutrients generates a flow of electrons to O_2 through protein complexes located in the mitochondrial inner membrane in eukaryotes, and in the cell intermembrane space in prokaryotes, that leads to the pumping of protons out of the matrix. The resulting uneven distribution of protons generates a pH gradient that creates a proton-motive force. This proton gradient is converted into phosphoryl transfer potential by ATP synthase which uses the energy stored in the electrochemical gradient to drive the synthesis of ATP from ADP and phosphate (P_i) [7]. Thus, oxidative phosphorylation is the culmination of a series of complex enzymatic transformations whose final phase is carried out by ATP synthase.

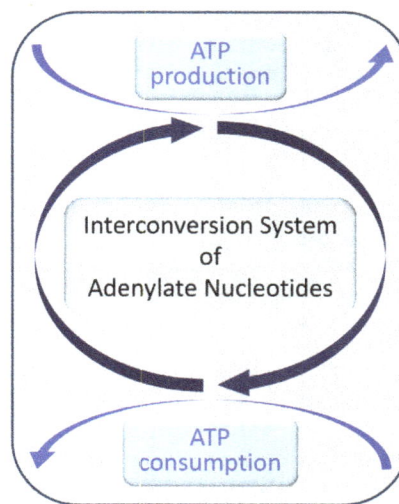

Figure 1. Elemental biochemical processes involved in the energy status of cells. The synthesis sources of ATP are coupled to energy-consumption processes through a network of enzymatic reactions which, interconverting ATP, ADP and AMP, shapes a permanent cycle of synthesis-degradation for the adenine nucleotides. This dynamic functional structure defines the elemental processes of the adenylate energy network, a thermodynamically open system able to accept, store, and supply energy to cells.

Experimental studies in non-pathologic cells have shown that ATP synthase generates the vast majority of cellular energy in the form of ATP (more than 90% in human cells) [136]; consequently, it is one of the central enzymes in energy metabolism for most cellular organisms, both prokaryotes and eukaryotes. This sophisticated rotatory macromolecular machine is embedded in the inner membrane of the mitochondria, the thylakoid membrane of chloroplasts, and the plasma membrane of bacteria [137].

The overall reaction sequence for the ATP synthase is:

$$ADP + P_i + nH^+_{\text{inter-membrane}} \rightarrow ATP + H_2O + nH^+_{\text{matrix}} \quad (1)$$

where n indicates the H^+/ATP ratio with values between 2 and 4 which have been reported as a function of the organelle under study [138].

II. Besides the oxidative phosphorylation, we have also considered that in optimal growth conditions a small part of ATP is generated through substrate-level phosphorylation [7].

III. Another essential metabolic process for cellular energy is the catalyzed phosphotransfer reaction performed by the enzyme adenylate kinase, which is required for interconversion of adenine nucleotides.

Almost since its discovery, about 60 years ago, adenylate kinase (phosphotransferase with a phosphate group as acceptor) has been considered to be a key enzyme in energy metabolism for all organisms [139–140]. This enzyme catalyzes the following reversible reaction for the interconversion of ATP, ADP and AMP:

$$2Mg^{2+}\cdot ADP \rightleftharpoons Mg^{2+}\cdot ATP + AMP \quad (2)$$

Adenylate kinase catalyzes the interconversion of the adenine nucleotides and so it is an important factor in the regulation of the adenine nucleotide ratios in different intracellular compartments, i.e. it contributes to regulate the adenylate energy charge in cells. The equilibrium will be shifted to the left or right depending on the relative concentrations of the adenine nucleotides. In contrast, ATP synthase catalyzes the *de novo* synthesis of the vast majority of ATP from ADP and Pi [136].

IV. The next catalytic process that we have considered corresponds to the enzymes implied in the hydrolysis of ATP to form ADP and orthophosphate (P_i). The chemical energy that is stored in the high-energy phosphoanhydridic bonds in ATP is released, ADP being a product of its catalytic activity.

The basic reaction sequence for the enzymatic process is:

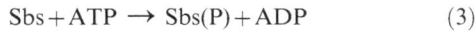

$$Sbs + ATP \rightarrow Sbs(P) + ADP \qquad (3)$$

where Sbs and Sbs(P) are the substrate and the product of the catalytic process, respectively. In this kind of metabolic reaction different groups of enzymes are involved, mainly kinases and ATPases. Particularly, kinases catalyze the transfer of a phosphoryl group from ATP to a different class of specific molecules, which may be also a protein. By adding phosphate groups to substrate proteins, the kinases enzymes shape the activity, localization and overall function of many proteins and pathways, which orchestrate the activity of almost all cellular processes. Up to 30% of all human proteins may be modified by a kinase activity, and they regulate the majority of cellular pathways, especially those involved in signal transduction [141]. These enzymes are fundamental for the functional regulation of the cellular metabolic network and they constitute one of the largest and most diverse gene families. The human genome contains about 500 protein kinase genes and they constitute about 2% of all human genes [141].

V. Finally, we have taken into account the ligase enzymes that catalyze the joining of smaller molecules to make larger ones, coupling the breakdown of a pyrophosphate bond in ATP to provide AMP and pyrophosphate as main products.

The basic reaction sequence for the ligases is:

$$A + B + ATP \rightarrow A\text{-}B + AMP + PP_i \qquad (4)$$

The enzymes belonging to the family of ligases involve different groups as DNA ligases, aminoacyl tRNA synthetases, ubiquitin ligase, etc. They are very important catalytic machines for anabolic processes and for the molecular architecture of the cell. Most ligases are mainly implied in the protein synthesis consuming a large part of the cellular ATP. Thus, for microbial cells, the protein synthesis accounts for 75% of the total energy during growth conditions [3,4].

Protein synthesis uses energy mainly from ATP at several stages such as the attachment of amino acids to transfer RNA, and the movement of mRNA through ribosomes, resulting in the attachment of new amino acids to the chain. In these processes, aminoacyl tRNA synthetases constitute an essential enzyme super-family, providing fidelity of the translation process of mRNA to proteins in living cells and catalyzing the esterification of specific amino acids and their corresponding tRNAs. They are common to all classes of organisms and are of utmost importance for all cells [142]. In the present model we have considered aminoacyl tRNA synthetase as a representative enzyme of the ligases group.

Figure 2 schematically shows the enzymatic processes of the ATP consuming-generating system. First, a permanent input of nutrients is considered to be the primary energy source. In the final phase of oxidative phosphorylation, the ATP synthase uses the energy stored in the proton gradient, generated by the enzymatic oxidation of nutrients, to drive the synthesis of ATP from ADP and phosphate (P_i). The flow of protons thus behaves like a gear that turns the rotary engine of ATP synthase. Likewise, a small part of ATP is also incorporated into the system via substrate-level phosphorylation. The ATP synthesized is fundamentally consumed by two different enzymatic reactions: (i) the ligase processes which provide the system with AMP molecules and (ii) the kinase and ATPase reactions which mainly generate ADP. The interconversion of ATP, ADP and AMP is performed by the enzyme adenylate kinase, which regenerates them according to the dynamic needs of the system.

The ATP consuming-generating system is open and consequently some AMP molecules are *de novo* biosynthesized [143]; whilst a part of AMP does not continue in the reactive system due to its hydrolysis, forming adenine and ribose 5-phosphate [144]. Finally, according to experimental observations, we have considered that a very small part of ATP does not remain in the system, but is drained out from the cell [145–149].

We want to emphasize that the biochemical energy system depicted in Figure 2 represents some key essential parts of the adenylate energy system (see for more details the end of the "Model Section"), which constitute a thermodynamically open system able to accept, store, and supply energy to cells.

This metabolic network of crucial biochemical processes for the cell can be rewritten in a simplified way to gain a better understanding about the dynamic behavior of the model:

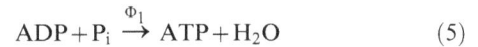

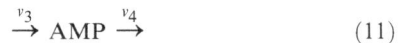

$$ADP + P_i \xrightarrow{\Phi_1} ATP + H_2O \qquad (5)$$

$$ATP + AMP \xrightarrow{\Phi_2} 2ADP \qquad (6)$$

$$2ADP \xrightarrow{\Phi_3} ATP + AMP \qquad (7)$$

$$ATP \xrightarrow{\Phi_4} ADP + P_i \qquad (8)$$

$$ATP \xrightarrow{\Phi_5} AMP + PP_i \qquad (9)$$

$$\xrightarrow{v_1} ATP \xrightarrow{v_2} \qquad (10)$$

$$\xrightarrow{v_3} AMP \xrightarrow{v_4} \qquad (11)$$

where Φ_i ($i = 1$–5) are the rates of the enzymatically-catalyzed reactions (5) to (9), v_1 is the rate of the ATP input into the system by substrate-level phosphorylation, v_2 is the rate of the ATP output from the cell [145–149], being $v_2 = k_2[ATP]$, v_3 is the rate of the biosynthesis *de novo* of AMP and v_4 is the rate of the sink of AMP, being $v_4 = k_4[AMP]$. The reversible adenylate kinase reaction (2) has been described by its corresponding reactions (6) and (7) linked by a control parameter (see below for more details) allowing to move the reactive process to either of the two reactions according

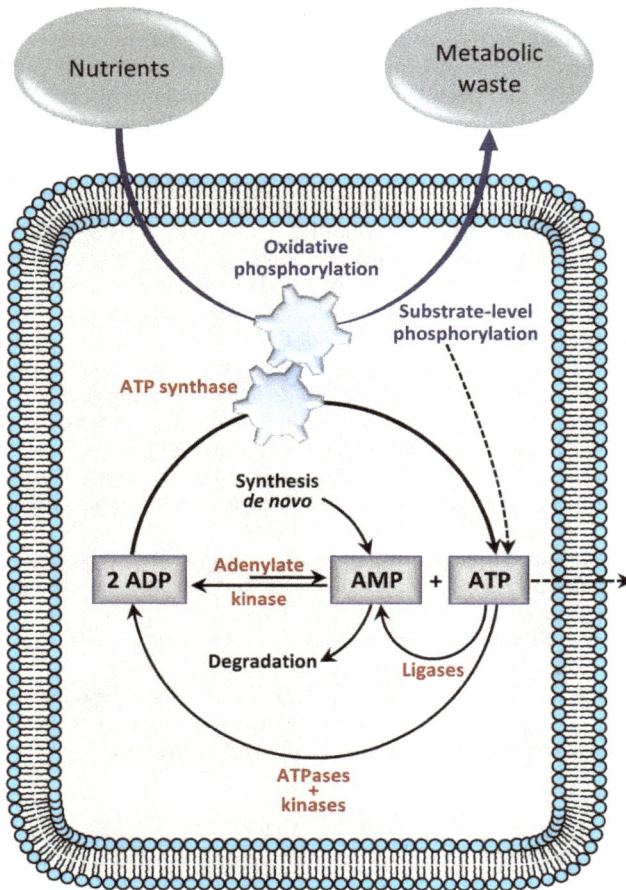

Figure 2. The Adenylate energy system. Oxidative phosphorylation and substrate-level phosphorylation generate ATP which is degraded by kinases (also ATPases) and ligases yielding ADP and AMP, respectively. The three adenine nucleotides are catalytically interconverted by adenylate kinase according to the needs of the metabolic system. AMP is also subjected to processes of synthesis-degradation, some AMP molecules are *de novo* biosynthesized, and a part of AMP is hydrolyzed. According to experimental observations, a very small number of ATP molecules may not remain in the adenylate reactive structure. The system (thermodynamically open) needs a permanent input of nutrients as primary energy source and a consequent output of metabolic waste. The biochemical energy system depicted in the figure represents some key essential parts of the adenylate energy system.

to the physiological needs of the system, i.e. the synthesis or the consumption of ATP or ADP. According to the stoichiometry of this set of chemical equations, there is a net consumption of ATP in the system, which can be regulated by reactions (6), (8), (9) and (10), as well as a production of AMP, which is regulated by steps (7), (9) and (11).

Although the kinetic behavior *in vivo* of most enzymes is unknown, *in vitro* studies can provide both adequate kinetic parameters and enzymatic rate functions. We have used this strategy to implement the dynamical model of the adenylate energy system. Thus, for ATP synthase we have assumed Michaelis–Menten kinetics with competitive inhibition by the product [150]. An iso-random Bi Bi mechanism has been reported for adenylate kinase kinetics [151–152]. We have also considered that a fraction of the adenylate kinases exhibit the balance shifted to the left and simultaneously the rest of the adenylate kinase

macromolecules present a balance shifted to the right, depending their catalytic activities on the system demand. For the kinase family we have selected phosphofructokinase, whose rate equation was developed in the framework of concerted transition theory of Monod and Changeux [153,154], and finally, for the ligase family we have chosen threonyl-tRNA synthetase, which shows Michaelis–Menten kinetics [155].

The time-evolution of the ATP consuming-generating system (Figure 2) can be described by the following three differential equations:

$$\frac{d\alpha}{dt} = v_1 + \lambda\sigma_1\Phi_1 - \Delta'\sigma_2\Phi_2 + \Delta''\sigma_3\Phi_3 - \sigma_4\Phi_4 - \sigma_5\Phi_5 - v_2,$$

$$\frac{d\beta}{dt} = -\lambda\sigma_1\Phi_1 + \Delta'\sigma_2\Phi_2 - \Delta''\sigma_3\Phi_3 + \sigma_4\Phi_4,$$

$$\frac{d\gamma}{dt} = v_3 - \Delta'\sigma_2\Phi_2 + \Delta''\sigma_3\Phi_3 + \sigma_5\Phi_5 - v_4 \qquad (12)$$

where the variables α, β and γ denote the ATP, ADP and AMP concentrations respectively, $\sigma_1,\ldots\sigma_5$ correspond to the maximum rates of the reactions (5) – (9), respectively, the nutrients are injected at a constant rate and λ is a control parameter related to the energy level stored in the proton gradient generated by the enzymatic oxidation of input nutrients. Δ' and Δ'' are also control parameters in the system regulating adenylate kinase activity towards the synthesis or the consumption of ADP, respectively, with $\Delta' = 2-\Delta''$.

The enzymatic rate functions are the following:

$$\Phi_1 = \frac{\beta}{\beta + K_{m,1}\left(1 + \dfrac{\alpha}{K_{I,1}}\right)} \qquad (13)$$

$$\Phi_2 = \frac{\alpha\gamma}{K_2 + K_{m,2}^{ATP}\gamma + K_{m,2}^{AMP}\alpha + \alpha\gamma} \qquad (14)$$

$$\Phi_3 = \frac{\beta^2}{K_3 + 2K_{m,3}^{ADP}\beta + \beta^2} \qquad (15)$$

$$\Phi_4 = \frac{\alpha(1+\alpha)(1+\beta)^2}{L_4 + (1+\alpha)^2(1+\beta)^2} \qquad (16)$$

$$\Phi_5 = \frac{\alpha}{K_{m,5} + \alpha} \qquad (17)$$

where $K_{m,1}, K_{m,2}^{ATP}$, $K_{m,2}^{AMP}$, $K_{m,3}^{ADP}$ and $K_{m,5}$ are the Michaelis constants for each respective enzyme, $K_{I,1}$ is the dissociation constant of the ADP-ATP synthase complex, K_2 and K_3 are kinetic parameters of the adenylate kinase, α and β in Eq. (16) are divided by 1 μM so that this equation is dimensionally homogeneous, and L_4 is the allosteric constant of phosphofructokinase. More details about the kinetic parameters and experimental references are given in Table 1.

Table 1. Values of the kinetic parameters used to simulate some of the dynamics of the adenylate energy system.

Parameter	Value	Reference
σ_1	7.14 µmol s^{-1}	[166]
$K_{m,1}$	30 µmol	[150]
$K_{I,1}$	25 µmol	[150]
σ_2	800 µmol s^{-1}	[167]
K_2	71000 µmol^2	[151]
$K_{m,2}^{ATP}$	25 µmol	[151]
$K_{m,2}^{AMP}$	110 µmol	[151]
σ_3	800 µmol s^{-1}	[168]
K_3	1360 µmol^2	[152]
$K_{m,3}^{ADP}$	29 µmol	[152]
σ_4	100 µmol s^{-1}	[132]
L_4	10^6	[169]
σ_5	0.43 µmol s^{-1}	[155]
$K_{m,5}$	100 µmol	[155]

These equations are simplified expressions, but they are particularly useful in the analysis of models of dynamic behavior [154]. For simplification, we do not consider orthophosphate molecules, nor the H_2O involved in the reaction (5), which has been omitted because the solvent has a standard state of 1M.

To study the system dynamics, the model here described has been analyzed by means of a system of delay differential equations accounting for the delays in the supplies of adenine nucleotides to the specific enzymes involved in the biochemical model.

Generally in the cellular metabolic networks the enzymatic processes are not coupled instantaneously between them. The metabolic internal medium is a complex, crowded environment [156], where the dynamic behavior of intracellular metabolites is controlled by a wide mixture of specific interactions and physical constraints mainly imposed by the viscosity of the cellular plasma, mass transport across membranes and variations in the diffusion times which are dependent on the physiological cellular context [157–160].

For example, there is a time-running from the instant in which ATP molecules are produced in the mitochondria until they come to the place where they are used by the target enzymes. Sometimes the spatial separations may involve long intracellular macroscopic distances. As a result of these intracellular phenomena (transport across membranes, diffusion, long macroscopic distances, interactions with the internal molecular crowded, etc.), the supply of metabolites to the enzymes (substrates and regulatory molecules) occurs in different time scales, and with different delays.

Time scales in biochemical systems mean an asynchronous temporal structure characterized by different magnitudes of metabolite supply delays associated to specific enzymatic processes.

Moreover, experimental studies have shown that metabolism exhibits complex oscillations in the concentrations of individual adenine nucleotides, with periods from seconds to several minutes [15,16], which shape a complex temporal structure for intracellular ATP/ADP/AMP concentrations. The phase shifts in this temporal structure also originate delays in the supplies of

substrates and regulatory molecules to the specific enzymes [161–165].

Consequently, metabolic reactions involving ATP/ADP/AMP may occur at different characteristic time scales, ranging from seconds to minutes, originating a temporal structure for intracellular ATP/ADP/AMP concentrations within the cell.

Dynamic processes with delay cannot be modeled using systems of ordinary differential equations. The different time scales can be considered with delay differential equations, which are not ordinary differential equations. In these systems, some dependent variables can be evaluated in terms of (t- r_i) where r_i are the delays and t the time, and consequently the metabolite supplies to the enzymes (substrates and regulatory molecules) are not instantaneous; other dependent variables may be evaluated in terms of t ($r_i = 0$), if metabolite supplies are considered instantaneous.

According to these regards, we have analyzed our system with three delayed variables $\alpha(t-r_1)$, $\beta(t-r_2)$ and $\gamma(t-r_3)$. r_1 models the delay in the supply of ATP to its specific enzymes; r_2 does the same for ADP and r_3 for AMP. Nevertheless, we have assumed that ATP concentration $(\alpha(t))$ in the equation corresponding to ATP synthase (Eq (18)) is not delayed, as this product formation can be considered instantaneous with respect to the competitive inhibition of the enzyme by the same ATP. Likewise, since the adenylate kinase enzyme is reversible, the ADP formed from ATP and AMP in the reaction (6) is used by the reaction (7) in the same place, and therefore, we have also considered that ADP concentration (β (t)) is not delayed in this process (Eq (20)).

Therefore, the adenylate energy system exhibits several time scales and we have used the system of delay-differential equations to model the asynchronous metabolite supplies to the enzymes. In some processes it can be considered that the substrate or regulatory molecules instantly reach the enzyme and in other processes there are delays for substrate supplies to them.

According to these kinds of dependent variables in the system, the enzymatic rate functions are written as follows:

$$\Phi_1 = \frac{\beta(t-r_2)}{\beta(t-r_2)+K_{m,1}\left(1+\dfrac{\alpha(t)}{K_{I,1}}\right)} \quad (18)$$

$$\Phi_2 = \frac{\alpha(t-r_1)\gamma(t-r_3)}{K_2+K_{m,2}^{ATP}\gamma(t-r_3)+K_{m,2}^{AMP}\alpha(t-r_1)+\alpha(t-r_1)\gamma(t-r_3)} \quad (19)$$

$$\Phi_3 = \frac{\beta(t)^2}{K_3+2K_{m,3}^{ADP}\beta(t)+\beta(t)^2} \quad (20)$$

$$\Phi_4 = \frac{\alpha(t-r_1)(1+\alpha(t-r_1))(1+\beta(t-r_2))^2}{L_4+(1+\alpha(t-r_1))^2(1+\beta(t-r_2))^2} \quad (21)$$

$$\Phi_5 = \frac{\alpha(t-r_1)}{K_{m,5}+\alpha(t-r_1)} \quad (22)$$

Our differential equations system with delay (12) takes the following particular form, up to a permutation of the indexes of the variables:

$$\begin{cases} y'_1(t) = f_1(y_1(t-r_1), y_1(t), ..., y_j(t-r_j), y_j(t), y_{j+1}(t), ..., y_n(t)) \\ \vdots \\ y'_n(t) = f_n(y_1(t-r_1), y_1(t), ..., y_j(t-r_j), y_j(t), y_{j+1}(t), ..., y_n(t)) \end{cases} \quad (23)$$

where the dependent variable is a n-dimensional vector of the form $y = (y_1, \cdots, y_n)$, t being the independent variable. In system (23), the derivatives of $y_1, \cdots, y_n$, evaluated in t, are related to the variables $y_1, \cdots, y_j$, where each y_i with $i \leq j$ appears evaluated in $t - r_i$, being r_i the corresponding delay, and might appear evaluated also in t, and the derivatives are also related to the variables $y_{j+1}, \cdots, y_n$ evaluated in t.

Unlike ODE systems, in delayed differential equations, in order to determine a particular solution, it is necessary to give the initial solution in the interval $[t_0, t_0 + \delta]$ with $\delta = \max\{r_1, ..., r_j\}$. That involves the consideration, in the solution of the system, of the function $f_0[t_0, t_0 + \delta] \rightarrow R^n$ called initial function. It can be observed therefore that infinite degrees of freedom exist in the determination of the particular solutions.

Since in our system simple oscillatory behavior of period 1 emerges from numerical integration, an acceptable approximation to the initial function is a periodic solution.

In the system described by (23), it is possible to take the initial function f_0 equal to any $y(t)$ and, in particular, it can be a periodic function.

With this type of systems, it is possible to take into account dynamic behaviours related to parametric variations linked to the independent variable. The parametric variations r_i affect the independent variable; they represent time delays and can be related to the domains of the initial functions.

Table 1 shows the values of the kinetic parameters involved in the system chosen to run the model. All of these values have been obtained from *in vitro* experiments reported in the scientific literature and they are within the range of the values published in the enzyme database Brenda (http://www.brenda-enzymes.info/). For these values, the preliminary integral solutions of the differential equations system (12) show a simple oscillatory behavior of period 1 and as an approximation we have assumed that the initial functions present simple harmonic oscillations in the following form:

$$\alpha_0(t) = C + D \sin(2\pi/P), \quad (24)$$

$$\beta_0(t) = E + F \sin(2\pi/P), \quad (25)$$

$$\gamma_0(t) = G + H \sin(2\pi/P), \quad (26)$$

with C = 6 µmol, D = 2 µmol, E = 4 µmol, F = 1 µmol, G = 7 µmol, H = 3 µmol and P = 200 s. The other parameter values used were $v_1 = 35 \times 10^{-3}$ µmol s^{-1}, $k_2 = 9 \times 10^{-5}$ s^{-1}, $v_3 = 1.4$ µmol s^{-1}, $k_4 = 0.69$ s^{-1}, $\Delta' = 1.98$, $r_1 = 5$ s, $r_2 = 27$ s and $r_3 = 50$ s.

In this paper, we have studied the dynamic behavior of the system under two parametric scenarios:

- In Scenario I, λ is the control parameter, which is related to the energy level stored in the proton gradient generated by the enzymatic oxidation of input nutrients. This scenario represents the main analysis of the paper, and the values used for the kinetic parameters involved in the model are those set out above.

- In Scenario II, the delay r_2 is the control parameter, modelfing the time constants for the time delays of ADP, with $v_1 = 3 \times 10^{-3}$ µmol s^{-1}, $k_2 = 2 \times 10^{-4}$ s^{-1}, $v_3 = 2.1$ µmol s^{-1}, $\lambda = 1.09$, $r_1 = 3$ s, and all other parameters as indicated in Scenario I.

The extracellular ATP concentration [145–149] is considerably much lower than its intracellular concentration [147], which makes accurate quantification of extracellular levels of ATP an extremely difficult task. Therefore, k_2 ($v_2 = k_2$ [ATP]) must be a sufficiently small value. The values of k_2 here used have been: 9×10^{-5} s^{-1} in Scenario I and 2×10^{-4} s^{-1} in Scenario II. If we now take an intermediate value for the ATP concentration, 10 nmol for example, the following data are obtained: $v_2 = 9 \times 10^{-7}$ µmol s^{-1} under Scenario I, and 2×10^{-6} µmol s^{-1} under Scenario II, which are significantly lower than the values considered for v_1 (the rate of the ATP input into the system by substrate-level phosphorylation): 3.5×10^{-2} µmol s^{-1} under Scenario I, and 3.0×10^{-3} µmol s^{-1} under Scenario II.

In this paper, we have studied the bifurcation analysis for the two control parameters (λ and r_2 here considered (Scenarios I and II, resp.). Further future studies, beyond the scope of the present work, might consider including other control parameters to understand how the stability of the solutions change along parameters space. Furthermore, the presence of "molecular noise" might also be included as a possibility to achieve non-periodic variability in the ATP/ADP/AMP oscillations.

An important feature of metabolism is the wide range of time scales in which cellular processes occur.

Generally enzymatic reactions take place at high speed e. g., carbonic anhydrase has a turnover number (k_{cat}) of 400,000 to 600,000 s^{-1} [170] and the turnover number for RNA polymerase II is less rapid, about 0.16 s^{-1}[171].

However, many cellular processes occur on a time scale of minutes. For instance, studies in glucose-limited cultures by up- and downshifts of the dilution rate in *Escherichia coli* K-12 have shown time delays of minutes in the metabolic mechanisms involved in the dynamics of the adenylate energy charge exhibiting drastic changes within 2 min after the nutrients dilution [109]. Intracellular concentrations of the adenine nucleotides and inorganic phosphate may present sustained oscillations in the concentrations of the adenine nucleotides with periods around a minute which can originate large temporal variations in the supplies of these substrates and regulatory molecules to the specific enzymes [30]. In addition to the temporal oscillations, sustained chemical redox waves (NAD(P)H − NAD(P)+) are a rather general feature of some cells [90] which may exhibit qualitative changes with wavefronts traveling in opposite directions within ≈2 min after the start [172].

It is also known that ATP can evoke fast currents by activation of different purinergic receptors expressed in the plasma membranes of many cells [173]. However, ATP exposure for several minutes can lead to the formation of a high conductance pore permeable for ions and molecules up to 900 Da [174,175]. The activation of some kinases, such as MAPK, occurs with a time scale of minutes [176,177]. Furthermore fructose-2,6-bisphosphate levels are also regulated through cyclic-AMP-based signalling, which occurs on the timescale of minutes [178].

According to these experimental observations, we have analyzed the dynamic behavior of the adenylate system taking into account both instantaneous substrate input conditions and delay times for metabolite supplies, between 1 to 120 seconds, which covers a wide range of cellular physiological processes.

The numerical integration of the system was performed with the package ODE Workbench, developed by Dr. Aguirregabiria

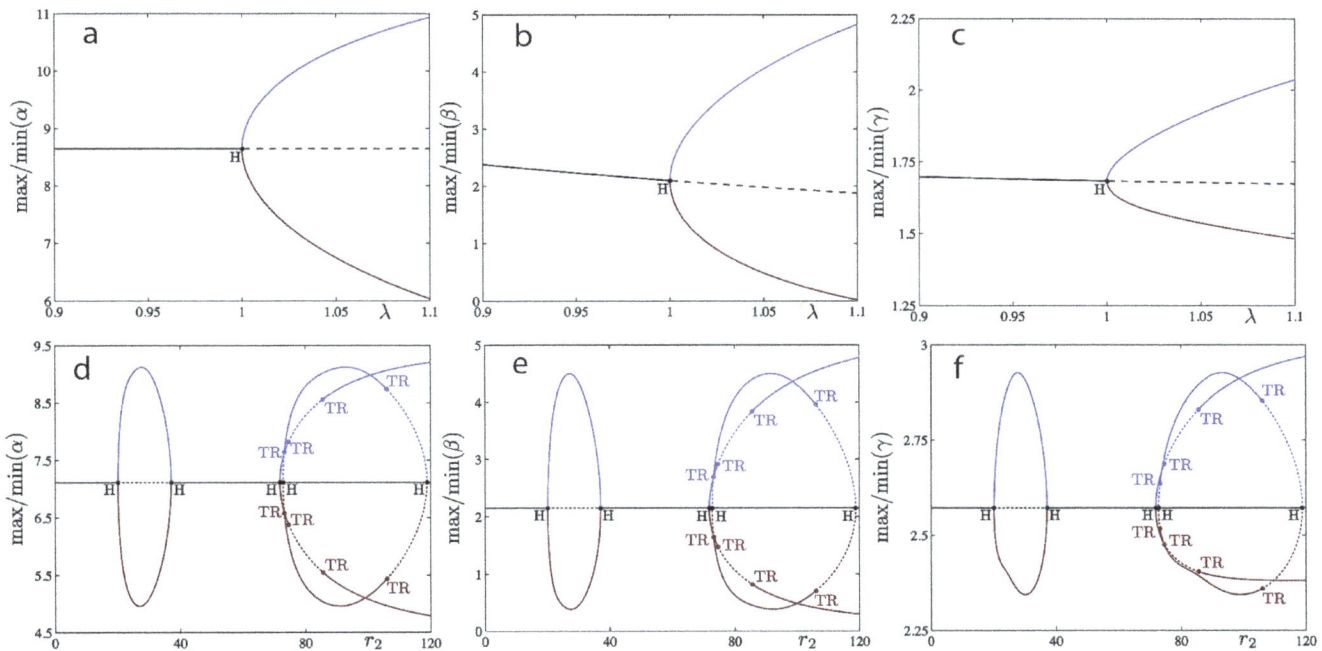

Figure 3. Numerical analysis for the model of the adenylate energy system. a–c: (cf. Scenario I in text) In y-axis we are plotting the max and the min of the different variables α, β and γ. For situations with no oscillations (stable fixed point colored in solid black lines) the max and the min are coincident. For situations with oscillations, the max and the min of the oscillations are plotted separately; in blue we are coloring the max of the oscillation, in red, its minimum value. λ is the control parameter. The numerical integration shows simple solutions. For small λ values ($0.9 \leq \lambda < 1$) the adenine nucleotide concentrations present different stable steady states which lose stability at a Hopf bifurcation at $\lambda \sim 1$. For $\lambda > 1$, the attractor is a stable limit cycle. d–f: (Scenario II) The delay r_2 is the control parameter. The numerical bifurcation analysis reveals that the temporal structure is complex, emerging 5 Hopf bifurcations as well as a secondary bifurcation of Neimark-Sacker type. Two pairs of Hopf bifurcations are connected in the parameter space. A third supercritical Hopf bifurcation occurs at $r_2 \sim 71.94$, rapidly followed by another Hopf bifurcation, subcritical, at $r_2 \sim 72.83$. This marks the beginning of the region where the system is multi-stable. The last Hopf bifurcation, born at $r_2 \sim 72.83$, which is subcritical exhibiting the presence of several Torus bifurcations, occurs on a branch of limit cycles when a pair of complex-conjugated Floquet multipliers, leave the unit circle. Branches of stable (resp. unstable) steady states are represented by solid (resp. dashed) black lines; branches of stable (resp. unstable) limit cycles are represented by the max of the oscillation in blue and the minimum in red and by solid (resp. dashed). Hopf bifurcation points are black dots labeled H; Torus bifurcation points are blue dots labeled TR. The bifurcation parameters λ (Scenario I) and r_2 (Scenario II) are represented on the horizontal axis. The max and min values of each variable are represented on the vertical axis.

which is part of the Physics Academic Software. Internally this package uses a Dormand-Prince method of order 5 to integrate differential equations (http://archives.math.utk.edu/software/msdos/diff.equations/ode_workbench/.html).

The use of differential equations in the study of metabolic processes is widespread nowadays and different biochemical regulation processes have been quantitatively analyzed using time delayed simulations, e.g., in the phosphorylation–dephosphorylation pathways [179], in the endocrine metabolism [180], in the Lactose Operon [181], in the regulation of metabolic pathways [182], in cell signaling pathways [183], and in metabolic networks [184].

Finally, we want to again emphasize that our model only represents some key essential parts of the adenylate energy system. As has previously been indicated, each living cell is essentially a sophisticated metabolic network characterized by hundreds to thousands of biochemical reactions, densely integrated, shaping one of the most complex dynamic systems in nature. The cellular metabolic network functionally integrates all their catalytic processes as a whole. For instance, in a cellular eukaryotic organism the systemic metabolic network includes the enzymatic reactions linked to the plasma membrane, the catabolic and anabolic processes of cytoplasm, the metabolism developed by organelles and subcellular structures, the processes of cell signaling, the adenylate energy system, the metabolism of the nuclear membrane and the nucleoplasm, the enzymatic processes for genetic expression, etc.

A fundamental property of this cellular metabolic network is their modularity. Metabolism is organized in a modular fashion and the emergence of modules is a genuine characteristic of the functional metabolic organization in all cells [185,186].

Energy is the essential element for the viability of the cellular metabolic network, and practically all bioenergetic processes are coupled with each other via adenosine nucleotides, which are consumed or regenerated by the different enzymatic reactions of the network.

The adenosine nucleotides also act as allosteric control of numerous regulatory enzymes allowing that changes in ATP, ADP and AMP levels can practically regulate the functional activity of the overall metabolic network of cell [10–13].

Accordingly, the cellular energetic system is an integral part of the systemic metabolic network and also shapes a super-complex dynamical system which consists of thousands of biochemical reactions.

In addition, the cellular energy system is involved as well in the set of catabolic and anabolic reactions of the systemic metabolism exhibiting specific processes, e.g., the oxidative phosphorylation, the glycolytic metabolism and other catalytic reactions of substrate-level phosphorylation, the regulatory modular sub-networks of adenosine nucleotide signals, the AMPK system

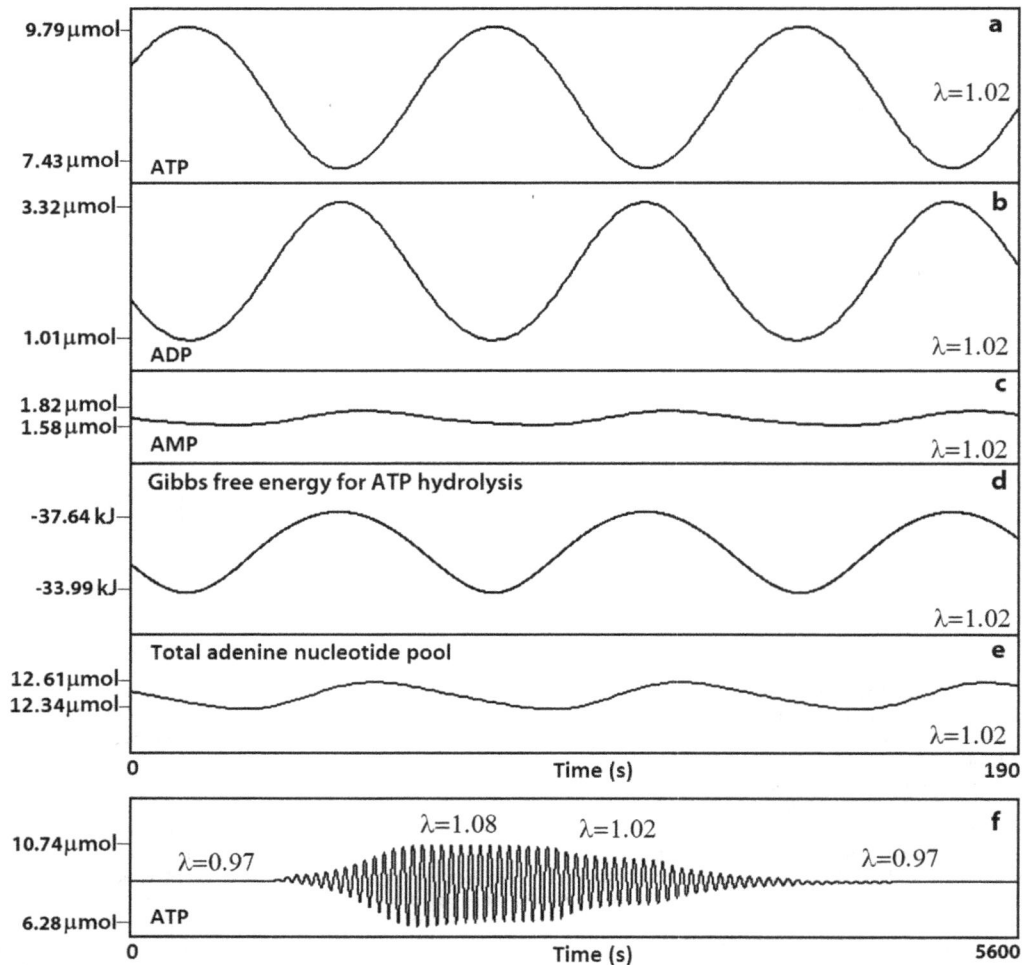

Figure 4. Dynamical solutions of Scenario I. For $\lambda = 1.02$ (normal activity for the ATP synthesis), periodic oscillations emerge. (a) ATP concentrations. (b) ADP concentrations. (c) AMP concentrations. (d) The Gibbs free energy change for ATP hydrolysis to ADP. (e) The total adenine nucleotide (TAN) pool. It can be observed that ATP and ADP oscillate in anti-phase (the ATP maximum concentration corresponds to the ADP minimum concentration). Likewise, it is noted that the total adenine nucleotide pool shows very small amplitude of only 0.27 µmol and a period around 65 s. (f) ATP transitions between different periodic oscillations and a steady state pattern for several values of λ(0.97, 1.08, 1.02, 0.97). Maxima and minima values per oscillation are shown in y-axis.

which acts as a metabolic master switch, the degradation processes of the adenosine nucleotides, the allosteric and covalent modulations of enzymes involved in bioenergetic processes, the role of AMP, AMPK and adenylate kinase in nucleotide-based metabolic signaling, the principles of dissipative self-organization of the bioenergetic processes and the significance of metabolic oscillations in the adenosine nucleotide propagation inside the cell.

Results

To understand the dynamics of the main enzymatic reactions interconverting the adenine nucleotides we have analyzed a biochemical model for the adenylate energy system using the system of delay differential equations (12) to account for the asynchronous conditions inside the cell.

Scenario I

Scenario I represents the fundamental analysis of the paper, being λ the main control parameter, which models the energy level stored in the proton gradient generated by the enzymatic

oxidation of input nutrients, and therefore, represents the modifying factor for the ATP synthesis in the system due to substrate intake.

The numerical integration illustrated in Figure 3a-c shows that the temporal structure of the biochemical model is simpler than Scenario II (see below). At small λ values, for $0.9 \leq \lambda < 1$ the adenine nucleotide concentrations display a family of stable steady states (notice that $\lambda = 0.9$ represents a 10% reduction of the ATP synthesis). These steady states lose stability at a Hopf bifurcation detected numerically for $\lambda \sim 1$ which corresponds to a normal activity of ATP synthase with a maximum rate of 7.14 µmol s^{-1} [166]. For values of λ bigger than 1 the attractor of the system is a stable limit cycle (therefore, the Hopf bifurcation is supercritical). Concretely, the amplitude of adenine nucleotide oscillations augments as λ increases, e. g., for $\lambda = 1.02$, which represents a 2% of increment in ATP synthesis, the adenine nucleotides exhibit new oscillations with amplitude values of 2.36 µmol (ATP), 2.21 µmol (ADP) and 0.24 µmol (AMP). With an 8% of increase in the ATP synthesis ($\lambda = 1.08$) the amplitudes show higher values, namely 4.47 µmol (ATP), 4.41 µmol (ADP) and 0.5 µmol (AMP).

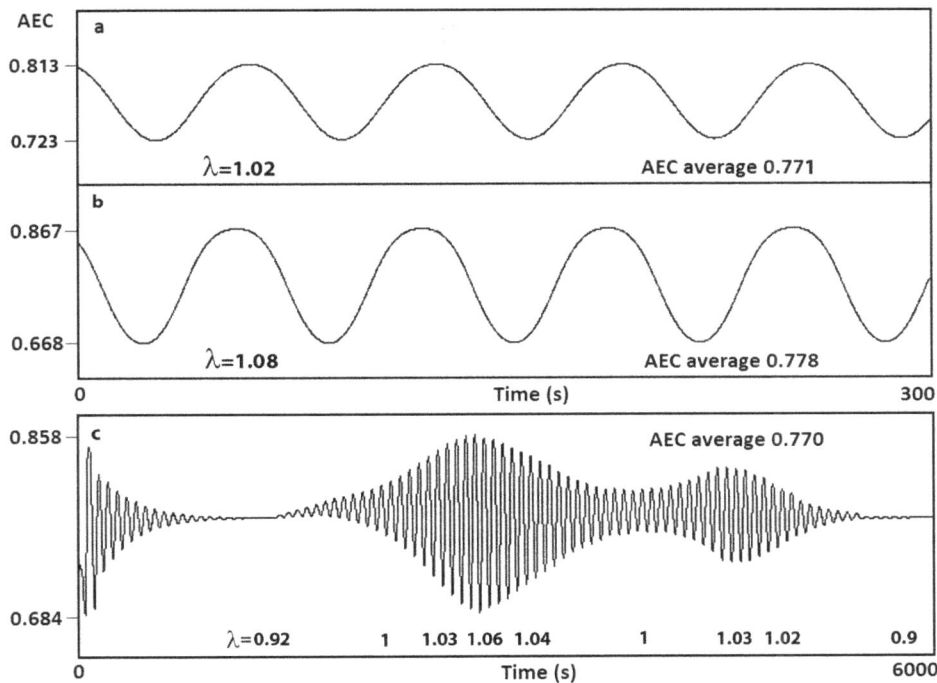

Figure 5. Emergence of oscillations in the AEC (Scenario I). Different oscillatory behavior appears when varying λ, the modifying factor for the ATP synthesis. (a) For $\lambda = 1.02$ (normal activity of ATP synthesis) the AEC periodically oscillates with a very low relative amplitude of 0.090. (b) At higher values of ATP synthesis ($\lambda = 1.08$) large oscillations emerge with a amplitude of 0.199. (c) AEC transitions between different periodic oscillations and steady state patterns for several values of λ (0.92, 1, 1.03, 1.06, 1.04, 1, 1.03, 1.02, 0.9).

Finally, when activity reaches a 10% increase ($\lambda = 1.1$) the three dependent variables of the metabolic system oscillate with higher amplitude concentrations: 4.92 µmol, 4.84 µmol and 0.55 µmol, respectively.

Figure 4 shows three time series belonging to ATP, ADP and AMP (panels a, b and c, respectively), for $\lambda = 1.02$. The largest oscillation values correspond to ATP (max = 9.79 µmol and min = 7.43 µmol) followed by ADP (max = 3.32 µmol and min = 1.01 µmol) and finally, AMP which oscillates with a low relative amplitude (max = 1.82 µmol and min = 1.58 µmol). We have also observed that ATP oscillates in anti-phase with ADP and consequently the maximum concentration of ATP corresponds to the minimum concentration of ADP.

In most metabolic processes, ATP (Mg-ATP) is the main energy source for biochemical reactions and its hydrolysis to ADP or AMP releases a large amount of energy. To this respect, we have estimated the Gibbs free energy change for ATP hydrolysis (to ADP) under an emergent oscillatory condition of the system, applying the known equation $\Delta G'_{reaction} = \Delta G'^0_{reaction} + RT \ln (\beta/\alpha)$. The change of the standard Gibbs free energy for this reaction was previously evaluated by Alberty and co-workers [187] obtaining a value of -32 kJmol^{-1} under standard conditions of 298 K, 1 bar pressure, pH 7, 0.25 M ionic strength and the presence of 1 mM Mg^{2+} ions forming the ATP.Mg^{2+} complex, which has different thermodynamic properties than free ATP and, it is closer to physiological conditions.

Under these conditions, Figure 4d shows the values of Gibbs free energy change of ATP hydrolysis for $\lambda = 1.02$ which corresponds to a normal activity for ATP synthesis. The resulting values for the oscillatory pattern were more negative than the standard value with a maximum and a minimum of -37.64 kJmol^{-1} and -33.99 kJmol^{-1}, meaning that the hydrolysis

of ATP releases a large amount of free energy that can be captured and spontaneously used to drive other energetically unfavorable reactions in metabolism.

The total of adenine nucleotides is another relevant element in the study of cellular metabolic processes. Different experimental observations have shown that changes in the size levels of the adenine nucleotide pool occur under different physiological conditions [188]. We have estimated the total adenine nucleotide (TAN) pool as [ATP] + [ADP] + [AMP], and Figure 4e shows for $\lambda = 1.02$ an emergent oscillatory behavior for TAN with a maximum of 12.61 µmol and a minimum of 12.34 µmol, i.e., a little amplitude of only 0.27 µmol and a period of 65 sec.

Likewise, we have observed that the sum of ATP and ADP concentrations exhibits very small range. So, for $\lambda = 1.02$, the amplitude is 93 nmol and for $\lambda = 1.1$ it is 202 nmol (data not shown in the Figure).

Figure 4f illustrates ATP transitions between different periodic oscillations and a steady state pattern for several values of λ (0.97, 1.08, 1.02, 0.97).

Next, to analyze the dynamics of the energetic status of the system we have calculated the energy charge level. Figure 5 shows different oscillatory patterns for AEC. For $\lambda = 1.02$ the AEC periodically oscillates with a low relative amplitude of 0.09 (max = 0.813 and min = 0.723) (Figure 5a). At higher values of ATP synthesis (an increment of 8%) larger oscillations emerge (max = 0.867 and min = 0.668) (Figure 5b).

Finally, Figure 5c illustrates AEC transitions between different periodic oscillations and steady state patterns for several arbitrary values of λ (0.92, 1, 1.03, 1.06, 1.04, 1, 1.03, 1.02, 0.9) and arbitrary integration times. All the oscillatory patterns for the energy charge maintain the AEC average within narrow physiological values between 0.7 and 0.9.

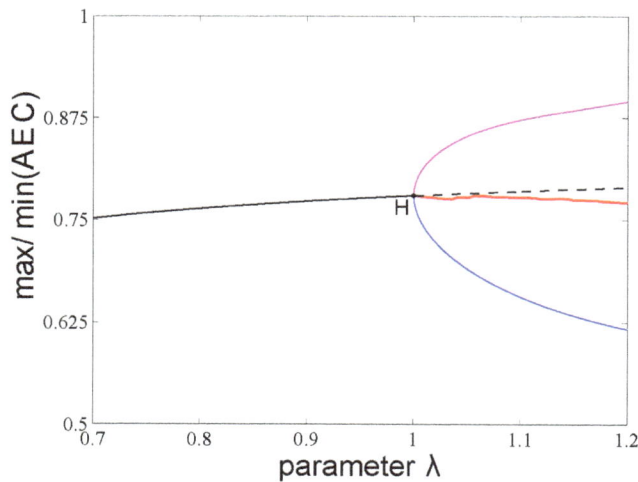

Figure 6. Robustness analysis for the adenylate energy charge (AEC) across different modeling conditions. In y-axis we have plotted the max and the min of the AEC. For situations with no oscillations stable fixed are point colored in solid black lines. In x-axis we have plotted the λ control parameter, which models the energy level stored in the proton gradient generated by the enzymatic oxidation of input nutrients. From left to right, we can see that the system has a fixed point solution which is stable for $\lambda < 1$ (black solid line) and becomes unstable for $\lambda > 1$ (black dashed line), i.e., there is a Hopf bifurcation (H) at $\lambda \sim 1$. For $\lambda > 1$, the limit cycle solution becomes stable, in magenta (blue) we have colored the max (min) of the oscillations. In red, we are coloring the average AEC value of the oscillations. For $\lambda < 1$, the AEC values range from 0.752 and 0.779, and for $\lambda > 1$, the AEC average value between the maximum and minimum per period range from 0.768 to 0.756. At very small λ values, for $\lambda \leq 0.45$ the AEC exhibits values below 0.6 (Figure 7). The AEC does not substantially change during the simulations indicating that it is strongly buffered against the changes of the main control parameter of the system.

Figure 6 shows a robustness analysis of the system in which the values of the adenylate energy charge (AEC) do not substantially change when λ, the main control parameter, is heavily modified (a 50% of its value) indicating that AEC is strongly buffered.

Thus, at small λ values, for $0.7 \leq \lambda \leq 0.99$, the AEC displays a family of stable steady states and the AEC values range from 0.752 to 0.779 (notice that $\lambda = 0.7$ represents a 30% reduction of the ATP synthesis). These steady states lose stability at a Hopf bifurcation for $\lambda \sim 1$ and the AEC exhibits oscillatory behaviors of period 1, being the average between the maximum and minimum of $\overline{AEC} = 0.769$. Notice that $\lambda = 1$ corresponds to an optimal activity of ATP synthase with a maximum rate of 7.14 μmols^{-1} [166].

As expected, the maximum and minimum per period get bigger as λ increases, and for $\lambda = 1.2$ the AEC maximum per oscillation reaches 0.896 and the $\overline{AEC}$ decreases to 0.769 ($\lambda = 1.2$ represents a 20% increase in activity of the optimal ATP synthesis).

This robustness analysis of the system for a perturbation of 50% in the λ values show that in the stable steady states the AEC values range from 0.752 to 0.779 and in the stable periodic behaviors the AEC average between the maximum and minimum per period ranges from 0.768 to 0.756.

At very small λ values, for $\lambda \leq 0.45$, the AEC exhibits values below 0.6, which are gradually descending up to reach very small energy values, when the system finally collapsed (Figure 7) [97–101].

During decades, experimental studies have shown that when yeast cells are harvested, starved and then supplemented they exhibit significant metabolic oscillations.

Following these observations, we have compared our results with a classical study for oscillations of the intracellular adenine nucleotides in a population of intact cells belonging to the yeast *Saccharomyces cerevisiae* [30]. These cells were quenched 5 min after adding 3 mM-KCN and 20 mM-glucose at time intervals of 5 s. Figure 8a shows the dynamics of adenine nucleotide concentrations experimentally obtained, exhibiting AEC rhythms between 0.6 and 0.9 values (in the first and second oscillation) and a period of around 50 s. In addition, Richard and colleagues attempted to fit a sinusoidal curve through the experimental points [30]. Figure 8b shows an AEC oscillatory pattern at high values of ATP synthesis ($\lambda = 1.1$), max = 0.873, min = 0.656 and a period of 65 s.

Scenario II

In this second Scenario we have considered r_2 as the control parameter, modeling time delays for ADP.

The numerical bifurcation analysis reveals that the temporal structure of the system (12) is complex, with several Hopf bifurcations emerging as well as secondary bifurcations of Neimark-Sacker type (torus), along two branches of limit cycles (Figure 3 d–f).

Concretely, using the numerical continuation package DDE-Biftools [189], we find 5 Hopf bifurcations. Two pairs of Hopf bifurcations are connected in parameter space, that is, the branch of limit cycles born at one, ends at the other, and the fifth Hopf bifurcation gives a branch that extends up to the upper limit of the interval considered, that is, $r_2 = 120$ s.

Gradually increasing r_2 from 1 s, we find that the branch of stable steady states that exists at $r_2 = 1$ s destabilizes at a first Hopf bifurcation occurring at $r_2 \sim 20.12$ s. This Hopf bifurcation is supercritical, which means that the emanating family of limit cycles is stable; this family remains stable until it disappears through the second (also supercritical) Hopf bifurcation at $r_2 \sim 37.04$ s, which allows the family of steady states to re-stabilize; it remains stable until a third supercritical Hopf bifurcation occurs at $r_2 \sim 71.94$ s, rapidly followed by another Hopf bifurcation, subcritical, at $r_2 \sim 72.83$ s.

This marks the beginning of the region where the system is multi-stable, with one stable steady state and (at least) one stable limit cycle. The last Hopf bifurcation, terminating the branch of limit cycle born at $r_2 \sim 72.83$ s, is subcritical.

The reason for the complex integral solutions in the Scenario II is the presence of several torus bifurcations detected along both branches of limit cycles in the region of r_2 between 71 s and 110 s. We recall that a Torus bifurcation occurs on a branch of limit cycles when a pair of complex-conjugated Floquet multipliers, leave the unit circle (in the complex plane). This corresponds to the fact that this branch of limit cycles becomes unstable and the stable solution starts winding on an invariant torus, periodic or quasi-periodic. We detect four Torus bifurcations, corresponding to the appearance and disappearance of multi-frequency oscillations, at the following values of r_2: $r_2 \sim 73.22$ s, $r_2 \sim 74.66$ s, $r_2 \sim 85.66$ s, and $r_2 \sim 106.06$ s. Note the following additional details about the Figure 3 d–f we made: branches of stable (resp. unstable) steady states are represented by solid (resp. dashed) black lines; branches of stable (resp. unstable) limit cycles are represented by the max of the oscillation in blue and the minimum in red and by solid (resp. dashed). Hopf bifurcation points are represented with black dots labeled H; Torus bifurcation points with blue dots labeled TR. The horizontal axis corresponds to the

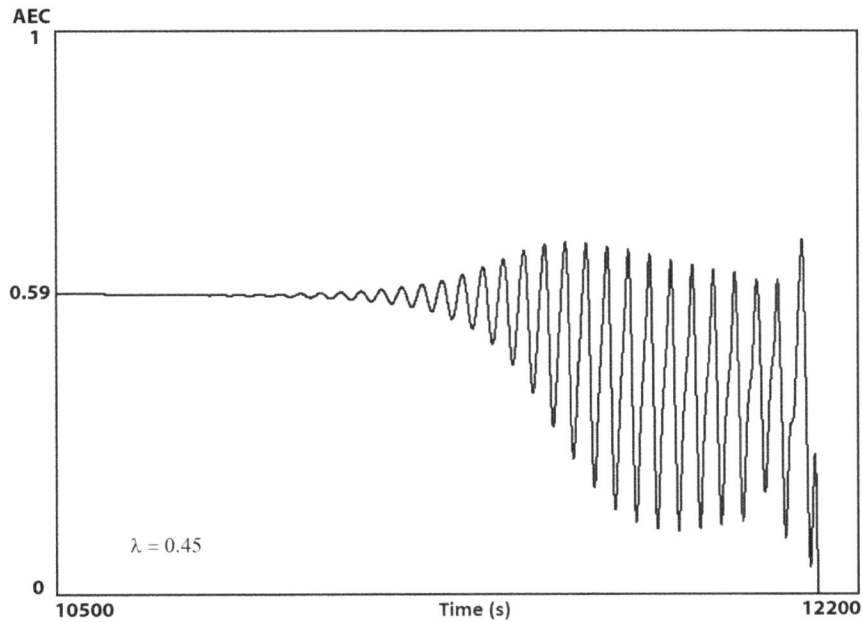

Figure 7. AEC dynamics under low production of ATP. AEC values as a function of time. At very small λ values ($\lambda \approx 0.45$), which represents a strong reduction of the ATP synthesis due to low substrate intake, the dynamic of the adenylate energy system shows a steady state behavior that slowly starts to descend, in a monotone way, up to reach the lowest energy values (AEC ~0.59) at which the steady state loses stability and oscillatory patterns emerge with a decreasing trend. Finally, when the maximum of the energy charge oscillations reaches a very small value (AEC ~0.28) the adenylate system suddenly collapses after 12,000 seconds of temporal evolution.

bifurcation parameters: λ (Scenario I) and r_2 (Scenario II). The vertical axis corresponds to the dependent variable's maxima along various computed branches.

Figure 9 illustrates several examples of oscillatory patterns for the adenylate energy charge under different delay times. For $r_2 = 37$ s the AEC periodically oscillates (Figure 9a). Increasing r_2 up to 72 s (Fig. 9b) and up to 94 s (Fig. 9c) there exist complex AEC oscillatory patterns. Finally, AEC transitions between different oscillatory behavior and steady state patterns are observed for several r_2 values (Figure 9d–e): (d) 50 s, 27 s, 30 s, 32 s, 33 s, 72 s, 52 s, (e) 50 s, 27 s, 30 s, 32 s, 34 s, 36 s, 33 s, 36 s, 38 s, 40 s. These r_2 values and the respective integration times have been arbitrarily taken.

Discussion

Energy is the fundamental element to maintain the turnover of the bio-molecular structures and the functional metabolic viability of all unicellular organisms.

The concentration levels of ATP, ADP and AMP reflect roughly the energetic status of cells, and a determined ratio between them was proposed by Atkinson as the adenylate energy charge (AEC) [96]. Under growth conditions, organisms seem to maintain their AEC within narrow physiological values, despite of extremely large fluctuations in the adenine nucleotide concentrations [96–101]. Intensive experimental studies have shown that the AEC ratio is preserved in a wide variety of organisms, both eukaryotes and prokaryotes (for details see Introduction section).

In order to understand some elements that determine the cellular energy status of cells we have analyzed a biochemical model conformed by some key essential parts of the adenylate energy system using a system of delay differential equations (12) in which the enzymatic rate equations of the main processes and all the corresponding physiological kinetic parameters have been explicitly considered and tested experimentally *in vitro* by other groups. We have used delay-differential equations to model the asynchronous metabolite supplies to the enzymes (substrates and regulatory molecules).

From the model results, the main conclusions are the following:

I. The adenylate energy system exhibits complex dynamics, with steady states and oscillations including multi-stability and multi-frequency oscillations. The integral solutions are stable, and therefore the adenine nucleotide concentrations (dependent variables of the system) can perform transitions between different kinds of oscillatory behavior and steady state patterns in a stabilized way, which is similar to that in the prevailing conditions inside the cell [15,16].

II. The model is in agreement with previous experimental observations [15,16,30], showing oscillatory solutions for adenine nucleotides under different ATP synthesis conditions, at standard enzymatic concentrations, and for different ADP delay times.

III. In all the numerical results, the order of concentration ratios between the adenine nucleotides is maintained in a way that the highest concentration values correspond to ATP, followed by ADP and AMP which displays the lowest values, in agreement with the experimental data obtained by other authors [30,109].

IV. During the oscillatory patterns, ATP and ADP exhibit anti-phase oscillations (the maxima of ATP correspond with the minima of ADP) also experimentally observed in [30].

V. As a consequence of the rhythmic metabolic behavior, the total adenine nucleotide pool exhibits oscillatory patterns (see experimental examples of this phenomenon in [188,190], as well as the Gibbs free energy change for ATP hydrolysis (see [30]). In agreement with these results, we have found that the oscillation for the Gibbs free energy has a maximum and minimum values per period of -37.64 kJmol^{-1} and -33.99 kJmol^{-1}, the same order

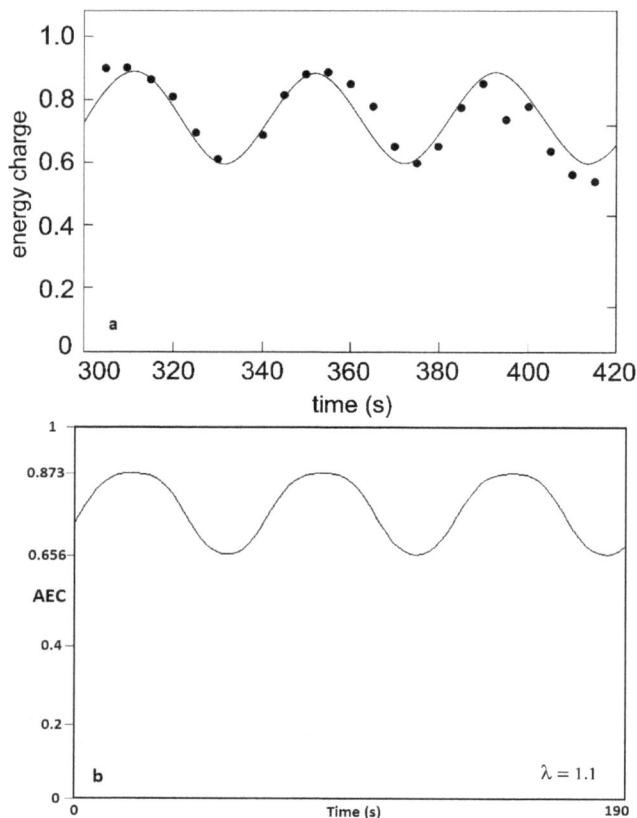

Figure 8. Experimental vs numerical results of AEC oscillations.
Figure 7a illustrates a classical study of the intracellular adenine nucleotides in a population of intact cells belonging to the yeast *Saccharomyces cerevisiae* [30] which exhibits AEC rhythms, with max = 0.9, min = 0.6 and a period around 50 s. The authors fitted the experimental points to a sinusoidal curve. Figure 7b shows AEC oscillations belonging to our model at high values of ATP synthesis ($\lambda = 1.1$), with max = 0.873, min = 0.656 and a period of 65 s.

of magnitude as in experimental observations (about -50 kJmol^{-1} in rat hepatocytes) [191].

VI. The adenylate energy charge shows transitions between oscillatory behaviors and steady state patterns in a stabilized way. We have compared an integral solution of our model with a classical study of intracellular concentrations for adenine nucleotides in a population of intact cells belonging to the yeast *Saccharomyces cerevisiae* and the model fits well with these data [30].

VII. The adenylate energy charge (AEC) does not substantially change during the simulations, indicating that is strongly buffered against the perturbations, in agreement with experimental data [97–101].

We want to remark that we have observed oscillatory patterns in the AEC, in the sum of ATP plus ADP and in the total adenine nucleotide pool but with very low amplitude, what might make difficult the experimental observation with traditional methods.

In fact, it is not clear yet what methodologies are the most appropriate to monitor the values of adenine nucleotides [192]. Although bioluminescence assays and high-performance liquid chromatography are the ones most commonly used for most of the studies [151,192], these procedures are discontinuous and do not allow to observe real-time variations at short temporal periods. Moreover, adenosine nucleoside levels are critically dependent on

sample manipulation and extraction by traditional methods. It has been demonstrated that even short lapses in sample preparation (2 min) can dramatically affect results [193].

It has been assumed for a long time that the temporal evolution of ATP, ADP and AMP concentrations present permanent steady state solutions and that, consequently, cells maintain the AEC as a constant magnitude (*homeostasis*). But this conservation is hard to be fulfilled for open systems.

Recently, the use of nanobiosensors has shown to be able to perform real-time-resolved measurements of intracellular ATP in intact cells; the ATP concentration is indeed oscillating, either showing a rhythmic behavior or more complex dynamics with variations over time, but importantly, the ATP concentration is never constant [15,16].

As a consequence of our analysis we suggest that the appropriate notion to describe the temporal behavior of ATP, ADP and AMP concentrations is *homeorhesis* i.e. the non-linear dynamics of the adenylate energy system shape in the phase space permanent transitions between different kinds of attractors including steady states (in cellular conditions correspond to quasi-steady states) and oscillating attractors, which represent the sets of the asymptotic solutions followed by the adenine nucleotide variables.

Homeorhesis is substantially different to *homeostasis*, which basically implies the ability of the system to maintain the adenine nucleotide concentrations in a constant state.

The concept of *homeostasis* was first suggested by the physiologist Walter Cannon [194] in 1932, but its roots are found back to the French physiologist Claude Bernard who argued that an alleged constancy of the internal medium for any organism results from regulatory processes in biological systems [195,196]. For a long time, the notable idea by Claude Bernard of constancy in the internal medium has paved the route of how cellular processes behaved. However, this constancy seems to be apparent.

In mid-twentieth century, the term *homeorhesis* was suggested to be a substitute of *homeostasis* by the prominent biologist Conrad Waddington [197,198] to describe those systems which return back to a specific dynamics after being perturbed by the external environment, thus opposite to *homeostasis*, in which the system returns back to a fixed state. Later, that concept of *homeorhesis* was mathematically applied in distinct biological studies [199–204].

Rather than a permanent physiological stable state (*homeostasis*), living systems seem to be characterized by changing energy dynamics (*homeorhesis*).

In our numerical study, the temporal dynamics for the concentrations of ATP, ADP and AMP are determined by the adenylate energy system, and these adenosine nucleotide dynamics present complex transitions across time evolution suggesting the existence of homeorhesis.

In addition, we have observed that the values of the AEC do not substantially change during the simulations indicating that is strongly buffered against the perturbations. Recall that the AEC represents a particular functional relationship between the concentrations of adenine nucleotides.

As indicated in the introduction section, intensive experimental measurements under growth cellular conditions have shown that AEC values between 0.7 and 0.95 are invariantly maintained in practically all classes of cells which seems to represent a common key feature to all cellular organisms.

Hence, there appear to be two essential elements in determining the cellular energy level: first, the adenylate energy system originates complex transitions over time in the adenosine nucleotide concentrations so that there is no homeostasis for energy; second, it emerges a permanent relationship among the

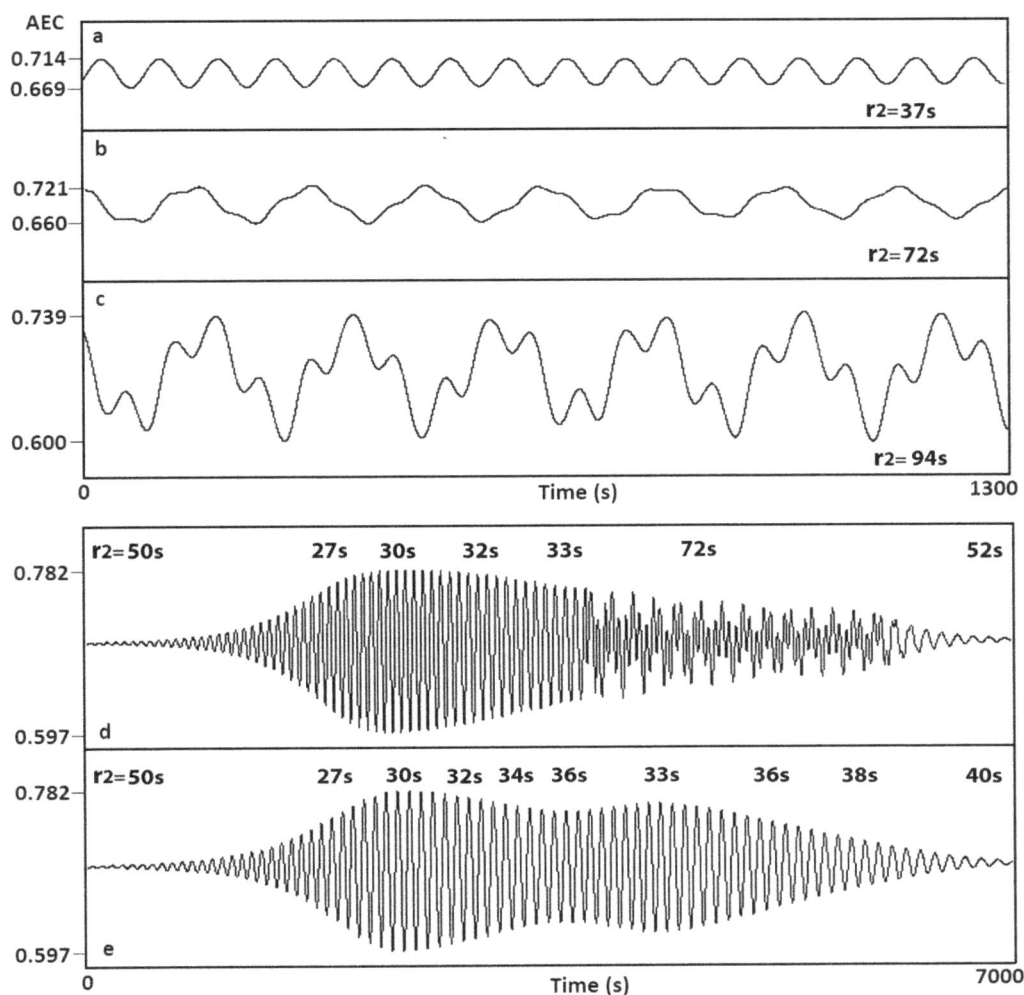

Figure 9. Emergence of oscillations in the AEC (Scenario II). Different oscillatory behavior appears when varying r_2, controlling the ADP time delays. (a) For $r_2 = 37$ s the AEC periodically oscillates with a very low relative amplitude of 0.045. (b–c) Existence of complex AEC oscillatory patterns for: (b) $r_2 = 72$ s and (c) r2 = 94 s. (d–e) AEC transitions between different oscillatory behavior and steady state patterns for several r_2 values. (d) 50 s, 27 s, 30 s, 32 s, 33 s, 72 s, 52 s. (e) 50 s, 27 s, 30 s, 32 s, 34 s, 36 s, 33 s, 36 s, 38 s, 40 s.

dynamics of adenine nucleotide concentrations (AEC values between 0.7 and 0.95), which seems to be strictly fulfilled during all the metabolic transformations that occur during the cell cycle.

These facts make possible to suppose that the cell is an open system where a given magnitude for energy is not conserved but there exists a functional restriction on the possible values that can adopt the adenine nucleotide concentrations.

At least, there seems to be a determinate function relating the adenine nucleotide values which appears to be invariant to all metabolic transformations occurring along the cell cycle. This invariant function, which it would define the real cellular energy state, might possibly have a complex attractor in the phase space since complex dynamic transitions in the adenine nucleotide concentrations have been observed *in vivo* [16], but these hypotheses need deserve further investigation.

Our interpretation to explain the essential elements of the cellular energy charge is that, in addition to the dynamical system which originates the complex transitions in the adenosine nucleotides, there exists an invariant of the energy function which restricts the values that adenylate pool dynamics can take, and the equation of Atkinson is the manifestation of that invariant function.

The main biological significance of the invariant energy function would be that under growth cellular conditions, the adenylate pool must be highly phosphorylated keeping the rate of adenylate energy production similar to the rate of adenylate energy expenditure.

Cell is a complex non-linearly open system where there is not a specific energy value which is conserved, but rather dynamic forms of change for energy. Unicellular organisms need energy to accomplish the fundamental tasks of the cell metabolism: today, in the post-genomic era, the understanding of the elemental principles and quantitative laws that govern the adenylate energy system is crucial to elucidate some of the fundamental dynamics of cellular life.

Acknowledgments

We acknowledge fruitful conversations and wise advice from Edison Lagares, Ricardo Grande, Josué Tonelli and Andoni Arteagoitia.

Author Contributions

Conceived and designed the experiments: IMDF. Performed the experiments: IMDF JMC EV MD SR IM LM. Analyzed the data: IMDF JMC EV MD SR IM LM. Contributed reagents/materials/analysis tools:

IMDF JMC EV MD SR IM LM. Wrote the paper: IMDF JMC EV MD SR LM. Supervised the research: IMDF. Developed the computational model: IMDF.

References

1. Jeong H, Tombor B, Albert R, Oltvai ZN, Barabási AL (2000) The large-scale organization of metabolic networks. Nature 407: 651–654.
2. Sear RP (2005) The cytoplasm of living cells: a functional mixture of thousands of components. J Phys Condens Matter 17: S3587–S3595.
3. Lane N, Martin W (2010) The energetics of genome complexity. Nature 467: 928–934.
4. Harold FM (1986) The vital force: A study of bioenergetics. New York: WH Freeman.
5. Xu Z, Spring DR, Yoon J (2011) Fluorescent sensing and discrimination of ATP and ADP based on a unique sandwich assembly of pyrene-adenine-pyrene. Chem Asian J 6: 2114–2122.
6. Knowles JR (1980) Enzyme-catalyzed phosphoryl transfer reactions. Annu Rev Biochem 49: 877–919.
7. Nelson DL, Cox MM (2008) Lehninger Principles of biochemistry. New York: WH Freeman.
8. Hardie DG (2011) Signal transduction: How cells sense energy. Nature 472: 176–177.
9. Khakh BS (2001) Molecular physiology of P2X receptors and ATP signalling at synapses. Nat Rev Neurosci 2: 165–174.
10. Cohen PF, Colman RF (1972) Diphosphopyridine nucleotide dependent isocitrate dehydrogenase from pig heart. Characterization of the active substrate and modes of regulation. Biochemistry 11: 1501–1508.
11. Ercan N, Gannon MC, Nuttall FQ (1996) Allosteric regulation of liver phosphorylase a: revisited under approximated physiological conditions. Arch Biochem Biophys 328: 255–264.
12. Ercan-Fang N, Gannon MC, Rath VL, Treadway JL, Taylor MR, et al. (2002) Integrated effects of multiple modulators on human liver glycogen phosphorylase. Am J Physiol Endocrinol Metab 283: E29–E37.
13. Nelson SW, Honzatko RB, Fromm HJ (2002) Hybrid tetramers of porcine liver fructose-1,6-bisphosphatase reveal multiple pathways of allosteric inhibition. J Biol Chem 277: 15539–15545.
14. Ataullakhanov FI, Vitvitsky VM (2002) What determines the intracellular ATP concentration. Biosci Rep 22: 501–511.
15. Ozalp VC, Pedersen TR, Nielsen LJ, Olsen LF (2010) Time-resolved measurements of intracellular ATP in the yeast Saccharomyces cerevisiae using a new type of nanobiosensor. J Biol Chem 285: 37579–37588.
16. Ytting CK, Fuglsang AT, Hiltunen JK, Kastaniotis AJ, Özalp VC, et al. (2012) Measurements of intracellular ATP provide new insight into the regulation of glycolysis in the yeast Saccharomyces cerevisiae. Integr Biol (Camb) 4: 99–107.
17. Yoshimoto Y, Sakai T, Kamiya N (1981) ATP oscillation in Physarum plasmodium. Protoplasma 109: 159–168.
18. Akitaya T, Ohsaka S, Ueda T, Kobatake Y (1985) Oscillations in intracellular ATP, cAMP and cGMP concentration in relation to rhythmical sporulation under continuous light in the myxomycete Physarum polycephalum. J Gen Microbiol 131: 195–200.
19. Ainscow EK, Mirshamsi S, Tang T, Ashford ML, Rutter GA (2002) Dynamic imaging of free cytosolic ATP concentration during fuel sensing by rat hypothalamic neurones: evidence for ATP-independent control of ATP-sensitive K(+) channels. J Physiol 544: 429–445.
20. Kwon HJ (2013) ATP oscillations mediate inductive action of FGF and Shh signalling on prechondrogenic condensation. Cell Biochem Funct 31: 75–81.
21. Kwon HJ, Ohmiya Y, Honma KI, Honma S, Nagai T, et al. (2012) Synchronized ATP oscillations have a critical role in prechondrogenic condensation during chondrogenesis. Cell Death Dis 3: e278.
22. Yang JH, Yang L, Qu Z, Weiss JN (2008) Glycolytic oscillations in isolated rabbit ventricular myocytes. J Biol Chem 283: 36321–36327.
23. Ainscow EK, Rutter GA (2002) Glucose-stimulated oscillations in free cytosolic ATP concentration imaged in single islet beta-cells: evidence for a Ca2+-dependent mechanism. Diabetes 51: S162–S170.
24. Kennedy RT, Kauri LM, Dahlgren GM, Jung SK (2002) Metabolic oscillations in beta-cells. Diabetes 51: S152–S161.
25. Dong K, Pelle E, Yarosh DB, Pernodet N (2012) Sirtuin 4 identification in normal human epidermal keratinocytes and its relation to sirtuin 3 and energy metabolism under normal conditions and UVB-induced stress. Exp Dermatol 21: 231–233.
26. MacDonald MJ, Fahien LA, Buss JD, Hasan NM, Fallon MJ, et al. (2003) Citrate oscillates in liver and pancreatic beta cell mitochondria and in INS-1 insulinoma cells. J Biol Chem 278: 51894–51900.
27. O'Neill JS, Reddy AB (2011) Circadian clocks in human red blood cells. Nature 469: 498–503.
28. Gilbert DA, Hammond KD (2008) Phosphorylation dynamics in mammalian cells. In: Lloyd D, Rossi E, editors. Ultradian rhythms from molecules to mind. Springer, pp.105–128.
29. Steven WE, LLoyd D (1978) Oscillations of respiration and adenine nucleotides in synchronous cultures of Acanthamoeba castellunii: mitochondrial respiratory control in vivo. Journal of General Microbiology 108: 197–204.
30. Richard P, Teusink B, Hemker MB, Van Dam K, Westerhoff HV (1996) Sustained oscillations in free-energy state and hexose phosphates in yeast. Yeast 12: 731–740.
31. Xu Z, Yaquchi S, Tsurugi K (2004) Gts1p stabilizes oscillations in energy metabolism by activating the transcription of TPS1 encoding trehalose-6-phosphate synthase 1 in the yeast Saccharomyces cerevisiae. Biochem J 383: 171–178.
32. Womac AD, Burkeen JF, Neuendorff N, Earnest DJ, Zoran MJ (2009) Circadian rhythms of extracellular ATP accumulation in suprachiasmatic nucleus cells and cultured astrocytes. Eur J Neurosci 30: 869–876.
33. Burkeen JF, Womac AD, Earnest DJ, Zoran MJ (2011) Mitochondrial calcium signaling mediates rhythmic extracellular ATP accumulation in suprachiasmatic nucleus astrocytes. J Neurosci 31: 8432–8440.
34. Getty-Kaushik L, Richard AM, Corkey BE (2005) Free fatty acid regulation of glucose-dependent intrinsic oscillatory lipolysis in perifused isolated rat adipocytes. Diabetes 54: 629–637.
35. Rosenspire AJ, Kindzelskii AL, Petty HR (2001) Pulsed DC electric fields couple to natural NAD(P)H oscillations in HT-1080 fibrosarcoma cells. J Cell Sci 114: 1515–1520.
36. Marquez S, Crespo P, Carlini V, Garbarino-Pico E, Baler R, et al. (2004) The metabolism of phospholipids oscillates rhythmically in cultures of fibroblasts and is regulated by the clock protein PERIOD 1. FASEB J 18: 519–521.
37. Holz GG, Heart E, Leech CA (2008) Synchronizing Ca2+ and cAMP oscillations in pancreatic beta cells: a role for glucose metabolism and GLP-1 receptors? Am J Physiol Cell Physiol 294: c4–c6.
38. Rengan R, Omann GM (1999) Regulation of oscillations in filamentous actin content in polymorphonuclear leukocytes stimulated with leukotriene B4 and platelet-activating factor. Biochem Biophys Res Commun 262: 479–486.
39. Shankaran H, Ippolito DL, Chrisler WB, Resat H, Bollinger N, et al. (2009) Rapid and sustained nuclear–cytoplasmic ERK oscillations induced by epidermal growth factor. Mol Syst Biol 5: 332.
40. Hans MA, Heinzle E, Wittmann C (2003) Free intracellular amino acid pools during autonomous ocillations in Saccharomyces cerevisiae. Biotechnol Bioeng 82: 143–151.
41. Hartig K, Beck E (2005) Endogenous cytokinin oscillations control cell cycle progression of tobacco BY-2 cells. Plant Biol 7: 33–40.
42. Hungerbuehler AK, Philippsen P, Gladfelter AS (2007) Limited functional redundancy and oscillation of cyclins in multinucleated Ashbya gossypii fungal cells. Eukaryot Cell 6: 473–486.
43. Shaul O, Mironov V, Burssens S, Van Montagu M, Inze D (1996) Two Arabidopsis cyclin promoters mediate distinctive transcriptional oscillation in synchronized tobacco BY-2 cells. Proc Natl Acad Sci USA 93: 4868–4872.
44. Chabot JR, Pedraza JM, Luitel P, van Oudenaarden A (2007) Stochastic gene expression out-of-steady-state in the cyanobacterial circadian clock. Nature 450: 1249–1252.
45. Tian B, Nowak DE, Brasier AR (2005) A TNF-induced gene expression program under oscillatory NF-κB control. BMC Genomics 6: 137.
46. Tonozuka H, Wang J, Mitsui K, Saito T, Hamada J, et al. (2001) Analysis of the upstream regulatory region of the GTS1 gene required for its oscillatory expression. J Biochem 130: 589–595.
47. Klevecz RR, Bolen J, Forrest G, Murray DB (2004) A genomewide oscillation in transcription gates DNA replication and cell cycle. Proc Natl Acad Sci USA 101: 1200–1205.
48. Lange G, Mandelkow EM, Jagla A, Mandelkow E (1988) Tubulin oligomers and microtubule oscillations. Antagonistic role of microtubule stabilizers and destabilizers. Eur J Biochem 178: 61–69.
49. Placantonakis D, Welsh J (2001) Two distinct oscillatory states determined by the NMDA receptor in rat inferior olive. J Physiol 534: 123–140.
50. Mellon D Jr, Wheeler CJ (1999) Coherent oscillations in membrane potential synchronize impulse bursts in central olfactory neurons of the crayfish. J Neurophysiol 81: 1231–1241.
51. García-Muñoz A, Barrio LC, Buño W (1993) Membrane potential oscillations in CA1 hippocampal pyramidal neurons in vitro: intrinsic rhythms and fluctuations entrained by sinusoidal injected current. Exp Brain Res 97: 325–333.
52. Sánchez-Armáss S, Sennoune SR, Maiti D, Ortega F, Martínez-Zaguilán R (2006) Spectral imaging microscopy demonstrates cytoplasmic pH oscillations in glial cells. Am J Physiol Cell Physiol 290: C524–C538.
53. Lloyd D, Eshantha L, Salgado J, Turner MP, Murray DB (2002) Respiratory oscillations in yeast: clock-driven mitochondrial cycles of energization. FEBS Lett 519: 41–44.

54. Danù S, Sùrensen PG, Hynne F (1999) Sustained oscillations in living cells. Nature 402: 320–322.

55. Ishii K, Hirose K, Iino M (2006) Ca2+ shuttling between endoplasmic reticulum and mitochondria underlying Ca2+ oscillations. EMBO Rep 7: 390–396.

56. Jules M, Francois J, Parrou JL (2005) Autonomous oscillations in Saccharomyces cerevisiae during batch cultures on trehalose. FEBS J 272: 1490–1500.

57. Getty L, Panteleon AE, Mittelman SD, Dea MK, Bergman RN (2000) Rapid oscillations in omental lipolysis are independent of changing insulin levels in vivo. J Clin Invest 106: 421–430.

58. Klevecz RR, Murray DB (2001) Genome wide oscillations in expression. Wavelet analysis of time series data from yeast expression arrays uncovers the dynamic architecture of phenotype. Mol Biol Rep 28: 73–82.

59. Brodsky VY, Boikov PY, Nechaeva NV, Yurovitsky YG, Novikova TE, et al. (1992) The rhythm of protein synthesis does not depend on oscillations of ATP level. J Cell Sci 103: 363–370.

60. Kindzelskii AL, Zhou MJ, Haugland RP, Boxer LA, Petty HR (1998) Oscillatory pericellular proteolysis and oxidant deposition during neutrophil locomotion. Biophys J 74: 90–97.

61. Fuentes JM, Pascual MR, Salido G, Soler G, Madrid JA (1994) Oscillations in rat liver cytosolic enzyme activities of the urea cycle. Arch Int Physiol Biochim Biophys 102: 237–241.

62. Wittmann C, Hans M, Van Winden WA, Ras C, Heijnen JJ (2005) Dynamics of intracellular metabolites of glycolysis and TCA cycle during cell-cycle-related oscillation in Saccharomyces cerevisiae. Biotechnol Bioeng 89: 839–847.

63. Aon MA, Roussel MR, Cortassa S, O'Rourke B, Murray DB, et al. (2008) The scale-free dynamics of eukaryotic cells. PLoS ONE 3: e3624.

64. Garmendia-Torres C, Goldbeter A, Jacquet M (2007) Nucleocytoplasmic oscillations of the yeast transcription factor Msn2: evidence for periodic PKA activation. Curr Biol 17: 1044–1049.

65. Baril EF, Potter VR (1968) Systematic oscillations of amino acid transport in liver from rats adapted to controlled feeding schedules. J Nutrition 95: 228–237.

66. Møller AC, Hauser MJ, Olsen LF (1998) Oscillations in peroxidase-catalyzed reactions and their potential function in vivo. Biophys Chem 72: 63–72.

67. Chiam KH, Rajagopal G (2007) Oscillations in intracellular signaling cascades. Phys Rev E 75: 061901.

68. Smrcinová M, Sørensen PG, Krempasky J, Ballo P (1998) Chaotic oscillations in a chloroplast system under constant illumination. Int J Bifurcation Chaos 8: 2467–2470.

69. Murray DB, Beckmann M, Kitano H (2007) Regulation of yeast oscillatory dynamics. Proc Natl Acad Sci USA 104: 2241–2246.

70. Allegrini P, Buiatti M, Grigolini P, West BJ (1988) Fractional Brownian motion as a non stationary process: An alternative paradigm for DNA sequences. Phys Rev E 57: 4558–4562.

71. Haimovich AD, Byrne B, Ramaswamy R, Welsh WJ (2006) Wavelet analysis of DNA walks. J Comput Biol 13: 1289–1298.

72. Ramanujan VK, Biener G, Herman B (2006) Scaling behavior in mitochondrial redox fluctuations. Biophys J 90: L70–L72.

73. Kazachenko VN, Astashev ME, Grinevitch AA (2007) Multifractal analysis of K+ channel activity. Biol Membrany 24: 175–182

74. De la Fuente IM, Martínez L, Benitez N, Veguillas J, Aguirregabiria JM (1998) Persistent behavior in a phase-shift sequence of periodical biochemical oscillations. Bull Math Biol 60: 689–702.

75. De la Fuente IM, Martínez L, Aguirregabiria JM, Veguillas J (1998) R/S analysis in strange attractors. Fractals 6: 95–100.

76. Eke A., Herman P, Kocsis L, Kozak LR (2002) Fractal characterization of complexity in temporal physiological signals. Physiol Meas 23: R1–R38.

77. De la Fuente IM, Martínez L, Aguirregabiria JM, Veguillas J, Iriarte M (1999) Long-range correlations in the phase-shifts of numerical simulations of biochemical oscillations and in experimental cardiac rhythms. J Biol Syst 7: 113–130.

78. Mahasweta D, Gebber GL, Barman SM, Lewis CD (2003) Fractal properties of sympathetic nerve discharge. J Neurophysiol 89: 833–840.

79. De la Fuente IM, Pérez-Samartín AL, Martínez L, García MA, Vera-López A (2006) Long-range correlations in rabbit brain neural activity. Ann Biomed Eng 34: 295–299.

80. Tu BP, Kudlicki A, Rowicka M, McKnight SL (2005) Cell Biology: Logic of the yeast metabolic cycle: Temporal compartmentalization of cellular processes. Science 310: 1152–1158.

81. Lloyd D, Murray DB (2006) The temporal architecture of eukaryotic growth. FEBS Lett 580: 2830–2835.

82. Oliva A, Rosebrock A, Ferrezuelo F, Pyne S, Chen H, et al. (2005) The cell cycle-regulated genes of Schizosaccharomyces pombe. PLoS Biol 3: e225.

83. Lloyd D, Murray DB (2005) Ultradian metronome: timekeeper for orchestration of cellular coherence. Trends Biochem Sci 30: 373–377.

84. De la Fuente IM, Benítez N, Santamaría A, Aguirregabiria JM, Veguillas J (1999) Persistence in metabolic nets. Bull Math Biol 61: 573–595.

85. De la Fuente IM, Martínez L, Pérez-Samartín AL, Ormaetxea L, Amezaga C, et al. (2008) Global self-organization of the cellular metabolic structure. PLoS ONE 3: e3100.

86. De la Fuente IM, Vadillo F, Pérez-Pinilla M-B, Vera-López A, Veguillas J (2009) The number of catalytic elements is crucial for the emergence of metabolic cores. PLoS ONE 4: e7510.

87. De la Fuente IM, Vadillo F, Pérez-Samartín AL, Pérez-Pinilla M-B, Bidaurrazaga J, et al (2010) Global self-regulations of the cellular metabolic structure. PLoS ONE 5: e9484.

88. De la Fuente IM, Cortes JM, Perez-Pinilla MB, Ruiz-Rodriguez V, Veguillas J (2011) The metabolic core and catalytic switches are fundamental elements in the self-regulation of the systemic metabolic structure of Cells. PLoS ONE 6: e27224.

89. De la Fuente IM, Cortes JM, Pelta DA, Veguillas J (2013) Attractor metabolic networks. PLoS ONE 8: e58284.

90. De la Fuente IM (2010) Quantitative analysis of cellular metabolic dissipative, self-organized structures. Int J Mol Sci 11: 3540–3599.

91. De la Fuente IM (2014) Metabolic dissipative structures. In: Aon MA, Saks V, Schlattner U, editors.Systems biology of metabolic and signaling networks: energy, mass and information transfer. Springer Berlin Heidelberg. pp. 179–212.

92. Edwards JM, Roberts TH, Atwell BJ (2012) Quantifying ATP turnover in anoxic coleoptiles of rice (Oryza sativa) demonstrates preferential allocation of energy to protein synthesis. J Exp Bot 63: 4389–4402.

93. Boender LGM, Almering MJH, Dijk M, van Maris AJA, de Winde JH, et al. (2011) Extreme calorie restriction and energy source starvation in Saccharomyces cerevisiae represent distinct physiological states. Biochim Biophys Acta 1813: 2133–2144.

94. Lim EL, Hollingsworth KG, Thelwall PE, Taylor R (2010) Measuring the acute effect of insulin infusion on ATP turnover rate in human skeletal muscle using phosphorus-31 magnetic resonance saturation transfer spectroscopy. NMR Biomed 23: 952–957.

95. Hochachka PW, McClelland GB (1997) Cellular metabolic homeostasis during large-scale change in ATP turnover rates in muscles. J Exp Biol 200: 381–386.

96. Atkinson DE, Walton GM (1967) Adenosine triphosphate conservation in metabolic regulation. Rat liver citrate cleavage enzyme. J Biol Chem 242: 3239–3241.

97. Chapman AG, Fall L, Atkinson DE (1971) Adenylate energy charge in Escherichia coli during growth and starvation. J Bacteriol 108: 1072–1086.

98. Ball WJ Jr, Atkinson DE (1975) Adenylate energy charge in Saccharomyces cerevisiae during starvation. J Bacteriol 121: 975–982.

99. Swedes JS, Sedo RJ, Atkinson DE (1975) Relation of growth and protein synthesis to the adenylate energy charge in an adenine-requiring mutant of Escherichia coli. J Biol Chem 250: 6930–6938.

100. Chapman AG, Atkinson DE (1977) Adenine nucleotide concentrations and turnover rates. Their correlation with biological activity in bacteria and yeast. Adv Microb Physiol 15: 253–306.

101. Walker-Simmons M, Atkinson DE (1977) Functional capacities and the adenylate energy charge in Escherichia coli under conditions of nutritional stress. J Bacteriol 130: 676–683.

102. Privalle LS, Burris RH (1983) Adenine nucleotide levels in and nitrogen fixation by the cyanobacterium Anabaena sp. strain 7120. J Bacteriol 154: 351–355.

103. Dai R, Liu H, Qu J, Zhao X, Ru J, et al. (2007) Relationship of energy charge and toxin content of Microcystis aeruginosa in nitrogen-limited or phosphorous-limited cultures. Toxicon 51: 649–658.

104. Beaman KD, Pollack JD (1981) Adenylate energy charge in Acholeplasma laidlawii. J Bacteriol 146: 1055–1058.

105. Holert J, Hahnke S, Cypionka H (2011) Influence of light and anoxia on chemiosmotic energy conservation in Dinoroseobacter shibae. Environ Microbiol Rep 3: 136–141.

106. Barrette WC Jr, Hannum DM, Wheeler WD, Hurst JK (1988) Viability and metabolic capability are maintained by Escherichia coli, Pseudomonas aeruginosa, and Streptococcus lactis at very low adenylate energy charge. J Bacteriol 170: 3655–3659.

107. Bulthuis BA, Koningstein GM, Stouthamer AH, van Verseveld HW (1993) The relation of proton motive force, adenylate energy charge and phosphorylation potential to the specific growth rate and efficiency of energy transduction in Bacillus licheniformis under aerobic growth conditions. Anton Leeuw Int J G 63: 1–16.

108. Kahru A, Liiders M, Vanatalu K, Vilu R (1982) Adenylate energy charge during batch culture of Thermoactinomyces vulgaris 42. Arch Microbiol 133: 142–144.

109. Weber J, Kayser A, Rinas U (2005) Metabolic flux analysis of Escherichia coli in glucose-limited continuous culture. II. Dynamic response to famine and feast, activation of the methylglyoxal pathway and oscillatory behaviour. Microbiology 151: 707–716.

110. Smith BA, Dworkin M (1980) Adenylate energy charge during fruiting body formation by Myxococcus xanthus. J Bacteriol 142: 1007–1009.

111. González F, Fernández-Vivas A, Muñoz J, Arias JM, Montoya E (1989) Adenylate energy charge during the life cycle of Myxococcus coralloides D. FEMS Microbiol Lett 58: 21–24.

112. Skjoldal HR (1981) ATP concentration and adenylate energy charge of tropical zooplankton from waters inside the great barrier reef. Mar Biol 62: 119–123.

113. Hünken M, Karsten U, Wiencke C (2005) Determination of the adenylate energy charge (AEC) as a tool to determine the physiological status of macroalgal tissues after UV exposure. Phycologia 44: 249–253.

114. Guimarães PMR, Londesborough J (2008) The adenylate energy charge and specific fermentation rate of brewer's yeasts fermenting high- and very high-gravity worts. Yeast 25: 47–58.

115. Chen Y, Xing D, Wang W, Ding Y, Du L (2007) Development of an ion-pair HPLC method for investigation of energy charge changes in cerebral ischemia of mice and hypoxia of Neuro-2a cell line. Biomed Chromatogr. 21: 628–634.

116. Derr RF, Zieve L (1972) Adenylate energy charge: relation to guanylate energy charge and the adenylate kinase equilibrium constant. Biochem Biophys Res Commun 49: 1385–1390.

117. Suska M, Skotnicka E (2010) Changes in adenylate nucleotides concentration and Na+, K+ - ATPase activities in erythrocytes of horses in function of breed and sex. Veterinary Medicine International 2010: ID 987309.

118. Bhatt DP, Chen X, Geiger JD, Rosenberger TA (2012) A sensitive HPLC-based method to quantify adenine nucleotides in primary astrocyte cell cultures. J Chromatogr B 889–890: 110–115.

119. Mills DCB, Thomas DP (1969) Blood platelet nucleotides in man and other species. Nature 222: 991–992.

120. Biegniewska A, Zietara MS, Rurangwa E, Ollevier F, Swierczynski J, et al. (2010) Some differences between carp (Cyprinus carpio) and African catfish (Clarias gariepinus) spermatozoa motility. J Appl Ichthyol 26: 674–677.

121. Plaideau C, Liu J, Hartleib-Geschwindner J, Bastin-Coyette L, Bontemps F, et al. (2012) Overexpression of AMP-metabolizing enzymes controls adenine nucleotide levels and AMPK activation in HEK293T cells. FASEB J 26: 2685–2694.

122. Rajab P, Fox J, Riaz S, Tomlinson D, Ball D, et al. (2000) Skeletal muscle myosin heavy chain isoforms and energy metabolism after clenbuterol treatment in the rat. Am J Physiol Regul Integr Comp Physiol 279: R1076–R1081.

123. Zubatkina IS, Emelyanova LV, Savina MV (2008) Adenine nucleotides and Atkinson energetic charge in liver tissue of cyclostomes and amphibians in ontogenesis. J Evol Biochem Phys+ 44: 763–765.

124. Rakotonirainy MS, Arnold S (2008) Development of a new procedure based on the energy charge measurement using ATP bioluminescence assay for the detection of living mould from graphic documents. Luminescence 23: 182–186.

125. Dinesh R, Chaudhuri SG, Sheeja TE (2006) ATP levels and adenylate energy charge in soils of mangroves in the Andamans. Curr Sci India 90: 1258–1263.

126. Pradet A, Raymond P (1983) Adenine nucleotide ratios and adenylate energy charge in energy metabolism. Annu Rev Plant Physiol 34: 199–224.

127. Singh J (1981) Isolation and freezing tolerances of mesophyll cells from cold hardened and nonhardened winter rye. Plant Physiol 67: 906–909.

128. Hanhijarvi AM, Fagerstedt KV (1995) Comparison of carbohydrate utilization and energy charge in the yellow flag iris (Iris pseudacorus) and garden iris (Iris germanica) under anoxia. Physiol Plantarum 93: 493–497.

129. McKee KL, Mendelssohn IA (1984) The influence of season on adenine nucleotide concentrations and energy charge in four marsh plant species. Physiol Plantarum 62: 1–7.

130. Pomeroy MK, Andrews CJ (1986) Changes in adenine nucleotides and energy charge in isolated winter wheat cells during low temperature stress. Plant Physiol 81: 361–366.

131. Sel'kov EE (1968) Self-oscillations in glycolysis. 1. A simple kinetic model. Eur J Biochem 4: 79–86.

132. Goldbeter A (1974) Modulation of the adenylate energy charge by sustained metabolic oscillations. FEBS Lett 43: 327–330.

133. Rapoport TA, Heinrich R, Rapoport SM (1976) The regulatory principles of glycolysis in erythrocytes in vivo and in vitro. A minimal comprehensive model describing steady states, quasi-steady states and time-dependent processes. Biochem J 154: 449–469.

134. Sel'kov EE (1975) Stabilization of energy charge, generation of oscillations and multiple steady states in energy metabolism as a result of purely stoichiometric regulation. Eur J Biochem 59: 151–157.

135. Reich JG, Sel'kov EE (1974) Mathematical analysis of metabolic networks. FEBS Lett 40: Suppl S119–S127.

136. Chen JQ, Cammarata PR, Baines CP, Yager JD (2009) Regulation of mitochondrial respiratory chain biogenesis by estrogens/estrogen receptors and physiological, pathological and pharmacological implications. Biochim Biophys Acta 1793: 1540–1570.

137. Weber J (2006) ATP synthase: subunit-subunit interactions in the stator stalk. Biochim Biophys Acta 1757: 1162–1170.

138. Steigmiller S, Turina P, Gräber P (2008) The thermodynamic H⁺/ATP ratios of the H⁺-ATPsynthases from chloroplasts and Escherichia coli. Proc Natl Acad Sci USA 105: 3745–3750.

139. Åden J, Weise CF, Brännström K, Olofsson A, Wolf-Watz M (2013) Structural topology and activation of an initial adenylate kinase-substrate complex. Biochemistry 52: 1055–1061.

140. Lange PR, Geserick C, Tischendorf G, Zrenner R (2008) Functions of chloroplastic adenylate kinases in Arabidopsis. Plant Physiol 146: 492–504.

141. Manning G, Whyte DB, Martinez R, Hunter T, Sudarsanam S (2002) The protein kinase complement of the human genome. Science 298: 1912–1934.

142. Gottlieb A, Frenkel-Morgenstern M, Safro M, Horn D (2011) Common peptides study of aminoacyl-tRNA synthetases. PLoS ONE 6: e20361.

143. Bønsdorff T, Gautier M, Farstad W, Rønningen K, Lingaas F, et al. (2004) Mapping of the bovine genes of the de novo AMP synthesis pathway. Anim Genet 35: 438–444.

144. Versées W, Steyaert J (2003) Catalysis by nucleoside hydrolases. Curr Opin Struct Biol 13: 731–738.

145. Erlinge D (2010) Purinergic and pyriminergic activation of the endothelium in regulation of tissue perfusion. In: Gerasimovskaya EV, Kaczmarek E, editors.Extracellular ATP and adenosine as regulators of endothelial cell function: Implications for health and disease. Springer Netherlands. pp. 1–13.

146. Tanaka K, Gilroy S, Jones AM, Stacey G (2010) Extracellular ATP signaling in plants. Trends Cell Biol 20: 601–608.

147. Falzoni S, Donvito G, Di Virgilio F (2013) Detecting adenosine triphosphate in the pericellular space. Interface Focus 3: 20120101.

148. Forsyth AM, Wan J, Owrutsky PD, Abkarian M, Stone HA (2011) Multiscale approach to link red blood cell dynamics, shear viscosity and ATP release. Proc Natl Acad Sci USA 108: 10986–10991.

149. Burnstock G (2012) Discovery of purinergic signalling, the initial resistance and current explosion of interest. Brit J Pharmacol 167: 238–255.

150. Nath S, Jain S (2000) Kinetic modeling of ATP synthesis by ATP synthase and its mechanistic implications. Biochem Biophys Res Commun. 272: 629–633.

151. Valero E, Varón R, García-Carmona F (2006) A kinetic study of a ternary cycle between adenine nucleotides. FEBS J 273: 3598–3613.

152. Sheng XR, Li X, Pan XM (1999) An iso-random Bi Bi mechanism for adenylate kinase. J Biol Chem 274: 22238–22242.

153. Goldbeter A, Lefeber R (1972) Dissipative structures for an allosteric model. Application to glycolytic oscillations. Biophys J 12: 1302–1315.

154. Goldbeter A (1990) Rythmes et chaos dans les systèmes biochímiques et cellulaires. Paris: Masson.

155. Curien G, Bastien O, Robert-Genthon M, Cornish-Bowden A, Cárdenas ML, et al. (2009) Understanding the regulation of aspartate metabolism using a model based on measured kinetic parameters. Mol Syst Biol 5: 271.

156. Ellis RJ (2001) Macromolecular crowding: an important but neglected aspect of intracellular environment. Curr Opin Struct Biol 11: 114–119.

157. Nenninger A, Mastroianni G, Mullineaux CW (2010) Size dependence of protein diffusion in the cytoplasm of Escherichia coli. J Bacteriol 192: 4535–4540.

158. Peiying Z (2007) Modeling the airway surface liquid regulation in human lungs. The University of North Carolina at Chapel Hill. Ed. ProQuest.

159. Mori Y, Matsumoto K, Ueda T, Kobatake Y (1986) Spatio-temporal organization of intracellular ATP content and oscillation patterns in response to blue light by Physarum polycephalum. Protoplasma 135: 31–37.

160. Ueda T, Mori Y, Kobatake Y (1987) Patterns in the distribution of intracellular ATP concentration in relation to coordination of amoeboid cell behavior in Physarum polycephalum. Exp Cell Res 169: 191–201.

161. De la Fuente IM (1999) Diversity of temporal self-organized behaviors in a biochemical system. BioSystems 50: 83–97.

162. De la Fuente IM, Martínez L, Veguillas J, Aguirregabiria JM (1996) Quasiperiodicity route to chaos in a biochemical system. Biophys J 71: 2375–2379.

163. De la Fuente IM, Martínez L, Veguillas J (1996). Intermittency route to chaos in a biochemical system. BioSystems 39: 87–92.

164. De la Fuente IM, Martínez L, Aguirregabiria JM, Veguillas J (1998) Coexistence of multiple periodic and chaotic regimes in biochemical oscillations. Acta Biotheor 46: 37–51.

165. De la Fuente IM, Cortes JM (2012) Quantitative analysis of the effective functional structure in yeast glycolysis. PLoS ONE 7: e30162.

166. Soga N, Kinosita K Jr, Yoshida M, Suzuki T (2011) Efficient ATP synthesis by thermophilic Bacillus F₀F₁-ATP synthase. FEBS J 278: 2647–2654.

167. Abrusci P, Chiarelli LR, Galizzi A, Fermo E, Bianchi P, et al. (2007) Erythrocyte adenylate kinase deficiency: characterization of recombinant mutant forms and relationship with nonspherocytic hemolytic anemia. Exp Hematol 35: 1182–1189.

168. Thuma E, Schirmer RH, Schirmer I (1972) Preparation and characterization of a crystalline human ATP:AMP phosphotransferase. Biochim Biophys Acta 268: 81–91.

169. Blangy D, Buc H, Monod J (1968) Kinetics of the allosteric interactions of phosphofructokinase from Escherichia coli. J Mol Biol 31: 13–35.

170. Hagen J (2006) Industrial catalysis: A practical approach. Weinheim, Germany: Wiley-VCH.

171. Jin J, Dong W, Guarino LA (1998) The LEF-4 subunit of Baculovirus RNA polymerase has RNA 5′-triphosphatase and ATPase activities. J Virol 72: 10011–10019.

172. Petty HR, Kindzelskii AL (2001) Dissipative metabolic patterns respond during neutrophil transmembrane signaling. Proc Natl Acad Sci USA 98: 3145–3149.

173. Burnstock G (1999) Current status of purinergic signalling in the nervous system. Prog Brain Res 120: 3–10.

174. Virginio C, MacKenzie A, Rassendren FA, North RA, Surprenant A (1999) Pore dilation of neuronal P2X receptor channels. Nat Neurosci 2: 315–321.

175. North RA (2002) The molecular physiology of P2X receptors. Physiol Rev 82: 1013–1067.

176. Blackwell KT (2013) Approaches and tools for modeling signaling pathways and calcium dynamics in neurons. J Neurosci Methods 220: 131–140.

177. Jacob T, Ascher E, Alapat D, Olevskaia Y, Hingorani A (2005) Activation of P38MAPK signaling cascade in a VSMC injury model: Role of P38MAPK inhibitors in limiting VSMC proliferation. Eur J Vasc Endovasc Surg 29: 470–478.

178. dos Passos JB, Vanhalewyn M, Brandao RL, Castro IM, Nicoli JR, et al. (1992) Glucose-induced activation of plasma-membrane H+-ATPase in mutants of the yeast Saccharomyces cerevisiae affected in cAMP metabolism, cAMP-dependent protein-phosphorylation and the initiation of glycolysis. Biochim Biophys Acta 1136: 57–67.

179. Srividhya J, Gopinathan MS, Schnell S (2007) The effects of time delays in a phosphorylation-dephosphorylation pathway. Biophys Chem 125: 286–297.

180. Li J, Kuang Y, Mason CC (2006) Modeling the glucose-insulin regulatory system and ultradian insulin secretory oscillations with two explicit time delays. J Theor Biol 242: 722–735.

181. Yildirim N, Mackey MC (2003) Feedback regulation in the lactose operon: a mathematical modeling study and comparison with experimental data. Biophys J 84: 2841–2851.

182. Locasale JW (2008) Signal duration and the time scale dependence of signal integration in biochemical pathways. BMC Syst Biol 2: 108.

183. Sung MH, Hager GL (2012) Nonlinear dependencies of biochemical reactions for context-specific signaling dynamics. Sci Rep 2: 616.

184. Chen BS, Chen PW (2009) On the estimation of robustness and filtering ability of dynamic biochemical networks under process delays, internal parametric perturbations and external disturbances. Math Biosci 222: 92–108.

185. Ravasz E, Somera AL, Mongru DA, Oltvai ZN, Barabási AL (2002) Hierarchical organization of modularity in metabolic networks. Science 297: 1551–1555.

186. Geryk J, Slanina F (2013) Modules in the metabolic network of E. coli with regulatory interactions. Int J Data Min Bioinform 8: 188–202.

187. Alberty RA, Goldberg RN (1992) Standard thermodynamic formation properties of adenosine 5′-triphosphate series. Biochemistry 31: 10610–10615.

188. Bonzon M, Hug M, Wagner E, Greppin H (1981) Adenine nucleotides and energy charge evolution during the induction of flowering in spinach leaves. Planta 152: 189–194.

189. Engelborghs K, Luzyanina T, Samaey G (2000) DDE-BIFTOOL: a Matlab package for bifurcation analysis of delay differential equations. TW Report 305.

190. Ching TM, Ching KK (1972) Content of adenosine phosphates and adenylate energy charge in germinating ponderosa pine seeds. Plant Physiol. 50: 536–540.

191. Moran LA, Horton RA, Scrimgeour G, Perry M (2011) Principles of biochemistry (5th Edition). New Jersey: Prentice Hall.

192. Manfredi G, Yang L, Gajewski CD, Mattiazzi M (2002) Measurements of ATP in mammalian cells. Methods 26: 317–326.

193. Buckstein MH, He J, Rubin H (2008) Characterization of nucleotide pools as a function of physiological state in Escherichia coli. J Bacteriol 190: 718–726.

194. Cannon WB (1932) The wisdom of the body. New York: WW Norton & Co.

195. Bernard C (1865) Introduction à l'étude de la médecine expérimentale. Paris: Flammarion.

196. Bernard C (1957) An introduction to the study of experimental medicine. New York: Dover.

197. Waddington CH (1957) The strategy of the genes. A discussion of some aspects of theoretical biology. London: George Allen and Unwin Ltd.

198. Waddington CH (1968) Towards a theoretical biology. Nature 218: 525–527.

199. Mamontov E. (2007). Modelling homeorhesis with ordinary differential equations. Math Comput Model 45: 694–707.

200. Mamontov E, Psiuk-Maksymowicz K, Koptioug A (2006) Stochastic mechanics in the context of the properties of living systems. Math Comput Model 44: 595–607.

201. Mamontov E, Koptioug A, Psiuk-Maksymowicz K (2006) The minimal, phase-transition model for the cell-number maintenance by the hyperplasia-extended homeorhesis. Acta Biotheor 54: 61–101.

202. Piotrowska MJ, Mamontov E, Peterson A, Koptyug A (2008) A model and simulation for homeorhesis in the motion of a single individual. Math Comput Model 48: 1122–1143.

203. Psiuk-Maksymowicz K, Mamontov E (2008) Homeorhesis-based modelling and fast numerical analysis for oncogenic hyperplasia under radiotherapy. Math Comput Model 47: 580–596.

204. Mamontov E (2011) In search for theoretical physiology–a mathematical theory of living systems: comment on "Toward a mathematical theory of living systems focusing on developmental biology and evolution: a review and perspectives" by N. Bellomo and B. Carbonaro. Phys Life Rev 8: 24–27.

8

Down-Regulation of Desmosomes in Cultured Cells: The Roles of PKC, Microtubules and Lysosomal/Proteasomal Degradation

Selina McHarg, Gemma Hopkins[¤a], Lusiana Lim[¤b], David Garrod*

Faculty of Life Sciences, University of Manchester, Manchester, United Kingdom

Abstract

Desmosomes are intercellular adhesive junctions of major importance for tissue integrity. To allow cell motility and migration they are down-regulated in epidermal wound healing. Electron microscopy indicates that whole desmosomes are internalised by cells in tissues, but the mechanism of down-regulation is unclear. In this paper we provide an overview of the internalisation of half-desmosomes by cultured cells induced by calcium chelation. Our results show that: (i) half desmosome internalisation is dependent on conventional PKC isoforms; (ii) microtubules transport internalised half desmosomes to the region of the centrosome by a kinesin-dependent mechanism; (iii) desmosomal proteins remain colocalised after internalisation and are not recycled to the cell surface; (iv) internalised desmosomes are degraded by the combined action of lysosomes and proteasomes. We also confirm that half desmosome internalisation is dependent upon the actin cytoskeleton. These results suggest that half desmosomes are not disassembled and recycled during or after internalisation but instead are transported to the centrosomal region where they are degraded. These findings may have significance for the down-regulation of desmosomes in wounds.

Editor: Michael Klymkowsky, University of Colorado, Boulder, United States of America

Funding: This work was supported by Wellcome Trust grant 08618/Z/08/Z, Medical Research Council Grant G0800004 and a MRC Research Studentship to GH. (http://www.wellcome.ac.uk/; http://www.mrc.ac.uk/index.htm) The funders had no role in study design, data collection and analysis, decision to publish, or preparation of the manuscript.

Competing Interests: The authors have declared that no competing interests exist.

* Email: d.garrod@manchester.ac.uk

¤a Current address: Paterson Institute for Cancer Research, University of Manchester, Manchester, United Kingdom
¤b Current address: Communicable Diseases, National Public Health Laboratory, Singapore, Singapore

Background

The intercellular adhesive strength of the epidermis and myocardium enables these tissues to withstand mechanical stress. Desmosomes are intercellular junctions that mediate this strong adhesion. By joining adjacent cells and binding to the keratin intermediate filament (IF) network, desmosomes act as linkers providing adhesion and great tensile strength. The importance of desmosomes is highlighted by the severe skin and cardiac defects that arise in autoimmune and genetic diseases [1–5].

Desmosomes are complex, transversely symmetrical structures composed of five main proteins. The desmosomal cadherins, desmoglein (Dsg) and desmocollin (Dsc), form the adhesive interface of the desmosome and their cytoplasmic tails bind to the armadillo proteins, plakoglobin (PG) and plakophilin (PKP), in the desmosomal plaque. The armadillo proteins in turn bind to desmoplakin (DP), which links the desmosome to the IFs [6–10].

Strong adhesion, though essential for tissue integrity, is incompatible with tissue remodelling such as takes place during epidermal wound healing and embryonic development. To facilitate remodelling, adhesion must be down-regulated but the mechanisms which govern down-regulation of desmosomes remain poorly understood. Ultrastructural studies of wound edge epidermis clearly show that entire desmosomes are internalised by cells [11,12]. Once internalised, they are presumably degraded.

Alternatively, they may be internally disassembled and their component proteins recycled.

In the context of tissue remodelling we have shown that desmosomes, both in culture and in vivo, can adopt two alternative adhesive states [12–15]. In normal tissues and confluent monolayers, desmosomes adopt calcium-independent adhesion, termed hyper-adhesion [12–13,15–16]. However, in subconfluent epithelial cultures [15], early embryogenesis and wound re-epithelialisation [12,14,16] desmosomal adhesion becomes calcium dependent. Hyper-adhesive desmosomes are more strongly adhesive than calcium dependent desmosomes [13]. The switch from hyper-adhesion to calcium dependence appears to be triggered by cell signalling since it (a) occurs without any qualitative or quantitative change in the major desmosomal components and (b) is triggered by activation of protein kinase C (PKC) or inhibition of protein phosphatases [13,15]. Moreover, the knockdown or knockout of PKCα promotes desmosomal hyper-adhesion [15–16].

On chelation of extracellular calcium, calcium dependent desmosomes have been shown by electron microscopy to split into half desmosomes that are rapidly internalised by the cells [17]. Calcium switching is widely regarded as an accepted method to study the assembly of desmosomes in tissue culture, and also, but perhaps less commonly, to study desmosome breakdown [17–20].

While calcium switching is unphysiological, in terms of desmosome breakdown it has the merit that it involves the vacuolar internalisation of complex structures, half desmosomes, and thus, to some extent reassembles the process that has been described in vivo. Half desmosomes are also produced by trypsinisation [21] and so is a daily occurrence when epithelia cells are passaged in culture. We have therefore used this model in order to attempt to provide novel information that may be relevant to the down-regulation of whole desmosomes.

We postulated that PKC signalling somehow primes the desmosomes in wounds for internalisation [12]. In the present study we test the role of PKC, and investigate both the role of the cytoskeleton in internalisation and internal transport and the fate of internalised desmosomal halves. Our results support a role for PKC and actin in internalisation. Once internalised, half desmosomes are transported to the centrosomal region by microtubules. Furthermore, internalised half desmosomes are not disassembled or recycled but are degraded by the combined action of lysosomes and the proteasome.

Materials and Methods

Cell culture

HaCaT cells [22] (a gift from Dr N.Fusenig), Madin Darby canine kidney type II cells (MDCK) [23] (ECACC, UK) and MDCK cells stably expressing Dsc2a-YFP (a gift from R.E.Leube) [24] were cultured in standard normal calcium medium (1.7 mM $CaCl_2$) (NCM) consisting of Dulbecco's Modified Eagle's Medium (DMEM) supplemented with 10% Foetal Calf Serum (FCS) (Sigma, Poole, UK) and 100 U/ml penicillin and 100 μg/ml streptomycin at 37°C in 5% humidified CO_2.

Low calcium medium and drug treatment of cells

Cells were seeded at a subconfluent density of 70,000 cells/cm^2 in NCM on 13 mm diameter coverslips in 24 well plates for 24 hours. They were then washed 3 times in calcium and magnesium free HBSS (CMF HBSS) and incubated in calcium free DMEM (Life Technologies, Paisley, UK) supplemented with 10% chelated FCS and 3 mM EGTA (LCM) for 60 minutes (unless specified otherwise) at 37°C in 5% humidified CO_2. To assess the role of actin, microtubules and conventional PKC isoforms in desmosome internalisation, MDCKs were pre-incubated with 5 μM latrunculin A, 33 μM nocodazole or 0.8 μM Gö6976 (all from Sigma) for 20–30 minutes. MDCK cells were then co-treated with LCM and the relevant inhibitor or vehicle (DMSO) alone for 60 minutes. To assess the role of kinesins in the internal transport of desmosomes, MDCK cells were treated in an identical manner with 500 μM adenylyl-imidodiphosphate (AMP-PNP) or 50 μM aurintricarboxylic acid (ATA) (Sigma) or vehicle alone (0.1% ethanol). Cells were then fixed in methanol and stained for immunoflourescence.

To investigate the degradation of desmosomal proteins cells were seeded at a subconfluent density of 70,000 cells/cm^2 in NCM and cultured for 24 hours. Cells were washed in CMF-HBSS and then treated with LCM for 16 or 24 hours in the presence of vehicle alone (0.1% DMSO) or 100 μM chloroquine (Fisher Scientific, Loughborough, UK), 250 nM bafilomycin A1, 100 μM leupeptin, 10 μM MG132 (all from Sigma), 20–200 nM bortezomib (a gift from Prof. S. High). Cells were then lysed in DTT sample buffer (100 mM DTT, 10% glycerol, 2% SDS, 80 mM Tris pH 6.8) and analysed by immunoblotting.

Antibodies

The following antibodies were used for immunoflourescence and western blotting: mouse monoclonal antibodies (mab) against desmoplakin I and II (11-5F) [25], desmoglein 2 (33-3D) [26], desmoglein 3 (32-2B) [27], pan-desmocollin (52-3D) [28], plakoglobin (Sigma) and plakophilin-2 (Progen, Heidelberg, Germany). For immunoflourescence only the following antibodies were used: rabbit polyclonal IgG against γ-tubulin (Abcam, Cambridge, UK), Rab11 (Life Technologies, Paisley, UK) and ninein (a gift from Dr C.Bierkamp), sheep polyclonal IgG against aurora A (a gift from Prof. S.S. Taylor), mouse mab against CLIP170 (Abcam), EB1 (Santa Cruz,) and Lamp1 (a gift from Prof P.G. Woodman). Rabbit antiserum against the cytoplasmic domain of Dsg2 was expressed as a His-tag fusion protein (details to be published at a later date). The following secondary antibodies were used: Alexaflour 488-conjugated goat anti-mouse IgG, Alexaflour 488-conjugated donkey anti-sheep IgG, Alexaflour 488-conjugated rabbit anti-goat IgG (all from Life Technologies), FITC-conjugated donkey anti-rabbit IgG, rhodamine-conjugated donkey anti-rabbit IgG and rhodamine-conjugated donkey anti-mouse IgG (all from Stratech Scientific Ltd., Newmarket, Suffolk, UK). For western blotting only, the following additional antibodies were used: mouse mabs against α-tubulin (Sigma), keratin 8 (LE41) (a gift from Prof. Birgit Lane) and peroxidase conjugated secondary antibodies including goat anti mouse IgG and IgM and goat anti rabbit IgG (Thermo Fisher Scientific, Loughborough, UK).

Immunoflourescence and image processing

For plasma membrane staining, cells were rinsed in ice-cold PBS and incubated with 40 μg/ml FITC-conjugated concanavalin A (con-A) (Sigma) for 30 minutes at 4°C prior to fixation. Cells were then either fixed in methanol for 10 minutes or in acetone:methanol for 5 minutes at −20°C and incubated with primary antibody for 1–3 hours at room temperature. Cells were rinsed with PBS and incubated with a flourophore conjugated secondary antibody for 30 minutes at room temperature. Cells were then rinsed extensively in PBS, and mounted in Vectashield antifade (Vector Laboratories ltd., Peterborough, UK). Flourescence was assessed either with a Zeiss Axioplan microscope using 63× plan-apochromat oil immersion objective (NA 1.4) an RTE/CCD-1300-Y camera (Princeton Instruments Inc., Trenton, NJ) and Metamorph software (Universal Imaging Corporation, West Chester, PA). Alternatively, for confocal fluorescence microscopy a Nikon Eclipse 90i (Nikon, Surrey, UK) was used to take z-stacks using a 60× oil immersion plan-apo objective (NA 1.4) and images acquired using Nikon EZC1 software. At least 5 random fields per condition were taken. For quantification of DP and γ-tubulin distribution in LCM- treated MDCK, 100–150 cells from several fields were assessed per time point. The degree of colocalisation in whole cell volumes was assessed by using the Image J intensity correlation analysis plugin to calculate the Pearson's correlation coefficient. For centrosomal analysis a threshold was set to restrict analysis to the subcellular region.

Live-imaging video of Dsc2a-YFP and pericentrin localisation

MDCK cells stably expressing Dsc2a-YFP were seeded at 70,000 cells/cm^2 in NCM for 24 hours. They were then transiently transfected with 0.4 μg pericentrin-RFP (a gift from Dr Sean Munro) per well of a 24 well plate using 2.4 μL Fugene 6 transfection reagent (Roche Diagnostics Ltd., West Sussex, UK). 24 hours post-transfection, cells were washed extensively with

CMF-HBSS and then treated with LCM for 3 hours during which images were taken every 2 minutes on a spinning disc confocal consisting of a Zeiss axio-observer with a 63× plan apo objective (NA 1.4), CSU-X1 spinning disc confocal (Yokagowa), an Evolve EMCCD camera (Photometrics, Tuscon, AZ) and XYpiezoZ stage (ASI, Eugene Oregon) at 37°C and images acquired using Slidebook 5.0 software (3i, Göttingen, Germany).

siRNA knockdown of desmoplakin

Human desmoplakin siRNA sequences were taken from Wan et al. [29] and are as follows: DP sense AACCCAGACUACA-GAAGCAAU, DP antisense AUUGCUUCUGUAGUCUGG-GUU (Dharmacon, Chicago, IL). Using Invivogen siRNA wizard v3.1 blast search, scrambled DP sequences were designed and are as follows: scrambled DP sense GCAGAACAACGCAUCAA-CAUA, scrambled DP antisense UAUGUUGAUGCGUUGUU-CUGC (Dharmacon). HaCaT cells were seeded in 6-well plates at 50,000 cells/cm^2 in NCM for 24 hours. They were then incubated in serum-free DMEM for 2 hours prior to transfection and oligonucleotides transfected at 50 nM using 2 µg/ml Lipofectamine 2000 (Life Technologies) diluted in serum-free DMEM. At 24 hours post-transfection the cells were either treated with LCM for 1 hour and stained for immunoflourescence or lysed in DTT sample buffer and immunoblotted to assess knockdown efficiencies.

Preparation of MDCK Triton-X100 soluble and insoluble fractions

MDCK II cells were seeded at 90,000 cells/cm^2 subconfluent density on 90 mm dishes for 24 hours in NCM. Cells were then washed extensively in CMF HBSS and incubated at 37°C in LCM for 90 minutes. Cells were then washed twice in PBS and treated with 0.1% Triton-X100 in PBS containing 1% protease inhibitor cocktail (Sigma) for 20 minutes at room temperature. The Triton-soluble fraction was collected and the remaining insoluble fraction dissolved in DTT sample buffer. DTT sample buffer was added to the Triton-soluble fractions, and insoluble and soluble fractions were equally loaded for protein amount onto polyacrylamide gels, separated by SDS-PAGE and then immunoblotted.

Desmoglein 2 recycling assay

MDCK cells were seeded at 70,000 cells/cm^2 subconfluent density in NCM on 60 mm dishes for 24 hours. Cells were then washed extensively with CMF HBSS and either treated with LCM or with a reverse calcium switch (REV) consisting of LCM for 1 hour followed by NCM for 1 hour to stimulate new desmosome formation. At the end of treatment, cells were treated with 0.25% TPCK treated trypsin (Sigma) and 1 mM EDTA (Sigma) for 15 minutes at 37°C. They were then centrifuged and the resultant pellet lysed in boiling DTT sample buffer. Protein was equally loaded on SDS-PAGE gels and immunoblotted.

SDS-PAGE & western blotting

Protein samples were all prepared in DTT sample buffer, equally loaded at concentrations of 50–100 µg protein per lane and then separated by SDS-PAGE. Protein was then transferred to nitrocellulose membrane, blocked in 5% milk dissolved in PBS/0.05% Tween-20 for 1 hour, then incubated in primary antibody for either 1 hour or overnight. Nitrocellulose membranes were washed, and probed with a secondary HRP -conjugated antibody for 30 minutes. Chemiluminescence (Supersignal West Pico CL kit, Thermo Fisher Scientific) was used to detect bound protein. Where necessary, densitometry was quantified using Adobe

Photoshop and normalised with respect to loading control proteins.

Online supplemental material

Video S1 shows internalised Dsc2a YFP surrounding and then co-localising with the centrosomal marker pericentrin-RFP. Figure S1 shows stills from Video S1 at time points 0 (A), 45 minute (B) and 1 hr 40minutes (C).

Results

Half desmosome internalisation is dependent on actin and PKC

We have used the calcium switch model to study desmosome internalisation. This involves chelating extracellular calcium with 3 mM EGTA from the medium surrounding cells in short term culture. Such cells have calcium dependent desmosomes so this process causes loss of desmosomal adhesion generating half desmosomes that are rapidly internalised by the cells [13,15,17,24]. This internalisation of intact plaque-bearing desmosomal structures is essentially similar to the internalisation of whole desmosomes that occurs in vivo and is currently the best culture model of desmosome down-regulation.

To test the hypothesis that conventional PKC isoforms are involved in internalisation of half desmosomes MDCK cells were co-treated with LCM and Gö6976 (0.8 µM), an inhibitor of these isoforms. Whilst LCM treatment alone elicited DP internalisation (Figure 1A,B), after co-treatment with Gö6976 the cells detached from each other but DP persisted at the cell surface (Figure 1C) indicating that the desmosomes had split through the extracellular domain but were not internalised. This suggests that conventional PKC isoform activity is required for the internalisation of half desmosomes. In MDCK cells the PKC isoform involved is PKCα [15].

(N. B. Since the effect of PKC signalling on desmosomal adhesion is rapid, this experiment cannot be done by PKCα depletion, because of the time required. Depletion renders desmosomes hyper-adhesive [15]. Such desmosomes do not split on treatment with LCM and are therefore not internalised. In fact we have done the depletion experiment and, as expected, the desmosomes remained intact at the cell surface.)

One of the cellular functions of PKC is regulation of the actin cytoskeleton [30] and both actin and PKC have been implicated in endocytic and phagocytic processes [31,32], as well as in the internalisation of desmosomes [33]. To confirm the latter result, cells were treated with LCM in the presence of 5 µM latrunculin A to depolymerise actin filaments. (Depolymerisation was confirmed by staining with fluorescent phalloidin (not shown).) This prevented the internalisation of half desmosomes, as demonstrated by the continued co-localisation of DP at the plasma membrane (Pearson's score ~0.6) (Figure 1F), thus confirming that the actin cytoskeleton is required for internalisation.

Microtubules and kinesins regulate internal transport of half desmosomes

Once internalised, half desmosomes continued to be transported to the perinuclear region of the cells (Figure 1B, E). It is well established that transport of endocytosed proteins and vesicles is carried out by microtubules [34,35]. To determine whether this is also the case for internalised desmosomal halves, cells cultured in NCM for 24 hours were pre-incubated for 20 minutes with 33 µM nocodazole to depolymerise microtubules and then with LCM in the presence of nocodazole. (Depolymerisation of microtubules was confirmed by immunofluorescence for tubulin (not shown).) In

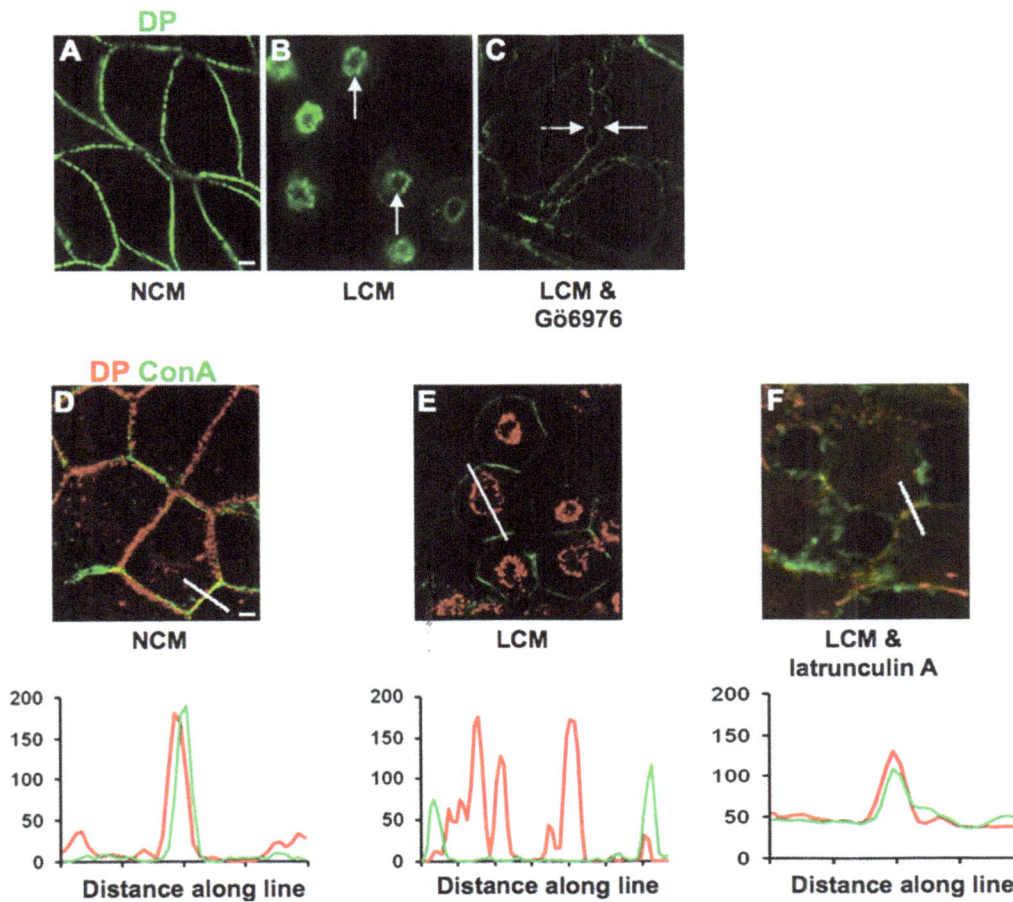

Figure 1. Half desmosome internalisation is cPKC and actin dependent. (A) MDCK cells cultured in NCM had desmosomes at cell-cell contacts as indicated by DP staining. (B) Internalised rings of DP were present in cells treated for 60 minutes with LCM (arrows). (All internalisation controls in LCM were carried out in the presence of the appropriate drug vehicle.) (C) Internalisation of desmosomes was prevented by co-treatment with LCM and Gö6976 (0.8 μM), as cell contact was lost but half desmosomes remained at the cell surface giving rise to the appearance of intercellular gaps (arrows). (D) In NCM, DP (red) was localised to the cell surface in association with the surface marker con-A (green). (E) LCM treatment caused internalisation of DP and separation from con-A, which remained at the surface. (F) Co-treatment with LCM and latrunculin A (5 μM) inhibited desmosome internalisation, as indicated by persistent association of DP with con-A. Fluorescence profiles depict the intensity of staining along the white line in the images.

cells thus treated DP was internalised but remained just beneath the cell surface rather than contracting into a tight ring (Figure 2A), suggesting that intracellular transport was impaired. Thus it appears that actin filaments are required for internalisation of half desmosomes but microtubules are responsible for their further transport within the cell.

The kinesins are microtubule motor proteins that bind their cargo to microtubules and are essential intermediaries in microtubular transport [36,37]. To determine whether kinesins might be involved in the transport of internalised desmosomal proteins MDCKs were pre-treated for 30 minutes with the broad-spectrum kinesin inhibitors adenylylimidodiphosphate (AMP-PNP) (500 μM) or aurintricarboxylic acid (ATA) (50 μM), followed by co-treatment with LCM and the inhibitor. Both inhibitors attenuated the transport of internalised DP, producing distributions that were identical to those elicited by nocodazole-induced microtubule disruption (Figure 2B, C). As the minus end-directed transport kinesin KIFC3 is expressed in MDCK cells [38], double immunoflourescence for DP and KIFC3 was carried out. This showed that whilst KIFC3 staining was diffuse throughout the

cytoplasm (Pearson's score ~0) (Figure 2D), there was co-localisation between KIFC3 and internalised rings of DP following 30 minutes LCM treatment (Pearson's score ~0.65) (Figure 2E). Furthermore, co-treatment with nocodazole reduced this co-localisation (Pearson's score ~0.2) (Figure 2F). Thus KIFC3 is a candidate kinesin for transport of internalised desmosomes.

Half desmosome internalisation does not require IF attachment

Internalisation of half desmosomes appears to depend upon the sequential action of the actin and microtubule cytoskeletons, but DP links the desmosomal plaque to IFs. To determine whether IFs are involved in half desmosome internalisation, we knocked down DP with siRNA, thus disrupting the desmosome-IF link. This experiment was done with HaCaT rather than MDCK cells as the human, but not the canine, sequence for DP is available. Immunoblotting of whole cell lysates showed substantial reduction of DP expression from cells transfected with 50 nM DP siRNA as compared to those transfected with scrambled siRNA or non-transfected samples (Figure 3A). Immunofluorescence for keratin 8

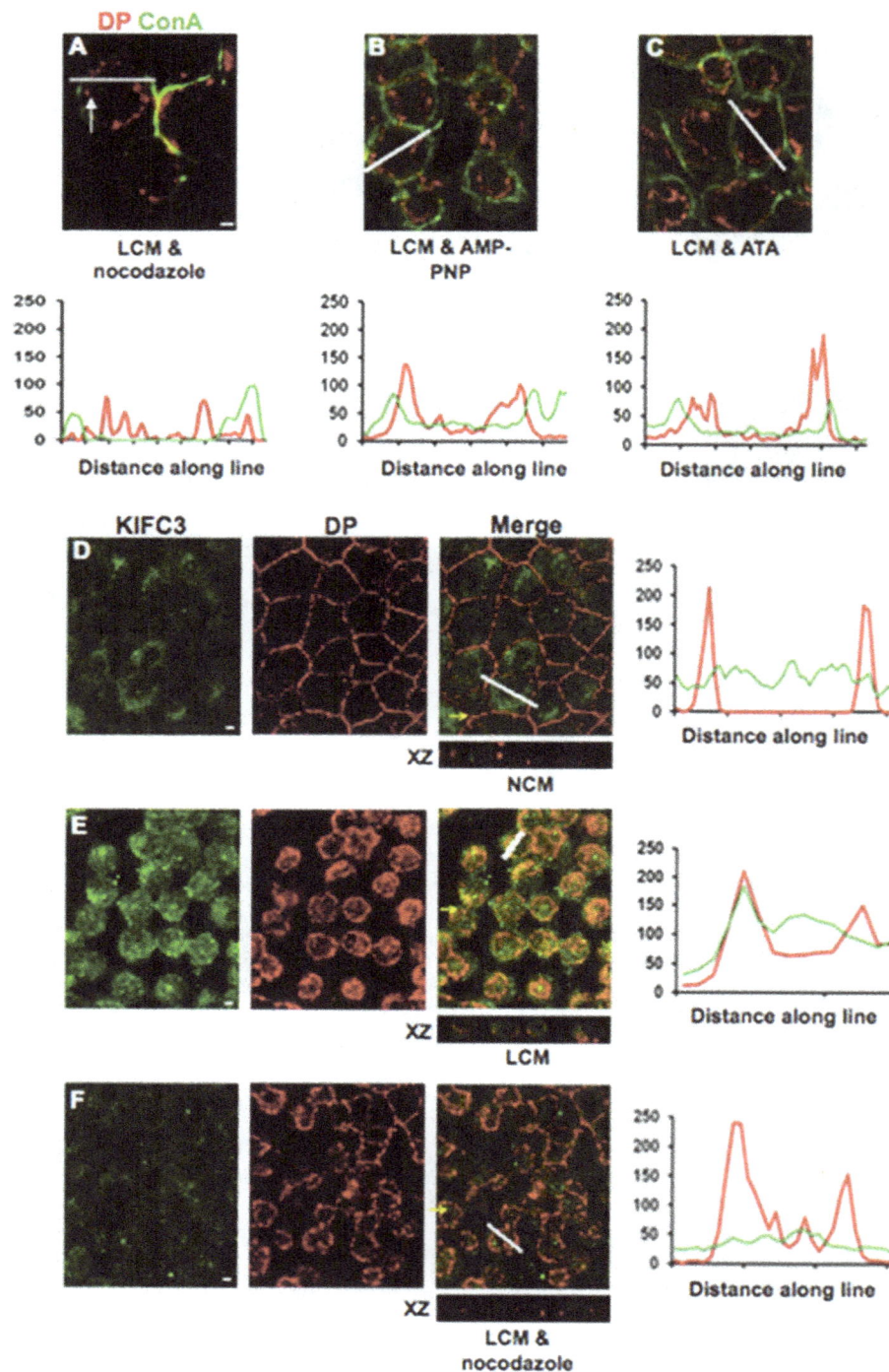

Figure 2. Half desmosome intracellular transport is regulated by microtubules. (A–C) Co-treatment with LCM and nocodazole (33 μM) (A), AMP-PNP (500 μM) (B) or ATA (50 μM) (C) caused DP to be internalised but to remain just beneath the cell surface (arrows). Confocal images of MDCK cultured in NCM and stained for KIFC3 (green) and DP (red) (D) which co-localise following 30minutes LCM treatment (E). This co-localisation was inhibited by LCM co-treatment with nocodazole (33 μM) (F). Yellow arrows indicate XZ axis. Bar, 5 μm. Fluorescence profiles depict the intensity of staining along the white line.

& 18 demonstrated a striking effect of DP knockdown on IF organisation. Cells lacking DP had diffuse, non-filamentous keratin staining that did not link to desmosomes, whilst cells transfected with scrambled siRNA had filamentous keratin (Figure 3B). Immunofluorescence revealed that LCM treatment of cells with DP knockdown were still able to internalise desmosomes; they had distinctive internalised rings of Dsg2 identical to those found in cells transfected with scrambled siRNA (Figure 3C). Thus it appears that the IF cytoskeleton is not required for internalisation of desmosomal proteins.

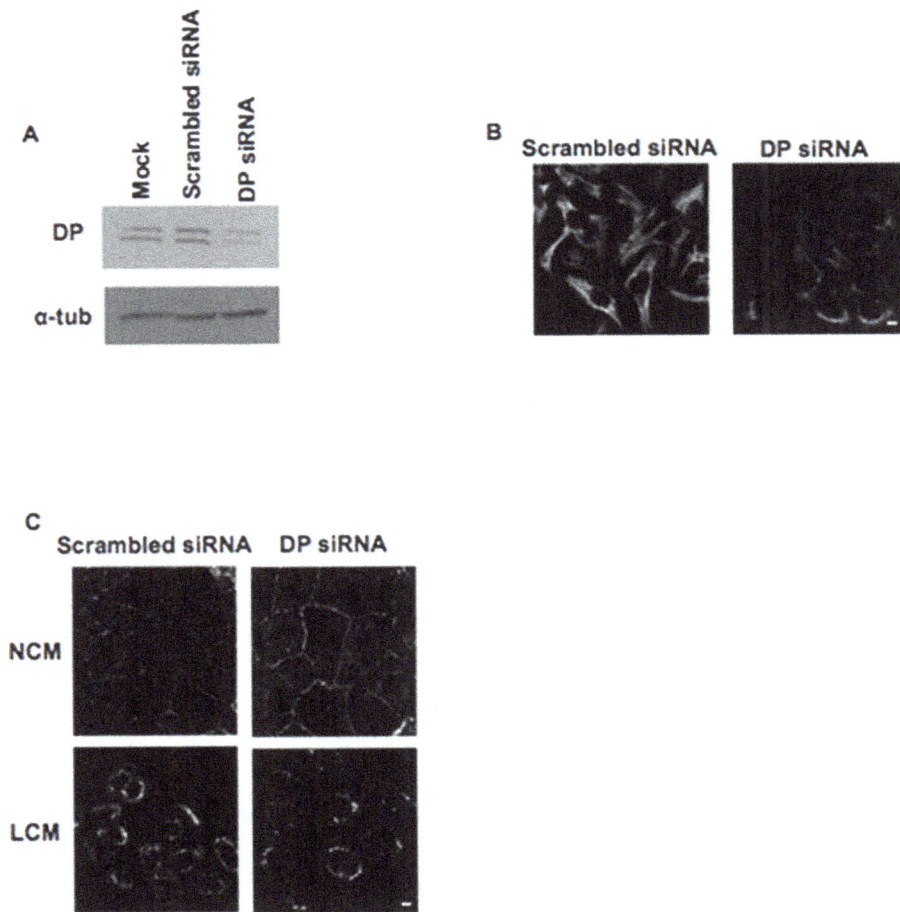

Figure 3. Intermediate filaments are not involved in internalisation of half desmosomes. (A) Western blot showing partial knockdown of DP in HaCaT cells transfected with 50 nM DP siRNA or compared to scrambled DP siRNA and mock transfected controls. (B) Single confocal slices of HaCaT cells or transfected with 50 nM scrambled or DP siRNA showing that the latter cells had disrupted intermediate filament organisation as indicated by keratin 8 & 18 staining. (C) HaCaT cells transfected with DP siRNA or scrambled siRNA were treated with either NCM or LCM for 1 hour and then stained for Dsg2. Half desmosomes were internalised in DP knockdown cells as in the controls. Bar, 5 μm.

Internalised half desmosomes co-localise with the centrosome

Because internalised half desmosomes are transported along microtubules and form a tight ring or dot adjacent to the nucleus, it seemed possible that they may be being transported to the centrosome, the cellular anchor for microtubule minus ends [39,40]. To test this, MDCK cells were treated with LCM for 2–3 hours to induce internalisation of half desmosomes, double immunofluorescence for DP and the centrosome markers aurora A, ninein or γ-tubulin carried out, and the results assessed by confocal microscopy. Analysis of individual confocal z-slices showed either rings of DP surrounding the centrosome or co-localisation of internalised DP with all 3 centrosome markers (Pearson's score ranged from ~0.5 to 0.8) (Figure 4A–C). It seemed possible that these alternative distributions represented different stages in the internalisation process. In order to determine whether this was the case, a time course was established by quantifying the distribution patterns at different times after LCM treatment. Since the distribution patterns of DP relative to all three centrosome markers were similar and as aurora A is only present at the centrosome for certain phases of the cell cycle, γ-tubulin staining was used for this study. Confocal z-slices were

analysed and cells counted as having one of three staining patterns: (i) a ring of DP surrounding the centrosome; (ii) a condensed spot of DP beside the centrosome; (iii) DP co-localised with the centrosome. After 1 hour of LCM treatment, 80% of cells had a ring of DP surrounding the centrosome (p<0.0286), but by 3 hours this had decreased to just 18% of cells (Figure 4D). However, as LCM treatment time progressed to 3 hours, the number of cells with DP in a dense spot beside the centrosome (Figure 4E) or co-localised with it (Figure 4F) significantly increased from 7% and 1% to 55% and 26% (p<0.0286), respectively. These data indicate that internalised DP is transported towards the centrosome.

To confirm the transport of internalised desmosome proteins to the centrosome, MDCK Dsc2a-YFP cells were transfected with a pericentrin-RFP construct (pericentrin is a centrosome matrix protein and centrosome marker) [41]. Live imaging over 3 hours of cells in LCM showed that the cells initially rounded up and detached from each other (Video S1). By 45 minutes an internalised ring of Dsc2a-YFP was seen surrounding the centrosome (Video S1 & Figure S1B). By 100 minutes, co-localisation of internalised Dsc2a and pericentrin occurred (Video

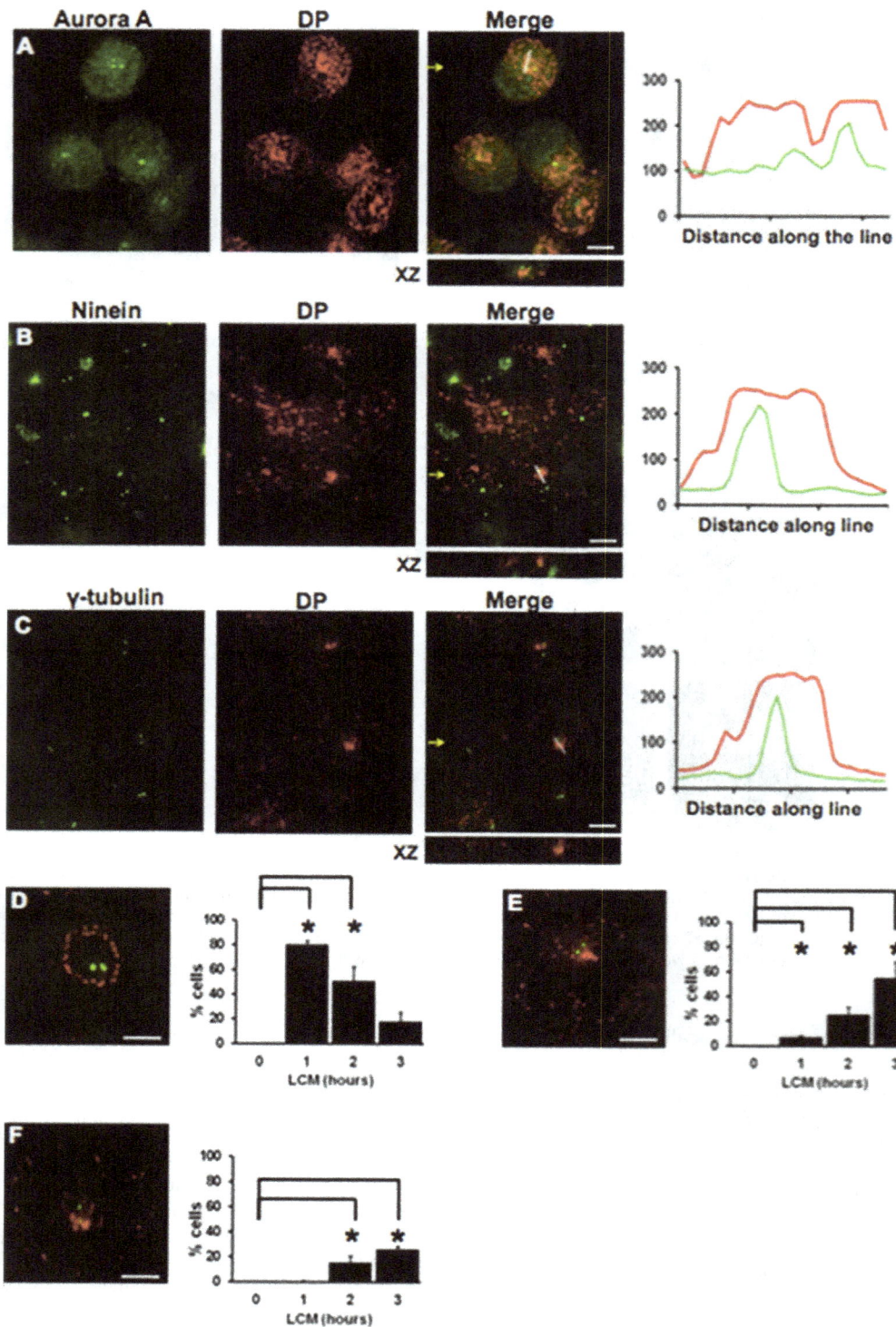

Figure 4. Internalised desmoplakin co-localises with the centrosome. (A–C) Co-localisation of DP with three centrosome markers. Single confocal slices of MDCK cells stained for DP (red) and the centrosome markers aurora A (A) ninein (B) and γ-tubulin (C) (all green) following 2-3 hours of LCM treatment. Fluorescence profiles depict the intensity of staining along the white line in the merged images. (D–F) The time course of co-localisation. Representative images of desmoplakin and γ-tubulin localisation demonstrate the 3 categories of staining pattern used for quantification with DP surrounding the centrosome (D), beside the centrosome (E) or co-localised with the centrosome (F) during LCM treatment, data are mean values ± s.e.m. Yellow arrows indicate XZ axis. Bar, 5 μm. Asterisk indicates statistical significance (p<0.0286, Mann-Whitney test).

S1 & Figure S1C). These results confirm that internalised desmosomal proteins are transported to the centrosome.

Internalised half desmosomes do not disassemble

To investigate whether internalised half desmosomes disassemble, LCM treated MDCK and MDCK Dsc2a-YFP cells were stained for a range of desmosome proteins. As expected, in NCM the desmosome proteins DP, PG, and Dsc2a were co-localised at points of cell-cell contact (Pearson's score ~0.9) (Figure 5A, B). Following LCM treatment co-localisation persisted with DP, PG and Dsc2a all continuing to associate with one another in internalised ring-like structures (Pearson's score ranged from ~0.8 to 0.9) (Figure 5A, B). This co-localisation persisted for up to 24 hours post internalisation (Pearson's score ~0.7) (Figure 5C, D), suggesting that desmosomal halves remain intact following internalisation.

To support this suggestion, LCM treated MDCK cells were separated into their Triton-X100-soluble and -insoluble fractions. Densitometry of immunoblots revealed that there was little or no change in DP, PKP2, PG, Dsg3 or Dsg2 levels in the insoluble fraction for up to 90 minutes post internalisation (Figure 5E, F). Only the pan-Dsc immunoblot showed a significant reduction of over 60% total protein from the insoluble fraction (p<0.0286). The reduction of Dsc from the insoluble fraction was not accompanied by a concomitant increase of Dsc in the soluble fraction. (We note that there were slight changes in the amounts of DP, PG and PKP2 in the soluble fraction but the significance of these is unclear since they were not accompanied by comparable changes in the insoluble fraction.) These data suggest that desmosomal halves do not disassemble upon internalisation, although there appears to be some loss of Dsc.

Internalised Dsg2 is not recycled

The recycling of both internalised adherens and tight junction proteins to the cell surface has been well documented [42–45]. However, previous work has suggested that this may not be the case for desmosomal proteins since those that are internalised as a result of LCM treatment appear to remain intracellular when new desmosomes are assembled [17,24]. We attempted to resolve this by biotinylation of cell surface desmosomal cadherins, but the results proved inconsistent, possibly because the molecules in the desmosomal intercellular space were difficult to biotinylate reliably. We therefore devised a method based on the trypsin-sensitivity of cell surface molecules and the trypsin-resistance of intracellular molecules.

Whole cell lysates of MDCK cells cultured in NCM showed a major band at 150 kDa on western blotting for Dsg2 (Figure 6A). Following whole cell trypsinisation, this band was completely absent, but replaced by a lower band, which represents the membrane-protected cytoplasmic domain of Dsg2 [26] (Figure 6A). However, if the cells were treated with LCM for 30 minutes and then trypsinised, western blotting revealed that some of the full length Dsg2 remained intact, indicating that it had become internalised and hence membrane-protected (Figure 6A). After 60 minutes in LCM, the amount of membrane-protected Dsg2 was increased showing that internalisation was a continuing process (Figure 6A).

To determine whether internalised Dsg2 was recycled to the cell surface, MDCK cells were either treated with LCM for 2 hours alone or for 1 hour, followed by NCM for 1 hour to stimulate new desmosome formation (confirmed by immunofluorescence, not shown. See [17]) followed by trypsinisation to remove residual Dsg2 on the cell surface. There were two possible outcomes of this experiment; either some of the internalised Dsg2 would be

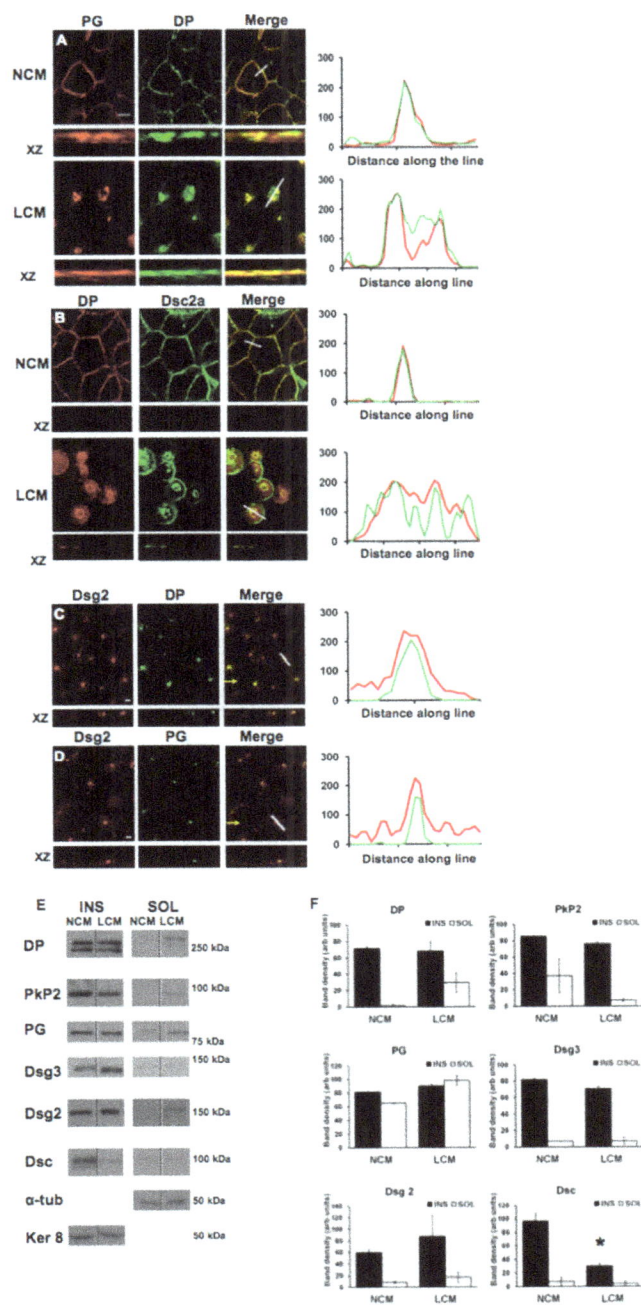

Figure 5. Desmosomal proteins remain co-localised following internalisation. (A–D) Co-localisation of the desmosomal proteins DP, PG, Dsc2a and Dsg2 in NCM persists following LCM-induced internalisation for 1 hour (A,B) and 24 hours (C,D). Yellow arrows indicate XZ axis. Bar, 5 μm. Fluorescence profiles depict the intensity of staining along the white line in the merged images. (E) Cells were treated with either NCM or LCM for 90 minutes and then separated into their insoluble (INS) and soluble (SOL) fractions. Western blots for desmosomal proteins (E) were quantified by densitometry (F) (dashed lines indicate lanes which have been re-ordered from the same western blot). Asterisk indicates statistical significance (p<0.0286, Mann-Whitney test). For further details see text.

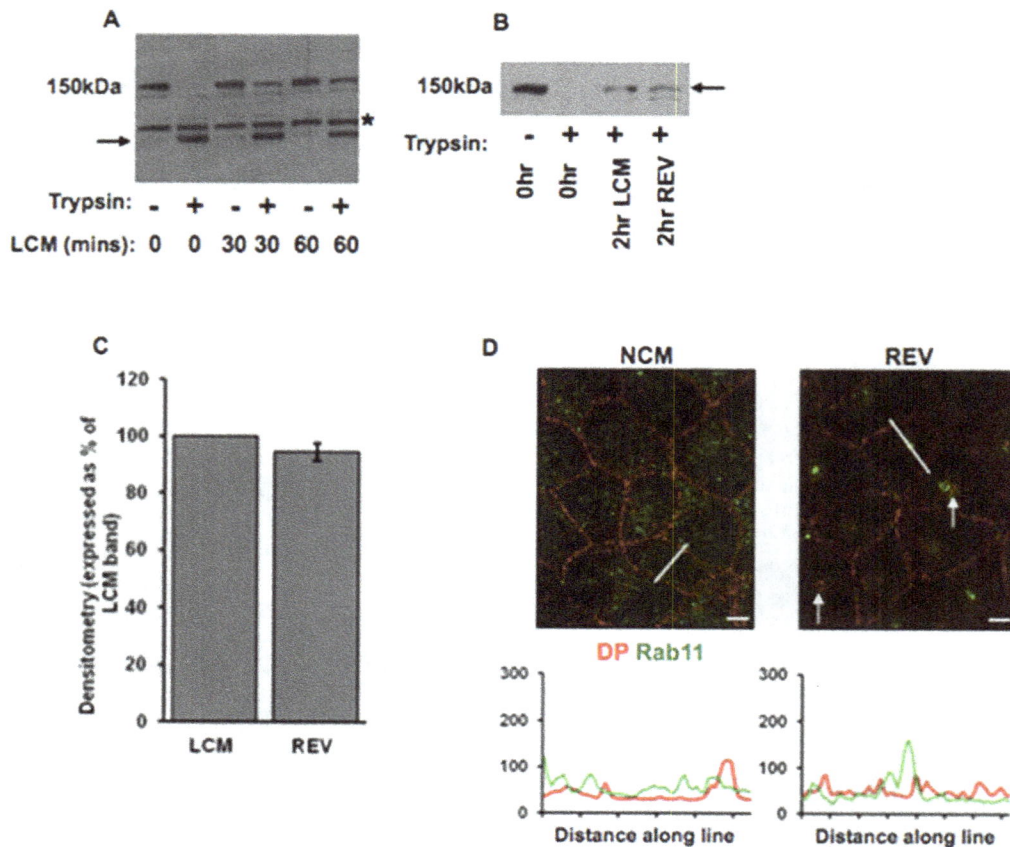

Figure 6. Internalised Dsg2 is not recycled to the cell surface. (A) LCM-induced internalisation protects Dsg2 from trypsin. MDCK cells were treated with trypsin/EDTA after incubation LCM for 0, 30 or 60 minutes. Western blots show that cell surface Dsg2 (150 KDa) was degraded by trypsin/EDTA generating a membrane-protected cytoplasmic fragment (arrow) (LCM (mins) 0). (The band above the trypsin fragment and present in each lane is believed to be a natural degradation product of Dsg2.) However, after LCM treatment a substantial amount of full length Dsg2 remained after trypsin/EDTA treatment (LCM (mins) 30 and 60) showing that it had become membrane-protected. (B, C) The amount of membrane-protected Dsg2 (arrow) remains the same after induction of new desmosome formation following LCM treatment. Cells were treated either with LCM for 2 hours, or with LCM for 1 hour followed by NCM for 1 hour (total time 2 hours) to induce new desmosome formation (verified by immunofluorescence but not shown). The latter treatment is referred to as reverse calcium switching (REV). Western blots (B, quantified in C) show that the amount of membrane protected Dsg2 was identical after both treatments (2 hr LCM, 2 hr REV). Bar in C indicates s.e.m. (D) Internalised DP (arrows) does not colocalise with Rab11 following a reverse calcium switch (REV). Bar, 5 μm. Fluorescence profiles depict the intensity of staining along the white line in the images.

recycled to the cell surface in the cells forming new desmosomes which would then be degraded by trypsin, or the internalised Dsg2 would remain internal. In the first case, the density of the internalised full length Dsg2 band should have decreased compared with that in cells that had remained in LCM for the full 2 hours, whereas in the second case, the density of the internalised Dsg2 band should have remained the same in both batches of cells. The result shows that that the densities of the Dsg2 bands were identical (Figure 6B, C). A similar result was observed for Dsg3, with densitometry indicating a 15% reduction of internalised Dsg3 in reverse switch conditions (data not shown). We conclude that the internalised desmosomal cadherins were not recycled to the cell surface. Furthermore, cells treated with a reverse calcium switch were stained for the recycling endosomal marker Rab11 and DP. DP was present both inside cells (Figure 6D, arrows) (caused by the initial LCM treatment) and at the plasma membrane (following subsequent NCM treatment) indicating new desmosome formation (Figure 6D). Internalised DP and Rab11 were not co-localised, substantiating the finding

that internalised desmosomal proteins are not subsequently recycled to the plasma membrane (Figure 6D).

Internalised desmosomal components are degraded by lysosomes and the 26S proteasome

If internalised desmosome proteins are, like Dsg2, not recycled, they must be degraded by the intracellular protein degradation apparatus. Previous reports have shown that internalised desmosomes associate with the late endosomal marker mannose 6-phosphate receptor [46]. Other research has shown association of the ubiquitin activating enzyme E1 with desmosome plaques [47]. However, definitive degradation pathway(s) for internalised desmosome components have not been identified.

We first showed that degradation of DP, Dsg2 and PG in LCM-treated cells proceeded gradually (Figure 7A). The rate of degradation varied slightly from one experiment to another but generally amounts had declined to low levels by 24 hours. In order to determine the degradation pathways, cells in LCM were treated with a range of lysosomal and proteasomal inhibitors. The lysosomal inhibitors chloroquine and bafilomycin A1 substantially

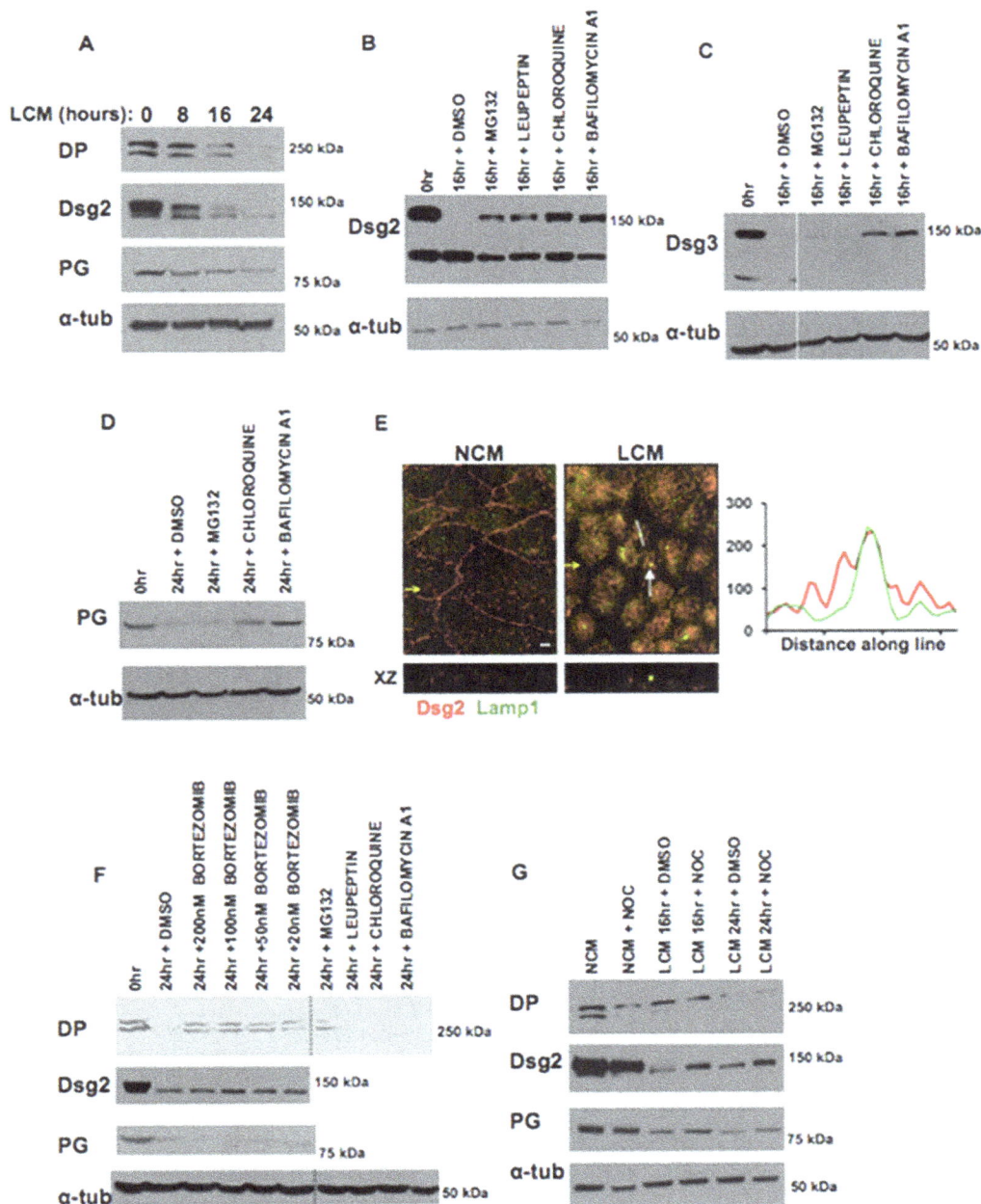

Figure 7. Desmosomal proteins are degraded by lysosomes and proteasomes, and co-localise with the lysosomal marker lamp1. (A) Internalised desmosomal proteins are gradually degraded. Western blots of whole MDCK cell lysates following LCM treatment for 0, 8, 16 or 24 hours show that Dsg2, DP and PG were gradually degraded. (B–D) Lysosomal and proteasomal degradation. Cells were treated with LCM for 16 or 24 hours in the presence of the noted inhibitors or vehicle alone (DMS0 or water). Western blots of whole cell lysates show that the lysosomal inhibitors bafilomycin A1 (250 nM), chloroquine (100 µM) and leupeptin (100 µM) and the proteasomal inhibitor MG132 (10 µM) inhibited degradation of Dsg2 (B) and chloroquine and bafilomycin A1 inhibited LCM-induced Dsg3 and PG degradation (C, D). (E) Internalised Dsg2 co-localises with the lysosomal marker lamp1. Cells cultured in NCM or treated with LCM for 16 hours stained for Dsg2 (red) and Lamp1 (green). Co-localisation in the latter cells is indicated by white arrows.Yellow arrows indicate XZ axis. Bar, 5 µm. Fluorescence profile depicts the intensity of staining along the white line in the image. (F) DP degradation was not inhibited by lysosomal inihibitors, but instead by the proteasomal inhibitors bortezomib (20–200 nM) and MG132 (10 µM) (dashed lines indicate lanes which have been re-ordered from the same western blot). Bortizomib had no effect on PG or Dsg2 degradation (F). (G) Western blots of whole cell lysates co-treated with LCM and nocodazole (33 µM) for 16 or 24 hrs shows degradation of DP is unaltered, whilst degradation of PG and Dsg2 is partially inhibited. Bar, 5 µm.

inhibited the degradation of Dsg2, Dsg3 and PG (Figure 7B–D). Lysosomal degradation of desmosomal proteins was substantiated by co-localisation of Dsg2 with the lysosomal marker Lamp1 in cells treated with LCM for 16 hours (Pearson's score ~0.7) (Figure 7E).

Intriguingly, no lysosomal inhibitor affected DP degradation. However, the specific 26S proteasome inhibitor bortezomib [48] blocked DP degradation substantially (Figure 7F). Bortezomib had no inhibitory effect on degradation of Dsg2 or PG (Figure 7F). MG132, a non-specific proteasome inhibitor, blocked degradation of both DP and Dsg2 (Figure7, F). These results suggest that (i) internalised desmosomal proteins are differentially degraded by lysosomes and proteasomes, and (ii) the centrosome is not essential in mediating the proteasomal degradation of DP, although disruption of the microtubule network partially disrupts lysosomal degradation.

Proteasomes have been shown to co-localise with the centrosome [49,50]. To determine whether the co-localisation of internalised desmosomal halves with the centrosome (see above) is essential for desmosome degradation cells were co-treated with nocodazole to block desmosome transport to the centrosome. Nocodazole treatment did not inhibit the degradation of DP, although there was a partial inhibition of PG and Dsg2 degradation (Figure 7G).

Discussion

Down-regulation of desmosomal adhesion is required to facilitate epithelial cell migration in embryonic development, wound healing and cancer invasion, but the mechanism of down-regulation is poorly understood. We have provided an outline of the process by which desmosomes are down-regulated in culture following chelation of extracellular calcium (Figure 8). We demonstrate that the internalisation of half desmosomes is dependent on the actin cytoskeleton and conventional PKC isoforms (Figure 8B, C). Once internalised desmosomal halves remain essentially intact and are conveyed to the region of the centrosome by microtubule transport (Figure 8D). Constituent proteins of internalised desmosomal halves are not recycled to the cell surface but instead are slowly degraded by lysosomes and proteasomes. We believe this is the first time that degradation of a cellular organelle has been shown to involve both lysosomal and proteasomal activity.

That the internalised structures remain intact is consistent with our observation that Dsg2 and Dsg3are not recycled; recycling would not be expected unless the desmosomal halves themselves were recycled. Demlehner et al. [51] showed that recycling of desmosomal halves may occur in HaCaT cells, but previous work by us and others suggests that this does not occur in MDCK cells, or at least is not a major mechanism [17,24]. The process of whole or half desmosome internalisation seems distinct from what is traditionally called "desmosome disassembly", which implies that desmosomes separate into their component molecules. While our results show that this does not occur in the model we have studied, they do not, of course, rule out the existence of such a process.

Because PKCα becomes associated with desmosomal plaques in wound edge epithelium in association with a weakening of desmosomal adhesion, and because no half desmosomes were detected, we speculated that PKCα might be involved in priming desmosomes for internalisation [12]. If desmosome internalisation is a type of phagocytosis, as postulated by Allen and Potten [11], a role for PKC might be suggested since much evidence suggests that PKC is involved in phagocytosis/endocytosis [52–54]. Our observation that inhibition of cPKC isoforms blocks internalisation of desmosomal halves is consistent with a role for PKC, and since PKCα was the only conventional isoform detectable in MDCK cells [15], this is likely to be the isoform involved. It will be interesting to discover whether the key targets for PKC are desmosomal proteins or components of the actin cytoskeleton, which is also involved in the internalisation process. It has recently been shown that loss of keratin from the epidermis activates PKCα, causing

phosphorylation of DP and destabilisation of desmosomes [55]. Actin, like PKC, plays a key role in phagocytosis [32,56] providing further support that desmosome internalisation resembles phagocytosis.

We demonstrate that intracellular transport of internalised half desmosomes is microtubule-dependent and probably involves kinesins. Microtubules have previously been associated with two aspects of desmosome function. Firstly, DP is involved in microtubule organisation in differentiated cells of the epidermis [57]. Here desmosomes associate with a number of microtubule end binding/centrosomal proteins including ninein, Lis1, Ndel1 and CLIP170 [58,59]. We found that neither ninein nor CLIP170 was associated with half desmosomes during the internalisation and intracellular transport processes (not shown). Secondly, microtubules and kinesins have been shown to function in the transport of desmosomal cadherins to the cell periphery during desmosome assembly in cultured cells, although there is some controversy about the involvement of microtubules in desmosome assembly [60–62].

Microtubules are organised in a stellate array with their negative ends anchored at the centrosome so that minus end-directed transport should logically end there. Indeed we have shown that internalised half desmosomes first surround, then co-localise with the centrosome. A number of cellular proteins have previously been shown to co-localise with the centrosome. These include IκBα, Dlc-1, hsp70 and p53 [50,63,64]. We believe that our work constitutes the first demonstration that an internalised cell surface structure localises to the centrosome.

The significance of localisation to the centrosome may lie in the observation that internalised half desmosomes are degraded rather than recycled to the cell surface. The centrosome has shown to be a site of proteasome localisation [49,50] where proteasomal activity appears to regulate both centrosome function and the degradation of cellular proteins [49,50,65,66]. We show that DP, the most peripheral desmosomal component, is proteasomally degraded. However, nocodazole-induced blocking of desmosome transport to the centrosome did not inhibit DP degradation possibly indicating shuttling of the 26S proteasome to a differing cellular location as has been well reported [67]. Partial inhibition of Dsg2 and PG degradation by microtubule disruption could be explained by the requirement of intact microtubules for final stages of lysosomal degradation following autophagy [68]. Alternatively, as lysosomes and endosomes are reportedly concentrated close to the microtubule organising centre [69], microtubule disruption may prevent lysosomal degradation there.

Our studies suggest that degradation of internalised half desmosomes is complex, involving both the proteasome and the lysosome. Thus degradation of the desmosomal proteins other than DP was blocked by lysosomal inhibitors and the internalised material co-localised with the lysosomal marker Lamp1. Lysosomal degradation was suggested previously by association of half desmosomes with late endosomes [46], structures which fuse with lysosomes [70].

Our work has revealed a number of novel aspects of desmosome down regulation in this model system. Calcium switching is clearly artificial; extracellular calcium concentrations have to be maintained relatively constant in vivo in order to preserve nerve and muscle function. Nevertheless calcium switching has been widely used in cell biology to study junction assembly. Virtually nothing is known about the mechanism of junction down-regulation in vivo. What little is known suggests that there are certain similarities to the model we have studied. Our overview provides a basis for the detailed mechanistic analysis of each of the steps - internalisation, intracellular transport and degradation - and possibly also for the

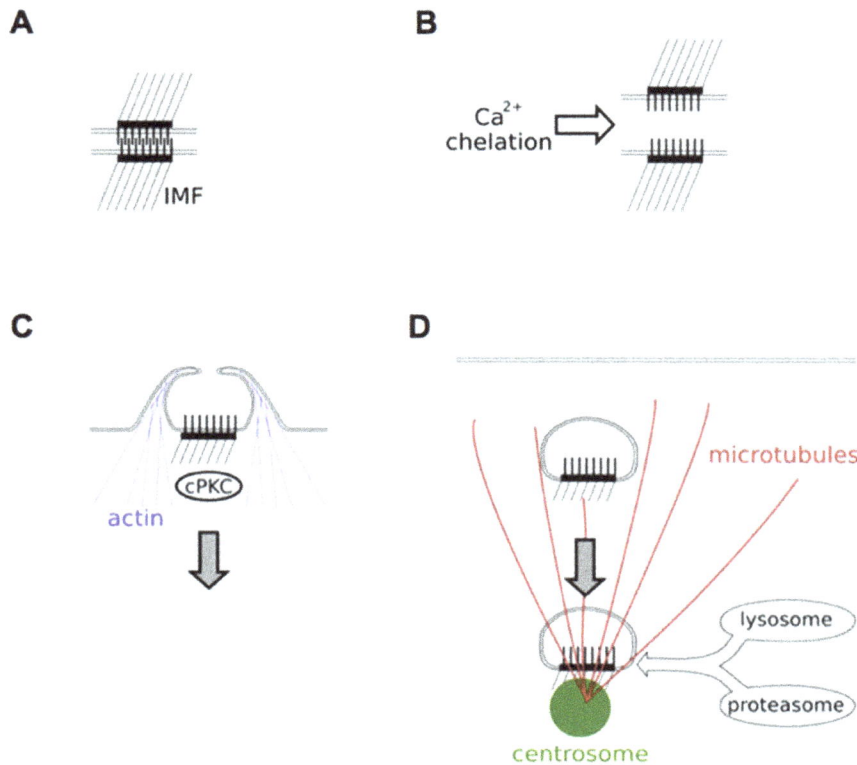

Figure 8. Model of the internalisation, transport and degradation of desmosomal halves. Treatment of calcium dependent desmosomes (A) with LCM causes loss of intercellular adhesion and formation of half desmosomes (B). These are internalised by a mechanism dependent on cPKC and actin filaments (C). Once internalised, desmosomal halves are transported in a microtubule/kinesin dependent manner to the centrosome and degraded by lysosomes and proteasomes (D).

understanding of the process in vivo. When they are induced to undergo epithelial-to-mesenchymal transition (EMT) by growth factors or artificial serum substitutes, epithelial cells in culture down-regulate desmosomes in the presence of calcium by a process that bears some resemblance to what we describe but which is slower and less synchronous [71,72].We are now investigating desmosome behaviour during EMT to determine whether it resemble that reveal by calcium chelation.

Supporting Information

Video S1 Live-imaging video of MDCK Dsc2a-YFP cells cultured at sub-confluence for 24 hours and then transfected with pericentrin-RFP for a further 24 hours using Fugene 6. LCM treatment induced detachment and rounding up of cells, and the internalisation of Dsc2a-YFP. Internalised Dsc2a-YFP initially surrounds pericentrin-RFP and is then transported towards it, with colocalisation occurring at around 1 hour 40 minutes LCM treatment. Images were taken every 2 minutes.

Figure S1   Single images from video S1 show Dsc2aYFP at the plasma membrane at time point 0 (A) surrounding pericentrin-RFP (B) at 45 minutes LCM treatment and co-localising with pericentrin-RFP (C) following 1 hour 40 minutes LCM treatment.

Acknowledgments

We thank the following for reagents: Birgit Lane, Stephen Taylor, Stephen High, Philip Woodman, Christiane Bierkamp, Sean Munro, Norbert Fusenig and Rudolph Leube. We also thank Stephen High and Philip Woodman for helpful advice and comments. We thank Peter March for help with microscopy.

Author Contributions

Conceived and designed the experiments: DG SM GH. Performed the experiments: SM GH LL. Analyzed the data: DG SM GH LL. Contributed reagents/materials/analysis tools: DG. Wrote the paper: DG SH.

References

1. Amagai M, Klaus-Kovtun V, Stanley JR (1991) Autoantibodies against a novel epithelial cadherin in pemphigus vulgaris, a disease of cell adhesion. Cell 67: 869–877.

2. McKoy G, Protonotarios N, Crosby A Tsatsopoulou A, Anastasakis A, et al. (2000) Identification of a deletion in plakoglobin in arrhythmogenic right ventricular cardiomyopathy with palmoplantar keratoderma and woolly hair (Naxos disease). Lancet 355: 2119–2124.

3. Stanley JR, Koulu L, Thivolet C (1984) Distinction between epidermal antigens binding pemphigus vulgaris and pemphigus foliaceus autoantibodies. J Clin Invest 74: 313–320.

4. Vasioukhin V, Bowers E, Bauer C, Degenstein L, Fuchs E (2001) Desmoplakin is essential in epidermal sheet formation. Nat Cell Biol 3: 1076–1085.

5. Yang Z, Bowles NE, Scherer SE, Taylor MD, Kearney DL, et al. (2006) Desmosomal dysfunction due to mutations in desmoplakin causes arrhythmogenic right ventricular dysplasia/cardiomyopathy. Circ Res 99: 646–655.

6. Delva E, Tucker DK, Kowalczyk AP (2009) The desmosome. Cold Spring Harb Perspect Biol 1: a002543.

7. Garrod D, Chidgey M (2008) Desmosome structure, composition and function. Biochim Biophys Acta 1778: 572–587.

8. Green KJ, Getsios S, Troyanovsky S, Godsel LM (2010) Intercellular junction assembly, dynamics, and homeostasis. Cold Spring Harb Perspect Biol 2: a000125.

9. Green KJ, Simpson CL (2007) Desmosomes: new perspectives on a classic. J Invest Dermatol 127: 2499–2515.

10. Thomason HA, Scothern A, McHarg S, Garrod DR (2010) Desmosomes: adhesive strength and signalling in health and disease. Biochem J 429: 419–433.

11. Allen TD, Potten CS (1975) Desmosomal form, fate, and function in mammalian epidermis. J Ultrastruct Res 51: 94–105.

12. Garrod DR, Berika MY, Bardsley WF, Holmes D, Tabernero L (2005) Hyper-adhesion in desmosomes: its regulation in wound healing and possible relationship to cadherin crystal structure. J Cell Sci 118: 5743–5754.

13. Kimura TE, Merritt AJ, Garrod DR (2007) Calcium-independent desmosomes of keratinocytes are hyper-adhesive. J Invest Dermatol. 127: 775–781.

14. Kimura TE, Merritt AJ, Lock FR, Eckert JJ, Fleming TP, et al. (2012) Desmosomal adhesiveness is developmentally regulated in the mouse embryo and modulated during trophectoderm migration. Dev Biol 369: 286–297.

15. Wallis S, Lloyd S, Wise I, Ireland G, Fleming TP, et al.(2000). The alpha isoform of protein kinase C is involved in signaling the response of desmosomes to wounding in cultured epithelial cells. Mol Biol Cell 11: 1077–1092.

16. Thomason HA, Cooper NH, Ansel DM, Chiu M, Merrit AJ, et al. (2012) Direct evidence that PKCalpha positively regulates wound re-epithelialization: correlation with changes in desmosomal adhesiveness. J Pathol 227: 346–356.

17. Mattey DL, Garrod DR (1986) Splitting and internalization of the desmosomes of cultured kidney epithelial cells by reduction in calcium concentration. J Cell Sci 85: 113–124.

18. Hennings H, Holbrook KA (1983) Calcium regulation of cell-cell contact and differentiation of epidermal cells in culture. An ultrastructural study. Exp Cell Res 143:127–142.

19. Watt FM, Mattey DL, Garrod DR (1984) Calcium-induced reorganization of desmosomal components in cultured human keratinocytes. J Cell Biol 99:2211–2215.

20. Kartenbeck J, Schmid E, Franke WW, Geiger B (1982) Different modes of internalization of proteins associated with adhaerens junctions and desmosomes: experimental separation of lateral contacts induces endocytosis of desmosomal plaque material. EMBO 1: 725–732.

21. Overton J (1962) Desmosome development in normal and reassociating cells of the chick blastoderm. Dev Biol 4: 532–548.

22. Boukamp P1, Petrussevska RT, Breitkreutz D, Hornung J, Markham A, et al. (1988) Normal keratinization in a spontaneously immortalized aneuploid human keratinocyte cell line. J Cell Biol 106: 761–771.

23. Madin SH, Darby NB Jr (1958) Established kidney cell lines of normal adult bovine and ovine origin. Proc Soc Exp Biol Med 98: 574–576.

24. Windoffer R, Borchert-Stuhltrager M, Leube RE (2002) Desmosomes: interconnected calcium-dependent structures of remarkable stability with significant integral membrane protein turnover. J Cell Sci 115: 1717–1732.

25. Parrish EP, Steart PV, Garrod DR, Weller RO (1987) Antidesmosomal monoclonal antibody in the diagnosis of intracranial tumours. J Pathol 153: 265–273.

26. Vilela MJ, Hashimoto T, Nishikawa T, North AJ, Garrod D (1995) A simple epithelial cell line (MDCK) shows heterogeneity of desmoglein isoforms, one resembling pemphigus vulgaris antigen. J Cell Sci 108: 1743–1750.

27. Vilela MJ, Parrish EP, Wright DH, Garrod DR (1987) Monoclonal antibody to desmosomal glycoprotein 1—a new epithelial marker for diagnostic pathology. J Pathol 153: 365–375.

28. Parrish EP, Marston JE, Mattey DL, Measures HR, Venning R, et al. (1990) Size heterogeneity, phosphorylation and transmembrane organisation of desmosomal glycoproteins 2 and 3 (desmocollins) in MDCK cells. J Cell Sci 96: 239–248.

29. Wan H, South AP, Hart IR (2007) Increased keratinocyte proliferation initiated through downregulation of desmoplakin by RNA interference. Exp Cell Res. 313: 2336–2344.

30. Larsson C (2006) Protein kinase C and the regulation of the actin cytoskeleton. Cell Signal 18: 276–284.

31. Lim J P, Gleeson PA (2011) Macropinocytosis: an endocytic pathway for internalising large gulps. Immunol Cell Biol 89: 836–843.

32. Swanson JA (2008) Shaping cups into phagosomes and macropinosomes. Nat Rev Mol Cell Biol 9: 639–649.

33. Holm PK, Hansen SH, Sandvig K, van Deurs B (1993) Endocytosis of desmosomal plaques depends on intact actin filaments and leads to a nondegradative compartment. Eur J Cell Biol. 62: 362–371.

34. Bomsel M, Parton R, Kuznetsov SA, Schroer TA,Gruenberg J (1990) Microtubule- and motor-dependent fusion in vitro between apical and basolateral endocytic vesicles from MDCK cells. Cell 62: 719–731.

35. Goltz JS, Wolkoff AW, Novikoff PM, Stockert RJ, Satir P (1992) A role for microtubules in sorting endocytic vesicles in rat hepatocytes. Proc Natl Acad Sci U S A 89: 7026–30.

36. Hirokawa N, Noda Y, Tanaka Y, Niwa S (2009) Kinesin superfamily motor proteins and intracellular transport. Nat Rev Mol Cell Biol 10: 682–696.

37. Loubery S, Wilhelm C, Hurbain I, Neveu S, Louvard D, et al. (2008) Different microtubule motors move early and late endocytic compartments. Traffic 9: 492–509.

38. Noda Y, Okada Y, Saito N, Setou M, Xu Y, et al. (2001) KIFC3, a microtubule minus end-directed motor for the apical transport of annexin XIIIb-associated Triton-insoluble membranes. J Cell Biol 155: 77–88.

39. Bornens M (2002). Centrosome composition and microtubule anchoring mechanisms. Curr Opin Cell Biol 14: 25–34.

40. Doxsey S, Zimmerman W, Mikule K (2005) Centrosome control of the cell cycle. Trends Cell Biol 15: 303–311.

41. Gillingham AK, Munro S (2000) The PACT domain, a conserved centrosomal targeting motif in the coiled-coil proteins AKAP450 and pericentrin. EMBO Rep 1: 524–529.

42. Dukes JD, Fish L, Richardson JD, Blaikley E, Burns S, et al. (2011) Functional ESCRT machinery is required for constitutive recycling of claudin-1 and maintenance of polarity in vertebrate epithelial cells. Mol Biol Cell 22: 3192–3205.

43. Le TL, Joseph SR, Yap AS, Stow JL (2002) Protein kinase C regulates endocytosis and recycling of E-cadherin. Am J Physiol Cell Physiol 283: C489–499.

44. Le TL, Yap AS, Stow JL (1999) Recycling of E-cadherin: a potential mechanism for regulating cadherin dynamics. J Cell Biol 146: 219–232.

45. Morimoto S, Nishimura N, Terai T, Manabe S, Yamamoto Y, et al. (2005) Rab13 mediates the continuous endocytic recycling of occludin to the cell surface. J Biol Chem 280: 2220–2228.

46. Burdett ID (1993) Internalisation of desmosomes and their entry into the endocytic pathway via late endosomes in MDCK cells. Possible mechanisms for the modulation of cell adhesion by desmosomes during development. J Cell Sci 106: 1115–1130.

47. Schwartz AL, Trausch JS, Ciechanover A, Slot JW, Geuze (1992) Immunoelectron microscopic localization of the ubiquitin-activating enzyme E1 in HepG2 cells. Proc Natl Acad Sci U S A 89: 5542–5546.

48. Adams J, Kauffman M (2004) Development of the proteasome inhibitor Velcade (Bortezomib). Cancer Invest 22: 304–311.

49. Fabunmi RP, Wigley WC, Thomas PJ, DeMartino GN (2000) Activity and regulation of the centrosome-associated proteasome. J Biol Chem 275: 409–413.

50. Wigley WC, Fabunmi RP, Lee MG, Marino CR, Muallem S, et al. (1999) Dynamic association of proteasomal machinery with the centrosome. J Cell Biol 145: 481–490.

51. Demlehner MP, Schäfer S, Grund C, Franke WW (1995) Continual assembly of half-desmosomal structures in the absence of cell contacts and their frustrated endocytosis: a coordinated Sisyphus cycle. J Cell Biol. 131:745–760.

52. Cheeseman KL, Ueyama T, Michaud TM, Kashiwagi K, Wang D, et al. (2006) Targeting of protein kinase C-epsilon during Fcgamma receptor-dependent phagocytosis requires the epsilonC1B domain and phospholipase C-gamma1. Mol Biol Cell 17: 799–813.

53. Jong A, Wu CH, Prasadarao NV, Kwon-Chung KJ, Chang YC, et al. (2008) Invasion of Cryptococcus neoformans into human brain microvascular endothelial cells requires protein kinase C-alpha activation. Cell Microbiol 10: 1854–1865.

54. Larsen EC, DiGennaro JA, Saito N, Mehta S, Loegering DJ, et al. (2000) Differential requirement for classic and novel PKC isoforms in respiratory burst and phagocytosis in RAW 264.7 cells. J Immunol 165: 2809–2817.

55. Kröger C, Loschke F, Schwarz N, Windoffer R, Leube RE, et al. (2013) Keratins control intercellular adhesion involving PKC-α-mediated desmoplakin phosphorylation. J Cell Biol. 201: 681–692.

56. Yamada H, Ohashi E, Abe T, Kusumi N, Li SA, et al. (2007). Amphiphysin 1 is important for actin polymerization during phagocytosis. Mol Biol Cell 18: 4669–4680.

57. Lechler T, Fuchs E (2007) Desmoplakin: an unexpected regulator of microtubule organization in the epidermis. J Cell Biol 176: 147–154.

58. Sumigray KD, Chen H, Lechler T (2011) Lis1 is essential for cortical microtubule organization and desmosome stability in the epidermis. J Cell Biol 194, 631–642.

59. Wacker IU, Rickard JE, De Mey JR, Kreis TE (1992) Accumulation of a microtubule-binding protein, pp170, at desmosomal plaques. J Cell Biol 117: 813–824.

60. Nekrasova OE, Amargo EV, Smith WO, Chen J, Kreitzer GE, et al. (2011) Desmosomal cadherins utilize distinct kinesins for assembly into desmosomes. J Cell Biol 195: 1185–1203.

61. Pasdar M, Krzeminski KA, Nelson WJ (1991) Regulation of desmosome assembly in MDCK epithelial cells: coordination of membrane core and cytoplasmic plaque domain assembly at the plasma membrane. J Cell Biol 113: 645–655.

62. Pasdar M, Li Z, Krzeminski KA (1992) Desmosome assembly in MDCK epithelial cells does not require the presence of functional microtubules. Cell Motil Cytoskeleton 23: 201–212.

63. Brown CR, Doxsey SJ, White E, Welch WJ (1994) Both viral (adenovirus E1B) and cellular (hsp 70, p53) components interact with centrosomes. J Cell Physiol 160: 47–60.

64. Crepieux P, Kwon H, Leclerc N, Spencer W, Richard S, et al. (1997) I kappaB alpha physically interacts with a cytoskeleton-associated protein through its signal response domain. Mol Cell Biol 17: 7375–7385.

65. Didier C, Merdes A, Gairin JE, Jabrane-Ferrat N (2008) Inhibition of proteasome activity impairs centrosome-dependent microtubule nucleation and organization. Mol Biol Cell 19: 1220–1229.

66. Ehrhardt AG, Sluder G (2005) Spindle pole fragmentation due to proteasome inhibition. J Cell Physiol 204, 808–18.

67. Voges D, Zwickl P, Baumeister W (1999) The 26S proteasome: a molecular machine designed for controlled proteolysis. Annu Rev Biochem 68: 1015–1068.

68. Yang Y, Feng LQ, Zheng XX (2011) Microtubule and kinesin/dynein-dependent, bi-directional transport of autolysosomes in neurites of PC12 cells. Int J Biochem Cell Biol 43: 1147–1156.

69. Luzio JP, Pryor PR, Bright NA (2007) Lysosomes: fusion and function. Nat Rev Mol Cell Biol 8: 622–632.

70. Huotari J, Helenius A (2011) Endosome maturation. Embo J 30: 3481–3500.

71. Stoker M, Gherardi E, Perryman M, Gray J (1987) Scatter-factor is a fibroblast-derived modulator of epithelial cell mobility. Nature 327: 239–242.

72. Boyer B, Tucker GC, Vallés AM, Franke WW, Thiery JP (1989) Rearrangement of desmosomal and cytoskeletal proteins during transition from epithelial to fibroblastoid organization in rat bladder carcinoma cells. J. Cell Biol. 109: 1495–1509.

9

Effects of Secondary Metabolite Extract from *Phomopsis occulta* on β-Amyloid Aggregation

Haiqiang Wu[1], Fang Zhang[1], Neil Williamson[2], Jie Jian[2,3], Liao Zhang[1], Zeqiu Liang[1], Jinyu Wang[1], Linkun An[4], Alan Tunnacliffe[2]*, Yizhi Zheng[1]*

1 College of Life Sciences, Shenzhen University, Shenzhen, China, 2 Department of Chemical Engineering and Biotechnology, University of Cambridge, Cambridge, United Kingdom, 3 College of Pharmacy, Guilin Medical University, Guilin, China, 4 School of Pharmaceutical Science, Sun Yat-sen University, Guangzhou, China

Abstract

Inhibition of β-amyloid (Aβ) aggregation is an attractive therapeutic and preventive strategy for the discovery of disease-modifying agents in Alzheimer's disease (AD). *Phomopsis occulta* is a new, salt-tolerant fungus isolated from mangrove *Pongamia pinnata* (L.) Pierre. We report here the inhibitory effects of secondary metabolites from *Ph. occulta* on the aggregation of Aβ42. It was found that mycelia extracts (MEs) from *Ph. occulta* cultured with 0, 2, and 3 M NaCl exhibited inhibitory activity in an *E. coli* model of Aβ aggregation. A water-soluble fraction, ME0-W-F1, composed of mainly small peptides, was able to reduce aggregation of an Aβ42-EGFP fusion protein and an early onset familial mutation Aβ42E22G-mCherry fusion protein in transfected HEK293 cells. ME0-W-F1 also antagonized the cytotoxicity of Aβ42 in the neural cell line SH-SY5Y in dose-dependent manner. Moreover, SDS-PAGE and FT-IR analysis confirmed an inhibitory effect of ME0-W-F1 on the aggregation of Aβ42 *in vitro*. ME0-W-F1 blocked the conformational transition of Aβ42 from α-helix/random coil to β-sheet, and thereby inhibited formation of Aβ42 tetramers and high molecular weight oligomers. ME0-W-F1 and other water-soluble secondary metabolites from *Ph. occulta* therefore represent new candidate natural products against aggregation of Aβ42, and illustrate the potential of salt tolerant fungi from mangrove as resources for the treatment of AD and other diseases.

Editor: Jaya Padmanabhan, University of S. Florida College of Medicine, United States of America

Funding: Work in China was supported by Shenzhen City, China (Grant No. JCYJ2012061408533365, http://www.szsti.gov.cn/, YZ). Work in Cambridge was funded by the European Research Council (Advanced Investigator Grant 233232, http://erc.europa.eu/, AT) and an anonymous donation. The funders had no role in study design, data collection and analysis, decision to publish, or preparation of the manuscript.

Competing Interests: Work in China was supported by Shenzhen City, China (Grant No. JCYJ2012061408533365, http://www.szsti.gov.cn/, YZ). Work in Cambridge was funded by the European Research Council (Advanced Investigator Grant 233232, http://erc.europa.eu/, AT) and an anonymous donation. The funders had no role in study design, data collection and analysis, decision to publish, or preparation of the manuscript.

* Email: at10004@cam.ac.uk (AT); yzzheng@szu.edu.cn (YZ)

Introduction

Alzheimer's disease (AD) is a devastating condition leading to progressive cognitive decline, functional impairment and loss of independence, and is the major cause of dementia in the elderly worldwide [1]. Its prevalence will continue to increase as life expectancy increases. AD therefore represents a major and rising public health concern. However, as none of the medicines currently in use are able to cure this neurodegenerative disorder [2], understanding its etiology and developing new protective medicines have become the primary research goals in AD research.

Many clinicopathological studies have demonstrated that the deposition of beta-amyloid (Aβ) peptides, fragments of the amyloid precursor protein (APP), in brain parenchyma and cerebral blood vessels is one of the hallmarks of AD [3,4]. Although the molecular mechanism of its involvement in the development and progression of AD is not clear, a critical role for Aβ is universally acknowledged [5]. Aβ fibrils were once thought to be the main molecular culprit in AD, but recent studies show a more decisive correlation between the levels of soluble, non-fibrillar Aβ

oligomers and the extent of synaptic loss and cognitive impairment [6–8]. Compared with Aβ fibrils and plaques, Aβ oligomers are more potent as neurotoxins that cause disruption of neuronal synaptic plasticity [9,10]. The relationships between Aβ peptides, oligomerisation, cellular dysfunction and AD suggest that inhibition of Aβ oligomerisation might lead to novel therapeutics for the treatment of AD [11].

In addition to chemical pharmacological agents, bioactive extracts derived from natural products are attracting increasing attention in the search for new effective agents for the treatment of AD. Examples of such extracts that, when administered, led to inhibition of Aβ aggregation and related downstream pathological responses include aged garlic extract (AGE) [12], *Ginkgo biloba* extract (EGb761) [13], fungal endophytic extracts of Malaysian medicinal plants [14], *Alpinia galanga* (L.) fractions [15], Yokukansan extract [16], coffee extract [17], Samjunghwan extract [18], *Paeonia suffruticosa* extract [19], GEPT (a combination of extracts of ginseng, *Epimedium*, *Polygala* and tubers of the *Curcuma* genus) [20].

Marine microorganisms are a source of potentially useful natural extracts for the treatment of multifaceted diseases such as AD [21,22], and we focus here on microbes associated with mangroves, which are salt-tolerant, woody trees that grow in coastal habitats. Recently, we isolated and identified a new salt-tolerant endophytic fungus, *Phomopsis occulta SN3-2* (CCTCC No. 2011044), from mangrove *Pongamia pinnata* (L.) Pierre, and have assessed water-soluble secondary metabolites from *Ph. occulta* for inhibitory effects on the aggregation of Aβ42 in mammalian cells and *in vitro*. Here we show that a bioactive fraction, ME0-W-F1, from *Ph. occulta* mycelia extract can reduce formation of high molecular weight (HMW) Aβ42 oligomer and tetramer *in vitro* by inhibiting the formation of β-sheet secondary structure. Moreover, ME0-W-F1 is able to reduce the neurotoxic effect of Aβ42 in SH-SY5Y cells.

Materials and Methods

Reagents

Phomopsis occulta SN3-2 is a new species of fungus, identified tentatively by the Institute of Microbiology, Chinese Academy of Sciences, and maintained at the Shenzhen Key Laboratory of Microbial & Genetic Engineering, Shenzhen University, Shenzhen, China and also at the China Center for Type Culture Collection (CCTCC No. 2011044). Synthetic Aβ42 peptide was purchased from GenScript USA Inc. (Piscataway NJ, USA). (−)-Epigallocatechin gallate (EGCG) was obtained from Sigma-Aldrich Company Ltd.; stock solutions (10 mM) were freshly prepared in water. Diaion-20 resin hexafluoro-2-propanol (HFIP; Sigma) and all other chemicals were of reagent grade and commercially available.

Culture of *Phomopsis occulta* and preparation of its secondary metabolite extracts

Axenic cultures of *Ph. occulta* were maintained on potato dextrose agar. The cultures were transferred to liquid medium LB for 5–7 days, and then incubated in LB medium containing 0, 1, 2 or 3 M NaCl at 28°C without shaking for 40 days. These cultures were separated by filtration into mycelia and filtrates. The filtrates were concentrated to 2 L below 45°C in the dark, and extracted five times by shaking with an equal volume of ethyl acetate (EtOAc). After drying using anhydrous Na_2SO_4, collection and evaporation of EtOAc at 50°C *in vacuo* using a rotary evaporator (RV06-ML 1-B, IKA, Germany) yielded the fermentation broth extracts BE0, BE1, BE2 and BE3 (corresponding to cultures at 0, 1, 2 or 3 M NaCl, respectively). The mycelia were dried under vacuum and extracted three times using 2 L methanol for 72 h. Combination and evaporation of methanol yielded the mycelia extracts ME0, ME1, ME2 and ME3 (corresponding to cultures at 0, 1, 2 or 3 M NaCl, respectively).

Escherichia coli cell model

E. coli cell models of Aβ aggregation have been developed by others previously [23–25]. Briefly, *E. coli* cultures capable of producing a secretable form of Aβ42 fused to β-lactamase were grown overnight in LB supplemented with chloramphenicol (Cam) and then diluted 1:100 and grown for another 3 h at 37°C. These exponential phase cultures were diluted 1:50 in 96-well plates containing LB supplemented with 12.5 μg/mL Cam, 1 mM isopropyl-β-D-thiogalactopyranoside, 50 μg/mL ampicillin (Amp) and, as required, 200 μg/mL test samples, and EGCG was used as positive control (100 μg/ml). The plates were incubated at 37°C for 20 h without shaking. The OD_{600} was read and relative growth rate (%) calculated according to the following formula.

$$\text{Relative Growth Rate}(\%) =$$

$$\left(\frac{\Delta_{E.coli + Sample + Amp} - \Delta_{Sample + Amp}}{\Delta_{E.coli + Sample} - \Delta_{Sample}} \right) \times 100$$

Here, $\Delta_{E.coli+Sample+Amp}$ and $\Delta_{E.coli+sample}$ represent the changes in OD_{600} in *E. coli* cell and sample interaction systems after 20 h in the presence of Amp or not, respectively. $\Delta_{Sample+Amp}$ and Δ_{Sample} represent the changes in OD_{600} in sample systems after 20 h in the presence of Amp or not, respectively. *E. coli* cells are normally killed by Amp because they are unable to export β-lactamase linked to aggregated Aβ42 peptide. If Aβ42 aggregation is inhibited, β-lactamase can be exported and degrade Amp, allowing cell growth.

Purification of active fractions and identification by TLC

For the active *Ph. occulta* secondary metabolite extract, extraction and column chromatography were used for further purification. The active fraction ME0 was distributed between n-butyl alcohol and water phases. The water soluble components, ME0-W, were separated by column chromatography filled with Diaion-20 resin. Methanol/water was used as mobile phase, and five fractions (ME0-W-F1 to F5; 0, 5, 10, 30 and 50% methanol/water respectively (v/v)) were collected when the gradient elution was finished. Components soluble in n-butyl alcohol were not separated because of the absence of bioactivity. The inhibitory effect of these fractions on Aβ42 aggregation was assessed using the *E. coli* model described above.

The bioactive fraction, ME0-W-F1 (10 μl), was applied to cellulose precoated (20×20 cm) thin layer chromatography (TLC) plates (Merck, Germany). TLC plates were developed in a chloroform:methanol:water system (1:3:1 v/v), then air dried and visualized with iodine. Dried TLC plates were sprayed with ninhydrin reagent and heated at 80°C for 6 min. Peptide complexes became visible as intensely pink and purple-coloured bands and spots [26].

Cell toxicity studies

SH-SY5Y cells were maintained in Ham's F12 and DMEM medium, mixed in a 1:1 ratio, containing 2 mM glutamine, 1% nonessential amino acids, 500 μg/mL penicillin/streptomycin and 15% FBS, in an atmosphere of 5% CO_2. Cells were transferred to a sterile 96-well plate with approximately 25000 cells per well and allowed to acclimatize for 48 h. The Ham's F12/DMEM medium was removed by suction and replaced with Optimem medium (100 μL/well) containing either no ME0-W-F1 or ME0-W-F1 (10, 100 and 200 μg/mL, in phosphate-buffered saline (PBS): 137 mM NaCl, 2.7 mM KCl, 6.5 mM Na_2HPO_4, 1.76 mM KH_2PO4, pH 7.4). The cells were left for 24 h and then assessed using the MTT assay. For the protective effects of ME0-W-F1 on SH-SY5Y cells against Aβ42 aggregation, the Ham's F12/DMEM medium was replaced with Optimem medium (100 μL/well) containing either no Aβ or Aβ42 (10 μM), with and without the ME0-W-F1 (10, 100 and 200 μg/mL).

Flp-In T-REx 293 anti-aggregation assay

The Flp-In T-REx 293 (Invitrogen) cell line, a derivative of HEK293 cells containing a stably integrated FRT site and a TetR repressor, was maintained in DMEM media (Sigma D6171) supplemented with 10% fetal bovine serum (FBS), 5 mM L-glutamine, 5 μg/ml blasticidin. T-REx 293 cells were grown at 37°C under a 5% CO_2 atmosphere. The anti-aggregation screen was performed essentially as described [27]. Briefly 20,000 cells per well were seeded into a 24-well plate and allowed to attach for 48 h. Transient transfections were performed using GeneJammer (Agilent Technologies) as per manufacturer's instructions with either 0.75 µg of pcDNA3-Aβ42–EGFP [27] or pATNRW20. The latter construct expresses an early onset familial form of Aβ42 (Aβ42E22G) fused with the fluorescent protein mCherry (pcDNA3.3-Aβ42E22G-mCherry, N. Williamson et al., in preparation). ME0-W-F1 was added three hours post-transfection at the indicated concentrations, and gene expression was allowed to proceed for a further 48 h. An equivalent volume of dimethyl sulfoxide (DMSO) was used as a negative control and 10 µM epigallocatechin gallate (EGCG), a compound known to inhibit amyloid formation, was used as a positive control.

Quantification of Aβ42 aggregates was performed as described previously [27]. Approximately 200 GFP (Aβ42) or mCherry (Aβ42E22G) positive cells were counted for each treatment and cells were scored as positive if they contained one or more aggregates. Images were acquired on an Olympus IX81 inverted wide field microscope and all experiments were performed in triplicate and odds ratio analysis of aggregation data was performed using the statistical package GraphPad Instat 3. The nature of Aβ42 aggregates was also demonstrated by confocal microscopy, performed as described [27].

SDS-PAGE analysis

Preparation of synthetic Aβ42 solution was carried out according as described [28,29]. Briefly, Aβ42 peptides were dissolved in hexafluoro-2-propanol (HFIP) for 10–12 h with shaking, sonicated for 15 min, lyophilized, and redissolved in DMSO. Aβ42 concentrations were determined by OD_{280} in a Nanodrop 8000 spectrophotometer (Thermo Fisher) after diluting with PBS (pH 7.4).

The inhibitory effect of ME0-W-F1 on Aβ42 fibril formation was monitored by sodium dodecyl sulfate polyacrylamide gel electrophoresis (SDS-PAGE) under reducing conditions on 15% Tricine gels (Invitrogen) followed by Coomassie blue staining. In each experiment, Aβ42 solution was incubated with ME0-W-F1 at 37°C and 8 µl samples were removed at various time points, then pooled and analyzed by SDS-PAGE. Gel band intensities were quantified using Quantitative One software (Bio-Rad).

Fourier transform infrared (FTIR) spectroscopy

Measurements and evaluation were as described [25]. Spectra were collected on a NICOLET-6700 (Thermo Nicolet, USA) spectrometer at room temperature using a CaF_2 cell with a 50 µm Teflon spacer. Aβ42 stock solution (10 mg/mL in DMSO) was prepared according to section 2.7. ME0-W-F1 was prepared in DMSO at a concentration of 1 mg/mL. Mixtures were prepared by addition of Aβ42 and ME0-W-F1 stock solutions to unbuffered D_2O in a mass ratio of 1:1 and measurements taken at various time points. IR spectra were collected at 2 cm^{-1} resolution. Electrode readings were uncorrected for deuterium effects. CO_2 was removed and the air moisture inside the chamber was reduced by flushing the chamber with nitrogen gas. During each experiment, spectra were scanned 32 times over the range $4000–400 \text{ cm}^{-1}$. In some cases, the residual overlapping band was eliminated by subtraction from the final spectrum. The OMNIC software package (Thermo Nicolet, USA) was used for analysis of FT-IR spectra. Second derivative spectra were generated by using a 9-data point (9 cm^{-1}) function included in the OMNIC software package.

Data analysis

The data were expressed as mean ± SD, or mean of means ± SE, and were evaluated by two-way analysis of variance (ANOVA) followed by a post hoc test, or t-test. P<0.05 was considered to be significant.

Results

Preparation of *Ph. occulta* secondary metabolite extracts and screening of bioactive fractions

Ph. occulta is a salt-tolerant fungus and we established LB cultures at various concentrations of NaCl, i.e. 0, 1, 2 or 3 M. However, fermentation was affected by salt concentration, with growth rate in the order: 1>0>2>3 M NaCl. After filtration, fermentation broth extracts (BEs) and mycelia extracts (MEs) were

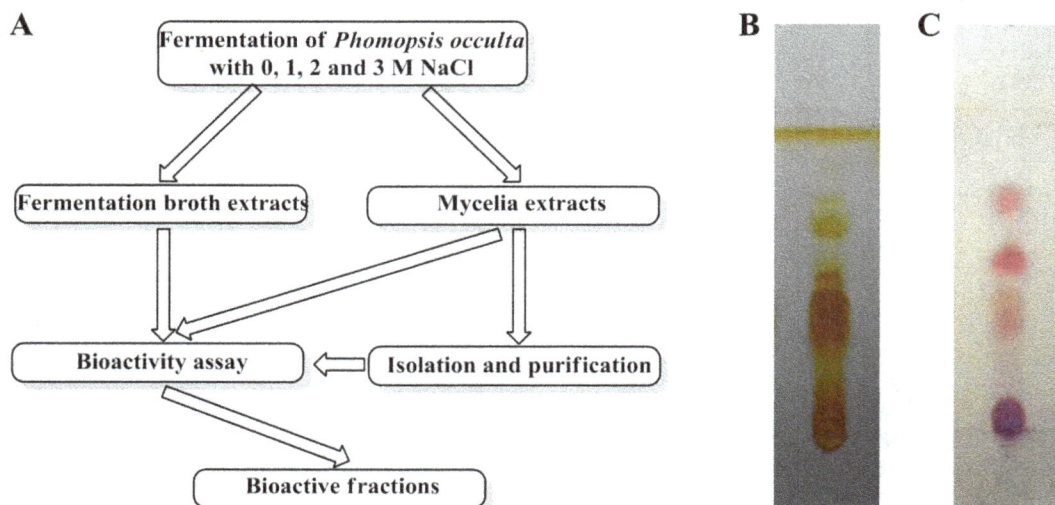

Figure 1. Preparation and analysis of secondary metabolites produced by *Phomopsis occulta*. A: Proposed strategy for preparation and screening of bioactive fractions from secondary metabolite extracts. B & C: Identification of ME0-W-F1 components by TLC analysis. B: visualized by iodide. C: visualized by ninhydrin.

prepared separately from each culture and labeled according to salt concentration (i.e. BE0, BE1, ME0, ME1 etc.), then purified as described in Materials and Methods. The strategy is outlined in Figure 1A. Peptides were the main components of MEs, as shown by TLC and stains such as iodide (Figure 1B) or ninhydrin (Figure 1C).

The effect of *Ph. occulta* secondary metabolites on the aggregation of Aβ42 was evaluated using an *E. coli* cell model. The fusion protein, ssTorA-Aβ42-Bla, was expressed in *E. coli*. In the presence of samples with Aβ42 aggregation inbitory effect, ssTorA-Aβ42-Bla can be transported into the extracellular space and degrade Amp. Thus, *E. coli* growth is proportional to the

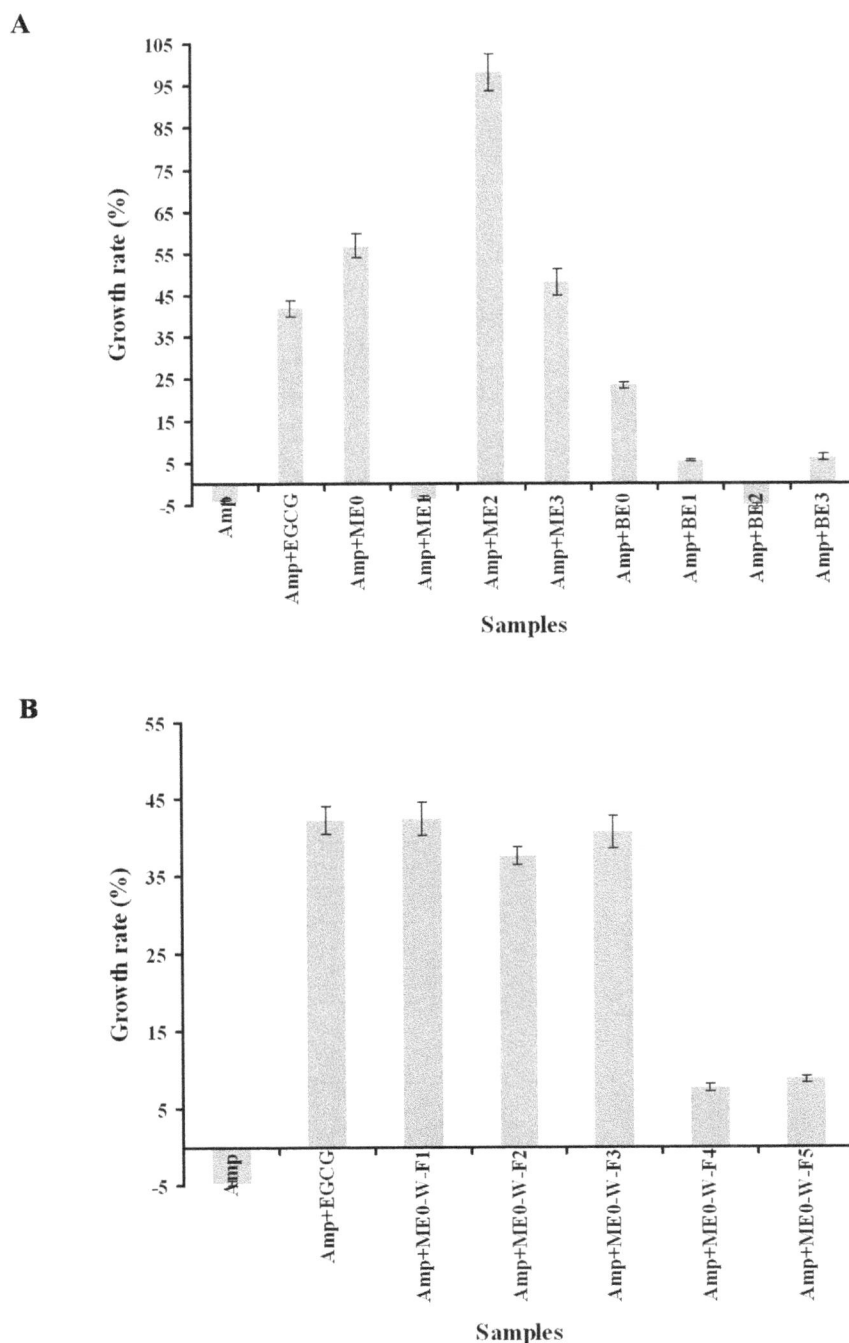

Figure 2. Screening of bioactive fractions from *Ph. occulta* secondary metabolite extracts using an *E. coli* cell model. A: Inhibitory effect on Aβ42 aggregation of *Ph. occulta* secondary metabolite extracts. ME: mycelia extracts; BE: broth extracts; 0, 1, 2 and 3 refer to molar salt concentrations in the cultures. B: Inhibitory effect on Aβ42 aggregation of ME0 fractions. ME0-W-F1 to ME0-W-F5 are water soluble fractions separated by column chromatography using Diaion-20 resin and a water/methanol mobile phase. Fraction concentrations were 200 μg/ml in each case; EGCG was used as positive control (100 μg/ml). Values represent mean of means ± SD of four separate experiments, each performed in triplicate.

inhibitory effect of samples on Aβ42 aggregation [25]. In most cases, growth rates were higher in the presence of MEs than in the presence of BEs. This indicated an inhibitory effect of MEs on the aggregation of Aβ42 in *E. coli*. Relative growth rates of *E. coli* cells with ME0, ME2 and ME3 were 57%, 98% and 48%, respectively, showing that all were at least as effective as the positive control, EGCG, which gave a relative growth rate of 42% (Figure 2A). The *E. coli* growth rate in the presence of ME1 was the same as that of the negative control (no additive), suggesting it had no effect on Aβ42 aggregation.

ME0 was selected for further study and was purified by column chromatography using a Diaion-20 resin with a water/methanol mobile phase. Fractions ME0-W-F1 to ME0-W-F5 were collected and re-tested in the *E. coli* assay: ME0-W-F1, ME0-W-F2 and ME0-W-F3 gave growth rates of 42.31%, 37.60% and 40.68%, respectively, similar to that of EGCG (42%), but ME0-W-F4 and ME0-W-F5 were less effective (Figure 2B). ME0-W-F1, which eluted with 100% H_2O, consisted largely of water-soluble peptides, and its proportion in ME0 was the highest (75% of the total mass). So, ME0-W-F1 was selected for further research.

Effect of ME0-W-F1 on Aβ42-induced cytotoxicity in SH-SY5Y cells

An MTT assay in the neuronal cell line SH-SY5Y was employed to explore the cytoprotective activity of ME0-W-F1. We showed that ME0-W-F1 did not affect the viability of SH-SY5Y cells, even at concentrations up to 200 μg/mL (Figure 3). In contrast, exposure to freshly prepared Aβ42 for 48 h was cytotoxic, producing a sharp decrease in SH-SY5Y viability, down to about 62% of control values. When ME0-W-F1 was added, however, the toxic effect of Aβ42 was significantly reduced in a dose-dependent manner, with cell viability of 77%, 84% and 89% at 10 μg/mL, 100 μg/mL and 200 μg/mL, respectively (Figure 3). Thus, ME0-W-F1 can reduce the cytotoxicity of Aβ42 significantly *in vitro*, in a similar fashion to the positive control, EGCG.

ME0-W-F1 reduces aggregation of fluorescently tagged Aβ42 in HEK293 cells

We have previously used Aβ42 aggregation in human cells as a screening tool to identify small molecules with anti-aggregation activity [27]. The effect of ME0-W-F1 was therefore tested in HEK293 cells transiently transfected with genes encoding either Aβ42-EGFP or Aβ42E22G-mCherry; the latter is a mutant form of Aβ42 associated with early onset AD. Both fluorescently tagged forms of Aβ42 aggregated in the human cell line (Figure 4; Figure S1). When ME0-W-F1 was added to cultures 3 h after transfection, however, the number of cells containing aggregates was reduced in an apparently dose-dependent manner. Therefore, ME0-W-F1 contains active components that can suppress aggregation of fluorescently tagged forms of Aβ42 in human cells, similar to the findings in bacteria.

Inhibitory effect of ME0-W-F1 on Aβ42 aggregation

SDS-PAGE was used to investigate the effect of ME0-W-F1 on Aβ42 aggregation *in vitro*. When Aβ42 was incubated at 37°C for 7 d in the absence of ME0-W-F1, four main forms of Aβ42 were visible on gels, i.e. monomer, dimer, tetramer and high weight molecular (HMW) oligomers, with the latter comprising about 50% of the material (Figure 5). However, the proportion of Aβ42 forming HMW oligomers was reduced to 32% and 7% after a 7 d incubation under the same conditions in the presence of low and high concentrations of ME0-W-F1, respectively (Figure 5; Figure S2 & S3). There was a corresponding, dose-dependent increase in the amount of Aβ42 monomer when ME0-W-F1 was present, suggesting that the water soluble fraction has an inhibitory effect on Aβ42 oligomerisation.

The formation of Aβ42 aggregates is characterised by a shift in conformation of the protein secondary structure from α-helix to β-sheet [30]. FT-IR spectroscopy allows this structural transition to be observed in the amide I band, 1600–1700 cm^{-1}, in which bands at ~1670 cm^{-1} and ~1627 cm^{-1} are characteristic of α-helix and β-sheet, respectively [31,32]. For Aβ42 alone, there is a progressive shift from α-helix to β-sheets over a 4 d period

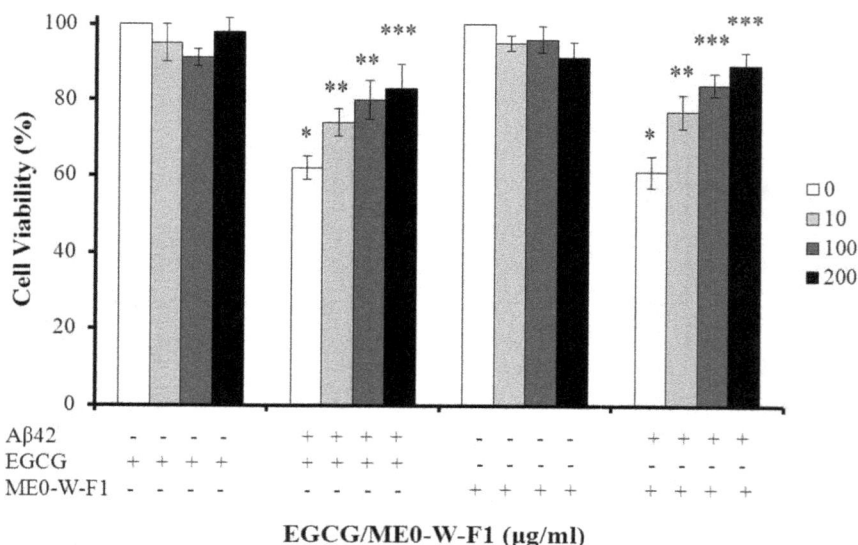

Figure 3. Protective effect of ME0-W-F1 in SH-SY5Y cells against cytotoxicity induced by aggregation of Aβ42 (10 μM), as shown by MTT analysis. Four concentrations (μg/ml) were used, with EGCG as positive control. Values represent mean of means ± SD of four separate experiments, each performed in triplicate (i.e. n = 12). The data were evaluated by two-way analysis of variance (ANOVA) followed by a post hoc test. *, **, ***, statistically significant from each other, $p < 0.05$. The treatments with EGCG or ME0-W-F1 were not significant.

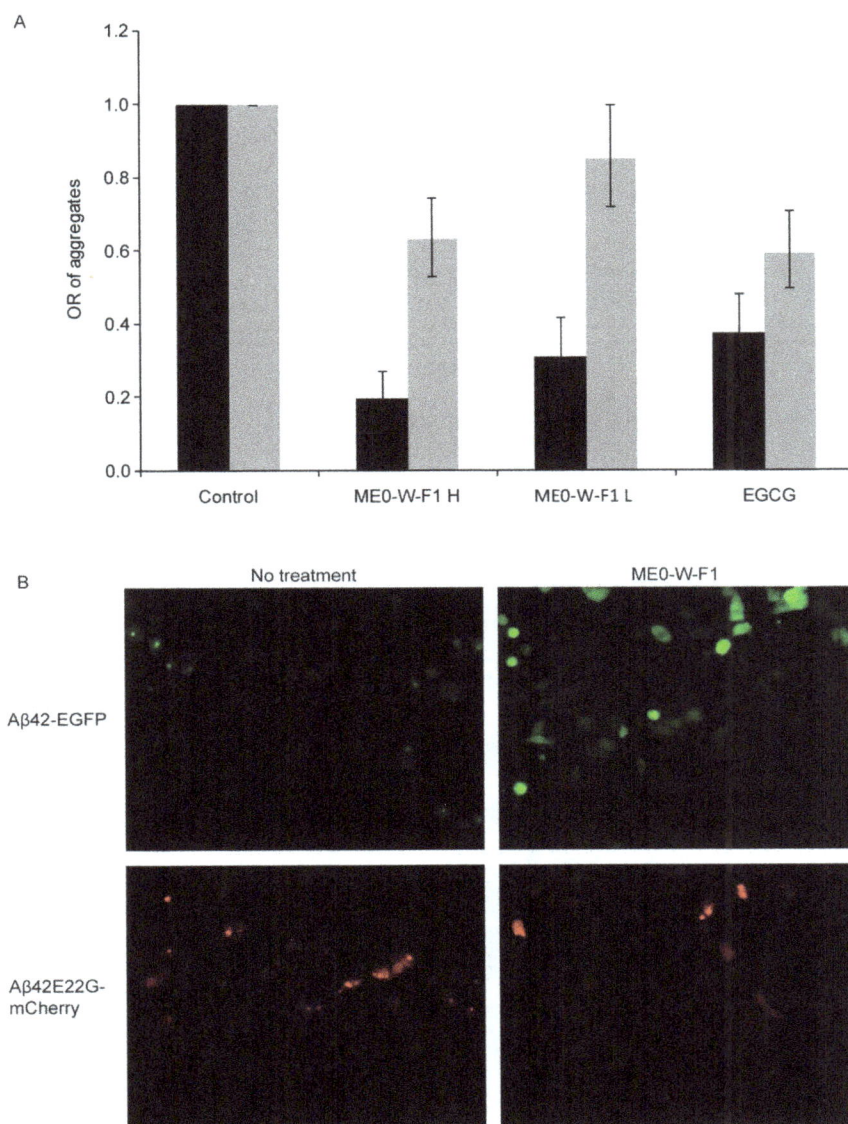

Figure 4. ME0-W-F1 reduces aggregation of Aβ42-EGFP and the early onset familial mutation Aβ42E22G-mCherry). (A, B) HEK293 transiently transfected with pcDNA3-Aβ42–EGFP (grey bars) and the Aβ42E22G-mCherry construct pATNRW20 (black bars) and treated with ME0-W-F1 at 17.5 μg/ml (H) and 1.75 μg/ml (L), a positive control (10 μM EGCG) and a negative control (DMSO only). For each construct and treatment, three fields of approximately 200 cells (i.e. n = 200 for each field) were counted for aggregates and odds ratios calculated. Error bars indicate 95% confidence interval for the odds ratio. Treatments of the Aβ42E22G-mCherry transfections with ME0-W-F1 (H and L) and EGCG were all statistically significant with a probability of P<0.0001. Treatments of the Aβ42-EGFP transient transfections with ME0-W-Fl (H) and EGCG were statistically significant with a probability of P<0.0001. The treatment with ME0-W-F1 (L) was not significant.

(Figure 6A). However, this transition is markedly reduced in the presence of ME0-W-Fl at 1 mg/mL (Figure 6B), with the proportion of β-sheet reducing from about 44% to 63% (Figure 6C). These data demonstrate that ME0-W-Fl can disrupt the transformation of α-helix to β-sheet associated with inhibition of the oligomerisation and aggregation of Aβ42.

Discussion

Aggregation of Aβ into plaques is a hallmark pathogenic feature of dementia and therefore is a primary target for amelioration of the disease [33]. Numerous chemical ligands have been developed as Aβ aggregation inhibitors in recent years including EGCG [34],

curcumin [35], scyllo-inositol [36] and LPFFD [37], but very few have progressed to clinical trials. In light of this disappointing situation, it is appropriate to search for alternative Aβ aggregation inhibitors among natural products. Some herb and fungal extracts have remarkable anti-AD activities in vivo and in vitro due to inhibition of Aβ aggregation [12–20]. Such studies justify further research on natural products, which could identify candidate lead compounds for AD treatment.

Evidence is accumulating that fungi are more likely to produce novel chemicals when they live in extreme environments [38]. Mangrove endophytic mycelium as a source of new microorganisms with potential pharmaceutical value has been intensively researched in recent years [39,40]. In this paper, the inhibitory

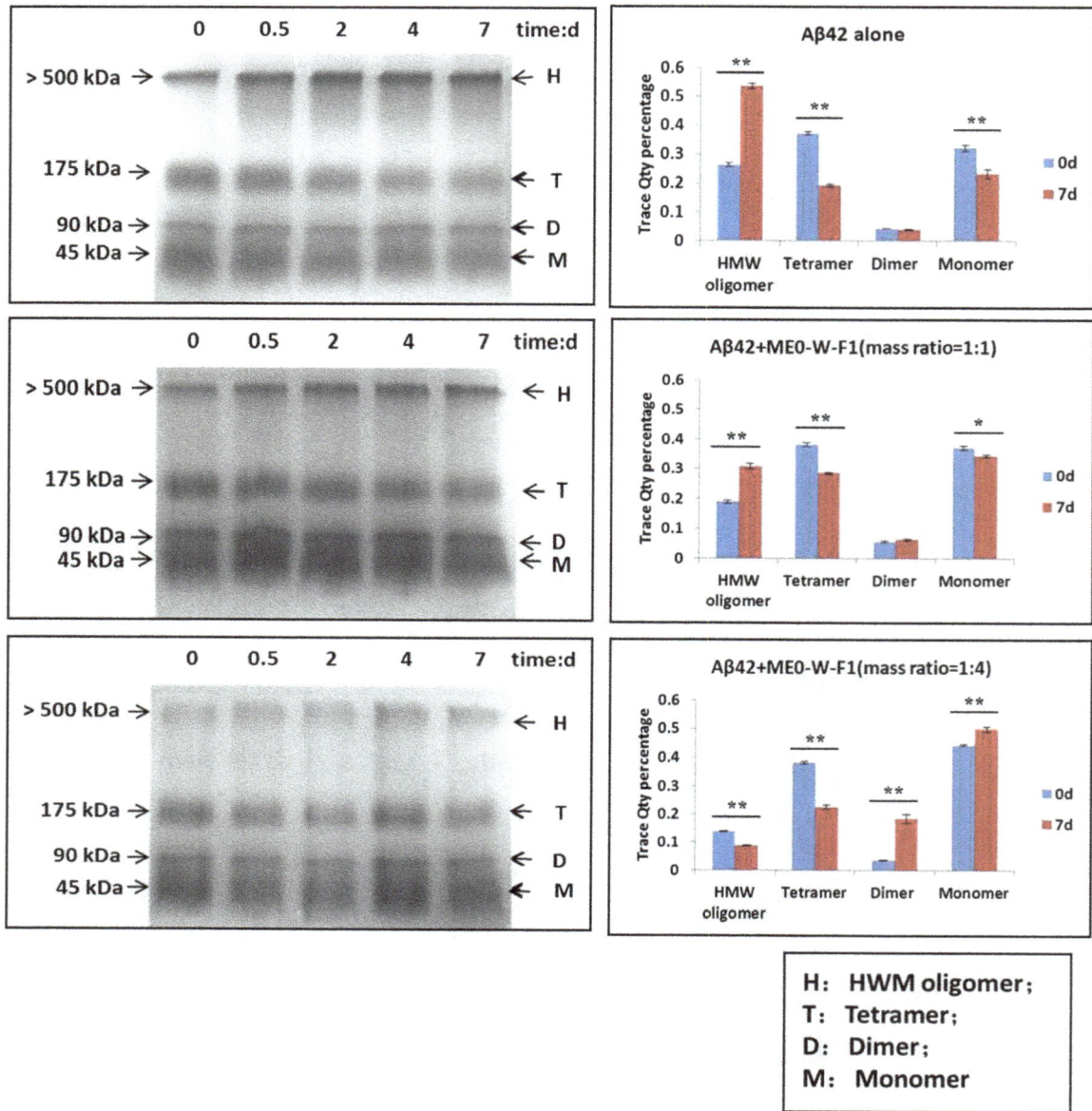

Figure 5. Effect of ME0-W-F1 on aggregation of Aβ42 analysed by SDS-PAGE and quantitative analysis with Quantitative One. Low (Aβ42: ME0-W-F1 = 1:1) and high (Aβ42: ME0-W-F1 = 1:4) concentrations of ME0-W-F1 were used. Values represent mean ± SD of three replicates. The data were evaluated by t-test, * $p<0.05$, ** $p<0.01$.

effects of secondary metabolites of *Ph. occulta*, a new endophytic fungus isolated from roots of the mangrove *Po. pinnata* (L.) Pierre, on the aggregation of Aβ42 *in vitro* and in cells were reported. Although its growth rate declined as the concentration of NaCl increased, *Ph. occulta* could be fermented in the presence of salt at various concentrations. This confirms *Ph. occulta* as a salt tolerant endophytic fungus.

The bioactivity of secondary metabolites from *Ph. occulta* is affected by the salt conditions during its growth. Thus, ME1, extracted from fungi grown at 1 M NaCl, has no clear growth-promoting, and hence anti-aggregation, effect in an *E. coli* model of Aβ42 aggregation. This might be because *Ph. occulta* grows naturally in sea water, which contains about 0.75 M NaCl, and is less stressed at 1 M NaCl than lower or higher salt concentrations. In contrast, MEs from *Ph. occulta* grown at 0 M, 2 M and 3 M

NaCl exhibited strong growth-promoting effects in the *E. coli* model, suggesting that certain secondary metabolites produced under such salt stress conditions have anti-aggregation activity. The water-soluble peptides in the selected bioactive fraction, ME0-W-F1, are candidates for such secondary metabolites. Because growth of *Ph. occulta* is very slow under high salt (2 M and 3 M NaCl) conditions, it was not possible to obtain sufficient quantities of material to test fractions such as ME2, which had strong bioactivity (Figure 2). Therefore, our analysis was limited to ME0 and related fractions. The BEs had no inhibitory effects on the aggregation of Aβ42.

ME0-W-F1 is active in human cells, as well as the *E. coli* model, reducing the cytotoxicity of Aβ42 in the SH-SY5Y cell line. Aβ oligomerisation and fibril formation are toxic to neurons, and these processes mediate Aβ toxicity mainly through interaction

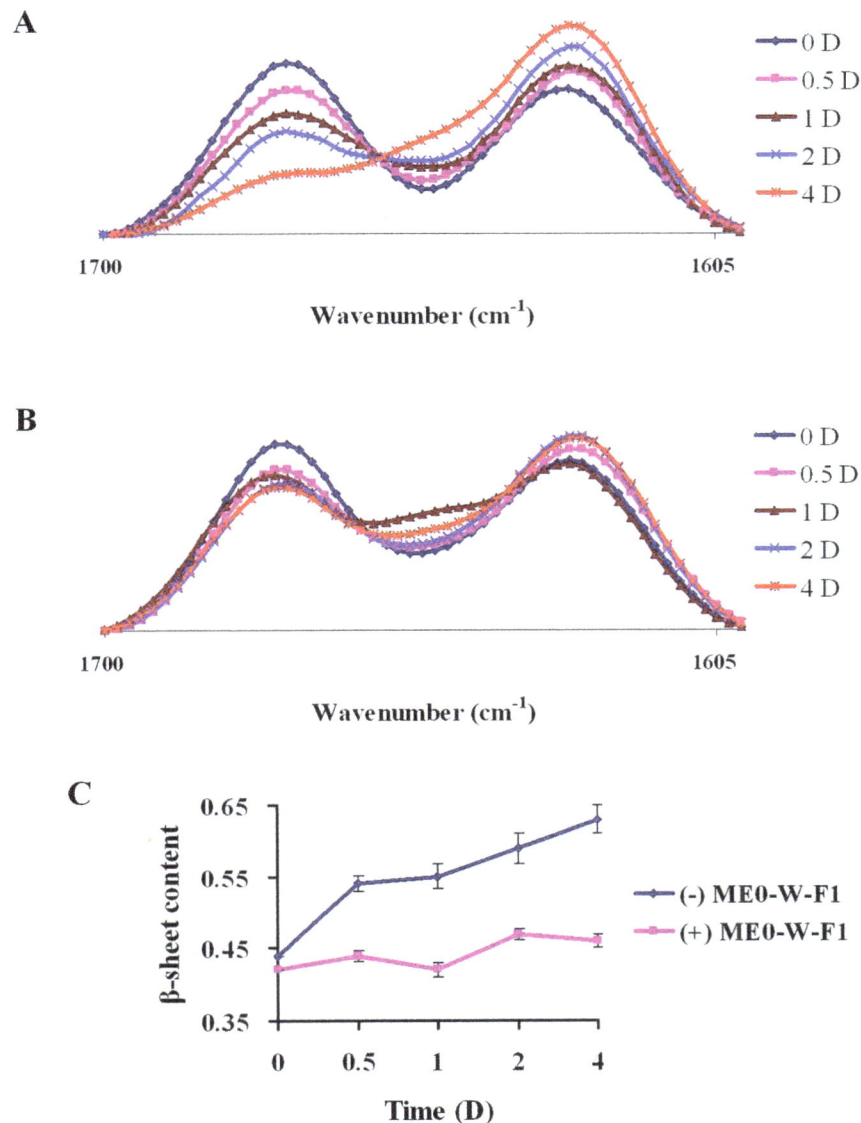

Figure 6. Inhibitory effects of ME0-W-F1 on the transformation of secondary structure of Aβ42 by FT-IR. A: Aβ42 alone; B: Aβ42 with ME0-W-F1; C: change in β-sheet content during incubation with (+) or without (−) ME0-W-F1. Time: 0, 0.5, 1, 2 and 4 days. Values represent mean ± SD of three separate experiments.

with other factors, e.g. Tau, in AD [5]. This suggests that ME0-W-F1 antagonises the oligomerisation and aggregation of Aβ42. An effect of ME0-W-F1 on intracellular Aβ42 aggregation was demonstrated in a HEK293 cell line, in which the water-soluble fraction reduced aggregation of both Aβ42 expressed as a fusion protein with EGFP and also an early onset form, Aβ42E22G, fused to mCherry; the fluorescent fusion partners allowed visualisation of aggregates within cells.

During the aggregation of Aβ *in vivo*, it is suggested that native Aβ peptides undergo conformational changes to form misfolded intermediates and various aggregated structures rich in β-sheet [41]. The transformation from α-helix to β-sheet is thought to be the rate-limiting step in the formation of soluble Aβ intermediates and oligomers, which are the most toxic Aβ species and are typically unstable, undergoing further aggregation to form higher-order oligomers and fibrillar deposits [7,42]. *In vitro*, ME0-W-F1

inhibits this structural transition from α-helix to β-sheet, as shown by both SDS-PAGE and FT-IR spectroscopy. Thus, SDS-PAGE demonstrated that the formation of tetramers and HMW oligomers of Aβ42 was disrupted in the presence of ME0-W-F1 in a dose- and time-dependent manner. Similarly, FT-IR spectroscopy showed that the shift from α-helix to β-sheet as Aβ42 aggregated was markedly reduced when ME0-W-F1 was present. These results suggest that ME0-W-F1 inhibits the oligomerisation and aggregation of Aβ42 through blocking the transformation of secondary structure and preventing subunit assembly.

Since ME0-W-F1 prevented or reduced the aggregation of intracellular Aβ42 fusion proteins in both bacteria and human cells, and since *in vitro* studies suggest an interaction with Aβ42 species, it seems likely that the active components of the water-soluble fraction must gain access to the intracellular space. The most likely mechanism of action is that these components interfere with Aβ42 aggregation within cells, as occurs *in vitro*, but we cannot rule out, for example, a stimulatory effect on molecular chaperone surveillance systems. The nature of the active components of ME0-W-F1 and the molecular mechanism of their action are currently under investigation.

In summary, water-soluble secondary metabolites from *Ph. occulta* exhibited inhibitory effects on the oligomerisation and aggregation of Aβ42 in cells and *in vitro*. Therefore, ME0-W-F1 and *Ph. occulta* are novel natural materials worthy of further investigation as potential therapeutic agents for AD.

Supporting Information

Figure S1 Confocal microscopy of Aβ42-EGFP and Aβ42E22G-mCherry aggregates in HEK293 cells. A: Aβ42-EGFP expression, showing DAPI staining of nuclei (blue), EGFP fluorescence (green) and a transmitted light image, together with a merged image. The main aggregate is visible as a bright fluorescent spot located adjacent to the nucleus in one cell. Non-aggregated Aβ42-EGFP can be seen as a less intense green fluorescence in the cytoplasm of this and other transfected cells. B: Aβ42E22G-mCherry expression, showing DAPI staining of nuclei (blue), mCherry fluorescence (red) and a transmitted light image, together with a merged image. Large aggregates are visible as bright fluorescent spots distributed around and within the nuclei of several cells. Non-aggregated Aβ42E22G-mCherry can be seen as a less intense red fluorescence in the cytoplasm of these cells.

Figure S2 Effect of ME0-W-F1 on aggregation of Aβ42 analysed by SDS-PAGE. Low (Aβ42: ME0-W-F1 = 1:1) concentration of ME0-W-F1 was used.

Figure S3 Effect of ME0-W-F1 on aggregation of Aβ42 analysed by SDS-PAGE. High (Aβ42: ME0-W-F1 = 1:4) concentration of ME0-W-F1 was used.

Author Contributions

Conceived and designed the experiments: HW AT YZ. Performed the experiments: HW FZ NW JJ LZ ZL JW LA. Analyzed the data: HW AT YZ. Contributed reagents/materials/analysis tools: HW FZ NW LZ ZL JW. Wrote the paper: HW AT YZ. Checking: HW.

References

1. Galimberti D, Scarpini E (2011) Disease-modifying treatments for Alzheimer's disease. Ther Adv Neurol Disord 4: 203–216.
2. Corbett A, Pickett J, Burns A, Corcoran J, Dunnett SB, et al. (2012) Drug repositioning for Alzheimer's disease. Nat Rev Drug Discov 11: 833–846.
3. Younkin SG (1995) Evidence that Abeta 42 is the real culprit in Alzheimer's disease. Ann Neurol 37: 287–288.
4. Haass C, Hung AY, Schlossmacher MG, Oltersdorf T, Teplow DB, et al. (1993) Normal cellular processing of the β-amyloid precursor protein results in the secretion of the amyloid β peptide and related molecules. Ann N Y Acad Sci 695: 109–116.
5. Ittner LM, Götz J (2011) Amyloid-β and tau-a toxic pas de deux in Alzheimer's disease. Nat Rev Neurosci 12(2): 65–72.
6. Hardy J, Selkoe DJ (2002) The amyloid hypothesis of Alzheimer's disease: Progress and problems on the road to therapeutics. Science 297: 353–356.
7. Walsh DM, Selkoe DJ. (2007) A beta oligomers - A decade of discovery. J Neurochem 101: 1172–1184.
8. Gessel MM, Wu C, Li H, Bitan G, Shea JE, et al. (2012) Aβ(39–42) modulates Aβ oligomerization but not fibril formation. Biochemistry 51: 108–117.
9. Bernstein SL, Dupuis NF, Lazo ND, Wyttenbach T, Condron MM, et al. (2009) Amyloid-β protein oligomerization and the importance of tetramers and dodecamers in the aetiology of Alzheimer's disease. Nat Chem 1: 326–331.
10. Selkoe DJ (2008) Soluble oligomers of the amyloid beta protein impair synaptic plasticity and behavior. Behav Brain Res 192: 106–113.
11. Amijee H, Bate C, Williams A, Virdee J, Jeggo R, et al. (2012) The N-methylated peptide SEN304 powerfully inhibits Aβ(1–42) toxicity by perturbing oligomer formation. Biochemistry 51: 8338–8352.
12. Ray B, Chauhan NB, Lahiri DK. (2011) The "aged garlic extract:" (AGE) and one of its active ingredients S-allyl-L-cysteine (SAC) as potential preventive and therapeutic agents for Alzheimer's disease (AD). Curr Med Chem 18: 3306–3313.
13. Luo Y, Smith JV, Paramasivam V, Burdick A, Curry KJ, et al. (2002) Inhibition of amyloid-beta aggregation and caspase-3 activation by the Ginkgo biloba extract EGb761. Proc Natl Acad Sci U S A 99: 12197–12202.
14. Harun A, James RM, Lim SM, Abdul Majeed AB, Cole AL, et al. (2011) BACE1 inhibitory activity of fungal endophytic extracts from Malaysian medicinal plants. BMC Complement Altern Med 11: 79.
15. Hanish Singh JC, Alagarsamy V, Diwan PV, Sathesh Kumar S, Nisha JC, et al. (2011) Neuroprotective effect of *Alpinia galanga* (L.) fractions on Aβ(25–35) induced amnesia in mice. J Ethnopharmacol 138: 85–91.
16. Fujiwara H, Takayama S, Iwasaki K, Tabuchi M, Yamaguchi T, et al. (2011) Yokukansan, a traditional Japanese medicine, ameliorates memory disturbance and abnormal social interaction with anti-aggregation effect of cerebral amyloid β proteins in amyloid precursor protein transgenic mice. Neuroscience 180: 305–313.
17. Dostal V, Roberts CM, Link CD (2010) Genetic mechanisms of coffee extract protection in a *Caenorhabditis elegans* model of β-amyloid peptide toxicity. Genetics 186: 857–866.
18. Kim HG, Ju MS, Park H, Seo Y, Jang YP, et al. (2010) Evaluation of *Samjunghwan*, a traditional medicine, for neuroprotection against damage by amyloid-beta in rat cortical neurons. J Ethnopharmacol 130: 625–630.
19. Fujiwara H, Tabuchi M, Yamaguchi T, Iwasaki K, Furukawa K, et al. (2009) A traditional medicinal herb *Paeonia suffruticosa* and its active constituent 1,2,3,4,6-penta-O-galloyl-beta-D-glucopyranose have potent anti-aggregation effects on Alzheimer's amyloid beta proteins in vitro and in vivo. J Neurochem 109: 1648–1657.
20. Tian J, Shi J, Zhang L, Yin J, Hu Q, et al. (2009) GEPT extract reduces Abeta deposition by regulating the balance between production and degradation of Abeta in APPV717I transgenic mice. Curr Alzheimer Res 6: 118–131.
21. Molinski TF, Dalisay DS, Lievens SL, Saludes JP (2009) Drug development from marine natural products. Nat Rev Drug Discov 8: 69–85.
22. Williams P, Sorribas A, Liang Z (2010) New methods to explore marine resources for Alzheimer's therapeutics. Curr Alzheimer Res 7: 210–213.
23. Fisher AC, Kim W, DeLisa MP (2005) Genetic selection for protein solubility enabled by the folding quality control feature of the twin-arginine translocation pathway. Protein Sci 15: 449–458.
24. Lee LL, Ha H, Chang YT, DeLisa MP (2009) Discovery of amyloid-beta aggregation inhibitors using an engineered assay for intracellular protein folding and solubility. Protein Sci 18: 277–286.
25. Wurth C, Guimard NK, Hecht MH (2002) Mutations that reduce aggregation of the Alzheimer's Abeta42 peptide: an unbiased search for the sequence determinants of Abeta amyloidogenesis. J Mol Biol 319: 1279–1290.
26. Ebada SS, Edrada RA, Lin W, Proksch P (2008) Methods for isolation, purification and structural elucidation of bioactive secondary metabolites from marine invertebrates. Nat Protoc 3: 1820–1831.
27. Chakrabortee S, Liu Y, Zhang L, Matthews HR, Zhang H, et al. (2012) Macromolecular and small-molecule modulation of intracellular Aβ42 aggregation and associated toxicity. Biochem J 442: 507–515.
28. LeVine H 3rd (2006) Biotin-avidin interaction-based screening assay for Alzheimer's beta-peptide oligomer inhibitors. Anal Biochem 356: 265–272.
29. Ying Z, Xin W, Jin-Sheng H, Fu-Xiang B, Wei-Min S, et al. (2009) Preparation and characterization of a monoclonal antibody with high affinity for soluble Abeta oligomers. Hybridoma (Larchmt) 28: 349–354.
30. Xu Y, Shen J, Luo X, Zhu W, Chen K, et al. (2005) Conformational transition of amyloid beta-peptide. Proc Natl Acad Sci U S A 102: 5403–5407.
31. Szabó Z, Klement E, Jost K, Zarándi M, Soós K, et al. (1999) An FT-IR study of the beta-amyloid conformation: standardization of aggregation grade. Biochem Biophys Res Commun 265: 297–300.
32. Lin SY, Chu HL (2003) Fourier transform infrared spectroscopy used to evidence the prevention of beta-sheet formation of amyloid beta(1–40) peptide by a short amyloid fragment. Int J Biol Macromol 32: 173–177.
33. Fändrich M, Schmidt M, Grigorieff N (2011) Recent progress in understanding Alzheimer's β-amyloid structures. Trends Biochem Sci 36: 338–345.
34. Ehrnhoefer DE, Bieschke J, Boeddrich A, Herbst M, Masino L, et al. (2008) EGCG redirects amyloidogenic polypeptides into unstructured, off-pathway oligomers. Nat Struct Mol Biol 15: 558–566.
35. Reinke AA, Gestwicki JE (2007) Structure-activity relationships of amyloid beta-aggregation inhibitors based on curcumin: influence of linker length and flexibility. Chem Biol Drug Des 70: 206–215.
36. Sun Y, Zhang G, Hawkes CA, Shaw JE, McLaurin J, et al. (2008) Synthesis of scyllo-inositol derivatives and their effects on amyloid beta peptide aggregation. Bioorg Med Chem 16(15): 7177–7184.

37. Bruce NJ, Chen D, Dastidar SG, Marks GE, Schein CH, et al. (2010) Molecular dynamics simulations of Aβ fibril interactions with β-sheet breaker peptides. Peptides 31: 2100–2108.

38. Jensen PR, Fenical W (1994) Strategies for the discovery of secondary metabolites from marine bacteria: ecological perspectives. Annu Rev Microbiol 48: 559–584.

39. Gutierrez RM, Gonzalez AM, Ramirez AM (2012) Compounds derived from endophytes: a review of phytochemistry and pharmacology. Curr Med Chem 19: 2992–3030.

40. Calcul L, Waterman C, Ma WS, Lebar MD, Harter C, et al. (2013) Screening mangrove endophytic fungi for antimalarial natural products. Mar Drugs 11: 5036–5050.

41. Schmidt M, Sachse C, Richter W, Xu C, Fändrich M, et al. (2009) Comparison of Alzheimer Abeta(1–40) and Abeta(1–42) amyloid fibrils reveals similar protofilament structures. Proc Natl Acad Sci U S A 106: 19813–19818.

42. Kayed R, Head E, Thompson JL, McIntire TM, Milton SC, et al. (2003) Common structure of soluble amyloid oligomers implies common mechanism of pathogenesis. Science 300: 486–489.

10

The Actin Binding Protein Adseverin Regulates Osteoclastogenesis

Siavash Hassanpour[1,9], Hongwei Jiang[1,2,9], Yongqiang Wang[1], Johannes W. P. Kuiper[1], Michael Glogauer[1]*

1 Matrix Dynamics Group, Faculty of Dentistry, University of Toronto, Toronto, Ontario, Canada, **2** Department of Operative Dentistry and Endodontics, Guanghua School of Stomatology, Sun Yat-sen University, Guangdong Provincial Key Laboratory of Stomatology, Guangzhou, P. R. China

Abstract

Adseverin (Ads), a member of the Gelsolin superfamily of actin binding proteins, regulates the actin cytoskeleton architecture by severing and capping existing filamentous actin (F-actin) strands and nucleating the assembly of new F-actin filaments. Ads has been implicated in cellular secretion, exocytosis and has also been shown to regulate chondrogenesis and megakaryoblastic leukemia cell differentiation. Here we report for the first time that Ads is involved in regulating osteoclastogenesis (OCG). Ads is induced during OCG downstream of RANK-ligand (RANKL) stimulation and is highly expressed in mature osteoclasts. The D5 isoform of Ads is not involved in regulating OCG, as its expression is not induced in response to RANKL. Three clonal Ads knockdown RAW264.7 (RAW) macrophage cell lines with varying degrees of Ads expression and OCG deficiency were generated. The most drastic OCG defect was noted in the clonal cell line with the greatest degree of Ads knockdown as indicated by a lack of TRAcP staining and multinucleation. RNAi mediated knockdown of Ads in osteoclast precursors resulted in distinct morphological changes characterized by altered F-actin distribution and increased filopodia formation. Ads knockdown precursor cells experienced enhanced migration while fusion of knockdown precursors cells was limited. Transient reintroduction of de novo Ads back into the knockdown system was capable of rescuing TRAcP expression but not osteoclast multinucleation most likely due to the transient nature of Ads expression. This preliminary study allows us to conclude that Ads is a RANKL induced early regulator of OCG with a potential role in pre-osteoclast differentiation and fusion.

Editor: Juha Tuukkanen, University of Oulu, Finland

Funding: This work is supported by a Canadian Institute of Health Research (CIHR) team research grant (MOP-384254). The funders had no role in study design, data collection and analysis, decision to publish, or preparation of the manuscript.

Competing Interests: The authors have declared that no competing interests exist.

* Email: michael.glogauer@utoronto.ca

9 These authors contributed equally to this work.

Introduction

The bone extracellular matrix (ECM) has historically been described as a static and protective scaffold [1]. Yet in reality, bone ECM is subjected to periodical remodeling to maintain its strength and integrity [2,3]. The task of skeletal remodeling falls in the domain of osteoclasts, which degrade the inorganic and organic phases of bone [1] and osteoblasts, which produce and secrete new matrix and regulate matrix mineralization [4]. Under normal circumstances bone destruction and formation are in steady state equilibrium. However, imbalances in bone remodeling result in perturbations of skeletal structure, integrity and function leading to diseases such as osteoporosis [5], osteopetrosis [6], inflammatory osteolysis such as rheumatic arthritis, periodontal disease [3,7] and Paget's bone disease [3]. Even though bone remodeling requires the collaborative action of osteoblasts and osteoclasts, the common thread to all the aforementioned disorders is abnormal bone resorption. Therefore, a thorough understanding of osteoclast formation or osteoclastogenesis (OCG) is crucial for development of novel drugs for treating bone-related diseases.

Osteoclasts are tissue specific multinuclear cells derived from hematopoietic stems cells [8] of the macrophage/monocyte lineage [9]. The intricate process of OCG, which involves coordinated cellular migration [10,11], adhesion and membrane fusion [12–14] is regulated by the critical hematopoietic cytokines, Macrophage Colony Stimulating Factor (M-CSF) and Receptor Activator of NF-κB Ligand (RANKL) [15]. OCG also requires dynamic regulation of cellular actin cytoskeleton. Cells organize their actin cytoskeleton through interactions with actin binding proteins [16–18] that control the length, flexibility and the viscosity of the actin network leading to changes in cell morphology and function. One such group of actin binding proteins is the Gelsolin superfamily, consisting of seven highly conserved members [17]. Adseverin (Ads), also known as Scinderin, is the closest homologue to the founding member of the Gelsolin superfamily. Like Gelsolin, Ads can promote F-actin depolymerization and nucleation depending on the intracellular conditions [19]. However, unlike Gelsolin, Ads's F-actin severing activity is inhibited by a wider range of membrane lipids [20–22], and its activation requires lower intracellular calcium concentrations [23]. Ads has been heavily implicated in regulating exocytosis through a rapid depolymerization of cortical actin in a number of biological systems, including chromaffin cells [24], airway goblet cells [25], murine pancreatic β-cell [26] and platelets [27]. In addition, Ads has been identified as a member of a multi-protein complex required for the trafficking of the water channel aquaporin-2 in rat renal collecting ducts [28] and has been shown to play a role in the differentiation of platelets [29] and chondrocytes [30]. A report by Robbens et al. [31] noted that Interleukin-9 stimulated T-helper lymphocytes express a splice variant of Ads, Ads D5, that is missing most of the fifth and a portion of the sixth Gelsolin like domains. Ads D5 has most of the typical characteristic common to all members of the Gelsolin family of actin binding proteins, with the exception of nucleation of filament assembly *in vitro*.

Surprisingly very little is known about the role of Ads during OCG. Unlike Gelsolin, which regulates osteoclast function and motility [32–34], a microarray analysis published by Yang et al. [35] is the only existing publication identifying Ads as a gene of interest in osteoclasts. A similar microarray performed in our lab also identified Ads as a gene of interest, which led us to test the hypothesis that Ads is a RANKL induced regulator of OCG with a potential functional role during osteoclast formation and function. The present study focused on investigating the role of Ads in OCG using two established *in vitro* model systems. It was shown for the first time that Ads is expressed during OCG in response to soluble RANKL (sRANKL) at both transcript and protein levels. Several Ads knockdown (KD) clonal cell lines with varying degrees of Ads expression reduction were generated. The clonal KD cell line with the greatest reduction in Ads expression failed to undergo OCG upon treatment with sRANKL, a phenotype most likely caused by a defect in osteoclast precursor fusion. The attenuation of Ads led to distinct morphological changes characterized by altered F-actin remodeling. The reintroduction of Ads was capable of rescuing the pre-osteoclast differentiation in the KD cells, in the form of TRAcP expression. Therefore, we concluded that Ads is a novel RANKL induced regulator of OCG with a functional role during the early stages of OCG.

Materials and Methods

2.1 Cell cultures

All procedures described were performed in accordance with the Guide for the Humane Use and Care of Laboratory Animals and were approved by the University of Toronto Animal Care Committee. Bone marrow monocytes (BMMs) were isolated from 6–12-week-old wild-type (WT) mice (SV129/BL6) as previously described [36]. OCG was induced by seeding 3×10^6 cells onto 60-mm² culture dishes in the Minimum Essential Medium – Alpha (α-MEM) (Gibco Life Technologies, Cat. No. 11095) containing 10% Fetal Bovine Serum (FBS) and 164 IU/mL of penicillin G, 50 μg/mL of gentamicin, and 0.25 μg/mL of fungizone (complete media), supplemented with 20 ng/ml M-CSF (Sigma, Cat No. M9170) for two days (unstimulated cells) or with 20 ng/ml M-CSF plus 30 ng/ml of sRANKL (Peprotech, Cat. No. 315-11) for up to 6 days. Fresh culture media supplemented with cytokines was added to the cells every two days. Cells were cultured at 37°C (5% CO_2). Murine RAW264.7 (RAW) macrophages (ATCC, provided by Keyin Li and Morris F Manolson at the University of Toronto) were cultured in the complete Dulbecco's Modified Eagle's Medium (DMEM) (Gibco Life Technologies, Cat. No. 11995). For osteoclast differentiation, 2.5×10^5 cells were seeded in 6-well culture dishes and cultured with 30 ng/ml sRANKL for up to 4 days. Unstimulated RAW cells were used as controls. Cells were cultured at 37°C (constant high humidity and 5% CO_2). All RAW cells used were between passage 5 and 12 to minimize the effects of passage on cell differentiation.

2.2 Microarray analysis

Total bone marrow cells from three mice were harvested, mixed and stimulated for 48 hours in complete α-MEM supplemented with 10 ng/ml M-CSF. Non-adherent cells were collected and centrifuged at 350×g for 30 minutes at 22°C over Ficoll-Paque PLUS (GE Healthcare, Cat No. 17-1440-02). A layer of mononuclear osteoclast precursors was obtained and split into two dishes, each containing 8×10^5 cells/cm². One dish was stimulated with 20 ng/ml M-CSF plus 200 ng/ml of purified sRANKL (as previously described [36]), and the other with 20 ng/ml M-CSF. After two days, the cells were washed with α-MEM and the total RNA was extracted (RNeasy Mini Kit, Qiagen, Cat. No. 74104). The quality of total RNA was determined using Agilent Technologies BioAnalyser. Microarray was performed using Affymetrix GeneChip Mouse Gene 1.0 ST array. Three independent experiments were performed making cells from three mice a single biological repeat. Gene expression profiles of cytoskeleton-associated genes were filtered. Gene normalization was performed prior to subsequent clustering computation using MultiExperiment Viewer 4.9.0 (MeV -TM4) [37]. MeV-HCL was run by defining Pearson Correlation metrics and an Average Linkage clustering method. Only genes with the statistically significant changes in expression of at least two-fold were reported. The data discussed in this publication have been deposited in NCBI's Gene Expression Omnibus and are accessible through GEO Series accession number GSE54779 (http://www.ncbi.nlm.nih.gov/geo/query/acc.cgi?acc=GSE54779).

2.3 Quantitative real-time PCR

Total RNA was extracted and residual genomic DNA was eliminated (RNase-Free DNase Kit, Qiagen, Cat. No. 79254). RNA concentration was determined using the Thermo Scientific Nanodrop 1000 Spectrophotometer. Ten random samples were analyzed for RNA quality using the Agilent Technologies Bioanalyzer. One μg of RAW and 120 ng BMM total RNA was reverse transcribed into cDNA using 200 units of Superscript II reverse transcriptase (Invitrogen, Cat. No. 18064-022) and 1 μM Oligo-dT$_{18}$ VN primers (ACGT Corp). A 1:32 cDNA dilution was used for all primer pairs to yield optimal PCR efficiency. Quantitative real-time PCR was performed in triplicate using the BioRad CFX96 real-time system. Each 20 μL reactions contained: 5 uL of 1:10 diluted cDNA, 500 nM of forward and reverse primers and 10 μL of SsoFast EvaGreen Supermix (Bio-Rad, Cat No. 172-5200). The PCR conditions were as follows:

95°C 2 min followed by 40 cycles of 95°C for 5 sec and 60°C/65°C for 5 sec. Melt curve analysis (95°C for 5 sec, 65°C for 5 sec, and 65° to 95°C with 0.5°C increase every 5 sec) of the amplified product was used to determine the specificity of the PCR reaction. The CFX Manger Software (Version 1.0) was used to analyze the PCR results. mRNA expression was normalized to the control, internal housekeeping gene GAPDH [38]. The following primers were used: *Ads* F: 5'-CCTATGGTGACTTTTACGTCGG-3', R: 5'-CTCATCCTGGGAACACTCCTT-3'; *GAPDH* F: 5'-CTTCCGTGTTCCTACCCC-3', R: 5'-GCCCAAGATGCCCTTCAGT-3'; *Gelsolin* F: 5'-CCTACCGCACATCCCCCAG-3', R: 5'-GTGATGGGGGTCCGCCTG-3', *RANK* F: 5'-CTAATCCAGCAGGGAAGCAAAT-3', R: 5'-GACACGGGCATAGAGTCAGTTC-3'; *SIRPα* F: 5'-TCGAGTGATCAAGGGAGCAT-3', R: 5'-CCTGGACACTAGCATACTCTGAG-3' and *CD44* F: 5'-TAGGAGAAGGTGTGGGCAG-3', R: 5'-AGGCACTACACCCCAATC-3'.

2.4 Western blot analysis

Immunoblotting was carried as previously described [36]. Equal amounts of total cell lysates (TCL) were resolved via SDS-PAGE on 8–12% polyacrylamide gels and transferred onto a nitrocellulose membrane (GE Healthcare, Cat. No. 8549062). Non-specific sites were blocked with Tris-buffered saline plus 0.05% v/v Tween 20 (TBS-T) with 5% nonfat milk powder for 1 hour. Incubations with primary antibodies were carried out overnight (O/N) at 4°C and secondary antibody for 1 hour at room temperature (RT). The following antibodies were used: Rabbit polyclonal anti-murine-Ads antibody (gift from Dr. C. Svensson [39], 1:2,000); Rabbit polyclonal anti-murine-Gelsolin antibody (gift from Dr. C. McCulloch, 1:2,000); Mouse monoclonal anti-murine ß-actin (Sigma, Cat No. A5316, 1:8,000); Rabbit polyclonal anti-murine Cathepsin K antibody (Abcam Inc, Cat No. ab19207, 1:1000); - conjugated donkey anti-rabbit IgG (GE Healthcare, Cat. No. NA934V); Sheep anti-mouse IgG-HRP (Amersham Pharmacia Biotech; Cat. No. NA931V). Immuno-reactive proteins were detected using chemiluminescence with Amersham ECL Plus Western Blotting Detection System (GE Healthcare, Cat. No. RPN2232), upon exposure to Bioflex MSI film (Clonex Corporation, Cat. No. CLMS810). Films were developed using the Kodak M35A X-OMAT Processor, scanned digitally using the Epson Perfection 1250 scanner and band intensities were quantified by densitometry using NIH ImageJ 1.41 software. Results are expressed as relative expression versus internal loading control, ß-actin.

2.5 Detection of Adseverin isoforms

Primers were designed to flank the fifth domain (D5) of Ads (esD5-F: 5'-GTGCAGGTCCGTGTCTCTC-3', esD5-R: 5'-GTGCAGGTCCGTGTCTCTC-3'). The design of the primers allowed for the specific detection of both Ads isoforms. Specifically, if the full-length Ads isoform was expressed, a 650 bp product was expected. However, if the smaller, Ads D5 isoform was expressed, a 350 bp product was expected (Fig. S1). Taq DNA polymerase (Qiagen, Cat. No. 201203) was used to non-quantifiably amplify the targeted sequences using cDNA from RAW macrophage and BMM-derived osteoclasts. PCR using 10 pg of pSE380-Ads and pSE380-Ads D5 (kind gift from Dr. J. Robbens [31]), and esD5 primers yielded 650 bp and 350 bp products respectively, thus allowing for the differentiation of two isoforms. To check for protein expression of Ads isoforms, 20 μg of TCL from unstimulated RAW and BMMs as well as TCL from Day 6 BMM cultures and Day 4 RAW cultures were resolved on an SDS-PAGE gel and immunoblotted against Ads. Lysates from BL21 E. coli (Agilent Technologies, Cat. No. 200132) transformed with pSE380-Ads or pSE380-Ads D5 and induced with 1 mM IPTG were used as positive controls. Two distinct bands at 85 kDa and 80 kDa are readily identifiable in lysates of BL21 E. coli induced to express exogenous Ads or Ads D5, respectively.

2.6 Retroviaral shRNA knockdown construct preparation

Ads specific (underlined) single stranded oligonucleotides were designed in accordance to the guidelines described by the Clontech RNAi systems (top strand: 5'-GATCCAACAAATAT-GAGCGTCTGATTCAAGAGATCAGACGCTCATATTTGT-TTTTTTTTACGCGTG-3', bottom strand: 5'-AATTCACGCG-TAAAAAAAACAAATATGAGCGTCTGATCTCTTGAATCA-GACGCTCATATTTGTTG-3'). The complementary oligonucleotides were annealed and ligated into the linearized RNAi-Ready-pSIREN-RetroQ-DsRed-Express vector (Clontech, Cat No. 632487) using T4 DNA ligase (New England Biolabs, Cat. No. M0202L) for 1 hour at RT. In addition, a separate ligation was carried out using the manufacturer provided Luciferase (Luc, underlined) control oligonucleotides (top strand 5'-GATCCG-TGCGTTGCTAGTACCAACTTCAAGAGATTTTTTACGC-GTG-3', bottom strand 5'-AATTCACGCGTAAAAAATCT-CTTGAAGTTGGTACTAGCAACGCACG-3'). The ligated construct was used to transform competent DH5 E. Coli (Invitrogen, Cat. No. 44-0098). The transformed E. Coli were allowed to recover for 1 hours at 37°C while shaking at 250 rpm. The transformed cultures were plated onto Ampicillin (100 μg/ml) selective media and incubated O/N at 37°C. A bacteria colony was selected and used to inoculate 2 mL of LB/Amp culture media, and incubated at 37°C for 8 hours while shaking at 250 rpm. The shRNA construct was extracted (Qiagen, QIAprep spin miniprep kit, Cat. No. 27108) and the DNA concentration was determined using the Nanodrop spectrophotometer. A fraction (1 μg) of the construct was digested with 20 U of MluI (New England Biolabs, Cat. No. R0198L) to confirm the presence of the insert. The sequence of the insert was confirmed by sequencing (ACGT Corp, Toronto, ON).

2.7 shRNA delivery and characterization of knockdown cell lines

GP-293 pantrophic packaging cells (gift from Drs. Helen Sarantis and Scott D. Gray-Owen, Department of Molecular and Medical Genetics, University of Toronto) were co-transfected (FuGENE HD, Roche, Indianapolis, Cat. No. 04709691001) with 2 μg of shRNA retroviral constructs (Ads-shRNA, Luc-shRNA) as well as 2 μg of the pVSV-G envelope protein-packaging vector (gift from Drs. Helen Sarantis and Scott D. Gray-Owen). Transfected GP-293 cells were incubated at 37°C for three days at which point viral containing supernatant was harvested and spun free of cells. The viral supernatant was filtered (Millex-HA 0.45 μm, Millipore, Cat. No. SLHA033SS), and residual cellular DNA degraded (Benzonase, Sigma, Cat. No. E1014), prior to RAW cell infections. Infected RAW cells were incubated for three days at 37°C and sorted based on the DsRed signal by fluorescence-activated cell sorting, using a Beckman-Coulter flow cytometer. 8.5×10^5 and 1.0×10^6 of Luc KD and Ads KD macrophages were sorted. DsRed positive cells were subjected to a standard limiting dilution. Briefly, 130 cells were resuspended in 6.5 mL cell culture media to get 2 cells/100 μL. One hundred μL cell suspension was added into each well of row A, B and C of a 96-well flat-bottomed plate. The remained cells were further diluted with 2.9 mL cell culture media to give a concentration of 1 cell/100 μL. One hundred μL cell suspension was added into each well of row D, E and F. Finally, the remained cells were further diluted with 2.2 mL cell culture media to reach 0.5 cell/100 μL, theoretically. One hundred μL of the cell suspension was seeded into each well of row G and H of the above plate. After one week incubation, colonies were examined under an inverted fluorescent microscope at 20× objective magnification and DsRed positive colonies were identified and expanded. Three Ads KD clonal cell lines (Ads KD Clone B, Clone C, Clone F), one Luc KD clonal cell line, as well as, non-infected WT RAW macrophages were expanded. A schematic outlining the generation of the various cell lines can be found in Fig. S2.

2.8 TRAcP staining and osteoclastogenesis index

BMMs and RAW macrophages were seeded in 6-well tissue culture plates at 2.5×10^5 cells/well and OCG was induced via 6-day and 4-day stimulations with sRANKL, respectively. Cells were washed twice with Phosphate Buffer Saline (PBS), fixed with 4% paraformaldehyde (PFA) and stained for tartrate-resistant acid phosphatase (TRAcP) as previously described [40]. Briefly, fixed cells were incubated in a solution of naphthol AS-BI phosphate and fast red TR salt in 0.2 M acetate buffer (pH 5.2) containing 100 mM sodium tartrate (Sigma, Cat No. S-4797) for 30 min at 37°C and washed with PBS. Cells were viewed with a Leitz Wetzlar microscope and images taken with a PixeLinK camera. The OCG index was determined by quantifying the number and size of TRAcP-positive osteoclasts in 10 random fields of view (FOVs) at 20× objective magnification. TRAcP positive osteoclasts with 3–10 nuclei were considered to be small osteoclasts, whereas TRAcP positive osteoclasts with 10–30, and 30+ nuclei were considered to be medium and large osteoclasts, respectively. The fusion index was quantified in a similar manner.

2.9 Differential seeding density assay

To evaluate the effect of initial cell seeding density on OCG, WT and KD cell lines were seeded onto 8-well slide chambers at three initial seeding densities: 1×10^4 cells/well (low density), 5×10^4 cells/well (normal density) and 1.0×10^5 cells/well (high density). Day 4 cultures were fixed and TRAcP stained as described above. The degree of osteoclast formation was determined by viewing strained cells at 20× objective magnification and representative images were taken with a PixeLink camera.

2.10 Transwell migration Assay

Two million cells were stimulated with sRANKL (60 ng/mL) in 10 mL of the complete DMEM for 30 hours in the 10-cm^2 Petri dishes. The cells were sRANKL starved O/N. The cells were scrapped and counted using a hemocytometer and resuspended in the reduced DMEM (0.5% FBS and antibiotics) to a final concentration was 1×10^6 cells/mL. Migration assays were carried out as previously described [41]. The transwells (8 μm) were pre-equilibrated and placed in the wells of a 24-well plate containing 600 μL reduced DMEM plus 60 ng/mL sRANKL. Two hundred microliters of cell suspension (containing 2×10^5 cells) were seeded into each well. The cells were incubated for 20 hours to allow for migration. After migration, non-migratory cells on the top of the membrane were removed with cotton swabs. The cells on the bottom of the membrane were fixed with 4% PFA and labeled with 0.165 μM 4′,6′-diamidino-2-phenylindole hydrochloride (DAPI) in PBS/0.1% Triton-100. The pictures were taken by using a Nikon Eclipse E1000 microscope. The experiment was repeated three times.

2.11 Functional resorption assay

WT, Luc KD and Ads KD cells were cultured for up to 6 days (+60 ng/mL sRANKL) in 96-well culture plates or 12-well cultures plates containing 18-CIR coverglass (Fisherbrand) or dentin slices prepared from narwhal tusks [42]. Cells cultured on the glass were fixed (4% PFA) and TRAcP stained. Pictures were taken by using a Nikon Eclipse E1000 microscope. To view resorption pits, cells were removed from the dentine slices by a 5 minute treatment in 6% NaOCl. The slices were painted with 1% toluidine blue (wt/vol) and 1% sodium borate (wt/vol) for 30 sec. Discs were viewed and representative images taken using a Nikon Elipse E400 microscope.

2.12 F-actin localization using confocal microscopy

Cells were cultured with sRANKL for two days. Cell culture medium was aspirated, cells were washed twice with pre-warmed PBS and permeabilized with 0.1% Triton-100 in PBS. Non-specific binding was blocked for 30 min at RT with 1% BSA/ 0.1% Triton-100/PBS. The cells were incubated in 100 mM glycine solution for 10 min and incubated in dark for 20 min in Alexa Fluor 488 phalloidin (0.15 μM in 1% BSA/0.1% Triton-100/PBS). A z-stack of 70 sections confocal images was taken (Leica, DMIRE2). A section at the glass was shown for each cell and the 3D reconstruction was construction using the maximizing projection method. Fluorescent intensity of two representative cells was measured by ImageJ. The number of membrane protrusions was enumerated in 30 representative cells, for each cell type respectively, and results are expressed as mean (+/- SD).

2.13 Adseverin rescue using site directed mutagenesis

As previously mentioned in the methods section 2.6, a 19 bp region of the Ads open reading frame (5′- nt. 547 AACAAATAT-GAGCGTCTGA nt. 565-3′) was targeted by the Ads specific shRNA. Using site-directed mutagenesis a new, mutated, Ads open reading frame was constructed where the same 19 bp region reads as follows: 5′- nt. 547 AACAA**G**TA**C**GA**A**CG**G**CT**C**A nt. 565-3′. These nucleotide substitutions (A^{552}-G, T^{555}-C, G^{558}-A, T^{561}-G, G^{564}-C) lead to five silent mutations such that the final amino acid sequence of both the original and mutated sequences was the same (N-K-Y-E-R-L). The newly constructed mut-Ads was cloned into the BamHI and BstBI restrictions site of pIRESpuro3 vector (Clontech, Mountain View, CA). Two million Ads KD Clone F cells were transiently transfected with 2 μg of pIRESpuro3-mut-Ads using Amaxa nucleatransfector and solution V (Program D-032). One fraction was stimulated for 4 days with sRANKL (60 ng/ml) and the cells were harvested for Western blot analysis. The other fraction was stimulated for 5 days with sRANKL (60 ng/ml) and the cells were processed for TRAcP staining.

2.14 Statistical analysis

For experiments where multiple observations were made per sample, numerical results were expressed as mean ± SEM or mean ± SD, as indicated. Each experiment had a sample size of $n \geq 3$, unless otherwise stated. Statistical analysis was performed using Student t-test and a p value of less than 0.05 was considered statistically significant.

Results

3.1 Adseverin expression is up-regulated during RANKL-induced OCG

An exploratory microarray was carried out to identify RANKL regulated genes during OCG. This microarray analysis (unpublished data) identified eight cytoskeleton associated genes of interest with statistically significant changes in gene expression of least 2-fold in BMMs stimulated for two days with M-CSF and sRANKL versus M-CSF alone (Fig. 1). Of these genes, Ads showed the second greatest change in expression with a 4.9-fold increase in expression in response to sRANKL. This finding prompted us to closely examine the temporal expression pattern of Ads during OCG.

In BMM-derived osteoclasts, Ads transcript was significantly up-regulated by 10-fold in Day 2 osteoclasts compared to basal Ads transcript expression in Day 0 BMMs. Day 4 cultures showed a significant 13-fold increase in expression when compared to Day 2 cultures. No statistically significant difference was noted between transcript levels of Day 4 and Day 6 cultures, both of which had 23-fold higher expression compared to Day 0 BMMs (Fig. 2A, black diamonds). RAW macrophage-derived osteoclasts also experienced significant 23 and 31-fold up regulation in Ads transcript in Day 2 and Day 4 cultures compared to Day 0

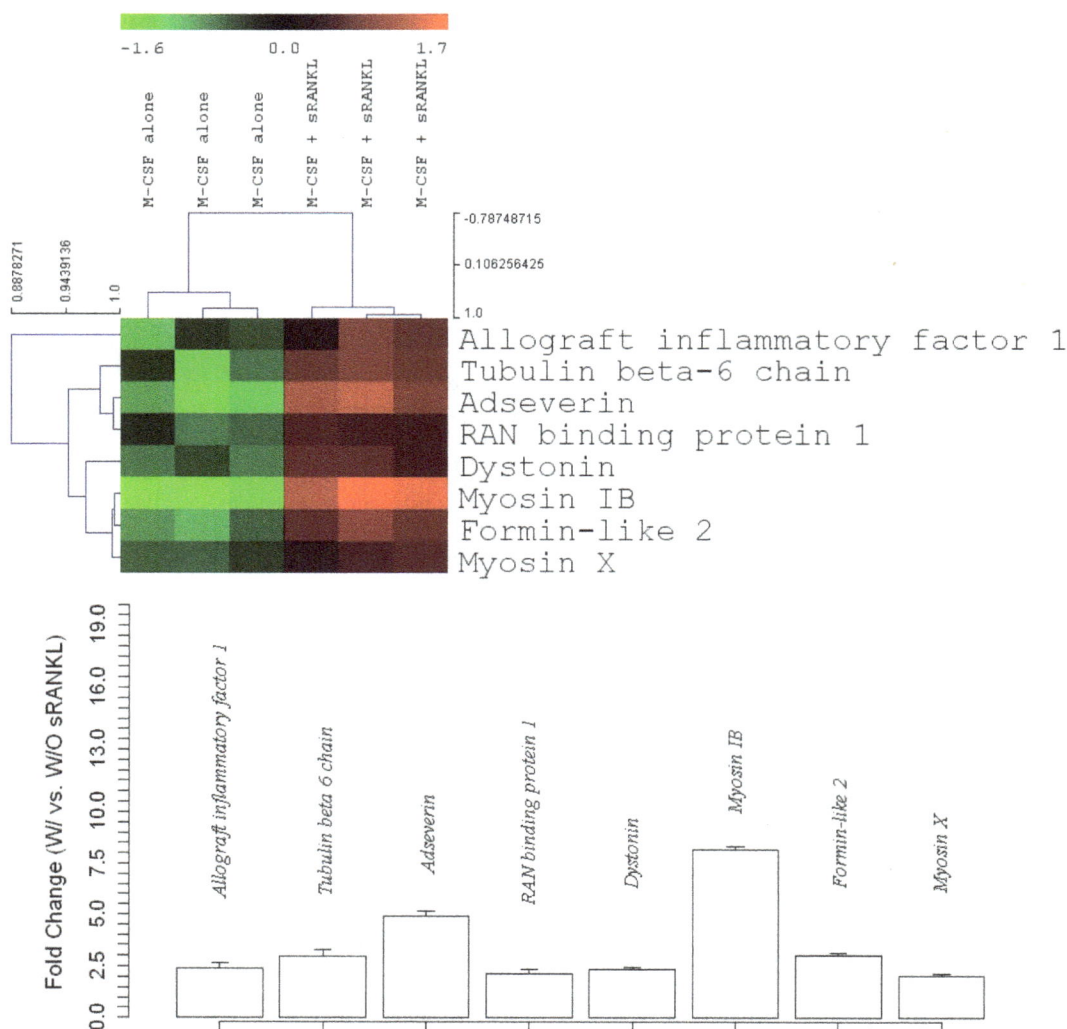

Figure 1. Microarray analysis identified Adseverin as a gene of interest during the early phases of OCG. Several genes were differentially expressed in osteoclast precursors stimulated for two days with sRANKL & M-CSF versus M-CSF alone. Eight cytoskeleton-associated genes with the greatest fold-change in expression (p<0.05) were reported. Specifically, Ads was up-regulated 4.9-fold in osteoclast precursors stimulated with sRANKL for two days. The microarray data has been deposited in NCBI's Gene Expression Omnibus and are accessible through GEO Series accession number GSE54779.

macrophages (Fig. 2A, black circles). The transcript levels were significantly higher in Day 4 cultures than in Day 2 cultures.

In BMM-derived osteoclasts, Ads protein expression was significantly up-regulated by 6-fold in Day 2 cultures versus Day 0 cultures. Ads protein expression remained elevated during osteoclast maturation. Ads protein expression in Day 4 and 6 cultures did not statistically differ from Day 2 (Fig. 2B & 2C). In RAW-derived osteoclast cultures (Fig. 2D & 2E), Ads protein was not detected in Day 1 or Day 2 cultures, while statistically significant 17 and 23-fold up regulation were noted in Day 3 and Day 4 cultures. Ads protein levels were significantly greater in Day 4 cultures compared to Day 3 cultures. In both model systems, Gelsolin protein levels did not change in response to sRANKL, while the expression osteoclast marker gene Cathepsin K was significantly increased by Day 4 in BMM cultures (Fig. 2B & 2C) and Day 3 in RAW macrophage cultures (Fig. 2D & 2E).

3.2 Adseverin D5 is not expressed during OCG

PCR analysis only detected the 650 bp product, representing full-length Ads, in Day 0 RAW macrophages and BMM as well as Day 4 RAW and Day 6 BMM-derived osteoclasts (Fig. 3A). This observation was validated at the protein level where only the 85 KDa full-length Ads was detected in lysates of mature osteoclasts derived from RAW macrophages and BMMs. Neither isoform was expressed in unstimulated, Day 0 RAW macrophages or BMMs (Fig. 3B).

3.3 Characterization of knockdown cell lines

To determine the functional role of Ads during OCG, RAW macrophage clonal cell lines with reduced endogenous Ads expression were generated. To exclude the possibility of pleiotropic effects of shRNA expression or retroviral transduction, a control clonal cell line was also generated. In addition, WT macrophages were used as no infection controls.

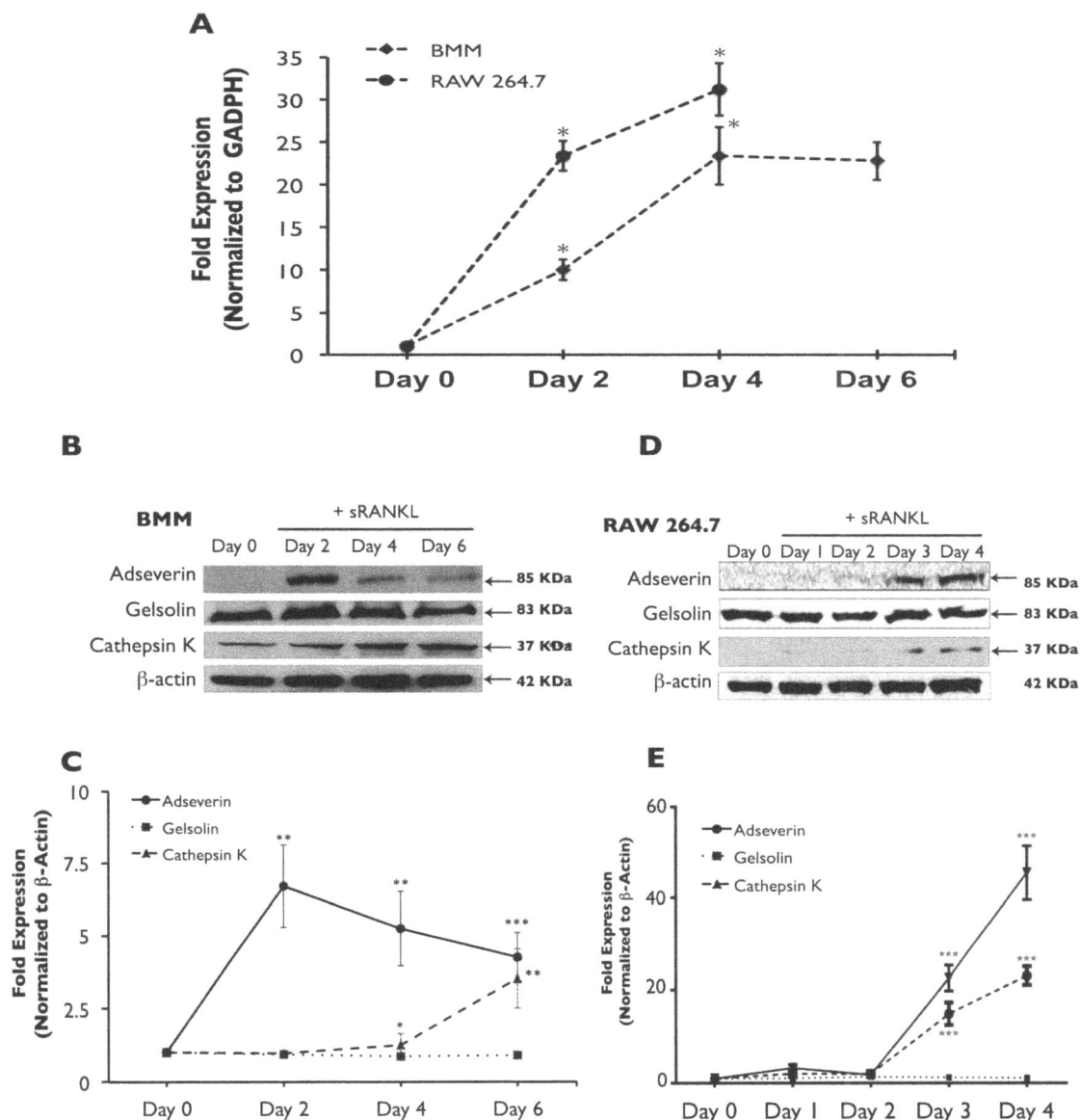

Figure 2. Adseverin expression is up-regulated during OCG in response to sRANKL. A) Quantitative real-time PCR analysis was used to quantify Ads gene expression in osteoclast cultures derived from BMMs (black diamonds) and RAW macrophages (black circles). Results are expressed as fold expression versus Day 0 and are normalized against GADPH ± SEM. In BMM-derived osteoclasts, Ads expression was significantly up-regulated by 10-fold after 2 days and 23-fold after 4 days. No difference was noted between Day 4 and Day 6 cultures. In RAW-derived osteoclasts, Ads gene expression was significantly up-regulated by 24-fold after 2 days and 32-fold after 4 days. **B)** Immunoblot analysis was used to quantify protein expression in BMMs and RAW macrophage-derived osteoclast cultures. Results are expressed as fold expression versus Day 0 and normalized against ß-actin ± SD. In BMM-derived osteoclasts, Ads protein expression was significantly up-regulated by 6-fold in Day 2 cultures. 5-fold and 4-fold increases in expression were noted in Day 4 and 6 cultures, respectively. The decrease in expression in days 4 and Day 6 was not found to be statistically significant. Gelsolin expression was not altered during OCG in response to sRANKL. Cathepsin K expression was significantly increased in response to sRANKL after a 4 day stimulation with sRANKL. **D)** Similarly in RAW macrophage-derived osteoclast cultures, Ads was up-regulated 17-fold and 23-fold in Day 3 and Day 4 cultures respectively. Gelsolin expression was not altered during OCG, and Cathepsin K expression was significantly increased in Day 3 and 4 cultures. **C & E)** Quantification of immunoblots (* $p<0.05$, ** $p<0.01$, *** $p<0.001$, n = 3).

Following expansion of the above mentioned clonal cell lines, quantitative real-time PCR was used to analyze the degree of Ads KD at the transcript level (Fig. 4A). Ads expression in WT Day 4 cultures was set at 100%. Ads gene expression in Day 0 WT cultures was 2% of Day 4 WT cultures. Ads gene expression was unchanged between Day 4 Luc KD and WT cultures. Day 4 osteoclasts derived from Ads KD Clone B, C and F cultures all experienced significant reductions in Ads expression. Ads transcript level in Day 4 cultures in Ads KD Clone B, C and F cells were 35%, 25% and 6% that of WT Day 4 cultures, respectively.

The degree of Ads attenuation was also determined at the protein level (Fig. 4B and 4C). Again, Ads expression in WT Day 4 cultures was set at 100%. Ads protein expression in Day 0, WT

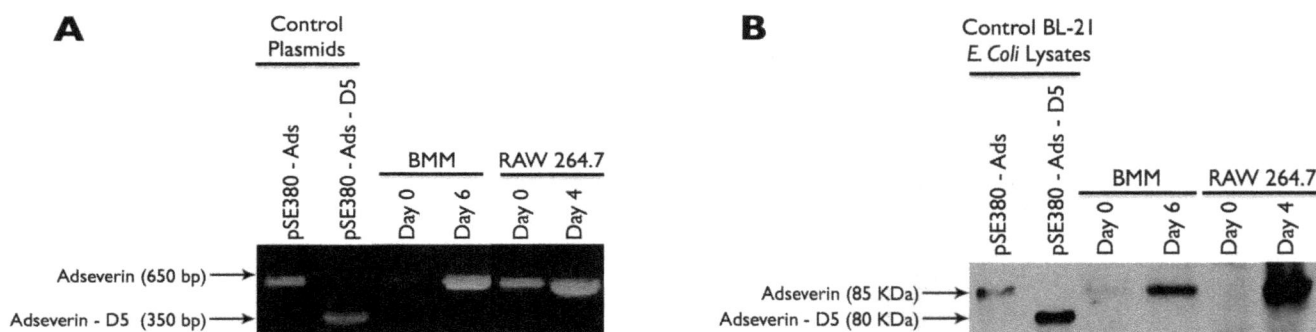

Figure 3. Adseverin D5 is not is expressed during *in vitro* OCG. A) PCR detected only the full-length Ads isoform in Day 0 and Day 6 BMMs and Day 0 and Day 4 RAW macrophage cultures. pSE380-Ads (full-length Ads) and pSE380-Ads-D5 (Ads D5 isoform) plasmids were used as positive controls. Equal volumes of cDNA from each representative samples were used as templates. The intensity of the signal is not a quantitative measure of Ads gene expression. **B)** Immunoblot analysis also failed to detect the Ads D5 in BMM and RAW macrophage-derived osteoclasts. Lysates from the transformed BL21 *E. coli* induced to express exogenous protein were used as positive controls. Only the 85 kDa full-length isoform was detected in 20 μg of TCL from Day 6 BMMs and Day 4 RAW macrophages cultures. Ads was not detected in 20 μg of TCL from unstimulated samples.

RAW macrophages was 30% of Day 4 WT cultures. Ads KD Clone B Day 4 cultures showed elevated (141%) albeit statistically non-significant increase in protein expression compared to Day 4 WT cultures. A slight, statistically non-significant, 2% reduction in protein expression was noted in Day 4 Ads KD Clone C and Luc KD cultures. Ads KD Clone F showed significant reductions in Ads protein levels after a 4-day stimulation with sRANKL. Ads protein expression in Day 4 Ads KD Clone F cultures was 28% that of the Day 4 WT cultures.

3.4 OCG is impaired in adseverin knockdown Clone F cells

OCG was induced in WT, Luc KD, and Ads KD clonal cell lines via a four day stimulation with sRANKL. Cultures were routinely examined for DsRed signal using an inverted fluorescent microscope to ensure that presence of the shRNA cassette. As expected, only WT macrophages were DsRed negative (results not shown). There were no statistically significant differences between WT, Luc KD, Ads KD Clones B and C with respect to OCG index (Fig 5). All four populations had similar number of small, medium and large osteoclasts. Furthermore, the total number of osteoclasts in each genotype was not significantly different. Although not significant, a trend towards fewer large osteoclasts was noted in Ads KD Clones B & C (Fig. 5B). TRAcP stain intensity was reduced in Ads KD Clone F cultures in comparison to all other cell lines (Fig. 5A). The reduction in staining translated to virtually no multinucleated osteoclasts. Prolonging the culture by an additional two days did not increase the size or number of Ads KD Clone F osteoclasts (results not shown). Quantitative real-time PCR was used to ensure that Ads KD did not alter the expression of other vital genes during OCG. Gene expression of RANK, CD44, S1PRα and Gelsolin was unchanged between Ads KD Clone F and WT cells (Fig. S3).

3.5 Increased seeding density failed to rescue OCG defect in adseverin knockdown Clone F cells

OCG was induced and TRAcP positive osteoclasts were counted in Day 4 cultures (Fig. 6). OCG in WT and Luc KD macrophages was not affected by differential seeding density, as mature osteoclasts of all sizes were seen in Day 4 cultures. Increasing seeding density failed to fully rescue OCG in Clone F Ads KD progenitors, as Clone F cells failed to form TRAcP positive multinuclear osteoclasts regardless of initial seeding density. However, increased seeding density did seem to have a positive effect on TRAcP staining in the KD population. This observation is due to increased cell density and excessive non-specific staining of cellular debris as opposed to restoration of TRAcP expression.

3.6 Adseverin attenuation enhances migration in response to sRANKL

Day 2 WT and Ads KD osteoclast cultures starved for sRANKL were incubated in transwell (8 μM) chambers conditioned with 60 ng/ml sRANKL, to analyze chemotaxis. Following the incubation period, Ads KD Clone F cells showed a statistically significant 2.4-fold increase in migratory cells when compared to WT cultures (Fig. 7).

3.7 Adseverin knockdown Clone F cells failed to resorb dentin discs substrates

The bone resorptive ability Day 4 Ads KD (Clone F) cultures were compared to Day 4 Luc KD cultures. Mature, Luc KD osteoclasts were capable of resorbing the dentin disc substrate at significantly higher rates than Ads KD cells (Fig. 8A, black arrow). Day 4 Ads KD Clone F osteoclast cultures had virtually no multinucleated osteoclasts, few mononuclear TRAcP positive osteoclasts and or detectable resorption pits (Fig. 8A). In order to assess the resorptive ability of premature, mononuclear osteoclasts, Day 4 Ads KD Clone F and Day 2 BMM-derived osteoclasts were also cultured on dentin discs. Nearly 80% of Day 2 WT mononuclear osteoclast precursors were TRAcP positive. In total 27 ± 1.41 resorptive pits/12 mm² disc were noted in Day 2 WT osteoclasts cultures, indicating that WT immature osteoclasts were capable of minimally resorbing dentin discs (Fig. 8B, black arrows). Conversely, no TRAcP staining or resorption pits were noted in Day 2 Ads KD Clone F cultures, indicating a lack of osteoclast differentiation and resorptive ability.

3.8 Adseverin knockdown disrupts F-actin architecture during OCG

F-actin, fluorescence intensity was plotted against the distance across a single cross section of a cell (Fig. 9A, white line). WT osteoclast precursors displayed well-defined cortical actin rings during the early phases of OCG as noted by high intensity fluorescence peaks at the cell boundaries, 0.5 μm and 6 μm

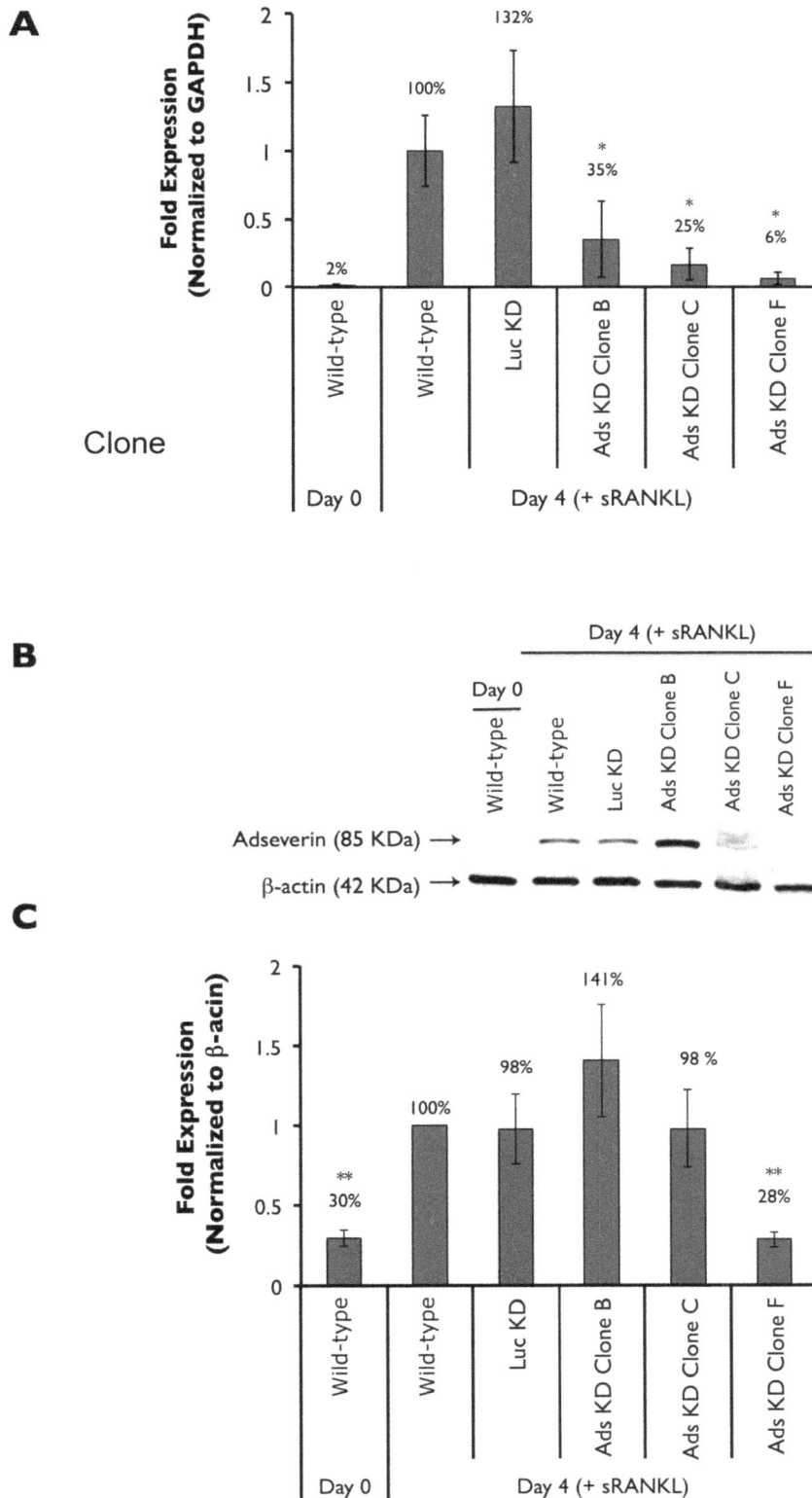

Figure 4. Adseverin knockdown clonal cell lines displayed varying degrees of Adseverin attenuation in response to sRANKL. A) Quantitative real-time PCR was used to quantify Ads gene expression in Day 0 and Day 4 osteoclast cultures. Results are expressed as fold expression versus WT Day 4 and are normalized against GAPDH ± SEM. Ads gene expression was not significantly different between Day 4 WT and Luc KD cultures. Ads transcript levels were significantly lower in all three Ads KD cell lines. Ads gene expression in Clones B and C were 35% and 25% that of Day 4 WT cultures, respectively. Ads KD Clone F cultures experienced the greatest reduction in Ads gene expression. Transcript levels in Day 4 Ads KD Clone F were similar to Day 0 WT cultures, which translated to 6% of Day 4 WT expression. **B)** Immunoblotting was used to quantify Ads protein expression in Day 0 and Day 4 osteoclast cultures. **C)** Results are quantified and expressed as fold expression versus WT Day 4 (±SD) and are

normalized against β-actin. Ads protein expression was not significantly different between Day 4 WT, Luc KD, and Ads KD Clone B and C cultures. Ads KD Clone F cultures experienced the greatest reduction in Ads expression. Protein levels in Day 4 Ads KD Clone F were similar to Day 0 WT cultures, which translated to 28% of Day 4 WT expression (**$p < 0.01$, n = 4).

(Fig. 9A & 9B). The KD of Ads protein expression resulted in significant morphological changes. The cortical actin rings commonly seen in the WT cells were lost in the Ads KD cells. The F-actin fluoresce intensity across Ads KD cells was ill defined and lacked the characteristic peaks seen in the WT cells (Fig. 9B). Dense F-actin islands and numerous actin projections resembling podosomes and filopodia characterized the F-actin architecture in the Ads KD cells. Although, fliopodial extensions were also noted in WT cells there was a significant 7-fold increase (Fig. 9C) in the Ads KD population giving the cells a more invasive and motile morphology (Fig. 9A).

3.9 Adseverin knockdown cell-rescue restores pre-osteoclast differentiation

A construct containing mutated Ads mRNA correlating to an Ads protein sequence identical to native Ads (pIRESpuro3-mut-Ads) was used to reintroduced Ads back into the Ads KD cells using transient transfections. An empty vector (pIRESpuro3-mut-Ads) was used as the control construct. Following transfections, the Ads protein expression level in cell lysates was measured after 4 days in culture (Fig. 10A). Ads was undetected in cell lysates from Ads KD Clone F transfected with the empty control vector. As expected these cells displayed a lack of OCG as noted by lack of TRAcP positive, multinucleated osteoclasts. Ads protein levels

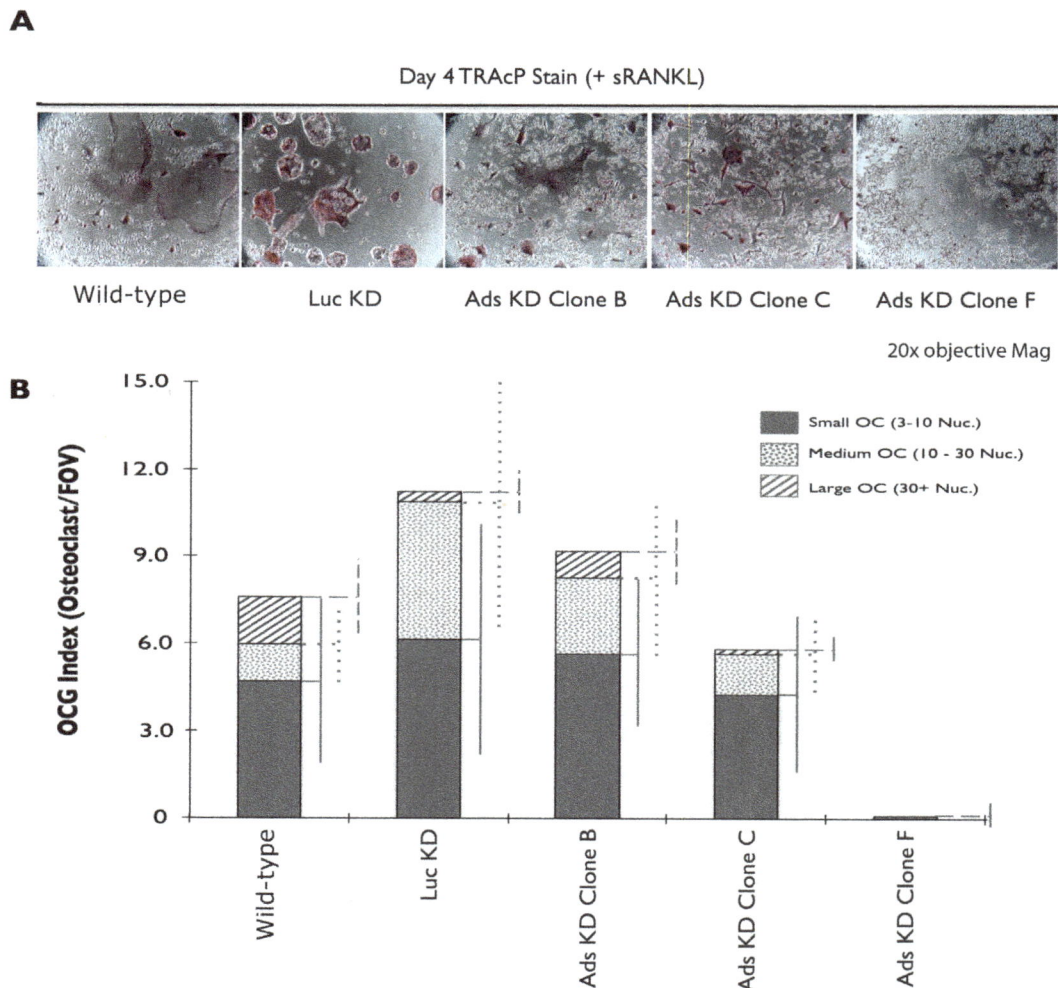

Figure 5. Adseverin knockdown Clone F displayed the greatest defect in osteoclast formation. A) 2.5×10^5 RAW macrophages were seeded onto 6-well tissue culture plates and stimulated for 4 days with sRANKL. Representative photomicrographs of TRAcP stained osteoclasts were taken at 20× objective magnification. A profound lack of osteoclasts was observed in Ads KD Clone F. **B)** The number of osteoclasts per random field of view (OCG index) was quantified. The number of osteoclasts is represented as three distinct populations of small (3–10 nuclei/osteoclast, solid grey column), medium (5–7 nuclei/osteoclast, dotted column) and large (30+ nuclei/osteoclast, dashed column) osteoclasts (n≥10). Ads KD Clones B and C experienced marginal and statically non-significant reductions in the number of large osteoclasts when compared to WT cultures. No statistically significant difference in OCG was observed between the WT and Luc KD cultures. Ads KD Clone F macrophages failed for form multinucleated osteoclasts.

Day 4 TRAcP Stain (+ sRANKL)

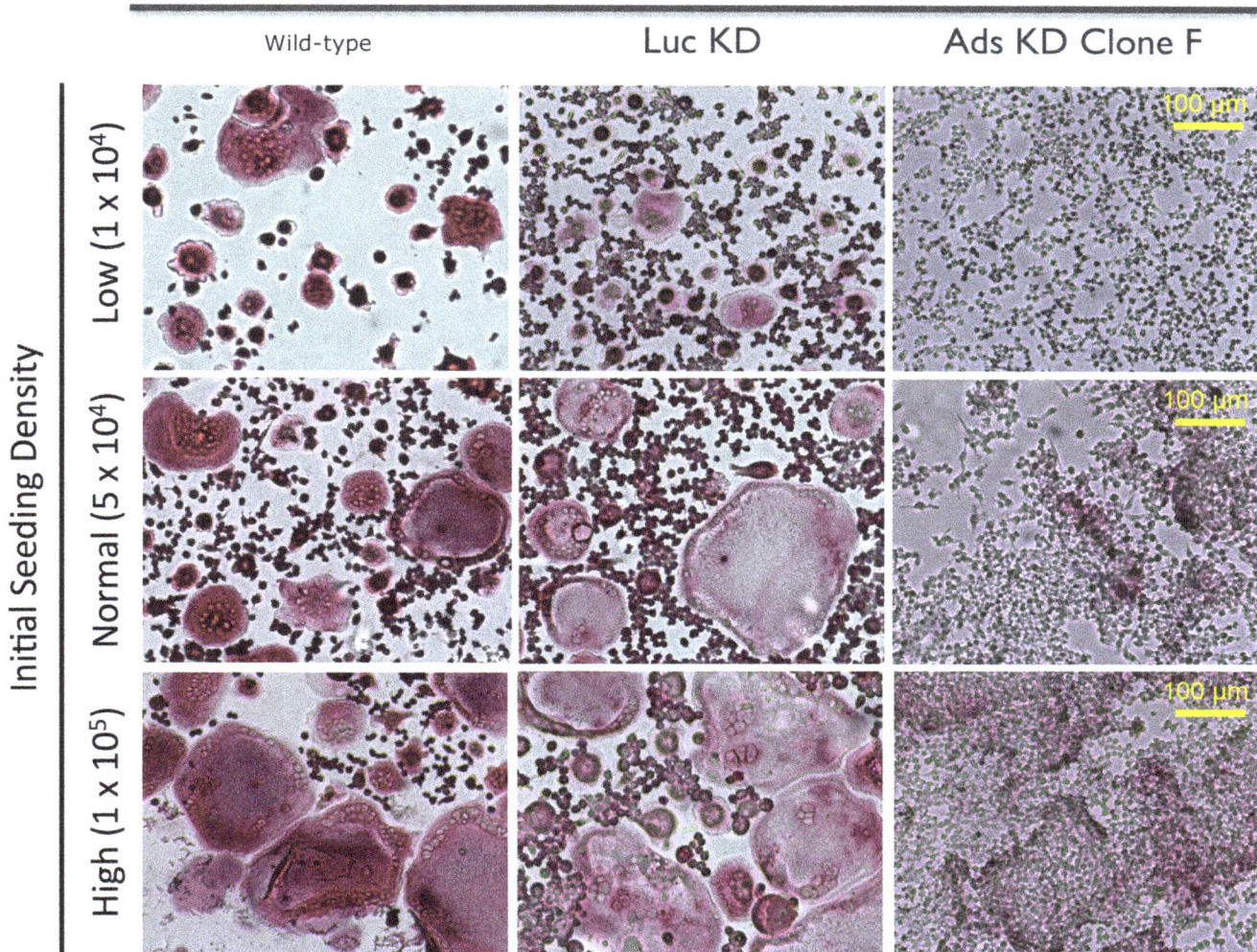

Figure 6. Increased seeding density failed to rescue OCG defect in Adseverin knockdown monocytes. WT, Luc KD and Ads KD (Clone F) monocytes were seeded at three initial plating densities in 12-well tissue culture plates, low (1×10^4 cells/well), normal (5×10^4 cells/well) and high (1×10^5 cells/well). Cells were stimulated for 4 days with sRANKL to stimulate OCG. Representative TRAcP stained photomicrographs are shown. Even at the 10-fold-higher initial plating density, Ads-KD cells failed to form multinucleated osteoclasts, suggestive of a fusion defect.

were slightly elevated (22% of WT) following transfection with the mutated Ads construct. The reintroduction of Ads was capable of restoring pre-osteoclast differentiation in the form of mononuclear TRAcP osteoclasts (Fig. 10B & C). It should be noted that no mature multinucleated osteoclasts were observed in culture following the transient reintroduction of Ads thereby eliminating the possibility of examining osteoclast function. The predominant reason for the lack of multinucleated osteoclasts is believed to be the transient nature of Ads rescue suggesting that Ads may be required for a prolonged period of time during OCG.

Discussion

OCG consists of a series of coordinated cellular events including migration of fusion competent osteoclast precursors [43], cell-cell adhesion and membrane fusion[12-14]. Upon contact with the bone matrix, multinucleated osteoclasts become polarized, undergo distinctive cell spreading and allocate their nuclei to the anti-

resorptive surface adhere to bone using actin rich structures and form a unique resorptive machinery [44,45]. All of the events described above are heavily dependent on the dynamic regulation of the actin cytoskeleton. This clearly illustrates a potential functional role for the actin binding protein, Ads, in the regulation of osteoclast formation and function. Much of what is known about Ads was discovered through detailed examination of exocytosis in bovine chromaffin cells of the adrenal medulla. Immunocytochemistry analysis of chromaffin cells co-localize Ads with a cortical F-actin ring [24,46], while numerous studies have demonstrated the role of Ads in depolymerizing the cortical F-actin ring and allowing for vesicular exocytosis [47-51]. To our knowledge, the microarray by Yang et al. [35] aimed at identifying genes differentially regulated by RANKL in mature osteoclasts was the first published report of Ads in the osteoclast biological system. Ads was up-regulated in mature osteoclasts and clustered with genes required for mature osteoclast attachment, movement and vesicular trafficking. This finding was validated in a

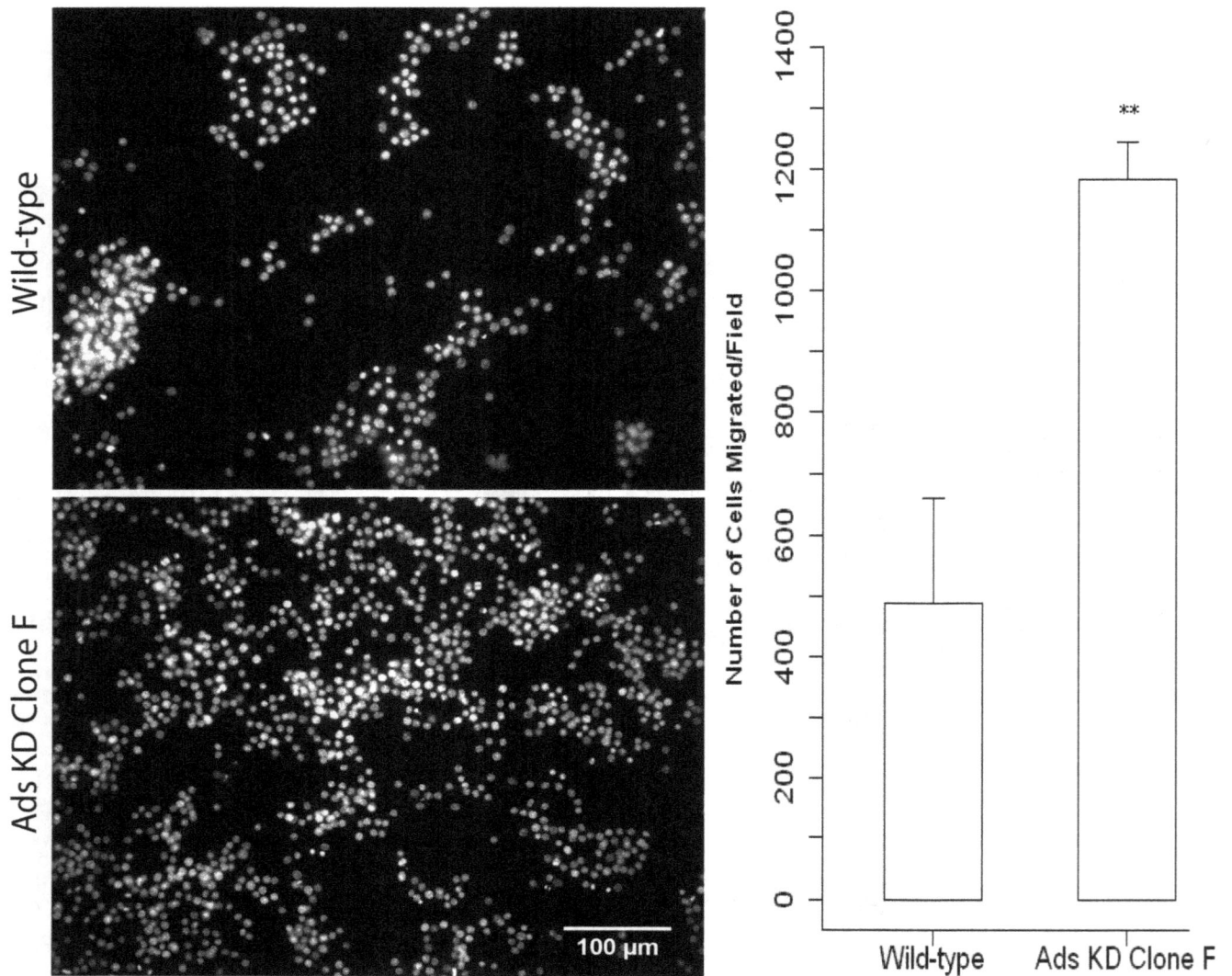

Figure 7. Adseverin knockdown osteoclast precursors display enhanced migration in response RANKL. 2×10^5 Day 2 sRANKL starved WT and Ads KD (Clone F) cells were seeded in transwells (8 μm) pre-equilibrated in reduced DMEM containing 60 ng/mL sRANKL. The cells were starved for RANKL and then incubated for 20 hours to allow for migration (chemotaxis). After migration, the non-migratory cells on the top of the membrane were removed and migration assessed by enumerating the cells on bottom of the membrane (n = 3). Significantly more Ads KD Clone F cells migrated towards the chemokine sRANKL suggestive of enhanced migration ability in the absence of Ads.

microarray performed in our lab (Fig. 1), which identified Ads as a gene of interest during the early phases of OCG. Taken together, these reports prompted us to investigate the hypothesis that Ads is a RANKL induced regulator of OCG with a potential functional role both during the early phases of OCG and in mature osteoclasts.

The temporal expression pattern of Ads was suggestive of a potential role during the earlier phases of OCG with a potential role in mature osteoclasts (Fig. 2). Further, it appears that any role for Ads D5 during OCG would be very limited, as only the full-length isoform of Ads was up-regulated during OCG in response to sRANKL (Fig. 3). To address the functional role of Ads during OCG, RNAi gene silencing was used to generate clonal Ads KD cell lines. The viral transduction and shRNA expression did not have any non-specific effects on OCG as indicated by the similarities between the non-infected WT and control, Luc KD cultures (Figs 4-6, 8, S3). Ads KD Clone F experienced the

greatest attenuation of Ads expression and the most distinct phonotypical variation from WT cells (Fig. 5) and was chosen as the experimental cell line for further experiments.

Considering the structural and functional similarities between Ads and Gelsolin, it is plausible to assume that these two proteins play similar or even redundant roles during OCG. Gelsolin, which has been implicated in osteoclast matrix adhesion, actin remodeling, motility and bone resorption [52] showed no changes in expression during OCG in response to RANKL. Ads on the other hand was significantly up-regulated in the early phases of OCG. Based on the differences in expression patterns, it is more likely that Ads and Gelsolin have divergent roles in the regulation of osteoclast formation and function. Strong evidence towards the divergent roles of Ads and Gelsolin during OCG comes from findings of Cellaiah et. al [52], where it was demonstrated that Gelsolin was not required for osteoclast precursor fusion or osteoclast formation. Instead, Gelsolin null mice experienced

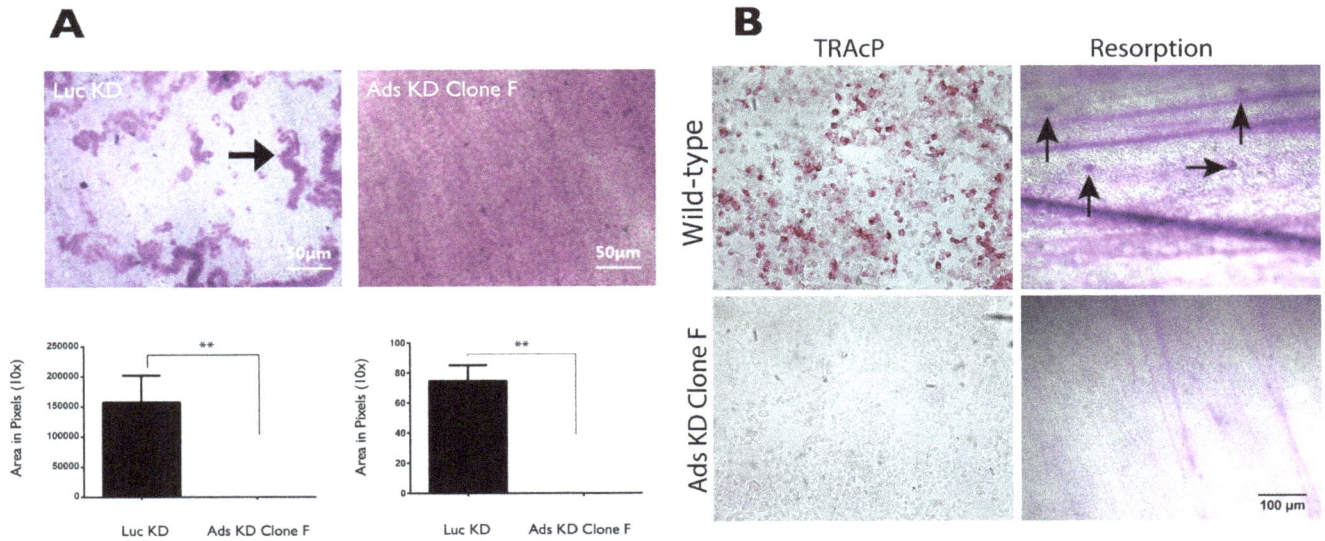

Figure 8. Adseverin knockdown osteoclasts fail to resorb mineralized matrix. A) Luc KD and Ads KD osteoclast precursors were cultured on dentin discs for 4 days. Mature Luc KD osteoclasts were capable of resorbing the dentin disc substrate (black arrow) significantly better than Ads KD cells, as indicated by the total resorbed area. Ads KD Clone F cultures had virtually no multinucleated osteoclasts or detectable resorption pits. **B)** WT and Ads KD (Clone F) cells were cultured for 2 days with sRANKL. More than 80% of WT cells were TRAcP. WT mononuclear TRAcP positive osteoclasts were capable of minimally resorbing the dentin slices as indicated by the small resorption pits (black arrows). Immature, Day 2 Ads KD cultures were TRAcP negative and displayed no resorptive ability.

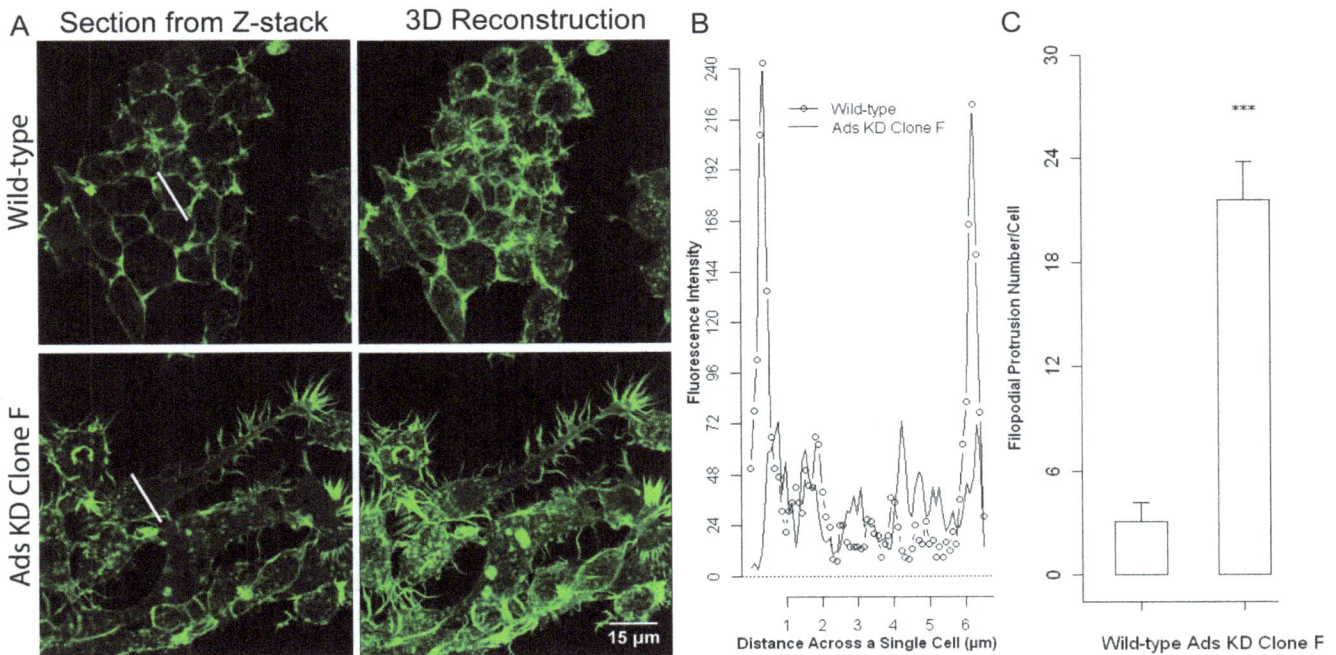

Figure 9. Adseverin knockdown alters F-actin architecture and cellular morphology in osteoclast precursors. A) WT RAW and Ads KD osteoclasts precursors were stimulated with sRANKL for two days, fixed and stained with green- fluorescent *phalloidin* to localize F-actin. **B)** F-actin intensity was plotted against the diameter of a cell (white line, Fig 9A). Well-defined cortical actin rings were noted in the periphery of WT cells as noted by high intensity fluorescence peaks at the cell boundaries, 0.5 μm and 6 μm. Fluoresce intensity across Ads KD cells was ill defined and lacked the characteristic peaks seen in the WT cells. Ads KD had a motile phenotype characterized by dense actin rich structure and significantly increased number of filopodial extensions at the cell periphery when compared to WT. **C)** The number of filopodial extensions of each cell type was determined. WT cells (leftmost column) had an average of 3 (+/−1) and Ads KD Clone F cells (rightmost column) had an average of 22 (+/−2) such membrane extensions.

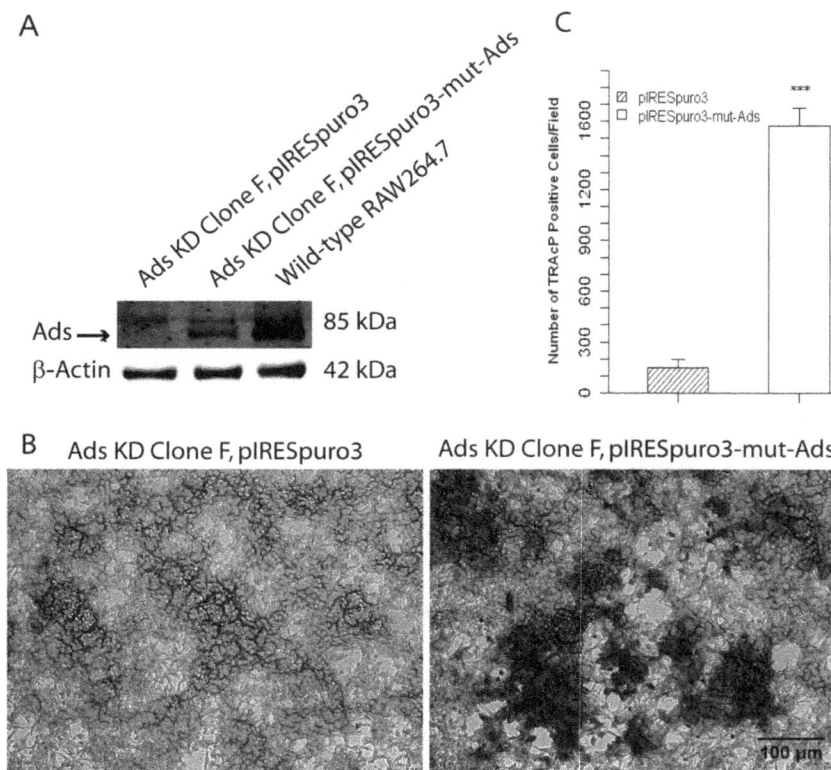

Figure 10. Adseverin knockdown cell-rescue restores pre-osteoclast differentiation. Two million Ads KD Clone F cells were transiently transfected with 2 µg of pIRESpuro3-mut-ads. The construct was constructed using site-directed mutagenesis such that the shRNA targeted sequence was mutated without changing the final protein product. A control transfection with an empty vector pIRES-puro3 was also carried out. **A)** Ads was not detected in the control transfection. Following transfection, Ads protein levels were increased to 22% of WT levels. **B & C)** The slight transient increase in Ads protein expression resulted in significantly more mononuclear, TRAcP positive ells. This phenotype was not found in the control transfection cultures.

abnormal actin cytoskeleton characterized by a loss of podosome assembly, reduced osteoclast motility caused by decreased podosome turnover, which contributed to decreased *in vivo* bone resorption.

The depletion of endogenous Ads (Clone F) resulted in a lack of multinucleated, TRAcP positive osteoclasts (Fig. 5), indicating a potential defect in early phases of OCG. The early phases of OCG can be divided into induction of fusion competency [9], migration [10-12] and membrane fusion [14]. A differential seeding density experiment was used as an indirect assay to tease out migration and/or fusion deficiencies in Ads KD osteoclast precursors. The rationale behind the differential seeding density experiments was as follows: 1) If fusion was compromised while migration was unaffected in Ads KD cell line, then OCG would be defective regardless of cell proximity; 2) If migration was compromised while fusion was unaffected in the Ads KD cell line, then the migration defect would be simply overcome by the increased initial proximity of precursors, resulting in large multinucleated osteoclasts at high densities; and 3) In the case of a concurrent defect in migration and fusion, this assay is incapable of determining the extent of the migration defect, but will still result in a lack of multinucleated osteoclasts due to a fusion defect. Large, TRAcP positive, multinucleated osteoclasts were observed regardless of seeding density in WT and Luc KD cell lines, while the phenotype of multinucleated osteoclasts was not rescued by increasing seeding density in the Ads KD (Clone F) cells (Fig. 6). Therefore, our findings favor the first or third possibilities, indicating a potential

fusion defect in the Ads KD (Clone F) osteoclast precursors. As mentioned before, the seeding density assay was unable to determine the extent of migration ability of the KD cells. However, pronounced cellular clumps observed within two days of differentiation at low and normal densities (results not shown) and transwell migration assays (Fig. 7) suggest enhanced migratory ability in KD cell lines. Further evidence supporting enhanced migratory of Ads KD cells ability comes from experiments that demonstrated the effect of Ads attenuation on morphological changes in cell shape and sub-cellular localization of F-actin (Fig. 9). Decreased Ads expression equated to the loss of cortical actin rings found in WT cultures. This was a surprising finding, as in most other cell systems, Ads induction results in the depolymerization of the cortical actin ring immediately prior to exocytosis. Therefore, a logical expectation would be to find dense, well-defined cortical actin rings in the absence of Ads. However, in the absence of Ads, the F-actin cytoskeleton is reorganized into dynamic F-actin rich structures resembling filopodia, and podosome cores. This finding corroborates with the finding of enhanced mobility and migration of Ads KD cells. Osteoclast migration is mediated through podosomes [53], while osteoclasts gain the ability to transmigrate through cell layers by arranging their cytoskeleton into long protrusive projections reminiscent of invadopodia [54]. The increased filopodia formation in the absence of Ads, is suggestive of a regulatory change in actin reorganization from fusion to migration.

As previously stated, the lack of multinucleated osteoclasts in the KD cell line implicates a fusion defect. That being said, a fusion defect does not equate to a defect in OCG [55,56]. Mononuclear, TRAcP positive osteoclasts are still capable of functioning and resorbing bone-like substrates, albeit less efficiently than multinucleated osteoclasts. This was clearly demonstrated by small resorption pits in dentin slices created by mononuclear TRAcP positive osteoclasts seen in two-day WT cultures (Fig. 8B). However, in the absence of Ads, there is also a marked defect in OCG, as noted by the lack TRAcP cells even after a four-day stimulation. As expected, the few TRAcP positive mononuclear osteoclasts noted in Ads KD (Clone F) cultures failed to resorb bone to any significant degree (Fig. 8). Therefore, in addition to a potential role in regulating fusion, Ads may also play a regulatory role during the mononuclear stages of OCG prior to pre-osteoclast fusion. It is unclear if the lack of fusion prevents differentiation or if altered differentiation program prevents fusion events. In order to illustrate the specific effects of targeted Ads attenuation, a mutated, shRNA insensitive Ads construct with altered message yet identical protein sequence was generated and used to transiently rescue the Ads defect in the KD cells (Fig. 10). Following the transient reintroduction of Ads, we were capable of rescuing TRAcP expression but not the phenotype of multinucleated osteoclasts. This observation can partly be explained by the transient nature of the rescue experiments, which only accommodate a 48-hour peak in Ads protein expression 24 to 48 hours post transfection. Earlier experiments have shown that although Ads expression peaks after two days of sRANKL stimulation, Ads levels remained elevated during OCG and in mature osteoclasts. Due to technical difficulties of stable transfection, we could not stably re-introduce Ads into the KD system. However, the transient rescue experiments are promising evidence depicting the importance of Ads for osteoclast differentiation. Presumably, mature, multinucleated osteoclast phenotype could be rescued if higher levels of Ads could be reintroduced into the system for a longer period of time. Taken together, we can conclude that Ads is critical in the regulation of in vitro OCG both in terms of differentiation and pre-osteoclast fusion. One possible mechanistic explanation is that in the absence of Ads, osteoclast precursors shift actin cytoskeletal organization to favor migration and inhibit fusion, resulting in a marked decrease in the number of multinucleated, TRAcP functioning osteoclasts. Further, the lack of Ads may have wider reaching implications downstream of the RANKL-RANK signal cascade resulting in altered osteoclast differentiation. The preliminary nature of this study has not focused on the intricate signaling cascade downstream of the RANKL-RANK in the Ads KD system. That being said, this proposed role of Ads in the regulation of osteoclast differentiation is consistent with the previously described cellular differentiation of role of Ads in platelet maturation and chondrocyte differentiation [29,30]. As such, we can conclude that a threshold, basal level of Ads may be required during the early mononuclear stages of OCG to promote TRAcP expression and osteoclast differentiation. Further, it is plausible that during the later stages of OCG, when cell-cell fusion is necessary to allow for the formation of large multinucleated osteoclasts, prolonged elevated levels of Ads expression may be required for the intricate regulation of pre-osteoclast fusion.

This project has already taken major strides towards determining the role of Ads in the osteoclast biological system. We have described the temporal expression pattern of Ads during OCG in response to RANKL and we have shown how attenuated Ads expression has deleterious impacts on the pre-osteoclast differentiation and the fusogenic ability of pre-osteoclasts. However, there are a number of important questions that still remain unanswered. First and foremost, we have yet to decipher the mechanism of Ads mediated osteoclast progenitor fusion. Alternatively, we plan to characterize the in vivo skeletal effect of Ads using newly developed transgenic mice containing a floxed Ads allele crossed this with LysM Cre transgenic mice to create viable conditional knockouts, in which Ads is deleted in osteoclast precursors and osteoclasts. Lastly, we plan on generating several Ads constructs with deletions of domains for either actin severing, actin nucleation or actin capping, to determine the mechanisms through which domain(s) regulates actin mediated pre-osteoclast fusion. All in all, the evidence presented in this paper represents an advancement in osteoclast biology. For the first time the role of Ads in osteoclast differentiation has been examined. Furthermore, although not unique to osteoclasts, Ads may also serve as a new marker for osteoclast differentiation.

Supporting Information

Figure S1 Detection of Adseverin D5 using primers flanking the 5th Domain. A schematic representation of the pSE380 bacterial expression vector with the full-length Ads (pSE380-A-FL) and Ads D5 (pSE380-A-D5) isoforms within its Nco I and Pst I sites. Primers flanking the 5th domain of Ads (esD5 forward and esD5 reverse) allow for the amplification of a 650 bp amplicon from Ads full-length plasmid and a 350 bp amplicon of the Ads D5 plasmid.

Figure S2 Schematic of knockdown clonal cell lines. Viral supernatant with the RNAi-Ready pSIREN-RetroQ-DsRed Express vector containing the Ads or Luc KD shRNA sequence was harvested and used to infect RAW macrophages. Infected cells were sorted based on the DsRed signal by fluorescence-activated cell sorting and subjected to a limiting dilution to obtain clonal cell lines. Colonies were examined under an inverted fluorescent microscope and DsRed positive colonies were identified and expanded. A single Luc KD clonal cell line (Luc KD) and three Ads KD clonal cell lines (Clones B, C, & F) were generated. WT RAW macrophages were used as non-infection controls. The schematic above shows that only cells stably infected cells were DsRed positive, while the WT cells showed no DsRed signal.

Figure S3 Adseverin knockdown does not alter the expression of multiple genes important in osteoclastogenesis. Quantitative real-time PCR was used to quantify gene expression on Days 0 and 4 of osteoclast cultures. Results are expressed as fold expression versus GAPDH used as internal control. There were no statistically significant differences in transcript levels of RANK, SIRPα, CD44 and Gelsolin between Ads KD and WT Day 0 and Day 4 osteoclast cultures (n = 3).

Author Contributions

Conceived and designed the experiments: SH YW JWPK MG. Performed the experiments: SH HJ YW. Analyzed the data: SH HJ YW. Contributed reagents/materials/analysis tools: SH HJ YW JWPK. Wrote the paper: SH.

References

1. Allori AC, Sailon AM, Warren SM (2008) Biological basis of bone formation, remodeling, and repair-part II: extracellular matrix. Tissue Eng Part B Rev 14: 275–283.
2. Boyle WJ, Simonet WS, Lacey DL (2003) Osteoclast differentiation and activation. Nature 423: 337–342.
3. Rodan GA, Martin TJ (2000) Therapeutic approaches to bone diseases. Science 289: 1508–1514.
4. Manolagas SC (2000) Birth and death of bone cells: basic regulatory mechanisms and implications for the pathogenesis and treatment of osteoporosis. Endocr Rev 21: 115–137.
5. Sambrook P, Cooper C (2006) Osteoporosis. Lancet 367: 2010–2018.
6. Tolar J, Teitelbaum SL, Orchard PJ (2004) Osteopetrosis. N Engl J Med 351: 2839–2849.
7. Novack DV, Teitelbaum SL (2008) The osteoclast: friend or foe? Annu Rev pathmechdis Mech Dis 3: 457–484.
8. Walker DG (1975) Bone resorption restored in osteopetrotic mice by transplants of normal bone marrow and spleen cells. Science.
9. Takahashi N, Yamana H, Yoshiki S, Roodman GD, Mundy GR, et al. (1988) Osteoclast-like cell formation and its regulation by osteotropic hormones in mouse bone marrow cultures. Endocrinology 122: 1373–1382.
10. Parent CA, Devreotes PN (1999) A cell's sense of direction. Science 284: 765–770.
11. Fuller K, Owens JM, Jagger CJ, Wilson A, Moss R, et al. (1993) Macrophage colony-stimulating factor stimulates survival and chemotactic behavior in isolated osteoclasts. J Exp Med 178: 1733–1744.
12. Helming L, Gordon S (2009) Molecular mediators of macrophage fusion. Trends in Cell Biology 19: 514–522.
13. Kyriakides TR, Foster MJ, Keeney GE, Tsai A, Giachelli CM, et al. (2004) The CC Chemokine Ligand, CCL2/MCP1, Participates in Macrophage Fusion and Foreign Body Giant Cell Formation. The American Journal of Pathology 165: 2157–2166.
14. Vignery A (2005) Macrophage fusion: the making of osteoclasts and giant cells. J Exp Med 202: 337–340.
15. Arai F, Miyamoto T, Ohneda O, Inada T, Sudo T, et al. (1999) Commitment and differentiation of osteoclast precursor cells by the sequential expression of c-Fms and receptor activator of nuclear factor kappaB (RANK) receptors. J Exp Med 190: 1741–1754.
16. Schafer DA, Cooper JA (1995) Control of actin assembly at filament ends. Annu Rev Cell Dev Biol 11: 497–518.
17. Silacci P, Mazzolai L, Gauci C, Stergiopulos N, Yin HL, et al. (2004) Gelsolin superfamily proteins: key regulators of cellular functions. Cell Mol Life Sci 61: 2614–2623.
18. Carlier MF, Pantaloni D (1997) Control of actin dynamics in cell motility. J Mol Biol 269: 459–467.
19. Sakurai T, Kurokawa H, Nonomura Y (1991) Comparison between the gelsolin and adseverin domain structure. J Biol Chem 266: 15979–15983.
20. Maekawa S, Sakai H (1990) Inhibition of actin regulatory activity of the 74-kDa protein from bovine adrenal medulla (adseverin) by some phospholipids. J Biol Chem 265: 10940–10942.
21. Janmey PA, Stossel TP (1987) Modulation of gelsolin function by phosphatidylinositol 4, 5-bisphosphate.
22. Chumnarnsilpa S, Lee WL, Nag S, Kannan B, Larsson M, et al. (2009) The crystal structure of the C-terminus of adseverin reveals the actin-binding interface. Proc Natl Acad Sci USA 106: 13719–13724.
23. Lueck A, Yin HL, Kwiatkowski DJ, Allen PG (2000) Calcium regulation of gelsolin and adseverin: a natural test of the helix latch hypothesis. Biochemistry 39: 5274–5279.
24. Trifaró JM, Gasman S, Gutiérrez LM (2007) Cytoskeletal control of vesicle transport and exocytosis in chromaffin cells. Acta Physiologica 192: 165–172.
25. Ehre C, Rossi AH, Abdullah LH, De Pestel K, Hill S, et al. (2005) Barrier role of actin filaments in regulated mucin secretion from airway goblet cells. Am J Physiol, Cell Physiol 288: C46–C56.
26. Bruun TZ, Høy M, Gromada J (2000) Scinderin-derived actin-binding peptides inhibit Ca(2+)- and GTPgammaS-dependent exocytosis in mouse pancreatic beta-cells. Eur J Pharmacol 403: 221–224.
27. Marcu MG, Zhang L, Nau-Staudt K, Trifaró JM (1996) Recombinant scinderin, an F-actin severing protein, increases calcium-induced release of serotonin from permeabilized platelets, an effect blocked by two scinderin-derived actin-binding peptides and phosphatidylinositol 4,5-bisphosphate. Blood 87: 20–24.
28. Noda Y, Horikawa S, Katayama Y, Sasaki S (2005) Identification of a multiprotein "motor" complex binding to water channel aquaporin-2. Biochemical and Biophysical Research Communications 330: 1041–1047.
29. Zunino R, Li Q, Rosé SD, Romero-Benítez MM, Lejen T, et al. (2001) Expression of scinderin in megakaryoblastic leukemia cells induces differentiation, maturation, and apoptosis with release of plateletlike particles and inhibits proliferation and tumorigenesis. Blood 98: 2210–2219.
30. Nurminsky D, Magee C, Faverman L, Nurminskaya M (2007) Regulation of chondrocyte differentiation by actin-severing protein adseverin. Developmental biology 302: 427–437.
31. Robbens J, Louahed J, De Pestel K, Van Colen I, Ampe C, et al. (1998) Murine adseverin (D5), a novel member of the gelsolin family, and murine adseverin are induced by interleukin-9 in T-helper lymphocytes. Mol Cell Biol 18: 4589–4596.
32. Beaulieu V, Da Silva N, Pastor-Soler N, Brown CR, Smith PJS, et al. (2005) Modulation of the actin cytoskeleton via gelsolin regulates vacuolar H+-ATPase recycling. J Biol Chem 280: 8452–8463.
33. Chellaiah M, Fitzgerald C, Alvarez U, Hruska K (1998) c-Src is required for stimulation of gelsolin-associated phosphatidylinositol 3-kinase. J Biol Chem 273: 11908–11916.
34. Wang Q, Xie Y, Du Q-S, Wu X-J, Feng X, et al. (2003) Regulation of the formation of osteoclastic actin rings by proline-rich tyrosine kinase 2 interacting with gelsolin. The Journal of Cell Biology 160: 565–575.
35. Yang G, Zaidi M, Zhang W, Zhu L-L, Li J, et al. (2008) Functional grouping of osteoclast genes revealed through microarray analysis. Biochemical and Biophysical Research Communications 366: 352–359.
36. Wang Y, Lebowitz D, Sun C, Thang H, Grynpas MD, et al. (2008) Identifying the relative contributions of Rac1 and Rac2 to osteoclastogenesis. J Bone Miner Res 23: 260–270.
37. Saeed AI, Sharov V, White J, Li J, Liang W, et al. (2003) TM4: a free, open-source system for microarray data management and analysis. BioTechniques 34: 374–378.
38. Leung R, Wang Y, Cuddy K, Sun C, Magalhaes J, et al. (2010) Filamin A regulates monocyte migration through Rho small GTPases during osteoclastogenesis. J Bone Miner Res 25: 1077–1091.
39. Svensson C, Silverstone AE, Lai Z-W, Lundberg K (2002) Dioxin-induced adseverin expression in the mouse thymus is strictly regulated and dependent on the aryl hydrocarbon receptor. Biochemical and Biophysical Research Communications 291: 1194–1200.
40. Kondo Y, Irie K, Ikegame M, Ejiri S, Hanada K, et al. (2001) Role of stromal cells in osteoclast differentiation in bone marrow. J Bone Miner Metab 19: 352–358.
41. Müller S, Quast T, Schröder A, Hucke S, Klotz L, et al. (2013) Salt-Dependent Chemotaxis of Macrophages. PLoS ONE 8: e73439.
42. Wang Y, Belsham DD, Glogauer M (2009) Rac1 and Rac2 in osteoclastogenesis: a cell immortalization model. Calcified Tissue International 85: 257–266.
43. Sheetz MP, Felsenfeld D, Galbraith CG, Choquet D (1999) Cell migration as a five-step cycle. Biochem Soc Symp 65: 233–243.
44. Takahashi N, Ejiri S, Yanagisawa S, Ozawa H (2007) Regulation of osteoclast polarization. Odontology 95: 1–9.
45. Roodman GD (1999) Cell biology of the osteoclast. Exp Hematol 27: 1229–1241.
46. Vitale ML (1991) Cortical filamentous actin disassembly and scinderin redistribution during chromaffin cell stimulation precede exocytosis, a phenomenon not exhibited by gelsolin. The Journal of Cell Biology 113: 1057–1067.
47. Trifaró J, Rosé SD, Lejen T, Elzagallaai A (2000) Two pathways control chromaffin cell cortical F-actin dynamics during exocytosis. Biochimie 82: 339–352.
48. Cheek TR, Burgoyne RD (1986) Nicotine-evoked disassembly of cortical actin filaments in adrenal chromaffin cells. FEBS Lett 207: 110–114.
49. Cuchillo-Ibáñez I, Lejen T, Albillos A, Rosé SD, Olivares R, et al. (2004) Mitochondrial calcium sequestration and protein kinase C cooperate in the regulation of cortical F-actin disassembly and secretion in bovine chromaffin cells. J Physiol (Lond) 560: 63–76.
50. Lejen T, Skolnik K, Rosé SD, Marcu MG, Elzagallaai A, et al. (2001) An antisense oligodeoxynucleotide targeted to chromaffin cell scinderin gene decreased scinderin levels and inhibited depolarization-induced cortical F-actin disassembly and exocytosis. J Neurochem 76: 768–777.
51. Zhang L, Rodríguez Del Castillo A, Trifaró JM (1995) Histamine-evoked chromaffin cell scinderin redistribution, F-actin disassembly, and secretion: in the absence of cortical F-actin disassembly, an increase in intracellular Ca2+ fails to trigger exocytosis. J Neurochem 65: 1297–1308.
52. Chellaiah M, Kizer N, Silva M, Alvarez U, Kwiatkowski D, et al. (2000) Gelsolin deficiency blocks podosome assembly and produces increased bone mass and strength. The Journal of Cell Biology 148: 665–678.
53. Gimona M, Buccione R (2006) Adhesions that mediate invasion. The International Journal of Biochemistry & Cell Biology 38: 1875–1892. doi:10.1016/j.biocel.2006.05.003.
54. Saltel F, Chabadel A, Bonnelye E, Jurdic P (2008) Actin cytoskeletal organisation in osteoclasts: a model to decipher transmigration and matrix degradation. European Journal of Cell Biology 87: 459–468.
55. Yagi M, Miyamoto T, Sawatani Y, Iwamoto K, Hosogane N, et al. (2005) DC-STAMP is essential for cell-cell fusion in osteoclasts and foreign body giant cells. J Exp Med 202: 345–351.
56. Yagi M, Miyamoto T, Toyama Y, Suda T (2006) Role of DC-STAMP in cellular fusion of osteoclasts and macrophage giant cells. J Bone Miner Metab 24: 355–358.

11

Human Equilibrative Nucleoside Transporter-1 Knockdown Tunes Cellular Mechanics through Epithelial-Mesenchymal Transition in Pancreatic Cancer Cells

Yeonju Lee[1], Eugene J. Koay[1,2], Weijia Zhang[1], Lidong Qin[1,4], Dickson K. Kirui[1], Fazle Hussain[3], Haifa Shen[1,4], Mauro Ferrari[1,5]*

1 Department of Nanomedicine, Houston Methodist Research Institute, Houston, Texas, United States of America, **2** Department of Radiation Oncology, M. D. Anderson Cancer Center, Houston, Texas, United States of America, **3** Department of Mechanical Engineering, Texas Tech University, Lubbock, Texas, United States of America, **4** Department of Cell and Developmental Biology, Weill Cornell Medical College, New York, New York, United States of America, **5** Department of Medicine, Weill Cornell Medical College, New York, New York, United States of America

Abstract

We report cell mechanical changes in response to alteration of expression of the human equilibrative nucleoside transporter-1 (hENT1), a most abundant and widely distributed plasma membrane nucleoside transporter in human cells and/or tissues. Modulation of hENT1 expression level altered the stiffness of pancreatic cancer Capan-1 and Panc 03.27 cells, which was analyzed by atomic force microscopy (AFM) and correlated to microfluidic platform. The hENT1 knockdown induced reduction of cellular stiffness in both of cells up to 70%. In addition, cellular phenotypic changes such as cell morphology, migration, and expression level of epithelial-mesenchymal transition (EMT) markers were observed after hENT1 knockdown. Cells with suppressed hENT1 became elongated, migrated faster, and had reduced E-cadherin and elevated N-cadherin compared to parental cells which are consistent with epithelial-mesenchymal transition (EMT). Those cellular phenotypic changes closely correlated with changes in cellular stiffness. This study suggests that hENT1 expression level affects cellular phenotype and cell elastic behavior can be a physical biomarker for quantify hENT1 expression and detect phenotypic shift. Furthermore, cell mechanics can be a critical tool in detecting disease progression and response to therapy.

Editor: Xin-Yuan Guan, The University of Hong Kong, China

Funding: The authors gratefully acknowledge funding support from the following sources: Department of Defense grants W81XWH-09-1-0212 and W81XWH-12-1-0414, National Institute of Health grants U54CA143837 and U54CA151668, the CPRIT grant RP121071 from the State of Texas, and the Ernest Cockrell Jr. Distinguished Endowed Chair. The funders had no role in study design, data collection and analysis, decision to publish, or preparation of the manuscript.

Competing Interests: The authors have declared that no competing interests exist.

* Email: mferrari@HoustonMethodist.org

Introduction

Pancreatic adenocarcinoma (PDAC) is one of the most lethal human cancers with an extremely poor prognosis [1]. PDAC has low survival rate, even after complete resection of the tumor, which is the only chance for cure. Unfortunately, most of tumors are unresectable and metastatic. Thus, chemotherapy and/or radiotherapy are the only options [2,3]. Gemcitabine (2′,2′-difluorodeoxycytidine) is one of efficient anticancer agents for pancreatic cancer [1]. It is a cytotoxic pyrimidine deoxynucleoside analogue that is transported into the cellular compartment through the primary transport protein, human equilibrative nucleoside transporter-1 (hENT1), and eventually inhibits DNA replication. The hENT1 expression level in pancreatic cancer cells has previously been correlated to therapeutic efficacy where cells with higher hENT1 expression were shown to respond better to gemcitabine. Furthermore, cellular level studies have also shown that pancreatic cancer cells with low hENT1 expression are highly resistant to gemcitabine [4]. Moreover, clinical studies have established that hENT1 expression affect how patients respond to

treatment where patients whose tumor expressed low hENT1 biomarker responded poorly to gemcitabine therapy [3,5].

Pancreatic cancer cells that acquire gemcitabine-resistance are characterized by epithelial-mesenchymal transition (EMT) phenotype and show distinct morphological changes from epithelial to spindle-shaped and increasing cellular motility [6,7]. EMT is a biological process that polarized epithelial cells shift to a mesenchymal-like phenotype through multiple biochemical changes. This phenotypic transition is characterized by loss of cell-cell adhesion and dynamic changes in the structure of the cytoskeleton which cause cells to detach from epithelium and to gain the ability to migrate to distant sites [8,9]. Thus, in this study, we hypothesized that modulation of hENT1 expression levels in pancreatic cancer cells may alter their physiological characteristics as it may induce phenotypic shift by inhibiting gemcitabine uptake. In addition to biochemical methods to identify cellular physiological changes, understanding cell mechanics can provide new biological insights. Recent studies reveal that mechanical properties of cells provide crucial information to understand various biophysical behaviors that include cell shape, motility, and cell adhesion that generate a cascade of biochemical signals that are critical for biological responses [10]. The mechanical signatures of cells can be an important tool in various aspects: (1) Identification of cancer cells from normal cells based on their relatively lower stiffness [11]; (2) Anticipation of a metastatic potential of cancer cells as cellular stiffness inversely correlates with migration and invasion [12–14]; (3) Recognition of phenotypic shifts associated with alteration in intracellular structure and motility in cancer cells by measurement of increases or decreases in elastic modulus [11,15–18]. Although there is no direct evidence that hENT1 is related to phenotypic events in pancreatic cancer cells, we can speculate that hENT1 expression may modulate cellular biophysical behaviors based on the close correlation between hENT1 expression and gemcitabine sensitivity. The cellular phenotypic shift from gemcitabine resistant cells established by culturing cells in serially increasing concentration of gemcitabine also supports our hypothesis.

In this study, two different methods, AFM and a microfluidic platform, were used to evaluate how modulation of hENT1 expression level influences on stiffness of pancreatic cancer cells. Then the accompanying morphological alterations, cytoskeleton rearrangements, cellular motility, and changes in expression levels of EMT markers to investigate cellular phenotypic shift were characterized (Figure 1). Together our results on hENT1 expression level and cell stiffness correlate very well with the mechanistic alterations of intracellular cytoskeletal structure, cellular motility, and suggest that cellular elastic properties can estimate hENT1 expression level as well as phenotypic shift.

Materials and Methods

Cell culture

All cell lines were purchased from American Type Culture Collection (ATCC, Manassas, VA). The AsPC-1, BxPC-3, MIA Paca 2, and Panc-1 cells were grown in Dulbeccos Modified Eagles Medium (DMEM, Thermo Scientific Hyclone, Waltham, MA) with 10% fetal bovine serum (FBS, ATLAS Biologicals, Fort Collins, CO), Capan-1 in DMEM with 20% FBS, and Panc 03.27 in RPMI 1640 (Thermo Scientific Hyclone, Waltham, MA) with 15% FBS and human recombinant insulin (10 units/ml, Sigma Aldrich, St. Louis, MO) in the presence of 5% CO_2 at 37°C.

hENT1 knockdown with siRNA transfection

The cells were cultured in 6 cm cell culture dish at a density of 5×10^5 cells/dish. After washing with PBS, cells were treated with INTERFERin (Polyplus transfection™) containing siRNA against hENT1 (SMARTpool: ON-TARGET plus SLC29A, Dharmacon, Inc., Pittsburgh, PA) or negative siRNA (Silencer Negative Control No. 1 siRNA, AM4635, Invitrogen, Grand Island, NY) at a concentration of 50 nM in Opti-MEM (Invitrogen, Grand Island, NY). After 24 hours transfection, cells were washed twice with PBS and incubated for 3 days.

Western blotting analysis

Cells were lysed with RIPA Pierce buffer (Thermo Scientific, Waltham, MA) with protease inhibitor cocktail (Thermo Scientific, Waltham, MA). Cell lysates (10 μg) with loading buffer (LDS sample buffer Non-reducing, Thermo Scientific, Waltham, MA) were heated for 5 min at 95°C. Cell lysates and protein ladder (Xpert 2 Prestained Protein Marker, GenDEPOT) loaded on polyacrylamide gels (Any kD Mini- PROTEAN TGX Precast Gel,

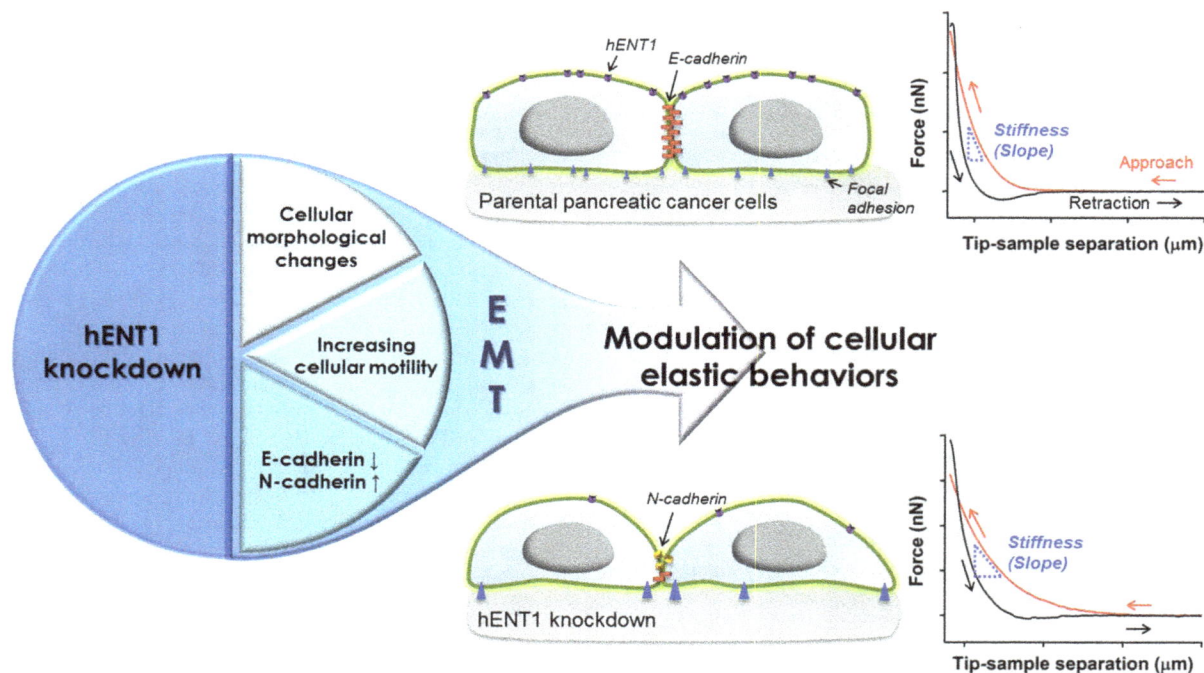

Figure 1. Schematic illustrations describing overall flow of this study. The hENT1 knockdown induces changes in cellular mechanics via EMT accompanied by alterations in E-cadherin and N-cadherin expression levels, cellular morphology, and motility of pancreatic cancer cells. Further, hENT1 knockdown induces decrease in cell stiffness as demonstrated on representative force separation curves obtained from Panc 03.27 cells (upper graph from a parent cell; the second graph from a hENT1 knockdown cell) using AFM.

Bio-Rad, Hercules, CA) and transferred to Nitrocellulose Membrane (Bio-rad, Hercules, CA). The blots were blocked with blocking buffer, TBST (20 mM TRIS pH 7.6/150 mM NaCl/ 0.05% Tween 20) containing 5% milk for an hour. The blots were incubated overnight with TBST and 5% milk containing primary antibodies at certain ratios: anti-hENT1 antibody (1:1000, Abcam, Cambridge, MA); anti-E-cadherin antibody (1:1000, Abcam, Cambridge, MA); anti-N-cadherin antibody (1:500, Abcam, Cambridge, MA); cytokeratin 18 (1:1000, Cell signaling technology, Danvers, MA); Lamin A/C (1:1000, Cell signaling technology, Danvers, MA); anti-GAPDH antibody (1:2000, Cell signaling technology, Danvers, MA). After washing three times with TBST, the blots were incubated with TBST and 5% milk containing secondary antibodies, anti-mouse IgG, HRP-linked antibody (1:5000, Cell signaling technology, Danvers, MA) or anti-rabbit IgG, HRP-linked antibody (1:5000, Cell signaling technology, Danvers, MA), for an hour at room temperature. Then, the blots placed onto a mixture of Pierce ECL western blotting substrate (Thermo Scientific, Waltham, MA) and Lumigen TMA-6 (Lumigen, Inc., Southfield, MI), and proteins were detected.

Immunofluorescence

Pancreatic cancer cells were seeded in 6-well plates containing sterilized, collagen I (Invitrogen, Grand Island, NY)-coated coverslips (22×22 mm, Corning) at a density of 2×10^5 cells/well. The cells were prepared with three groups: 1) control (without treatment); 2) scramble (transfected with negative control siRNA); and 3) hENT1 knockdown. The cells were fixed with 4% paraformaldehyde (Affymetrix, Santa Clara, CA) for 30 min at room temperature and then permeabilized with 2% Triton X-100 (Sigma Aldrich, St. Louis, MO) for 10 min. Following blocking with 2% bovine serum albumin (Calbiochem) for 1 hour, the cells were incubated overnight with anti-Vinculin antibody (1:200, Abcam, Cambridge, MA), anti-E-cadherin (1:200), anti-N-cadhrein (1:100), anti-cytoketarin 18 (1:200), or anti-Lamin A/C (1:200) at 4°C with gentle shaking. Then cells were washed twice with PBS for 10 min and incubated with Alexa 488 or 594-conjugated secondary antibodies for an hour at room temperature. After washing with PBS, for F-actin staining, cells were incubated with Alexa Fluor 488 Phalloidin (1:250, Invitrogen, Grand Island, NY) for 30 min at room temperature. Then, the nuclei were stained using Hoechst 33342 (Molecular probes, Life Technologies, Grand Island, NY) for 10 min. The cells were monitored by confocal laser scanning microscope (CLSM, Olympus FluoView FV1000).

AFM measurements

Cellular stiffness was measured by the force-curve technique on a Bioscope (Bruker Corporation, Billerica, MA). The cells were cultured on 6 cm cell culture dish and all measurements were performed in culture medium at 37°C. The AFM was equipped with an inverted light microscope (Olympus IX81) so that the cells were constantly monitored. To minimize cell damage, silica microparticle (diameter: 5 μm) modified silicon nitride cantilevers (Novascan Technologies, Inc., Ames, IA) with approximate spring constant values of ∼0.06 N/m were employed in all AFM experimentations. The exact spring constant value was measured by the thermal tuning method. Probes were positioned at the cells' nuclei proximities under optical control, and force curves were acquired at a sampling rate of 1 Hz. The Young's modulus, E, was calculated from obtained force curves based on the Hertz model (Eq. 1) using Nanoscope analysis program from Bruker corporation.

$$F = \frac{4}{3} \frac{E}{(1-\nu^2)} \sqrt{R} \delta^{\frac{3}{2}} \qquad (1)$$

where F = force, E = Young's modulus, ν = Poisson's ratio ($\nu = 0.5$, in this study), R = radius of the indenter (R = 2500 nm, in this study), and δ = indentation depth.

To obtain Young's modulus, two independent experiments were performed and force curves from at least 50 cells were collected in each experiment.

MS-chip design and fabrication

Silicon wafers (4 inch) from Corning Inc. SU-8 2015 photoresist, and SU-8 developer from MicroChem Corp. and poly-dimethylsiloxane (PDMS RTV615) from Momentive Performance Materials Inc. were used. The microchip pattern was designed with AutoCAD software (Autodesk Inc.) and then printed out as 10-μm resolution chrome masks by Photo Science Inc. The photomask pattern was first translated into a microstructure on a 4-in silicon wafer using SU-8 2015 photoresist, which is a mold for casting PDMS materials. Briefly, the mold was prepared by spin coating SU-8 2015 photoresist onto a silicon wafer and cross-linking by UV for 180 seconds. Subsequently, the designed pattern was developed using SU-8 developer (Microchem Corp.) and cleaned with isopropyl alcohol and nitrogen gas. The holes for the inlets and outlets were punched using needles. The PDMS layer was cleaned by rinsing with isopropyl alcohol and deionized water, and dried with nitrogen gas. After treatment with oxygen plasma, the PDMS layer was bonded immediately to a 75×50-mm glass slide. Finally, the bonded device was baked for 2 h at 80°C.

On-chip cell separation

The channels in MS-chip and Tygon tube connected to the chip inlet were wetted with PBS and then kept with 0.5% BSA in PBS for 1 hour. BSA blocks the surface and further prevents nonspecific adhesion of cells to PDMS. Membranes of control and hENT1 knockdown cells were stained using Alexa Fluor 594 or 488-conjugated wheat germ agglutinin (Invitrogen, Grand Island, NY), respectively. The cell mixture in equal amounts at a final density of 1×10^5 cells/ml was prepared. Suspended cells were then applied to the MS-chip via a Tygon tube. During the experiment, compressed nitrogen gas was applied to the cell suspension at a pressure of ∼10 psi (69×10^3 Pa). A typical separation lasted ∼15 min, and average flow rate was controlled at 1–2 mL/h. Cells applied to the MS-chip were imaged by fluorescence microscopy (IX81, Olympus). Number of cells retained on chip after separation was counted by Image J.

In vitro scratch assay

The scratch assay was performed on either native cells (control) or transfected cells (scramble and siRNA against hENT1) to study the effect of hENT1 knockdown on cell migration. Capan-1 or Panc 03.27 cells were seeded into 24-well plate. When the cells are approximately 50–60% confluent, cells were washed with PBS and then transfected with INTERFERin containing siRNA against hENT1 or negative siRNA at a concentration of 50 nM in Opti-MEM. After 6 hours transfection, cells were washed twice with PBS and incubated for 2 days. A scratch of the cell monolayer was created by using a pipet tip. Then, the plates was washed and replaced with the desired medium. The time-lapse microscope (EVOS FL Auto Imaging System, Life Technologies) with chamber (95% Air and 5% CO_2) and temperature control (37°C) was used for acquiring the images from the same field

Capan-1

Panc 03.27

(A)

(B)

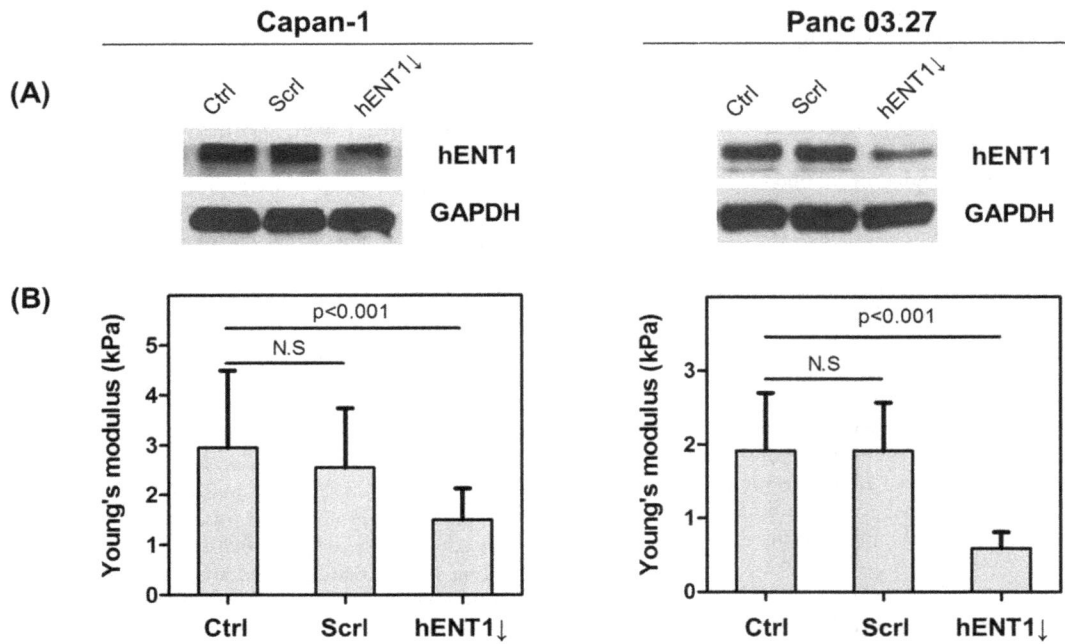

Figure 2. Establish hENT1 knockdown pancreatic cancer cell lines. (A) Western blots of hENT1 (55 kDa) and GAPDH (37 kDa) in Capan-1 (left) and Panc 03.27 cells (right) without treatment (ctrl) or after treatment of negative siRNA (scrl) or hENT1 siRNA (hENT1 ↓), (B) Bar histograms show Young's modulus of control, negative siRNA transfected, and hENT1 knockdown cells. (N.S: statistically not significant).

automatically for 18 hours. The images acquired were analyzed quantitatively by using Image J.

Statistical analysis

The data are expressed as mean ± standard deviation. Statistical significance was identified by two-way ANOVA analysis using GraphPad Prism 5.0. A P value<0.05 was considered to be statistically significant.

Results and Discussion

To investigate the correlation between hENT1 expression level and cell mechanics, two types of pancreatic cancer cell lines, Capan-1 and Panc 03.27 with relatively higher hENT1 expression (Figure S1 in File S1), were chosen. By siRNA transfection, hENT1 downregulated cells were established (Figure 2). We then measured changes in cell mechanics by two different methods, AFM and microfluidic separation. AFM measures elastic properties of living cells through a direct mechanical interaction between a probe and cell surface. Cell stiffness affects the extent of cantilever deflection upon interaction with the surface of adherent cells [19]. A live cell indentation induces deformation of cell compartments that include membrane, cytoskeletons, nucleus, and various organelles. Thus, the elastic properties of cells result from the aggregated effects of deforming numerous cellular components. We used a silica microparticle (diameter: 5 μm) modified cantilever (Figure S2 in File S1) to reduce cell membrane damage during contact and skew the dataset differently than a sharp tip. To optimize indentation force, we applied force up to 10 nN as shown in Figure S2 in File S1. A constant Young's modulus of Panc 03.27 cells was obtained within the indentation forces up to 200 pN, indicating the measurement of cellular stiffness using AFM is only valid at small deformations of the living cells. Both cell lines showed a similar trend under the same indentation force, i.e. 100 pN. The control cells and cells transfected by negative

siRNA were significantly stiffer than hENT1 knockdown cells (Figure 2 and Figure S3 in File S1). The average cell stiffness of both control cells is 2.95±1.55 kPa (Capan-1) and 1.91±0.78 kPa (Panc 03.27); however, that of hENT1 knockdown cells showed significantly decreased Young's moduli with values of 1.50±0.63 kPa (Capan-1) and 0.59±0.22 kPa (Panc 03.27).

Since AFM is limited to local measurement of cell mechanics, we also evaluated cellular deformability using microfluidic separation method [20] to corroborate AFM findings. A mechanical separation chip (MS-chip, Figure 3A) was designed with artificial microbarriers in combination with hydrodynamic force to separate deformable from stiff cells [21]. To demonstrate the capacity of the MS-chip to separate cells based on deformability, we tested the separation of a mixture containing two different cells: control and hENT1 knockdown. Membrane of control and hENT1 knockdown cells was stained by using Alexa Fluor 594 and 488 conjugated wheat germ agglutinin (WGA), respectively. As shown in Figure S4 in File S1, the effect of WGA on cell stiffness is negligible. The proportion of the two cell lines varied along the length of the chip. As shown in Figure 3D and 3G, diameters of both cells, control and hENT1 knockdown, were similar. The cell mixture in equal amounts at a density of 1×10^5 cells/mL was injected to MS-chip and flowed for 15 min. The gaps between posts array in this designed MS-chip range from 22 μm to 2 μm. The channels avoid obstruction that occurs in repeating arrays of posts which regulate and equalize hydrodynamic pressure throughout the chip. It is assumed that there is no physical contact between PDMS pillars and cells because the PDMS is coated with BSA. The shear stress is a main factor to interact with cells. When the gap size of post arrays is bigger than cell diameter, pressure force (F_p, 10 psi in this study) is the only force acting on cells. If cell diameter is the same or larger than the gap size, friction force (F_f) opposes pressure force (Figure 3B). If cells are stiffer, they push harder on the post arrays, and then F_f is

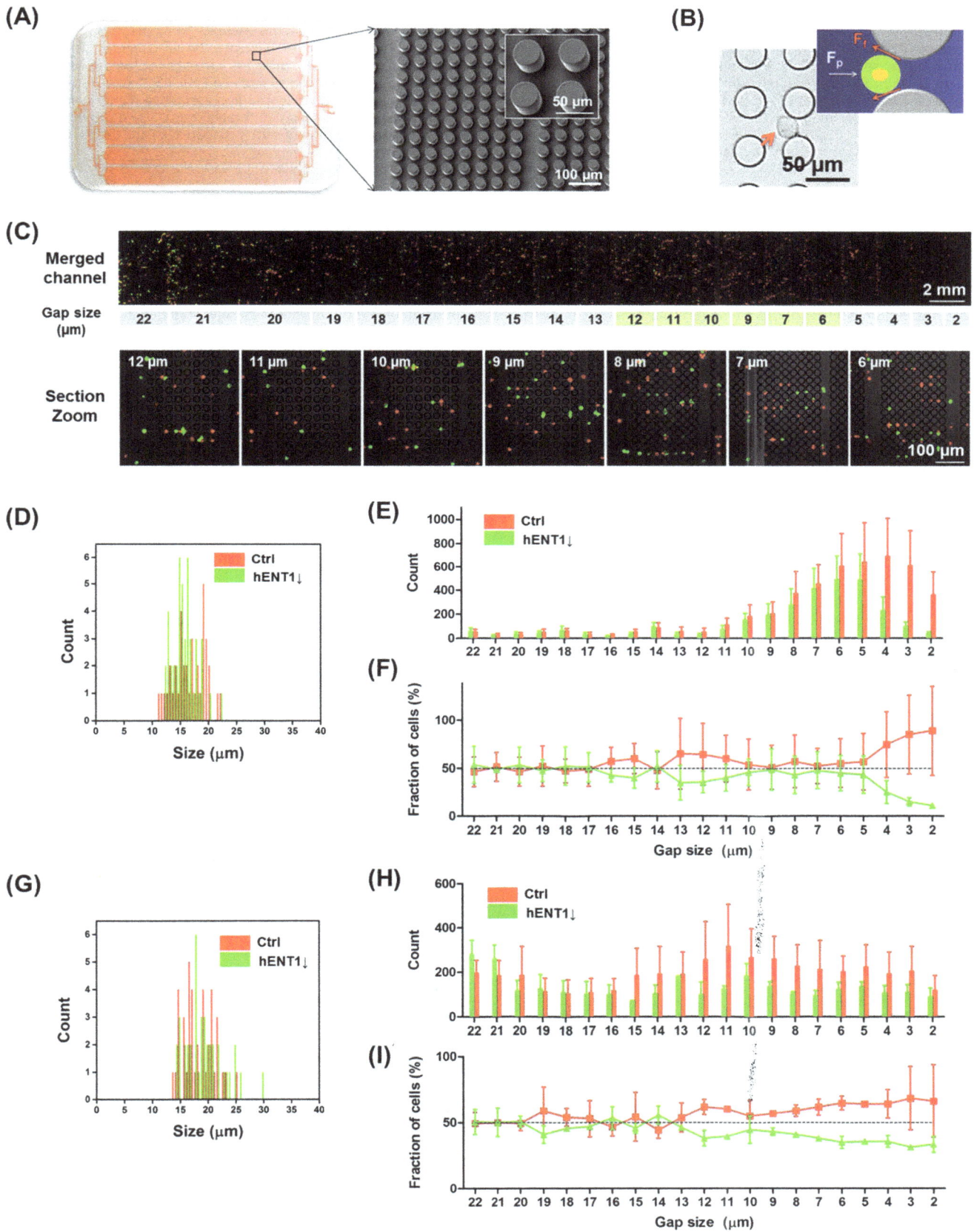

Figure 3. Compare deformability of control and hENT1 knockdown cells using a microfluidic separation chip. (A) Photograph (left) of microfluidic separation chip (MS-chip) visualized by red dye inside and a representative scanning electron microscopic (right) image of the post array (diameter of pillar: 35 μm, height of pillar: 30 μm). (B) Bright field image and scheme of cells flowing in MS-chip when the cell diameter is smaller than pillar gap size (red arrow indicates cells). It is assumed here that friction is due only to shear stress (F_p: pressure force against the applied pressure (10 psi in this study), F_f: friction force). (C) The fluorescence image of one of channels among eight shows retained Panc 03.27 cells in the MS-chip after separation of Panc 03.27-ctrl (red fluorescence) and Panc 03.27-hENT1 knockdown (green fluorescence). Representative higher magnification confocal microscopic images of the MS chip (gap size from 12 μm to 6 μm) show the efficiency of separation through gaps. Size distribution of cells (D: Capan-1, G: Panc 03.27 cells). Statistical analysis of cells (E: Capan-1, H: Panc 03.27) and fraction ratio of cells (F: Capan-1, I: Panc 03.27) retained on chip. Values represent mean ± standard deviation.

significantly increased. Figure 3C shows Panc 03.27 cells trapped on MS-chip imaged by fluorescence microscopy; red fluorescence shows control cells and green fluorescence shows hENT1 knockdown cells. At the inlet, equal number of red control Panc 03.27 (or Capan-1) cells and green hENT1 knockdown Panc 03.27 (or Capan-1) cells were observed (Figure 3C and 3E–I). In this region, channels were much wider that the diameter of cells, resulting in no significant separation of flexible and stiff cells. Due to smaller diameter of Capan-1 cells (average diameter: 15 μm), the separation was achieved with 8 μm gap size of post arrays. But, in case of Panc 03.27 cells (average diameter: 19 μm), the major separation occurs at 12 μm. In these cell lines, cells in the control group were consistently trapped while hENT1 knockdown cells passed through the gaps more frequently. These experiments demonstrate the overall efficiency of separation of the two different phenotypes with similar diameter in the MS-chip. Thus, we conclude both control cells are relatively stiffer than hENT1 knockdown cells, and these results are in agreement with AFM findings.

To understand how hENT1 influences cell mechanics, physiological changes in cells after hENT1 knockdown were studied by confocal microscopy. We observed that cell morphological changes associated with hENT1 knockdown. As shown in Figure 4, cells were visualized using confocal laser scanning microscope and the cell roundness was quantitatively analyzed by calculating form factors (FF). The FF is obtained by the equation, $4\pi(\text{area})/(\text{perimeter})^2$, which gives a value of 1 for a perfectly circular perimeter and decreasingly smaller positive values for less circular perimeters [22]. Capan-1 and Panc 03.27 cells are epithelial type and they are closely attached to each other via intercellular adhesion complexes. They have squarish shapes with FF values of 0.74±0.08 and 0.77±0.06, respectively (Figure 4B). In contrast, both cells after hENT1 knockdown became elongated with much lower FF with values of 0.44±0.14 and 0.52±0.16, respectively.

Further, the effect of hENT1 knockdown on organization of cytoskeletons was evaluated. We stained for vinculin, which controls the formation of focal adhesions (FAs) by directly interacting with actin and other proteins that are involved in FA formation [23]. FAs, sites of adhesions between the cell and the extracellular matrix (ECM), and stress fibers at the margin have clear roles in moving the cell forward [23,24]. Studies have reported that FA size is correlated with cell speed; larger FAs found in more elongated and faster-moving cells [25,26]. In this study, we analyzed redistribution of F-actin and changes in focal adhesion area of hENT1 knockdown cells (Figure 5). Actin fibers in both control groups were mostly concentrated around the cell periphery, and small, nascent FAs were observed. But, after

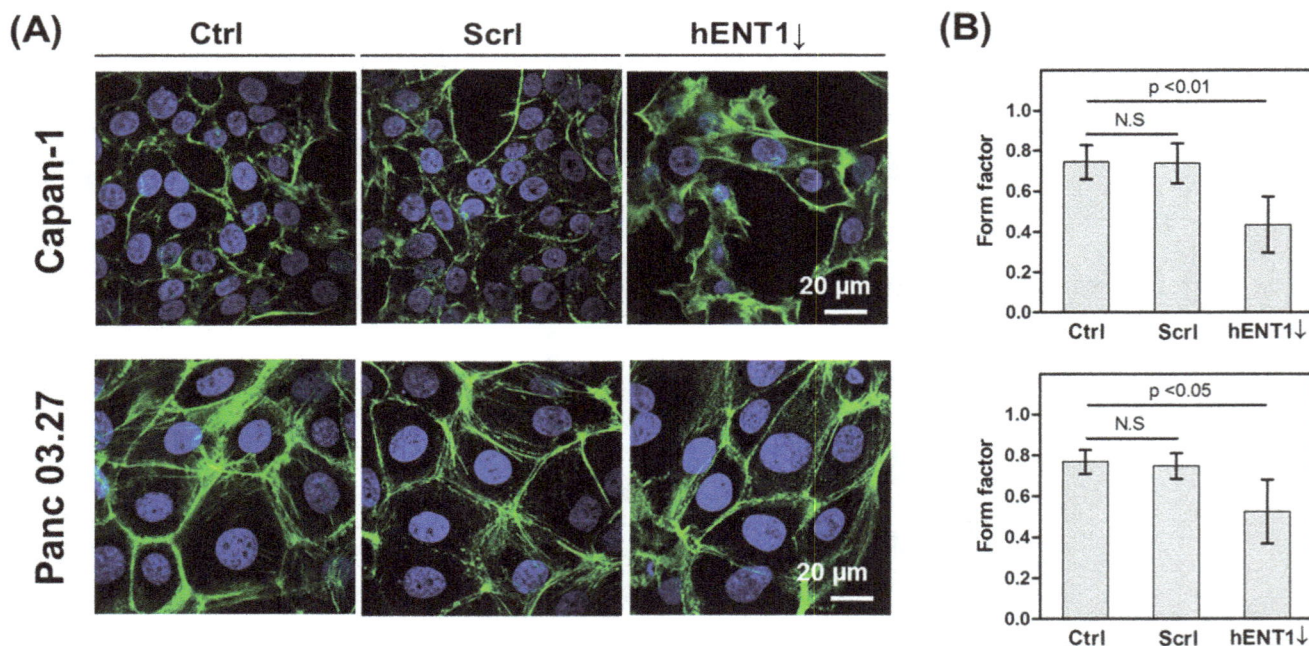

Figure 4. Quantify cell roundness. (A) Representative confocal micrographs of pancreatic cells (Top panel: Capan-1, Bottom panel: Panc 03.27) showing F-actin distribution, (B) form factor ($f = 4\pi a/p^2$; a: area; p: perimeter) analyzed by Image J (Capan-1 cells, ctrl: n = 37, scrl: n = 59, hENT1 ↓: n = 29; Panc 03.27, ctrl: n = 36, scrl: n = 28, hENT1 ↓: n = 28, N.S: statistically not significant).

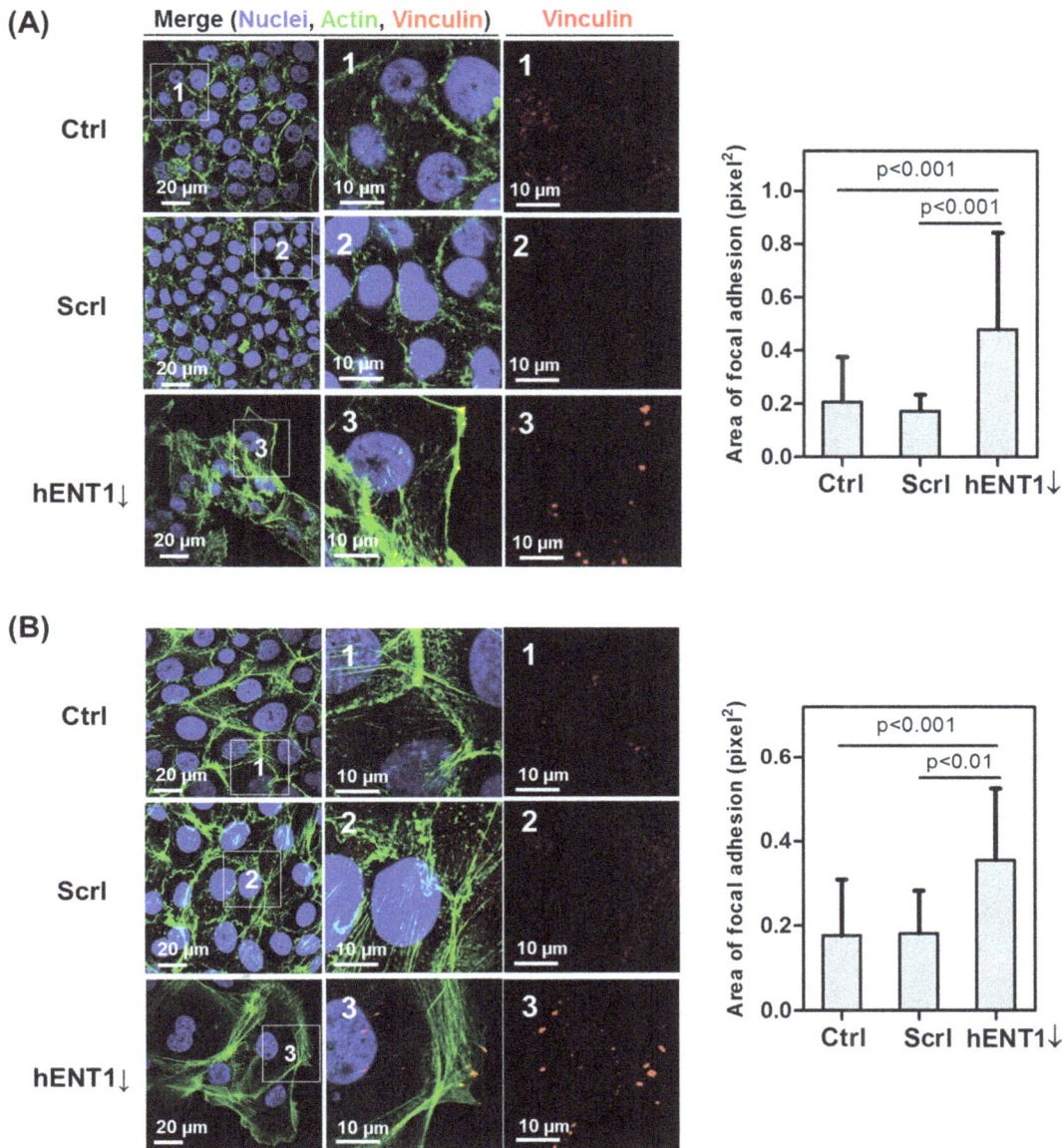

Figure 5. Analysis of focal adhesion area. Confocal micrographs (z-stacks of basal plane, 0–1.5 μm) showing vinculin (red) and F-actin (green) in (A) Capan-1 and (B) Panc 03.27 cells. Area of focal adhesion (pixel^2) is analyzed by Image J.

hENT1 knockdown, there was an increase in the number of stress fibers as well as measurable focal adhesion areas. These results suggest increased cellular motility in both cell lines.

In addition, in order to understand effect of hENT1 knockdown on cell motility, we examined the effect of downregulation of hENT1 on cell migration. A wound scratch was introduced in confluent cell monolayer, and then the wound sealing was observed for 18 hours (Figure 6). The area of wound closure was calculated by using Image J and the migration speed (area of wound closure (μm^2)/hour) was calculated as shown in Figure 6. The wound sealing area and migration speed of both of control cells and cells transfected with scramble siRNA are similar. But hENT1 suppressed Capan-1 or Panc 03.27 cells migrate 1.8 (1.7) or 1.5 (1.5) –fold faster than control (scramble siRNA transfected) cells, respectively. Thus, we confirm that downregulation of hENT1 induces formation of larger focal adhesions and promotes faster cell migration. Based on cell morphological changes and increased cell motility, it is possible that both of cells are undergoing phenotypic changes, which is associated with loosened adhesion between cells and cytoarchitectural rearrangement. Upon EMT, reorganization of cytoskeletons along with increase in FA dynamics is crucial for cells to leave the epithelium and begin migrating through the ECM [22,27,28]. For example, a recent study showed increased migration of post-EMT SKOV-3 cells which was transforming growth factor-β (TFG-β) that generated larger FA mechanical forces than that of pre-EMT cells [16]. TGF-β is reported as EMT inducer in different epithelial cell lines including renal proximal tubular and alveolar epithelial cells [16,29,30]. Increased interaction of actin cytoskeleton and intracellular molecules with FAs initiated by the association of integrin α5β1 and fibronectin enhanced cellular motility [16].

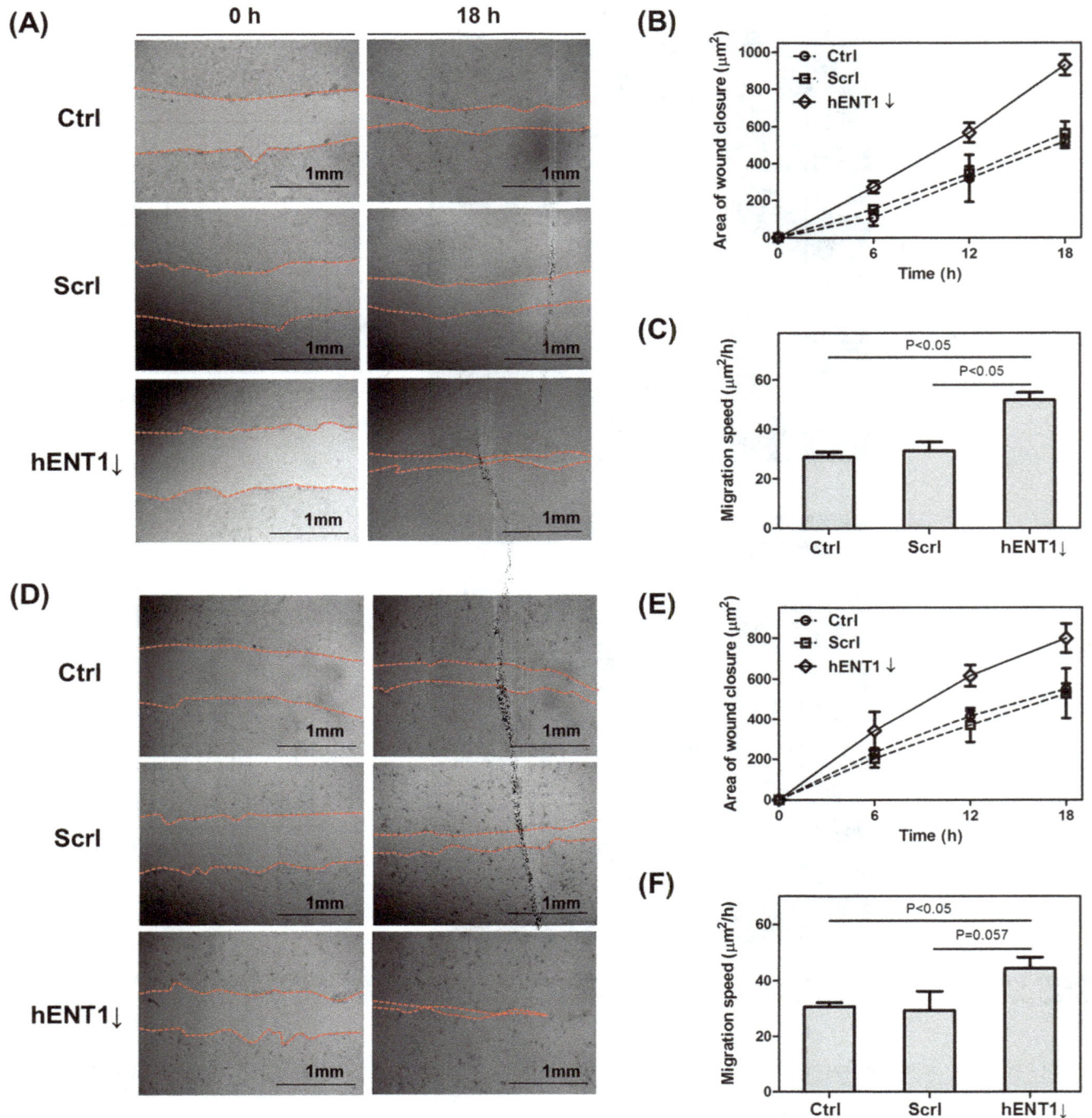

Figure 6. Downregulation of hENT1 promote cell migration. The scratches were introduced to monolayer (A) Capan-1 and (D) Panc 03.27 cells. Photographs were taken immediately after wound induction for 18 hours. The wound sealing areas of (B) Capan-1 and (E) Panc 03.27 cells were calculated using Image J and compared with control, scramble siRNA transfected, and hENT1 knockdown cells. Migration speed of (C) Capan-1 and (F) Panc 03.27 cells was calculated.

In order to confirm those physiological changes induced by hENT1 knockdown are related to EMT, we analyzed changes in EMT marker proteins, such as E-cadherin and N-cadherin, after hENT1 knockdown. E-cadherin is one of cell-cell adhesion molecules. That maintains cell-cell contacts and epithelial architecture. Alternation of expression of cadherins from E-cadherin to N-cadherin, which is primarily expressed in mesen-

chymal cells, occurs during EMT. This alternation leads to a drastic change in the adhesive properties of cells, as it loses its affinity for epithelial neighbors and gains affinity for mesenchymal cells, which are nonpolarized, lack intercellular junctions, and have a unique spindle-like shape [27,31,32]. After hENT1 downregulation, both cell lines showed low expression of E-cadherin and high expression of N-cadherin, which are a major

(A)

(B)

(C)

(D)

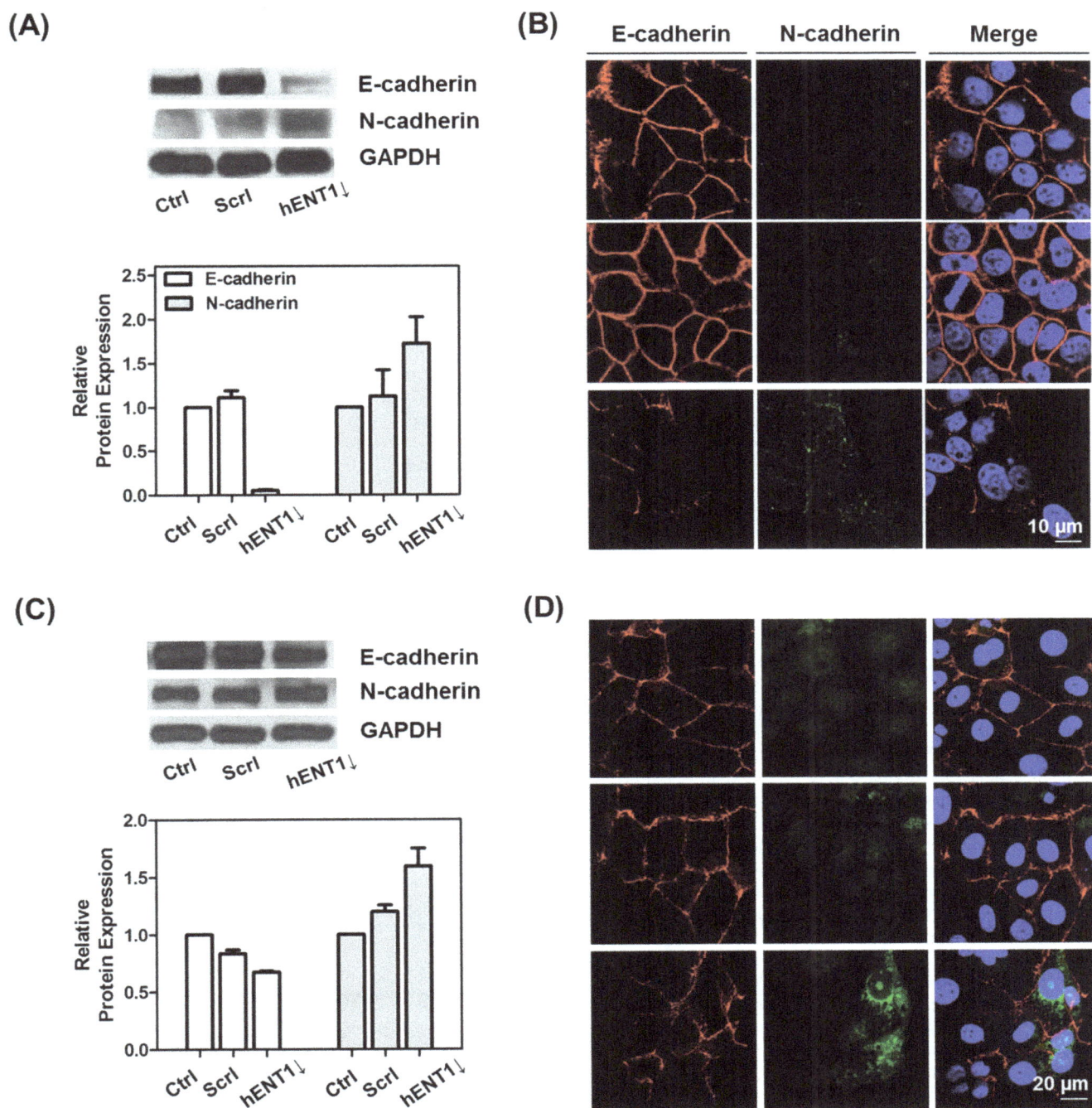

Figure 7. hENT1 knockdown induces changes in EMT markers. Western blots and relative expression levels of E-cadherin (110 kDa), N-cadherin (140 kDa) in (A) Capan-1 and (C) Panc 03.27 cells after treatment with hENT1 siRNA. Columns in the histograms are the mean of two independent experiments (intensity ratio of GAPDH to E-cadherin or N-cadherin). Representative confocal images of E-cadherin (red fluorescence) and N-cadherin (green fluorescence) expressions in (B) Capan-1 and (D) Panc 03.27 cells.

hallmark of EMT (Figure 7). The loss of E-cadherin and gain of N-cadherin increases cell motility and metastatic potential. Several studies have shown that reduced cell stiffness directly correlates with increased metastatic potential using various in vitro biomechanical analysis methods [11,17,33]. Reduction in cellular stiffness modulates that cells are undergoing EMT after hENT1 knockdown as it is demonstrated in Figure 2. However, there are negligible difference in expression levels of intermediate filaments,

cytokeratin 18, and nuclear cytoskeleton, lamin A/C in both of cell lines (Figure S5 in File S1).

To confirm our forementioned studies which suggest that EMT can be characterized by changes in cell stiffness, we evaluated phenotypic changes of pancreatic cancer cells after treatment with TGF-β. The TGF-β has been reported in several studies to induce EMT [15,16,30]. Therefore, we further examined these results in another experiment using EMT-induced pancreatic cancer cells.

Panc 03.27 cells were exposed to TGF-β for 2 days at a concentration of 10 ng/ml to induce EMT, and then measured cellular stiffness. Slightly suppressed E-cadherin and elevated N-cadherin and vimentin expressions indicate that EMT is induced in Panc 03.27 cells (Figure S6A in File S1). Panc 03.27 cells treated with TGF-β were demonstrated by decreased Young's modulus (1.42±0.53 kPa) versus untreated cells (1.91±0.78 kPa). This result suggests that cell stiffness decreases when the cells are undergoing EMT. Recently, there is a report concerning cellular stiffness changes in EMT-induced A549 cells with TGF-β treatment [15]. In their study, they measured local mechanical properties of A549 cells using a sharp DNP cantilever (20 nm tip radius) which has a smaller contact area compared to a microparticle-modified cantilever. Even though there is crucial evidence of cell phenotypic change from epithelial to mesenchymal, i.e., cellular shape changes and increasing stress fibers, are still consistent with our work. The A549 cells treated with TGF-β became stiffer than untreated cells, which is different from our findings. Probably, this conflict result is because they used a sharp tip and the stiffness value of these cells reflects structural rearrangements of F-actin stress fibers in the cells. To demonstrate whether different tip geometry is a critical factor to distinguish cellular compartments, additional experiment with two different AFM cantilevers, a pyramid tip cantilever with 10 nm nominal radius (Figure S7A in File S1) and 5 μm silica bead-modified cantilever (Figure S7D in File S1), were performed. Sequential stiffness maps of Panc 03.27 cells demonstrated that a sharp tip is capable of distinguishing between different cellular compartments (Figure S7B, C in File S1). However, stiffness of overall cells obtained by microparticle-modified cantilever is similar (Figure S7E, F in File S1). The stiffness of cytoplasm region is 2.2-fold higher than nucleus region, when a sharp tip was used; Young's moduli from cytoplasm and nucleus are 68.31±28.00 kPa and 31.10±6.29 kPa, respectively. These results suggest that stiffness obtained from cell membrane and cytoskeleton (cytoplasm region) is higher than that from combination of cell membrane, cytoskeleton nucleus cytoskeleton, and nuclei (nucleus region). A sharp AFM cantilever creates 9.3-fold (nucleus region) and 17-fold (cytoplasm region) increases in the value of the Young's modulus relative to that of the microparticle-modified cantilever under same indentation force at the same loading rate. The Young's moduli from cytoplasm and nucleus areas obtained by using microparticle-modified cantilever are 2.30±0.65 kPa and 1.91±0.82 kPa, respectively. The small contact area of the pyramidal tip is able to identify the mechanical properties of a single cytoskeleton filament, especially in the periphery of a cell. But, the microparticle-modified cantilever has larger contact area compared to a sharp cantilever. Thus, it can indent several cytoskeleton filaments simultaneously and well-organized filaments under cell membrane dissipate the indenting force. These results confirm our hypothesis that different tip geometry influences on detecting specific cellular compartments. The microparticle-modified cantilever is not able to distinguish cellular compartments, however, consistent and reproducible results are obtained from overall cells suggesting this cantilever is suitable to measure cell stiffness compared to a sharp tip.

To further demonstrate that changes in cell mechanical properties upon EMT are universal features of both parental and transformed cells, we evaluated stiffness in six different types of pancreatic cancer cells that classifies as either epithelial-like or mesenchymal-like cells (Figure S8 in File S1). Three cell lines which include AsPC-1, MIA Paca 2 and Panc-1 have vimentin expression and less or no expression of E-cadherin indicating mesenchymal-like cells. As measured by AFM, these cells are relatively softer and have lower hENT1 expression than other three epithelial-like cell lines, BxPC-3, Capan-1, and Panc 03.27. These findings comparing cell stiffness in parental cell lines are consistent to the results where siRNA was used to knockdown hENT1 and induce phenotypic shift. Also, it is consistent with the findings from siRNA transfected cells that mesenchymal-like cells are softer than epithelial-like cells. It is still difficult to fully explain how hENT1 regulates E-cadherin or N-cadherin expression and further cellular stiffness; however, we can conclude that hENT1 expression level is somehow related to cellular stiffness based on results. Also, our results establish a relationship between cell stiffness and EMT whereby cells are undergoing EMT showed reduced stiffness. Those findings are consistent with other studies which have shown that cancer cells from body fluids from patients diagnosed with metastatic tumor are more than 70% softer than the benign cells [11]. Therefore, we suggest that cellular mechanical properties are a critical marker to estimate hENT1 expression and identify phenotypic shift, which is a hallmark in cancer metastasis.

Conclusions

In this study, we described the role that hENT1 plays in modulating physiological and mechanical properties of pancreatic cancer cells. After downregulation of hENT1, cells became elongated, migrated faster with larger focal adhesion area, and showed altered EMT marker expressions that included downregulation of E-cadherin and upregulation of N-cadherin. These properties are primary signatures associated with phenotypic shift, EMT. In addition, hENT1 knockdown also decreased cellular stiffness which was evaluated with AFM and confirmed by microfluidic platform. These findings suggest that hENT1 knockdown induces EMT in pancreatic cancer cells, results that we corroborated with further measurements in parental pancreatic cancer cells. In conclusion, we have established a novel method to evaluate alterations in cellular biophysical behaviors that result from hENT1 knockdown – a critical drug transporter that has been correlated to patients' response to chemotherapeutic treatment. The ability to predict drug response by evaluating cellular or tumor biophysical properties is a potentially powerful tool that can aid clinicians monitor how patients respond to therapy and may provide predictive capability for personalized treatment.

Supporting Information

File S1 Figure S1, Western blots of hENT1 (55 kDa) and GAPDH (37 kDa) in pancreatic cancer cells. Figure S2, Young's modulus of Panc 03.27 cells at different indentation force from 10 pN to 10000 pN (inset: Bright-field image showing AFM tip approaching cells and schematic diagram of indentation of microparticle-modified cantilever and cell) (A), Calculated average indentation depth (nm) corresponding to Figure S2A using Eq. (1), and representative force-displacement (f-d) curve from a Panc 03.27 control cell when the indentation force is (C) 100 pN and (D) 1 nN (red: approach curve, black: retract curve) (B). The calculated Young's moduli from (C) and (D) are 1.88 and 5.21 kPa, respectively. Figure S3, Stiffness distribution of (A) Capan-1 and (B) Panc 03.27 cells corresponding to bar histograms shown in Figure 1. The solid line shows Lorentzian distribution. Figure S4, Cellular stiffness of Panc 03.27 cells: Ctrl (without treatment); WGA treated (cell membrane is stained by Alexa Fluor 488 Conjugated wheat germ agglutinin). Young's modulus of cells measured by AFM under same indentation force at 100 pN. Figure S5, Representative confocal micrographs of pancreatic cancer Capan-1 and Panc 03.27 cells showing cytokeratin 18 (green, top panel), Lamin A/C (green, middle and bottom panels), and nuclei (blue) (A). (B) Western blots of Lamin A/C (74, 63 kDa), cytoketarin 18 (46 kDa) and GAPDH (37 kDa) in control, scramble siRNA transfected, and hENT1 knockdown Capan-1 and Panc 03.27 cells. Figure S6, Western blots of E-cadherin (110 kDa), N-cadherin (140 kDa), vimentin (57 kDa), and GAPDH (37 kDa) in untreated and TGF-β treated Panc 03.27 cells (A), (B) Young's modulus of untreated and TGF-β treated Panc 03. 27 cells (concentration of TGF-β: 10 ng/ml, exposure for 2 days). Figure S7, Representative AFM topographic image, deflection image, and corresponding stiffness map of Panc 03.27 cells obtained by using sharp MSCT-C (A, B) and microparticle modified cantilever (D, E). Histograms show (C, F) corresponding stiffness from force volume map (B, E). Figure S8, Young's modulus of cells measured by AFM under same indentation force at 100 pN (A), (B) Western blots of E-cadherin

(110 kDa), vimentin (57 kDa), and GAPDH (37 kDa) expressed in six different pancreatic cancer cells.

Acknowledgments

The authors gratefully acknowledge funding support from the following sources: Department of Defense grants W81XWH-09-1-0212 and W81XWH-12-1-0414, National Institute of Health grants U54CA143837 and U54CA151668, the CPRIT grant RP121071 from the State of Texas, and the Ernest Cockrell Jr. Distinguished Endowed Chair.

Author Contributions

Conceived and designed the experiments: YL MF. Performed the experiments: YL WZ. Analyzed the data: YL WZ LQ. Contributed reagents/materials/analysis tools: LQ HS. Wrote the paper: YL. Revised the article critically for important intellectual content: EK DK FH. Final approval of manuscript: HS MF.

References

1. Burris HA, Moore MJ, Andersen J, Green MR, Rothenberg ML, et al. (1997) Improvements in survival and clinical benefit with gemcitabine as first-line therapy for patients with advanced pancreas cancer: a randomized trial. J Clin Oncol 15: 2403–2413.
2. Li D, Xie K, Wolff R, Abbruzzese JL (2004) Pancreatic cancer. Lancet 363: 1049–1057.
3. Giovannetti E, Del Tacca M, Mey V, Funel N, Nannizzi S, et al. (2006) Transcription analysis of human equilibrative nucleoside transporter-1 predicts survival in pancreas cancer patients treated with gemcitabine. Cancer Res 66: 3928–3935.
4. Mori R, Ishikawa T, Ichikawa Y, Taniguchi K, Matsuyama R, et al. (2007) Human equilibrative nucleoside transporter 1 is associated with the chemosensitivity of gemcitabine in human pancreatic adenocarcinoma and biliary tract carcinoma cells. Oncol Rep 17: 1201–1205.
5. Spratlin J, Sangha R, Glubrecht D, Dabbagh L, Young JD, et al. (2004) The absence of human equilibrative nucleoside transporter 1 is associated with reduced survival in patients with gemcitabine-treated pancreas adenocarcinoma. Clin Cancer Res 10: 6956–6961.
6. Shah A, Summy J, Zhang J, Park S, Parikh N, et al. (2007) Development and characterization of gemcitabine-resistant pancreatic tumor cells. Ann Surg Oncol 14: 3629–3637.
7. Wang R, Cheng L, Xia J, Wang Z, Wu Q, et al. (2014) Gemcitabine resistance is associated with epithelial-mesenchymal transition and induction of HIF-1α in pancreatic cancer cells. Curr Cancer Drug Targets 14: 407–417.
8. Lee JM, Dedhar S, Kalluri R, Thompson EW (2006) The epithelial-mesenchymal transition: new insights in signaling, development, and disease. J Cell Biol 172: 973–981.
9. Kalluri R, Weinberg RA (2009) The basics of epithelial-mesenchymal transition. J Clin Invest 119: 1420–1428.
10. Kamm R, Lammerding J, Mofrad M (2010) Cellular Nanomechanics. In: B . Bhushan, editor. Springer handbook of nanotechnology: Springer Berlin Heidelberg. pp. 1171–1200.
11. Cross SE, Jin Y-S, Rao J, Gimzewski JK (2007) Nanomechanical analysis of cells from cancer patients. Nat Nanotechnol 2: 780–783.
12. Xu W, Mezencev R, Kim B, Wang L, McDonald J, et al. (2012) Cell stiffness is a biomarker of the metastatic potential of ovarian cancer cells. PLoS ONE 7: e46609.
13. Indrajyoti I, Vishnu U, Casey K, Umadevi T, Micah D, et al. (2011) An in vitro correlation of mechanical forces and metastatic capacity. Phys Biol 8: 015015.
14. Swaminathan V, Mythreye K, O'Brien ET, Berchuck A, Blobe GC, et al. (2011) Mechanical stiffness grades metastatic potential in patient tumor cells and in cancer cell lines. Cancer Res 71: 5075–5080.
15. Buckley ST, Medina C, Davies AM, Ehrhardt C (2012) Cytoskeletal re-arrangement in TGF-β1-induced alveolar epithelial-mesenchymal transition studied by atomic force microscopy and high-content analysis. Nanomedicine: NBM 8: 355–364.
16. Lee S, Yang Y, Fishman D, Banaszak Holl MM, Hong S (2013) Epithelial-mesenchymal transition enhances nanoscale actin filament dynamics of ovarian cancer cells. J Phys Chem B 117: 9233–9240.
17. Guck J, Schinkinger S, Lincoln B, Wottawah F, Ebert S, et al. (2005) Optical deformability as an inherent cell marker for testing malignant transformation and metastatic competence. Biophys J 88: 3689–3698.
18. Suresh S, Spatz J, Mills JP, Micoulet A, Dao M, et al. (2005) Connections between single-cell biomechanics and human disease states: gastrointestinal cancer and malaria. Acta Biomater 1: 15–30.
19. Sirghi L, Ponti J, Broggi F, Rossi F (2008) Probing elasticity and adhesion of live cells by atomic force microscopy indentation. Eur Biophys J 37: 935–945.
20. Hou HW, Li QS, Lee GYH, Kumar AP, Ong CN, et al. (2009) Deformability study of breast cancer cells using microfluidics. Biomed Microdevices 11: 557–564.
21. Zhang W, Kai K, Choi DS, Iwamoto T, Nguyen YH, et al. (2012) Microfluidics separation reveals the stem-cell-like deformability of tumor-initiating cells. Proc Natl Acad Sci USA 109: 18707–18712.
22. Mendez MG, Kojima S-I, Goldman RD (2010) Vimentin induces changes in cell shape, motility, and adhesion during the epithelial to mesenchymal transition. FASEB J 24: 1838–1851.
23. Yaron RS, Gleb EY, Michael AH, Andrew EP (2009) Cell nanomechanics and focal adhesions are regulated by retinol and conjugated linoleic acid in a dose-dependent manner. Nanotechnology 20: 285103.
24. Izzard CS, Lochner LR (1980) Formation of cell-to-substrate contacts during fibroblast motility: an interference-reflexion study. J Cell Sci 42: 81–116.
25. Kim D-H, Wirtz D (2013) Focal adhesion size uniquely predicts cell migration. FASEB J 27: 1351–1361.
26. Parsons JT, Horwitz AR, Schwartz MA (2010) Cell adhesion: integrating cytoskeletal dynamics and cellular tension. Nat Rev Mol Cell Biol 11: 633–643.
27. Acloque H, Adams MS, Fishwick K, Bronner-Fraser M, Nieto MA (2009) Epithelial-mesenchymal transitions: the importance of changing cell state in development and disease. J Clin Invest 119: 1438–1449.
28. Carragher NO, Frame MC (2004) Focal adhesion and actin dynamics: a place where kinases and proteases meet to promote invasion. Trends Cell Biol 14: 241–249.
29. Singh A, Settleman J (2010) EMT, cancer stem cells and drug resistance: an emerging axis of evil in the war on cancer. Oncogene 29: 4741–4751.
30. Willis BC, Borok Z (2007) TGF-β-induced EMT: mechanisms and implications for fibrotic lung disease. Am J Physiol-Lung C 293: L525–L534.
31. Gjorevski N, Boghaert E, Nelson C (2012) Regulation of epithelial-mesenchymal transition by transmission of mechanical stress through epithelial tissues. Cancer Microenviron 5: 29–38.
32. Kalluri R (2009) EMT: When epithelial cells decide to become mesenchymal-like cells. J Clin Invest 119: 1417–1419.
33. Suresh S (2007) Biomechanics and biophysics of cancer cells. Acta Biomater 3: 413–438.

12

Regulation of the Tyrosine Phosphorylation of Phospholipid Scramblase 1 in Mast Cells That Are Stimulated through the High-Affinity IgE Receptor

Asma Kassas[1,2], Ivan C. Moura[1,2], Yumi Yamashita[3], Jorg Scheffel[3], Claudine Guérin-Marchand[1,2], Ulrich Blank[1,2], Peter J. Sims[4], Therese Wiedmer[4], Renato C. Monteiro[1,2], Juan Rivera[3], Nicolas Charles[1,2*⑨], Marc Benhamou[1,2*⑨]

1 INSERM U1149, Faculté de Médecine Xavier Bichat, Paris, France, 2 University Paris-Diderot, Sorbonne Paris Cité, Laboratoire d'excellence INFLAMEX, DHU FIRE, Paris, France, 3 Laboratory of Molecular Immunogenetics, Molecular Immunology and Inflammation Branch, NIAMSD, NIH, Bethesda, Maryland, United States of America, 4 Department of Medicine, University of Rochester School of Medicine and Dentistry, Rochester, New York, United States of America

Abstract

Engagement of high-affinity immunoglobulin E receptors (FcεRI) activates two signaling pathways in mast cells. The Lyn pathway leads to recruitment of Syk and to calcium mobilization whereas the Fyn pathway leads to phosphatidylinositol 3-kinase recruitment. Mapping the connections between both pathways remains an important task to be completed. We previously reported that Phospholipid Scramblase 1 (PLSCR1) is phosphorylated on tyrosine after cross-linking FcεRI on RBL-2H3 rat mast cells, amplifies mast cell degranulation, and is associated with both Lyn and Syk tyrosine kinases. Here, analysis of the pathway leading to PLSCR1 tyrosine phosphorylation reveals that it depends on the FcRγ chain. FcεRI aggregation in Fyn-deficient mouse bone marrow-derived mast cells (BMMC) induced a more robust increase in FcεRI-dependent tyrosine phosphorylation of PLSCR1 compared to wild-type cells, whereas PLSCR1 phosphorylation was abolished in Lyn-deficient BMMC. Lyn association with PLSCR1 was not altered in Fyn-deficient BMMC. PLSCR1 phosphorylation was also dependent on the kinase Syk and significantly, but partially, dependent on detectable calcium mobilization. Thus, the Lyn/Syk/calcium axis promotes PLSCR1 phosphorylation in multiple ways. Conversely, the Fyn-dependent pathway negatively regulates it. This study reveals a complex regulation for PLSCR1 tyrosine phosphorylation in FcεRI-activated mast cells and that PLSCR1 sits at a crossroads between Lyn and Fyn pathways.

Editor: David Holowka, Cornell University, United States of America

Funding: This work was supported in part by a grant from the Fondation de France (2002004544). N.C. was supported in part by a grant from the Ligue Nationale Contre le Cancer. The work of Y.Y., J.S. and J.R. was supported by the Intramural Research Program, NIAMS, NIH. A.K. was supported by Investissements d'Avenir programme ANR-11-IDEX-0005-02, Sorbonne Paris Cite, Laboratoire d'excellence INFLAMEX. The funders had no role in study design, data collection and analysis, decision to publish, or preparation of the manuscript.

Competing Interests: The authors have declared that no competing interests exist.

* Email: nicolas.charles@inserm.fr (NC); marc.benhamou@inserm.fr (MB)

⑨ These authors contributed equally to this work.

Introduction

High-affinity receptors for IgE (FcεRI) expressed on mast cells promote, after their aggregation by IgE and antigen, the release of preformed mediators stored in cytoplasmic granules and of newly synthesized lipid mediators and cytokines [1]. Engagement of FcεRI leads to the activation of at least two signaling pathways. One is initiated by the tyrosine kinase Lyn [2] and leads to recruitment of another tyrosine kinase, Syk, to the receptor and to activation of the signaling complex recruited by the protein adaptor LAT [3], resulting in calcium mobilization [4]. The other pathway, initiated by the tyrosine kinase Fyn [4], leads to phosphatidylinositol 3-kinase recruitment [4,5]. Both pathways cooperate to determine the extent of degranulation and of cytokine and lipid inflammatory mediator production. It has been demonstrated that the Lyn-initiated pathway negatively regulates the Fyn-initiated pathway through recruitment of the kinase Csk [6]. Since the FcεRI-dependent cell activation combines these

pathways into one coherent signal, mapping of their connections is an important task that remains to be completed to fully understand signal integration.

Recently, we reported that phospholipid scramblase 1 (PLSCR1) is phosphorylated on tyrosine after aggregation of FcεRI on mast cells [7]. PLSCR1 is a multi-function protein. It was originally identified based on its capacity to accelerate transbilayer migration of phospholipids upon interaction with calcium, thereby collapsing the lipid asymmetry existing between inner and outer leaflets of plasma membranes [8,9]. Activation of scrambling leads to increased cell surface exposure of phosphatidylserine and other aminophospholipids. This has been implicated in the recognition of apoptotic cells by phagocytes and in the cell surface expression of procoagulant activity by activated platelets and perturbed endothelium [10,11]. Interestingly, activated mast cells also demonstrate transient exposure of phosphatidylserine [12,13]. However, studies with knock-out mice questioned the involvement of PLSCR1 alone in phospholipid scrambling [14,15]. Recently, several reports have implicated the Ca^{2+}-activated ion channels belonging to the TMEM16 family in phospholipid scrambling induced by a calcium ionophore [16–18]. By contrast, phospholipid scrambling following caspase activation during apoptosis was shown to be promoted by Xkr8, a putative transporter [19]. Therefore, depending on the triggering signal, phospholipid scrambling now appears to result from a variety of alternative mechanisms, in which the specific role of plasma membrane PLSCR1 remains to be resolved. In addition to its putative role in mediating transbilayer movement of plasma membrane phospholipids that accompanies PS exposure at the cell surface, there is now also considerable evidence that: i) PLSCR1 serves as a signaling intermediate for the Epidermal Growth Factor (EGF) receptor promoting optimal activation of p60c-Src [20,21]; ii) PLSCR1 contains a nuclear localisation signal domain that mediates nuclear trafficking of the unpalmitoylated form of the protein [22,23]; iii) *de novo* synthesis of PLSCR1 is induced by interferon-α (IFNα) and results in its nuclear trafficking and binding to chromosomal DNA [23–25]. In this setting, PLSCR1 may serve as a transcription factor since it amplifies the expression of IFNα/β-stimulated genes [26] and promotes the transcription of the inositol 1, 4, 5-trisphosphate receptor gene [27]; iv) PLSCR1 potentiates granulopoiesis by prolonging expansion of granulocyte precursors presumably through its role in transcriptional regulation [15]; v) Expression of PLSCR1 has been shown to be tumor suppressive, and its level of expression in bone marrow cells to correlate with long-term survival in acute myelogenous leukemia, whereas mutations affecting PLSCR1 appear to promote the leukemogenic potential of myeloid progenitors [28–31]; vi) PLSCR1 regulates compensatory endocytosis in neuroendocrine cells [32]; vii) PLSCR1 is capable of potentiating a select set of mast cell responses following FcεRI aggregation [33]. In this study, we observed that endogenous expression of PLSCR1 in RBL-2H3 mast cells doubles VEGF production and the degranulation response to FcεRI engagement as compared to PLSCR1-knock-down RBL-2H3 cells, without any detectable impact on MCP-1 production and release of arachidonic acid metabolites. In PLSCR1-knocked-down RBL-2H3 cells the LAT-PLCγ-calcium axis initiated by Lyn was inhibited [33]. Interestingly, Lyn was found to colocalize with PLSCR1 at the cell membrane and to coprecipitate with it. Syk, which is downstream of Lyn activation, was also found to interact with PLSCR1 [33].

Whereas the kinase responsible for the tyrosine phosphorylation of PLSCR1 after engagement of the EGF receptor has been identified as p60c-Src [20,21], and as c-Abl after induction of apoptosis [34], the regulatory pathway leading to PLSCR1 tyrosine phosphorylation after FcεRI aggregation remains to be elucidated. The present study was carried out to identify this pathway. We found that the Lyn and Fyn pathways cooperate in FcεRI-dependent tyrosine phosphorylation of PLSCR1. Indeed, PLSCR1 phosphorylation was dependent on FcRγ chain, on Lyn and on Syk and partially on intracellular calcium mobilization, suggesting direct involvement of both kinases through a direct association with PLSCR1, and indirectly as a result of calcium mobilization. We also provide evidence that Fyn negatively regulates PLSCR1 tyrosine phosphorylation.

Materials and Methods

Ethics statement

Mice were maintained and used in accordance with NIH guidelines and Animal Study Proposals approved by the NIAMS Animal Care and Use Committee (ASP number AO12-11-01).

Antibodies and reagents

The anti-DNP monoclonal IgE DNP-48 [35] was a kind gift of Dr. Reuben Siraganian (NIH, Bethesda, MD). The monoclonal anti-rat PLSCR1 antibody 129.2 was produced in the laboratory [7,36]. Monoclonal anti-mouse PLSCR1 antibody 1A8 has been described elsewhere [20]. Anti-Lyn and anti-Fyn polyclonal antibodies were from Santa-Cruz Biotech (Santa-Cruz, CA). Anti-phosphotyrosine monoclonal antibody 4G10 used as a culture supernatant of the 4G10 hybridoma or as horseraddish peroxidase (HRP)-coupled antibody (Calbiochem, La Jolla, CA). HRP-labeled secondary antibodies (rabbit anti-mouse and goat anti-rabbit) were from Sigma-Aldrich (St-Louis, MO). Cell culture medium (D-MEM and E-MEM), fetal calf serum and other culture reagents were from Gibco-BRL-Life Technologies (Gaithersburg, MD). Triton X-100 and sodium dodecyl sulfate (SDS), bovine serum albumin (BSA), DNP-human serum albumin (DNP-HSA), proteases and phosphatases inhibitors (leupeptin and aprotinin, NaF and Na_3VO_4) were from Sigma-Aldrich (St-Louis, MO). G418 was from PAA Laboratories (Les Mureaux, France).

Cells and cell lines

The RBL-2H3 [37] or the β-hexosaminidase high-releaser subclone 9 of this cell line were maintained in culture as described [13]. The Syk-deficient RBL-2H3 variant and the Syk reconstituted clone have been described elsewhere [38] and were generously provided by Dr. Reuben P. Siraganian (NIDCR, NIH, Bethesda, MD). The Syk reconstituted clone was maintained in selective medium containing 1 mg/ml G418.

RBL-2H3 transfectants expressing either the wild-type FcαRI or the mutated R209L FcαRI have been described elsewhere and express equivalent levels of the transfected receptor [39]. Whereas the wild-type FcαRI does associate with endogenous FcRγ chain, the mutated R209L FcαRI does not, as determined by co-immunoprecipitation experiments [39]. These transfectants were maintained in selective medium containing 1 mg/ml G418.

Bone marrow-derived mast cells (BMMC) were obtained from Lyn-deficient, Fyn-deficient, Lyn/Fyn double deficient or wild-type control mice [4]. After sacrifice of the mice by CO_2 inhalation, bone marrow cells were cultured in RPMI-1640 containing 16% fetal calf serum, glutamax, non-essential amino acids, streptomycin and penicillin, and supplemented with 20 ng/ml interleukin-3 and 20 ng/ml Stem Cell Factor (both from Peprotech, Rocky Hill, NJ). After four weeks in culture more than 95% of the cells were mast cells as assessed by FcεRI and c-kit expression and toluidine blue staining.

Reconstitution experiments

A lentiviral approach was used that was essentially as previously described [40]. The lentiviral vectors carried Lyn kinase or the catalytically inactive form of Lyn (K172R). The forward and reverse primer sequences used to insert a point mutation in the kinase domain of Lyn (Lyn K172R) were as follows: 5′-GTGGCTGTGAGGACCCTCA-3′ (forward), 5′-CTTGAGGGTCCTCACAGCC-3′ (reverse).

Cell stimulation and lysis

One million RBL-2H3 cells, variants or transfectants in 2 ml or BMMC at 1×10^6/ml were plated overnight with or without anti-DNP IgE ascitic fluid (1:10^3 dilution for RBL-2H3 cells and 1:10^2 dilution for BMMC). After several washes in Tyrode's solution (Hepes 10 mM pH 7.3, NaCl 135 mM, KCl 5 mM, glucose 5.6 mM, $CaCl_2$ 1.8 mM, $MgCl_2$ 1 mM, BSA 0.5 mg/ml), cells (1×10^6/ml RBL-2H3 cells, variants or transfectants, or 2×10^6/ml

BMMC) were stimulated at 37°C for the time indicated with 1 μg/ml DNP-HSA for RBL-2H3 cells, variants or transfectants and with 0.1 μg/ml DNP-HSA for BMMC.

For stimulation of FcαRI transfectants (wild-type and R209L mutated proteins), cells were first incubated with 10 μg/ml F(ab')2 fragment of the anti-FcαRI monoclonal antibody A77 [41] for 1 hr at 4°C in Tyrode's buffer. Cells were then washed twice and stimulated for the indicated time at 37°C with 20 μg/ml F(ab')2 fragment of rabbit anti-mouse IgG antibody.

In all experiments, the supernatant was recovered for control of degranulation and cells were washed in ice-cold PBS. Cells were lysed on ice in 200 μl lysis solution (Hepes 50 mM pH 7.2, 50 mM NaF, 50 mM NaCl, 1 mM Na_3VO_4, 1% Triton X-100, 0.1% SDS, 10 μg/ml leupeptin, 10 μg/ml aprotinin). After 10 min on ice, the wells were scraped, their content transferred to microtubes and the soluble cell lysates were recovered following centrifugation at 14,000 g for 10 minutes at 4°C for immunoblotting or immunoprecipitation. For co-immunoprecipitation purposes, 1×10^7 cells were lysed in 500 μl lysis buffer containing 0.5% Triton X-100.

Imumunoprecipitation and immunoblotting

Proteins in the soluble cell lysates were immunoprecipitated with either Sepharose-4B (Pharmacia, Uppsala, Sweden) coupled antibodies (for 129.2, and control IgG) or with anti-mouse PLSCR1 monoclonal antibody 1A8 adsorbed to protein G beads (Pharmacia, Uppsala, Sweden), by incubation in a rotating wheel for 2 hours (hr) at 4°C. Beads were washed 6 times with 1 ml ice-cold lysis buffer and material was eluted by boiling for 5 min in Laemmli sample buffer [42].

Immunoblots were performed after resolution of proteins by SDS-PAGE and their transfer onto PVDF membranes. After blocking non specific sites by incubation for 1 hr in 10 mM Tris pH 7.2 containing 150 mM NaCl, 0.05% Tween 20 and 4% BSA, membranes were incubated with the desired first antibody for 1 hr, washed 3 times for 10 min, incubated with the relevant secondary antibody and washed 3 times for 10 min. Blots were revealed by incubation in SuperSignal West Pico solution from Pierce (Rockford, IL) and exposure to X-OMAT films from Kodak (Rochester, NY). In some cases, membranes were stripped by several washes in methanol and reblotted with control antibodies. Blots were quantified by densitometry using the National Institutes of Health Software *Image J 1.46r* after scanning of the films. In the case of PLSCR1 the band corresponding to phosphorylated PLSCR1 was normalized to the total PLSCR1 protein recovered in the precipitates for each experimental point. Fold phosphorylation is the ratio of the corrected value of phospho-PLSCR1 obtained with stimulated cells to that obtained with the non-stimulated cells.

Degranulation measurements

Degranulation was assessed by measurement of the release of the granule marker β-hexosaminidase as described [43].

Calcium measurement

RBL-2H3 cells were trypsinized and washed two times in culture medium. Cells were then sensitized in the presence $1:10^3$ dilution of anti-DNP IgE ascitic fluids for two hours in a 37°C waterbath with frequent agitation and washed in Tyrode's buffer with or without Ca^{++} and Mg^{++}. Cells were incubated or not with 40 μM of the intracellular calcium chelator BAPTA-AM (Calbiochem, La Jolla, CA) for 15 minutes at 37°C protected from light. Treated cells were washed two times in Tyrode's buffer with or without Ca^{++} and Mg^{++}. All cells were resuspended at 1×10^6/ml

in Tyrode's buffer with or without Ca^{++} and Mg^{++}. Cells were loaded with 5 μM Fura-2-AM (Calbiochem, La Jolla, CA) for 30 minutes at 37°C protected from light, washed two times in Tyrode's buffer with or without Ca^{++} and Mg^{++}, and resuspended at 1.5×10^6/ml. The fluorescence of free intracellular Ca^{++}, after stimulation of the cells with 1 μg/ml DNP-HSA, was measured at 37°C using two excitation wavelengths at 340 and 380 nm and an emission wavelength of 510 nm in an Hitachi H-2000 spectrofluorimeter (Sciencetec, Les Ulis, France). Ca^{++} concentrations were calculated using a Kd of 224 nM for the interaction between Fura-2 and Ca^{++}.

Statistical analyses

All experiments were conducted at least three times (see figure legends). Statistical analyses were performed using GraphPad Prism 5.0 as indicated in figure legends.

Results

IgE-mediated PLSCR1 tyrosine phosphorylation is dependent on the FcRγ chain

The tetrameric form of FcεRI is composed of one FcεRIα chain responsible for IgE binding, and of the signal transducing chains FcεRIβ and a dimer of FcRγ $(\alpha\beta(\gamma)_2)$. The FcRγ chain is also mandatory for FcεRI cell membrane expression. We first examined whether this chain was required for the signal leading to PLSCR1 tyrosine phosphorylation. The FcRγ chain is shared by many Fc receptors [44] including the receptor for IgA (FcαRI) [45,46]. Interestingly, this receptor cannot associate with the FcRγ chain when an arginine-to-leucine (R209L) mutation is introduced in its transmembrane domain but is readily expressed at the plasma membrane [39]. Since mast cells do not express FcαRI, we took advantage of RBL-2H3 rat mast cells that had been transfected with wild-type or mutant (R209L) FcαRI. Of note, these transfectants expressed comparable levels of FcαRI at the plasma membrane and engagement of wild-type FcαRI induced degranulation of the transfectants whereas aggregation of of the R209L mutant did not ([47] *and data not shown*). As in the parental cell line, aggregation of FcεRI in these transfectants resulted in PLSCR1 tyrosine phosphorylation (figure 1). Yet, whereas tyrosine phosphorylation of PLSCR1 was readily observed after aggregation of wild-type FcαRI, no such phosphorylation could be detected after engagement of the FcαRI$_{R209L}$ mutated form of FcαRI (figure 1). We conclude that PLSCR1 phosphorylation required the FcRγ chain.

IgE-mediated PLSCR1 tyrosine phosphorylation depends on Lyn

It is well established that Lyn [2] and Fyn [4], both Src-family kinase members, function immediately downstream of the receptor. To determine which one of these kinases was important for PLSCR1 tyrosine phosphorylation, we used bone marrow-derived mast cells (BMMC) from Lyn-deficient and from Fyn-deficient mice. As a first step, FcεRI-dependent tyrosine phosphorylation of PLSCR1 was examined in wild-type BMMC. An increased phosphorylation was observed as soon as 30 sec after FcεRI engagement and reached a plateau after 9 min stimulation (Figure 2A) confirming that the phosphorylation observed in RBL-2H3 rat mast cells was not the result of the tumoral phenotype of this cell line and that it was not restricted to rat mast cells. Therefore in all BMMC experiments, subsequent analyses were performed after at least 9 minutes of stimulation. As in RBL-2H3 cells [7] phospho-PLSCR1 migrated as a heterogeneous band between 37 and 48 kDa (Figure 2A). We first assessed the

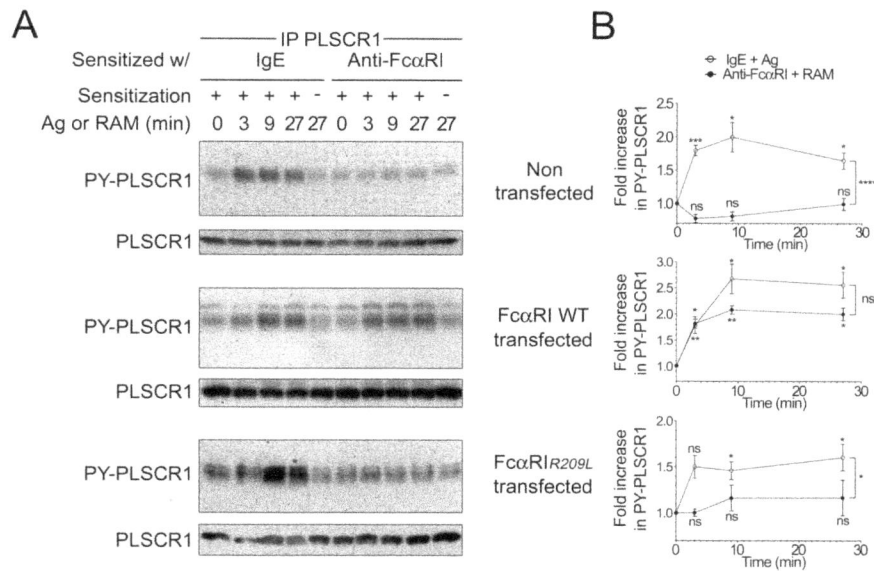

Figure 1. FcεRI-mediated PLSCR1 tyrosine phosphorylation depends on the FcRγ chain. (A) Non-transfected, FcαRI-transfected or FcαRI_R209L_-transfected RBL-2H3 cells were sensitized for 1 hr with anti-DNP IgE or F(ab')2 fragment of the anti-FcαRI monoclonal antibody A77, as indicated. After washes, IgE-sensitized cells were stimulated with specific antigen (*Ag*), whereas A77-F(ab')2-sensitized cells were stimulated with F(ab')2 fragment of rabbit anti-mouse IgG (*RAM*) for the indicated time. PLSCR1 in cell lysates was immunoprecipitated with anti-rat PLSCR1 monoclonal antibody 129.2 (*IP PLSCR1*). Eluates were analyzed by immunoblotting with anti-phosphotyrosine monoclonal antibody to detect PY-PLSCR1 and, after stripping of the membranes, with 129.2 to detect total PLSCR1, as indicated. (B) Quantification of PLSCR1 tyrosine phosphorylation relative to immunoprecipitated PLSCR1. Shown are the fold increases relative to basal PLSCR1 phosphorylation. Two-way ANOVA was used to compare the two kinetics observed for each cell line (bracket); and two-tailed unpaired student t test was used to compare stimulated conditions to unstimulated condition. Data are presented as mean ± s.e.m. of at least three independent experiments. ns: not significant; *: $p < 0.05$; **: $p < 0.01$; ***: $p < 0.001$.

involvement of Lyn in the PLSCR1 tyrosine phosphorylation by using BMMC from $Lyn^{-/-}$ mice. Interestingly, no increase in tyrosine phosphorylation of PLSCR1 was observed in these cells after FcεRI aggregation (Figure 2B). Reconstitution of the $Lyn^{-/-}$ cells with kinase-sufficient Lyn restored PLSCR1 tyrosine phosphorylation. By contrast, reconstitution with kinase-deficient Lyn was unable to restore this phosphorylation. These results show that Lyn is mandatory and its kinase activity is required for FcεRI-dependent increase in PLSCR1 phosphorylation.

IgE-mediated PLSCR1 tyrosine phosphorylation is negatively regulated by Fyn

Surprisingly, the increased tyrosine phosphorylation of PLSCR1 after FcεRI engagement was higher in Fyn-deficient BMMC (Figure 3), revealing that Fyn is a negative regulator for this event. This more robust increase in PLSCR1 phosphorylation was also dependent on Lyn. No increased PLSCR1 phosphorylation was observed in BMMC deficient for both Lyn and Fyn (Figure 4) suggesting that Fyn impedes the Lyn-dependent PLSCR1 phosphorylation rather than targeting an altogether different pathway. This led us to examine whether Fyn deficiency had an impact on the expression of Lyn or of PLSCR1, or on their association. There was no difference in Lyn or PLSCR1 expression between wild-type and Fyn-deficient BMMC (Figure 5A). As well, co-immunoprecipitation experiments demonstrated that the absence of Fyn did not result in a detectable increase in Lyn association with PLSCR1 (Figure 5B), although it confirmed in BMMC an association between Lyn and PLSCR1 that was not modulated by FcεRI aggregation as we previously reported in RBL-2H3 cells [33]. We conclude that Lyn-dependent

and Fyn-dependent signals converge as antagonists in the regulation of FcεRI-mediated PLSCR1 tyrosine phosphorylation.

IgE-mediated PLSCR1 tyrosine phosphorylation depends on Syk

Syk is a tyrosine kinase that is activated downstream of Lyn [1] and that coprecipitates with PLSCR1 [38]. Therefore, we used a variant of RBL-2H3 that is deficient in Syk [38] to evaluate its involvement in PLSCR1 phosphorylation. Aggregation of FcεRI on these cells resulted in no significant increase in tyrosine phosphorylation of PLSCR1 (Figure 6). Reconstitution of these cells with Syk restored in great part the FcεRI-induced tyrosine phosphorylation of PLSCR1 (Figure 6). These data demonstrate that Syk is required for optimal FcεRI-dependent phosphorylation of PLSCR1.

IgE-mediated PLSCR1 tyrosine phosphorylation is partially dependent on intracellular calcium mobilization

Calcium mobilization occurs downstream of Lyn and of Syk activation in FcεRI-activated mast cells [4,38], and FcεRI engagement induces both early (calcium independent) and late (calcium dependent) tyrosine phosphorylations [48]. PLSCR1 has an EF-hand-like domain allowing calcium binding that results in a conformational change of the molecule [49–51]. In addition, calcium ionophores can induce PLSCR1 tyrosine phosphorylation [52]. Therefore, we examined whether FcεRI-dependent phosphorylation of PLSCR1 is modulated by calcium mobilization that occurs in response to activation of this receptor. To that effect, RBL-2H3 cells were activated in the presence extracellular calcium, or in cells loaded with the intracellular calcium-chelator BAPTA-AM and suspended in calcium-free medium. In the

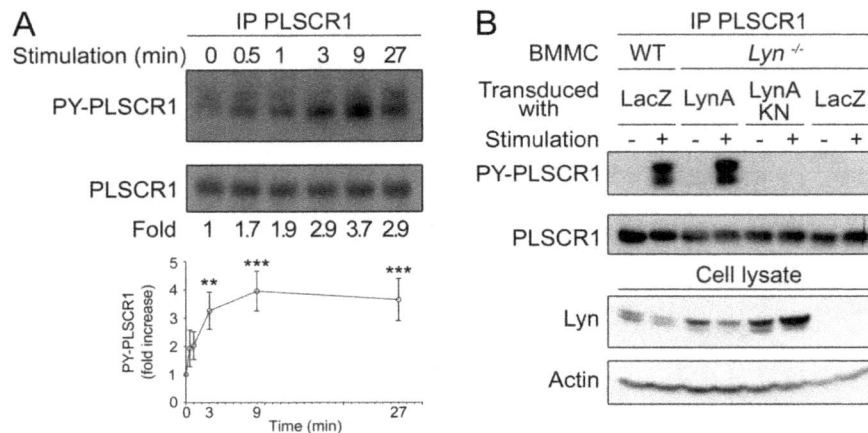

Figure 2. Time- and Lyn-dependency of FcεRI-mediated PLSCR1 tyrosine phosphorylation. (A) Time-dependent FcεRI-mediated PLSCR1 tyrosine phosphorylation in WT mouse BMMC. IgE-sensitized cells were stimulated for the indicated length of time with antigen. PLSCR1 was immunoprecipitated from cell lysates and tyrosine phosphorylated PLSCR1 (PY-PLSCR1) was analyzed by immunoblotting with anti-phosphotyrosine antibody (4G10). After stripping of the membrane, total PLSCR1 was analyzed by immunoblotting with anti-mouse PLSCR1 1A8 antibody. Lower panel: quantification. Statistical analysis was done by a one-way ANOVA followed by a Tukey's multiple comparison test. Data are presented as mean ± s.e.m. of six independent experiments. **: p<0.01; ***: p<0.001. (B) FcεRI-mediated PLSCR1 tyrosine phosphorylation is dependent on Lyn. Wild-type or Lyn−/− BMMC were transduced either with empty vector (LacZ), LynA containing vector (LynA) or dead-kinase LynA (LynAKN). After reconstitution, IgE-sensitized cells were stimulated for 30 minutes with antigen. Upper panels: PLSCR1 was immunoprecipitated from cell lysates and tyrosine phosphorylated PLSCR1 (PY-PLSCR1) was analyzed by immunoblotting with anti-phosphotyrosine antibody. After stripping of the membrane, total PLSCR1 was analyzed by immunoblotting with anti-mouse PLSCR1 1A8 antibody. Lower panels: Controls for the presence of Lyn in the reconstituted cells were performed by immunoblotting cell lysates with anti-Lyn antibody and with anti-actin antibody for loading control.

absence of calcium (and in the presence of BAPTA-AM), aggregation of FcεRI induced a detectable increase in tyrosine phosphorylation of PLSCR1. Yet, this increase amounted to approximately 25% of that observed after FcεRI aggregation in the presence of calcium (figure 7A). Under these conditions, and as previously reported [53], no degranulation (figure 7B) and no detectable calcium signal (figure 7C) were observed. Although one cannot completely exclude that infinitesimal or highly localised variations of free cytosolic calcium could account for the residual tyrosine phosphorylation of PLSCR1 in the absence of external calcium and in the presence of BAPTA, our data strongly indicate

collectively that IgE-mediated PLSCR1 phosphorylation involves calcium-dependent and calcium-independent signals.

Discussion

We previously reported that PLSCR1 is phosphorylated on tyrosine after aggregation of FcεRI expressed on the RBL-2H3 rat mast cell line [7]. Background levels of PLSCR1 tyrosine phosphorylation before receptor engagement varied from experiment to experiment but could be seen in all cases upon sufficient exposure time of immunoblots. This was true not only for RBL cells, but also for BMMC. We believe that these variations might

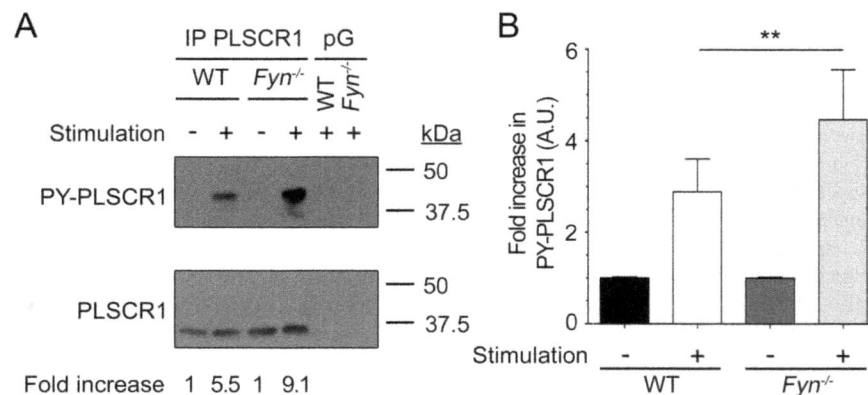

Figure 3. FcεRI-mediated PLSCR1 tyrosine phosphorylation is negatively regulated by Fyn. (A) Five million IgE-sensitized BMMC from wild-type (WT) or Fyn knock-out (Fyn−/−) mice were stimulated or not with antigen for 10 min. BMMC cell lysates were subjected to immunoprecipitation with anti-mouse PLSCR1 mAb 1A8 or protein G beads alone (pG), and eluates were analyzed by immunoblotting with anti-phosphotyrosine (upper panel) and, after stripping, anti-muPLSCR1 (lower panel) monoclonal antibodies. Fold increase in phosphorylation corresponds to the ratio of the value of phospho-PLSCR1 obtained for stimulated and non-stimulated cells for each condition normalized with the corresponding value of recovered total PLSCR1. (B) Quantification. Statistical analysis was done by a one-way ANOVA followed by a two tailed paired student t test. Data are presented as mean + s.e.m. of four independent experiments. **: p<0.01.

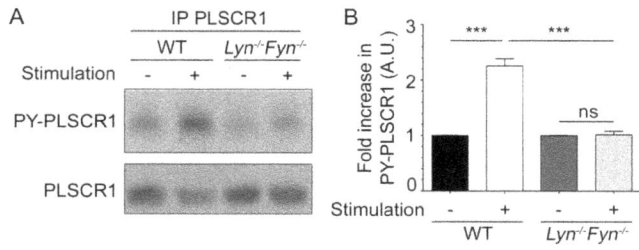

Figure 4. The Fyn-dependent negative regulation of FcεRI-mediated PLSCR1 tyrosine phosphorylation targets Lyn-dependent PLSCR1 phosphorylation. (A) Five million IgE-sensitized BMMC from wild-type (WT) or Lyn/Fyn-double deficient ($Lyn^{-/-}Fyn^{-/-}$) mice were stimulated or not with antigen for 10 min and PLSCR1 was immunoprecipitated from the lysates with anti-muPLSCR1 antibody 1A8. Tyrosine phosphorylated PLSCR1 (PY-PLSCR1) recovered from the lysates was detected by immunoblotting with anti-phosphotyrosine antibody and, after stripping of the membrane, the total amount of recovered muPLSCR1 was analyzed. (B) Quantification. Statistical analysis was done by a one-way ANOVA followed by a Tukey's multiple comparison test. Data are presented as mean + s.e.m. of three independent experiments. ns: not significant; ***: $p<0.001$.

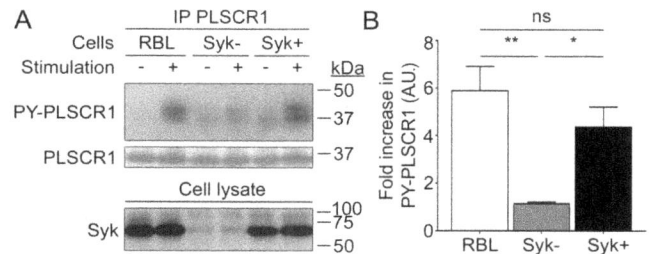

Figure 6. FcεRI-mediated PLSCR1 tyrosine phosphorylation is dependent on Syk. (A) One million IgE-sensitized RBL-2H3 cells (RBL) and Syk-deficient RBL-2H3 cells before (Syk–) or after (Syk+) reconstitution with Syk, were stimulated for 30 minutes with antigen. PLSCR1 was immunoprecipitated from cell lysates with anti-rat PLSCR1 monoclonal antibody 129.2. Tyrosine phosphorylated PLSCR1 (PY-PLSCR1) in eluates was detected by immunoblotting with phosphotyrosine-specific antibodies and, after stripping of the membrane, PLSCR1 was detected by immunoblotting with 129.2. Lower panel: the presence and absence of Syk was confirmed in cell lysates by immunoblotting. (B) Quantification of the increase in PLSCR1 tyrosine phosphorylation after cell stimulation relative to PLSCR1 basal phosphorylation. Statistical analysis was done by a one-way ANOVA followed by a Tukey's multiple comparison test. Data are presented as mean + s.e.m. of four independent experiments. ns: not significant; *: $p<0.05$; **: $p<0.01$.

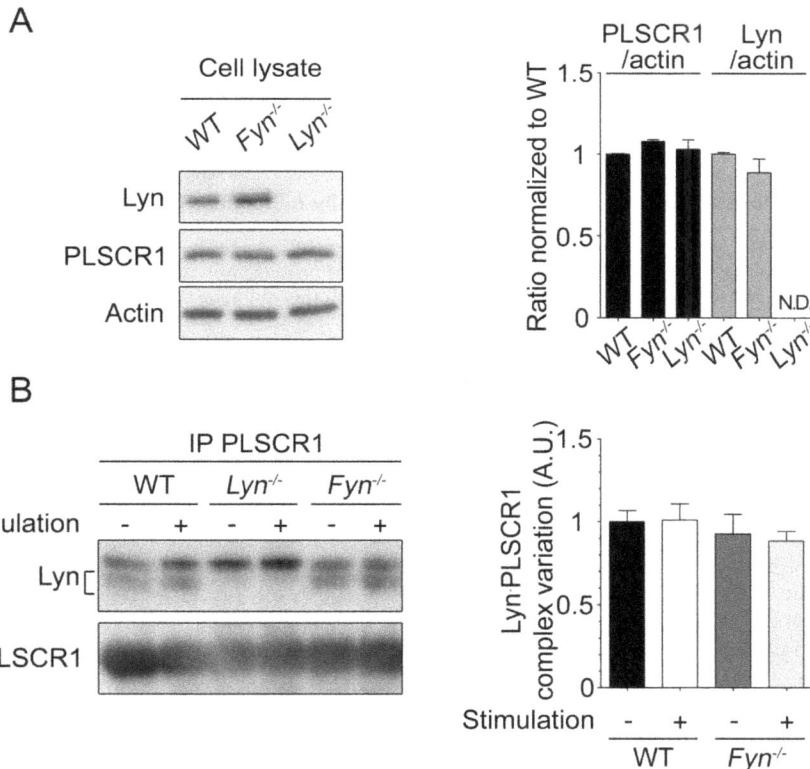

Figure 5. Lyn: PLSCR1 association is not modulated by Fyn in mast cells. A) Quantification of PLSCR1 and Lyn relative to actine in wild-type (WT), $Lyn^{-/-}$ and $Fyn^{-/-}$ BMMC. Left: representative immunoblots for the indicated proteins from one analysis. Right: relative quantifications of three independent experiments. The relative amount of the indicated protein was standardized as 1 in WT BMMC, and served as reference for all other relative quantifications. N.D., not detected. Data are presented as mean + s.e.m. One-way ANOVA analysis showed no significant difference. B) Sensitized BMMC from wild-type (WT), from Fyn-deficient ($Fyn^{-/-}$) and from Lyn-deficient ($Lyn^{-/-}$) mice were stimulated or not stimulated with antigen for 10 min. BMMC cell lysates were immunoprecipitated with anti-mouse PLSCR1 mAb 1A8-coupled beads (IP PLSCR1). Eluates were analyzed by immunoblotting with anti-Lyn antibodies (Lyn) and, after stripping, with anti-muPLSCR1 monoclonal antibody (PLSCR1). Left: one experiment representative of four. Bracket: Lyn doublet. The upper band is the heavy chain of the immunoprecipitating antibody. Right: Quantification of the amounts of Lyn co-immunoprecipitated with PLSCR1. Data were normalized to values obtained with WT unstimulated BMMC. Data are presented as mean + s.e.m. of four independent experiments. One-way ANOVA analysis showed no significant difference.

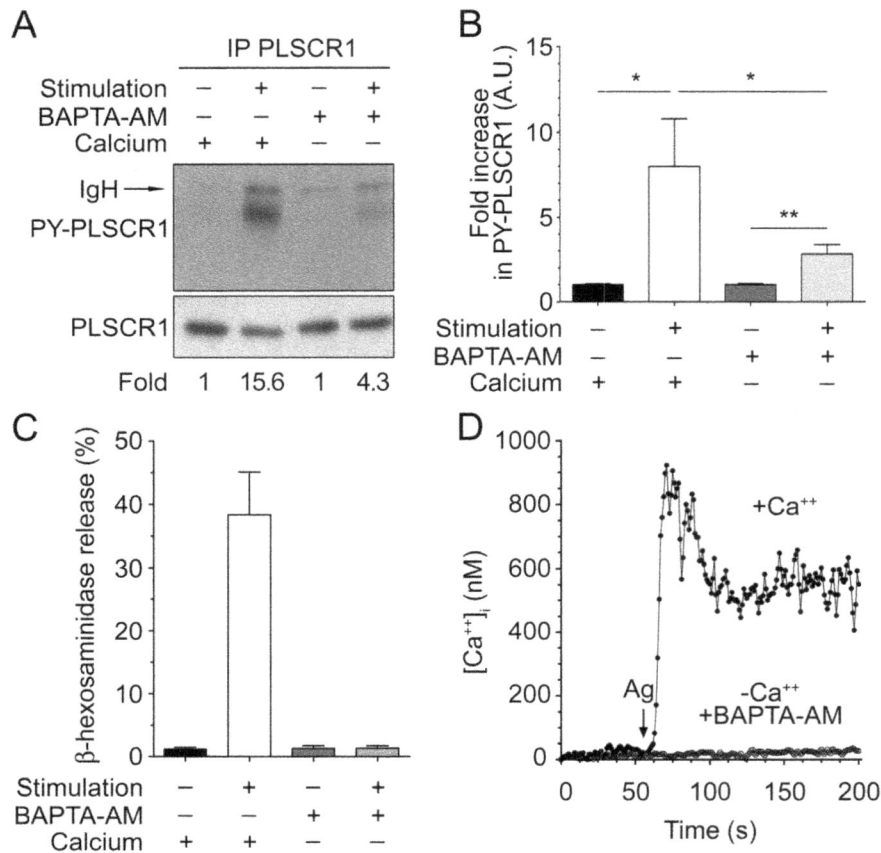

Figure 7. FcεRI-mediated PLSCR1 tyrosine phosphorylation is partially dependent on calcium. (A) Phosphorylation of PLSCR1 in the absence of intra- and extra-cellular calcium. RBL-2H3 cells were stimulated or not stimulated with IgE and antigen for 30 min in the presence or absence of calcium and of the intracellular calcium chelator BAPTA-AM as indicated and as described in the Materials and Methods Section. RBL-2H3 cell lysates were immunoprecipitated with anti-rat PLSCR mAb 129.2, and tyrosine phosphorylated PLSCR1 (PY-PLSCR1) in the eluates was analyzed in immunoblotting with anti-phosphotyrosine antibodies. After stripping of the membrane, the presence of total PLSCR1 was analyzed by immunoblotting with anti-rat PLSCR1 monoclonal antibodiy 129.2. Quantification of the PLSCR1 phosphorylation was performed using the NIH Image J software. Fold phosphorylation corresponds to the ratio of the value of phospho-PLSCR1 obtained for stimulated and non-stimulated cells (*Fold*, see Materials and Methods section) for each condition. One experiment representative of three is shown. IgH: Heavy chain of immunoprecipitating antibody. (B) Quantification. Statistical analysis was done by a one-way ANOVA followed by a two tailed paired student t test. Data are presented as mean + s.e.m. of three independent experiments. *: $p<0.05$; **: $p<0.01$. (C) Degranulation. The supernatants of the cells shown in the panel (A) were tested for the percent of β-hexosaminidase released. (D) Fluorescence measurement of intracellular calcium of sensitized RBL-2H3 cells loaded with Fura-2-AM and stimulated as in (A) with antigen (Ag) in the presence (+Ca^{++}) and absence (−Ca^{++} + BAPTA-AM) of intra- and extra-cellular calcium. This experiment is representative of three.

be related to variations in batch-to-batch cell cultures as we routinely observed variations in cell growth intensity and in degranulation responses of BMMC. However, a statistical increase in PLSCR1 tyrosine phosphorylation was reproducibly observed after aggregation of the receptor in WT BMMC. The data presented here demonstrate that this phosphorylation is subject to regulation that involves calcium mobilization as well as both positive and negative regulatory mechanisms mediated by Lyn and Syk, and Fyn, respectively (summarized in Figure 8). These data suggest that the phosphorylation of PLSCR1 is pivotal to the cross-interaction of the Lyn- and Fyn-initiated signaling pathways. Whereas initial analyses of FcεRI-dependent signaling pathways concentrated on defining specific effectors for each pathway, it becomes increasingly evident that multiple signals converge and that cross-talk of signals is key for an integrated cellular response. Example of this is provided by the tyrosine kinase Csk and by the adaptor Cbp that are recruited by Lyn and that negatively regulate the Fyn signaling in FcεRI-mediated mast cell activation [6].

Therefore, whereas Lyn can negatively regulate Fyn-initiated signals [6], herein we report that, conversely, Fyn can negatively regulate at least some Lyn-initiated signals, demonstrating that both pathways have the capacity to control each other. Thus, particular substrates could function at such crossroads promoting integration of signaling pathways and allowing fine-tuning and regulation. Interestingly, we previously demonstrated that PLSCR1 acts as an amplifier of the LAT-PLCγ-calcium axis thus modulating degranulation and VEGF production [33]. This axis depends on the Lyn-initiated signaling pathway. The association between Lyn and PLSCR1 is reminiscent of the association between PLSCR1 and Src in EGF receptor signaling that potentiates Src kinase activity [20,21]. Our present study demonstrates that whereas PLSCR1 can modulate the Lyn initiated pathway, this pathway controls PLSCR1 tyrosine phosphorylation, revealing a particular partnership between PLSCR1 and Lyn in FcεRI-induced mast cell activation.

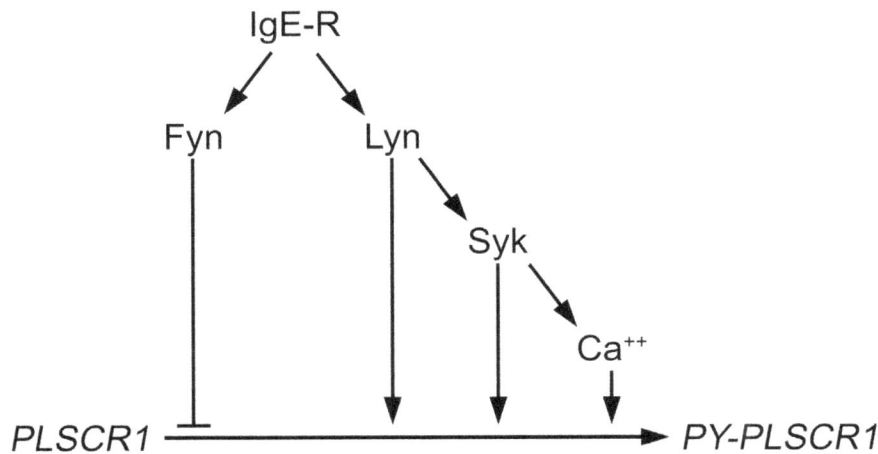

Figure 8. Complex regulation of FcεRI-dependent tyrosine phosphorylation of PLSCR1. Based on the data presented herein, we propose that tyrosine phosphorylation of PLSCR1 following FcεRI engagement is regulated on at least three levels in mast cells. Positive regulation is mediated by the Lyn-dependent pathway, whereas negative regulation is mediated by the Fyn-dependent pathway. After FcεRI aggregation, PLSCR1 can be phosphorylated on tyrosine directly either by Lyn or by Syk and indirectly as a result of Lyn/Syk-dependent activation of subsequent calcium mobilization. The other activation pathway initiated by FcεRI, that is dependent on Fyn, negatively regulates tyrosine phosphorylation of PLSCR1. Whether it acts by directly modulating Lyn-mediated or calcium-dependent tyrosine phosphorylation of PLSCR1, by controlling the PLSCR1 cellular localization required for its optimal phosphorylation by the Lyn pathway or by promoting its dephosphorylation is still unresolved.

The absolute requirement of Lyn and Syk is demonstrated by the absence of FcεRI-dependent increase in PLSCR1 phosphorylation in mast cells deficient for either kinase. Of note, absence of either kinase abolishes the mobilization of calcium [4,38]. Yet, we observed a residual, but significant, increase in tyrosine phosphorylation of PLSCR1 in the absence of calcium after FcεRI aggregation. This indicates that PLSCR1 could be phosphorylated on multiple tyrosines (calcium-dependent and calcium-independent), which would be in agreement with the heterogeneous molecular weight observed for phospho-PLSCR1 in mast cells, although the extent of this heterogeneity varies somehow from experiment to experiment for reasons that remain so far unclear. The increase in PLSCR1 tyrosine phosphorylation observed in the presence of calcium may be due to its phosphorylation on additional phosphorylation sites accessible after the conformational change induced by calcium [49,50]. Alternatively, new molecules of PLSCR1 could be recruited for tyrosine phosphorylation by the same or other kinases.

That Lyn and Syk could be directly involved in the phosphorylation of PLSCR1 is supported by several observations. Lyn and Syk were found to physically interact with PLSCR1 [33]. In addition, the FcεRI-dependent tyrosine phosphorylation of PLSCR1 was fully ablated in the absence of FcRγ, and partially in the absence of calcium, mapping the occurrence of the residual increase in the phosphorylation of PLSCR1 upstream of the calcium signal but downstream of the FcRγ chain, i.e. where Lyn and Syk are thought to function in IgE-mediated signaling [1]. Moreover, tyrosine phosphorylation of PLSCR1 was recovered when *Lyn*$^{-/-}$ BMMC and Syk-deficient RBL-2H3 cells were reconstituted with kinase competent Lyn or Syk, respectively, demonstrating that the defect in PLSCR1 phosphorylation was not due to aberrant maturation of these cells but to a kinase-related defect in FcεRI signaling. After FcεRI aggregation no significant increase in Lyn association with PLSCR1 was observed indicating that tyrosine phosphorylation of PLSCR1 by Lyn is not solely regulated by the interaction between both proteins but might be also regulated by Lyn activation. Previous studies have demonstrated that PLSCR1 also interacts with the prototypic kinase

p60c-src within the EGF receptor signaling pathways [20,21], thus serving as a substrate for p60c-src and, in turn, amplifying the activation of this kinase [20]. These data together with the ones collected herein, suggest a preferred connection between PLSCR1 and this family of tyrosine kinases that might be due to their common localization at the plasma membrane and particularly in lipid rafts [21,33].

This preferred connection is also highlighted by the negative regulation of PLSCR1 tyrosine phosphorylation by Fyn, another member of the Src family. This was revealed by the more robust increase in FcεRI-dependent phosphorylation of PLSCR1 in Fyn-deficient BMMC. The increased phosphorylation of PLSCR1 was not due to an increased calcium signal because Fyn deficiency does not increase this signal [4]. Neither was it due to an overall increase in tyrosine phosphorylation of cellular proteins since tyrosine phosphorylation of proteins in whole cell lysates from activated cells is significantly lower in Fyn-deficient BMMC compared to their wild-type counterparts ([4] and *data not shown*).

The data presented here extend our original observation that rat PLSCR1 is phosphorylated on tyrosine after FcεRI aggregation in the RBL-2H3 mast cell line [7] to non-tumoral mast cells derived in culture (BMMC) and to another species. Furthermore, the demonstration that it is initiated by the FcRγ chain and by another FcRγ-associated receptor (the transfected FcαRI) suggests that other Fc receptors should be able to promote the tyrosine phosphorylation of PLSCR1. The complexity of the mechanisms regulating PLSCR1 tyrosine phosphorylation suggests that this phosphorylation might play an important role in the regulation of PLSCR1 amplifier function. Studies by others have reported that there is no correlation between mast cell degranulation, phosphatidylserine externalization and tyrosine phosphorylation of PLSCR1 when comparing mast cells stimulated through FcεRI, Thy-1 and calcium ionophores [52]. The evidence presented here suggesting that PLSCR1 is phosphorylated on multiple tyrosines raises the possibility that positive regulation of degranulation by PLSCR1 may be associated with its phosphorylation on particular tyrosines whereas phosphorylation of other tyrosines may be involved in down-regulation of PLSCR1 function as is known for many receptors. Studies are underway to clarify this question.

Acknowledgments

We greatly acknowledge Reuben P. Siraganian (NIDCR, NIH, Bethesda, MD) for the generous gift of the Syk-deficient and Syk-reconstituted RBL-2H3 mast cell lines and of the DNP-48 anti-DNP IgE hybridoma.

References

1. Blank U, Rivera J (2004) The ins and outs of IgE-dependent mast-cell exocytosis. Trends Immunol 25: 266–273.
2. Eiseman E, Bolen JB (1992) Engagement of the high-affinity IgE receptor activates src protein-related tyrosine kinases. Nature 355: 78–80.
3. Saitoh S, Arudchandran R, Manetz TS, Zhang W, Sommers CL, et al. (2000) LAT is essential for Fc(epsilon) RI-mediated mast cell activation. Immunity 12: 525–535.
4. Parravicini V, Gadina M, Kovarova M, Odom S, Gonzalez-Espinosa C, et al. (2002) Fyn kinase initiates complementary signals required for IgE-dependent mast cell degranulation. Nat Immunol 3: 741–748.
5. Furumoto Y, Gonzalez-Espinosa C, Gomez G, Kovarova M, Odom S, et al. (2004) Rethinking the role of Src family protein tyrosine kinases in the allergic response: new insights on the functional coupling of the high affinity IgE receptor. Immunol Res 30: 241–253.
6. Odom S, Gomez G, Kovarova M, Furumoto Y, Ryan JJ, et al. (2004) Negative regulation of immunoglobulin E-dependent allergic responses by Lyn kinase. J Exp Med 199: 1491–1502.
7. Pastorelli C, Veiga J, Charles N, Voignier E, Moussu H, et al. (2001) IgE receptor type I-dependent tyrosine phosphorylation of phospholipid scramblase. J Biol Chem 276: 20407–20412.
8. Zhao J, Zhou Q, Wiedmer T, Sims PJ (1998) Level of expression of phospholipid scramblase regulates induced movement of phosphatidylserine to the cell surface. J Biol Chem 273: 6603–6606.
9. Zhou Q, Zhao J, Stout JG, Luhm RA, Wiedmer T, et al. (1997) Molecular cloning of human plasma membrane phospholipid scramblase. A protein mediating transbilayer movement of plasma membrane phospholipids. J Biol Chem 272: 18240–18244.
10. Bevers EM, Comfurius P, Dekkers DW, Harmsma M, Zwaal RF (1998) Transmembrane phospholipid distribution in blood cells: control mechanisms and pathophysiological significance. Biol Chem 379: 973–986.
11. Fadok VA, Bratton DL, Rose DM, Pearson A, Ezekewitz RA, et al. (2000) A receptor for phosphatidylserine-specific clearance of apoptotic cells. Nature 405: 85–90.
12. Demo SD, Masuda E, Rossi AB, Throndset BT, Gerard AL, et al. (1999) Quantitative measurement of mast cell degranulation using a novel flow cytometric annexin-V binding assay. Cytometry 36: 340–348.
13. Martin S, Pombo I, Poncet P, David B, Arock M, et al. (2000) Immunologic stimulation of mast cells leads to the reversible exposure of phosphatidylserine in the absence of apoptosis. Int Arch Allergy Immunol 123: 249–258.
14. Wiedmer T, Zhao J, Li L, Zhou Q, Hevener A, et al. (2004) Adiposity, dyslipidemia, and insulin resistance in mice with targeted deletion of phospholipid scramblase 3 (PLSCR3). Proc Natl Acad Sci U S A 101: 13296–13301.
15. Zhou Q, Zhao J, Wiedmer T, Sims PJ (2002) Normal hemostasis but defective hematopoietic response to growth factors in mice deficient in phospholipid scramblase 1. Blood 99: 4030–4038.
16. Malvezzi M, Chalat M, Janjusevic R, Picollo A, Terashima H, et al. (2013) Ca2+-dependent phospholipid scrambling by a reconstituted TMEM16 ion channel. Nat Commun 4: 2367.
17. Suzuki J, Umeda M, Sims PJ, Nagata S (2010) Calcium-dependent phospholipid scrambling by TMEM16F. Nature 468: 834–838.
18. Yang H, Kim A, David T, Palmer D, Jin T, et al. (2012) TMEM16F forms a Ca2+-activated cation channel required for lipid scrambling in platelets during blood coagulation. Cell 151: 111–122.
19. Suzuki J, Denning DP, Imanishi E, Horvitz HR, Nagata S (2013) Xk-related protein 8 and CED-8 promote phosphatidylserine exposure in apoptotic cells. Science 341: 403–406.
20. Nanjundan M, Sun J, Zhao J, Zhou Q, Sims PJ, et al. (2003) Plasma membrane phospholipid scramblase 1 promotes EGF-dependent activation of c-Src through the epidermal growth factor receptor. J Biol Chem 278: 37413–37418.
21. Sun J, Nanjundan M, Pike LJ, Wiedmer T, Sims PJ (2002) Plasma membrane phospholipid scramblase 1 is enriched in lipid rafts and interacts with the epidermal growth factor receptor. Biochemistry 41: 6338–6345.
22. Ben-Efraim I, Zhou Q, Wiedmer T, Gerace L, Sims PJ (2004) Phospholipid scramblase 1 is imported into the nucleus by a receptor-mediated pathway and interacts with DNA. Biochemistry 43: 3518–3526.
23. Chen MH, Ben-Efraim I, Mitrousis G, Walker-Kopp N, Sims PJ, et al. (2005) Phospholipid scramblase 1 contains a nonclassical nuclear localization signal with unique binding site in importin alpha. J Biol Chem 280: 10599–10606.
24. Wiedmer T, Zhao J, Nanjundan M, Sims PJ (2003) Palmitoylation of phospholipid scramblase 1 controls its distribution between nucleus and plasma membrane. Biochemistry 42: 1227–1233.
25. Zhou Q, Zhao J, Al-Zoghaibi F, Zhou A, Wiedmer T, et al. (2000) Transcriptional control of the human plasma membrane phospholipid scramblase 1 gene is mediated by interferon-alpha. Blood 95: 2593–2599.
26. Dong B, Zhou Q, Zhao J, Zhou A, Harty RN, et al. (2004) Phospholipid scramblase 1 potentiates the antiviral activity of interferon. J Virol 78: 8983–8993.
27. Zhou Q, Ben-Efraim I, Bigcas JL, Junqueira D, Wiedmer T, et al. (2005) Phospholipid scramblase 1 binds to the promoter region of the inositol 1,4,5-triphosphate receptor type 1 gene to enhance its expression. J Biol Chem 280: 35062–35068.
28. Kasukabe T, Okabe-Kado J, Honma Y (1997) TRA1, a novel mRNA highly expressed in leukemogenic mouse monocytic sublines but not in nonleukemogenic sublines. Blood 89: 2975–2985.
29. Silverman RH, Halloum A, Zhou A, Dong B, Al-Zoghaibi F, et al. (2002) Suppression of ovarian carcinoma cell growth in vivo by the interferon-inducible plasma membrane protein, phospholipid scramblase 1. Cancer Res 62: 397–402.
30. Yokoyama A, Yamashita T, Shiozawa E, Nagasawa A, Okabe-Kado J, et al. (2004) MmTRA1b/phospholipid scramblase 1 gene expression is a new prognostic factor for acute myelogenous leukemia. Leuk Res 28: 149–157.
31. Zhao KW, Li X, Zhao Q, Huang Y, Li D, et al. (2004) Protein kinase Cdelta mediates retinoic acid and phorbol myristate acetate-induced phospholipid scramblase 1 gene expression: its role in leukemic cell differentiation. Blood 104: 3731–3738.
32. Ory S, Ceridono M, Momboisse F, Houy S, Chasserot-Golaz S, et al. (2013) Phospholipid scramblase-1-induced lipid reorganization regulates compensatory endocytosis in neuroendocrine cells. J Neurosci 33: 3545–3556.
33. Amir-Moazami O, Alexia C, Charles N, Launay P, Monteiro RC, et al. (2008) Phospholipid scramblase 1 modulates a selected set of IgE receptor-mediated mast cell responses through LAT-dependent pathway. J Biol Chem 283: 25514–25523.
34. Sun J, Zhao J, Schwartz MA, Wang JY, Wiedmer T, et al. (2001) c-Abl tyrosine kinase binds and phosphorylates phospholipid scramblase 1. J Biol Chem 276: 28984–28990.
35. Stracke ML, Basciano LK, Fischler C, Berenstein EH, Siraganian RP (1987) Characterization of monoclonal antibodies produced by immunization with partially purified IgE receptor complexes. Mol Immunol 24: 347–356.
36. Charles N, Monteiro RC, Benhamou M (2004) p28, a novel IgE receptor-associated protein, is a sensor of receptor occupation by its ligand in mast cells. J Biol Chem 279: 12312–12318.
37. Barsumian EL, Isersky C, Petrino MG, Siraganian RP (1981) IgE-induced histamine release from rat basophilic leukemia cell lines: isolation of releasing and nonreleasing clones. Eur J Immunol 11: 317–323.
38. Zhang J, Berenstein EH, Evans RL, Siraganian RP (1996) Transfection of Syk protein tyrosine kinase reconstitutes high affinity IgE receptor-mediated degranulation in a Syk-negative variant of rat basophilic leukemia RBL-2H3 cells. J Exp Med 184: 71–79.
39. Launay P, Patry C, Lehuen A, Pasquier B, Blank U, et al. (1999) Alternative endocytic pathway for immunoglobulin A Fc receptors (CD89) depends on the lack of FcRgamma association and protects against degradation of bound ligand. J Biol Chem 274: 7216–7225.
40. Furumoto Y, Brooks S, Olivera A, Takagi Y, Miyagishi M, et al. (2006) Cutting Edge: Lentiviral short hairpin RNA silencing of PTEN in human mast cells reveals constitutive signals that promote cytokine secretion and cell survival. J Immunol 176: 5167–5171.
41. Monteiro RC, Cooper MD, Kubagawa H (1992) Molecular heterogeneity of Fc alpha receptors detected by receptor-specific monoclonal antibodies. J Immunol 148: 1764–1770.
42. Laemmli UK (1970) Cleavage of structural proteins during the assembly of the head of bacteriophage T4. Nature 227: 680–685.
43. Razin E, Mencia-Huerta JM, Stevens RL, Lewis RA, Liu FT, et al. (1983) IgE-mediated release of leukotriene C4, chondroitin sulfate E proteoglycan, beta-hexosaminidase, and histamine from cultured bone marrow-derived mouse mast cells. J Exp Med 157: 189–201.
44. Ravetch JV, Kinet JP (1991) Fc receptors. Annu Rev Immunol 9: 457–492.
45. Monteiro RC, Van De Winkel JG (2003) IgA Fc receptors. Annu Rev Immunol 21: 177–204.

Author Contributions

Conceived and designed the experiments: MB NC. Performed the experiments: AK ICM YY JS CGM. Analyzed the data: MB NC JR RCM TW PJS UB. Contributed reagents/materials/analysis tools: MB RCM YY JS PJS TW JR. Wrote the paper: MB NC JR RCM TW PJS UB CGM JS YY ICM AK.

46. Pfefferkorn LC, Yeaman GR (1994) Association of IgA-Fc receptors (Fc alpha R) with Fc epsilon RI gamma 2 subunits in U937 cells. Aggregation induces the tyrosine phosphorylation of gamma 2. J Immunol 153: 3228–3236.

47. Pasquier B, Launay P, Kanamaru Y, Moura IC, Pfirsch S, et al. (2005) Identification of FcalphaRI as an inhibitory receptor that controls inflammation: dual role of FcRgamma ITAM. Immunity 22: 31–42.

48. Benhamou M, Stephan V, Robbins KC, Siraganian RP (1992) High-affinity IgE receptor-mediated stimulation of rat basophilic leukemia (RBL-2H3) cells induces early and late protein-tyrosine phosphorylations. J Biol Chem 267: 7310–7314.

49. Sahu SK, Aradhyam GK, Gummadi SN (2009) Calcium binding studies of peptides of human phospholipid scramblases 1 to 4 suggest that scramblases are new class of calcium binding proteins in the cell. Biochim Biophys Acta 1790: 1274–1281.

50. Stout JG, Zhou Q, Wiedmer T, Sims PJ (1998) Change in conformation of plasma membrane phospholipid scramblase induced by occupancy of its Ca2+ binding site. Biochemistry 37: 14860–14866.

51. Zhou Q, Sims PJ, Wiedmer T (1998) Identity of a conserved motif in phospholipid scramblase that is required for Ca2+-accelerated transbilayer movement of membrane phospholipids. Biochemistry 37: 2356–2360.

52. Smrz D, Lebduska P, Draberova L, Korb J, Draber P (2008) Engagement of phospholipid scramblase 1 in activated cells: implication for phosphatidylserine externalization and exocytosis. J Biol Chem 283: 10904–10918.

53. Beaven MA, Rogers J, Moore JP, Hesketh TR, Smith GA, et al. (1984) The mechanism of the calcium signal and correlation with histamine release in 2H3 cells. J Biol Chem 259: 7129–7136.

13

Establishment of *Myotis myotis* Cell Lines - Model for Investigation of Host-Pathogen Interaction in a Natural Host for Emerging Viruses

Xiaocui He[1], Tomáš Korytář[1], Yaqing Zhu[1], Jiří Pikula[2], Hana Bandouchova[2], Jan Zukal[3,4], Bernd Köllner[1]*

1 Institute of Immunology, Friedrich-Loeffler-Institute (FLI), Federal Research Institute for Animal Health, Greifswald- Insel Riems, Germany, 2 Department of Ecology and Diseases of Game, Fish and Bees, Faculty of Veterinary Hygiene and Ecology, University of Veterinary and Pharmaceutical Sciences Brno, Brno, Czech Republic, 3 Institute of Vertebrate Biology, Academy of Sciences of the Czech Republic, Brno, Czech Republic, 4 Department of Botany and Zoology, Masaryk University, Brno, Czech Republic

Abstract

Bats are found to be the natural reservoirs for many emerging viruses. In most cases, severe clinical signs caused by such virus infections are normally not seen in bats. This indicates differences in the virus-host interactions and underlines the necessity to develop natural host related models to study these phenomena. Due to the strict protection of European bat species, immortalized cell lines are the only alternative to investigate the innate anti-virus immune mechanisms. Here, we report about the establishment and functional characterization of *Myotis myotis* derived cell lines from different tissues: brain (*Mm*Br), tonsil (*Mm*To), peritoneal cavity (*Mm*Pca), nasal epithelium (*Mm*Nep) and nervus olfactorius (*Mm*Nol) after immortalization by SV 40 large T antigen. The usefulness of these cell lines to study antiviral responses has been confirmed by analysis of their susceptibility to lyssavirus infection and the mRNA patterns of immune-relevant genes after poly I:C stimulation. Performed experiments indicated varying susceptibility to lyssavirus infection with *Mm*Br being considerably less susceptible than the other cell lines. Further investigation demonstrated a strong activation of interferon mediated antiviral response in *Mm*Br contributing to its resistance. The pattern recognition receptors: RIG-I and MDA5 were highly up-regulated during rabies virus infection in *Mm*Br, suggesting their involvement in promotion of antiviral responses. The presence of CD14 and CD68 in *Mm*Br suggested *Mm*Br cells are microglia-like cells which play a key role in host defense against infections in the central nervous system (CNS). Thus the expression pattern of *Mm*Br combined with the observed limitation of lyssavirus replication underpin a protective mechanism of the CNS controlling the lyssavirus infection. Overall, the established cell lines are important tools to analyze antiviral innate immunity in *M. myotis* against neurotropic virus infections and present a valuable tool for a broad spectrum of future investigations in cellular biology of *M. myotis*.

Editor: Matthias Johannes Schnell, Thomas Jefferson University, United States of America

Funding: Bundesministerium für Forschung und Technologie German Ministry for Research and Technology Grant 01KI1016A. The funders had no role in study design, data collection and analysis, decision to publish, or preparation of the manuscript.

Competing Interests: The authors have declared that no competing interests exist.

* Email: bernd.koellner@fli.bund.de

Introduction

Bats belong to one of the most abundant, diverse and widely distributed mammalian groups. In the order of *Chiroptera* which is divided into two suborders *Megachiroptera* and *Microchiroptera*, a total of 1,240 species have been yet described [1]. Bats evolved early and changed very little over the past 52 million years [2]. Their wide distribution and migratory behaviour favour bats as vectors for viruses and raise concerns over their role in zoonotic diseases [3–5]. Among the large number of viruses detected in bats, some like Hendra virus, Nipah virus, severe acute respiratory syndrome coronavirus (SARS), Ebola virus, West Nile virus were reported to be zoonotic [4,6–11]. Also 13 of the 15 lyssaviruses, except Mokola virus and Ikoma lyssavirus, were detected in bats. In North America bats host RABV, whereas in Europe European Bat lyssavirus type 1 and 2 (EBLV-1 and EBLV-2) are found in different bat species [12,13]. Annually, there are approximately

55,000 human deaths caused by rabies, especially in the developing countries of Asia and Africa [14]. Despite most human rabies deaths are associated with dog rabies, some of them can be directly linked to the contact with bats, such as 8 out of 226 human rabies cases were of bat origin in the Americas in 1983 and a few human cases caused by EBLVs were reported in Europe to date [15–18]. Although bat associated viruses can cause severe diseases in various mammals, they seem to be less pathogenic for bats [3,19–25]. After experimental infection with Hendra or Nipah virus, bats showed no clinical disease, while guinea pigs succumbed to the same dose of virus [21,22]. Similar situation was also observed in Hendra virus infection in horses and bats [24]. Lyssaviruses are the only viruses that were reported to cause clinical disease in bats [26]. However, only a small proportion of bats develop clinical symptoms after experimental infection [25,27]. This indicates a critical difference in the development of viral disease between bats and other mammals and requires

Table 1. Primers used in this study.

Name*	Sequence (5'→3')
[a]SV40T-F	GGGTCTTCTACCTTTCTCTTCTTT
[a]SV40T-R	GCAGTGGTGGAATGCCTTT
ND1-F	TATTAGCCCTATCAAGTTTAGC
ND1-R	GGATGCTCGGACCCATAA
β-actin-F	GCGCAAGTACTCTGTGTGGA
β-actin-R	ATCTCGTTTTCTGCGCAAGT
[a] Mx1-F	TCTACTGCCAAGACCAAGCGT
[a] Mx1-R	CGAGGGAGCAAGTCAAAGGA
[a] IFIT3-F	AGCAGAGGAGCTTGCAGAAG
[a] IFIT3-R	CCGGAAAGCCATAAACAAGA
[a] ISG56-F	CAGGCTAAATCCAGAAGATG
[a] ISG56-R	TTCCAGAGCAAATTCAAAAT
ISG43-F	CATGATGCTGCTCAACTCTA
ISG43-R	TAAGGTGGATTGTCAAGGTC
TLR3-F	TCTCGCTCCTTCTATGGG
TLR3-R	TGCCTGGAAAGTTGTTATCG
RIG-1-F	GAAGAGCAAGAGGTAGCAAA
RIG-1-R	CCTTTGCTTTCTTCTCAAAA
MDA5-F	TCCGAATGATTGATGCCTAT
MDA5-R	ATTATCCCTCTTGCTGACCC
CD14-F	GCTCTCTTAACCTGTCCTCCG
CD14-R	CTCTGTTCAGCCGGTTGTTG
CD68-F	GCCCTGGTGCTTGTTATCCT
CD68-R	GAGGCAGCTGAGTGGTTCAG
[a] EBLV1-F	GAAAGGKGACAAGATAACACC
[a] EBLV1-R	ARAGAAGAAGTCCAACCAGAG
[a] EBLV2-F	GGTGTCTGTAAAGCCAGAAG
[a] EBLV2-R	TTATAAGCTCTGTTCAAG
[a] RABV-F	GATCCTGATGAYGTATGTTCCTA
[a] RABV-R	GATTCCGTAGCTRGTCCA

* F indicates forward primer, R indicates reverse primer. [a]Primers are from previous studies [33,35].

profound investigation of bat immunology and host-virus interactions.

Since all of 52 identified European bats species are endangered and strictly protected, the use of animal trials for the investigation of immune mechanisms in bats is not possible. Thus, development of stable cell lines for *in vitro* studies derived from European bat species is desirable. So far, several bat cell lines were reported in previous studies, but most of them were established from non-European bats, like Tb1-Lu from *Tadarida brasiliensis*, Mvi/It from *Myotis velifer incautus*, and several primary immortalized cell lines from *Pteropus alecto, Carollia perspicillata, Eidolon helvum* and *Rousettus aegyptiacus* [28–31]. Viral infection studies have been carried out in the fruit bat cell lines to investigate the susceptibility, infection kinetics of henipavirus as well as the host innate immunity [28,32]. However, the susceptibility to lyssavirus has not yet been examined in these cell lines. Additionally, except for a brain cell line from *Eptesicus serotinus* employed to investigate the type I interferon (IFN) response after lyssavirus infection [33], the use of a bat cell line as a tool for studies into lyssavirus infection in its natural reservoir host is rare. A broader

variety of bat cell lines, particularly European bat cell lines from tissues of immune relevance, is therefore urgently in demand for lyssavirus-host studies.

In this study, we established different cell lines from the European bat *M. myotis*, evaluated their susceptibility to EBLV-1, EBLV-2 and RABV infection and investigated innate immune gene responses after the polyinosinic:polycytidylic acid (poly I:C) stimulation. The established *M. myotis* cell lines present a valuable *in vitro* model to study the interactions between lyssaviruses and their natural host, and to shed light on the mechanisms of resistance in bat's central nervous system (CNS).

Materials and Methods

Ethics statement

Ethical approval for all of the capturing and sampling were confirmed by the competent authorities in the respective Federal Republic of Germany and Czech Republic. The Czech Academy of Sciences Ethics Committee reviewed and approved the animal use protocol No. 169/2011 in compliance with Law No. 312/

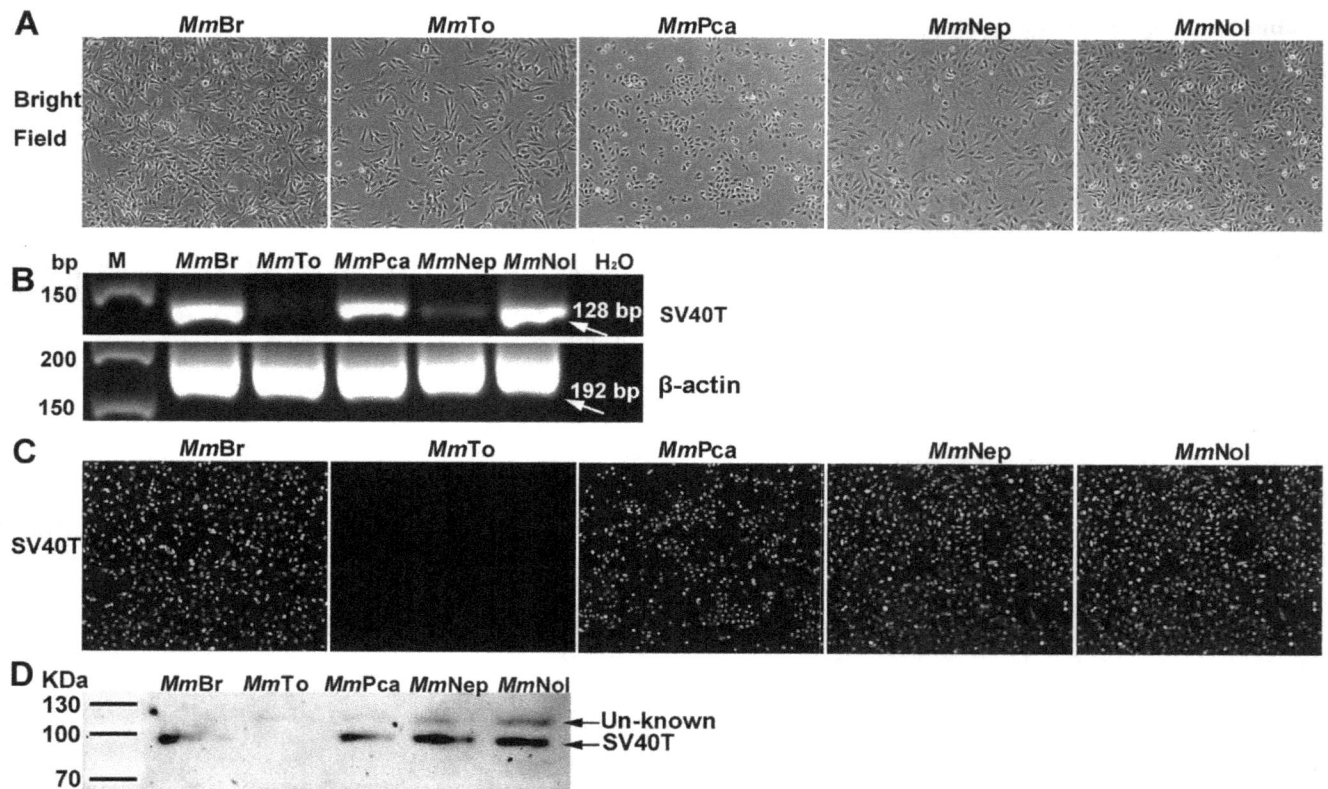

Figure 1. Newly established immortalized *M. myotis* cell lines from different tissues *MmBr* - brain; *MmTo* - tonsil; *MmPca* - peritoneal cavity; *MmNep* - nasal epithelium; *MmNol* - nervus olfactorius. (**A**) Morphology of a 24 h culture of immortalized *M. myotis* cell lines. (**B**) Expression of SV40T transcripts in different *M. myotis* cell lines. Note the very low expression in *MmTo*. (**C**) Expression of SV40T protein visualized by immunofluorescence and (**D**) by western blot using anti-SV40T monoclonal antibody. Note the absence of SV40T protein in *MmTo* cell line.

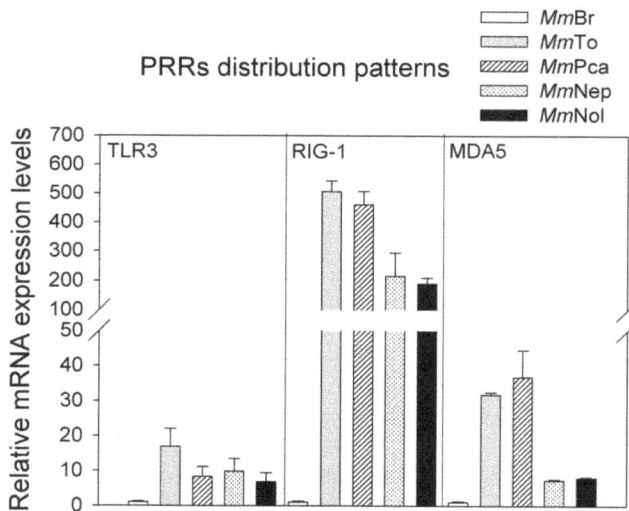

Figure 2. PRRs distribution patterns in the established unstimulated *M. myotis* cell lines. The mRNA expression levels of TLR3, RIG-1 and MDA5 in *MmBr*, *MmTo*, *MmPca*, *MmNep* and *MmNol* were determined by qRT-PCR (n = 3). The expression level was shown as a related fold and normalized against β-actin. The expression level of different genes in *MmBr* showed the lowest expression and was presented as one fold.

2008 on Protection of Animals against Cruelty adopted by the Parliament of the Czech Republic. The capture and sampling of a *M. myotis* specimen in the Moravian Karst in November 2012 was in compliance with Law No. 114/1992 on Nature and Landscape Protection, and was based on permit 01662/MK/2012S/00775/ MK/2012 issued by the Nature Conservation Agency of the Czech Republic. Established *M. myotis* cell lines from the single sacrificed specimen have been used to examine bat responses to the infection by *Pseudogymnoascus destructans* (un-published data) as well as for the present study of rabies. Three co-authors of the present manuscript concerning establishment of *M. myotis* cell lines to investigate lyssavirus infection, i.e. Hana Bandouchova, Jiri Pikula, and Jan Zukal, examine white-nose syndrome in the Czech Republic and hold the necessary permits. A paper based on these permits and excemption from Law No. 114/1992 on Nature and Landscape Protection of the Czech Republic allowing euthanasia of up to 10 *M. myotis* bats has already been published [34].

Primary cell culture and immortalization

A single *M. myotis* male was captured in Sloupsko-Sosuvske caves of the Moravian Karst (Czech Republic, coordinates 49° 24′ 40.88″ and 16° 44′ 20.54″). The bat was kept to minimize stress and handling between capture and euthanasia in a clean plastic box with soft mesh to enable roosting under temperature of hibernation torpor of 6°C and transferred to our laboratory at Veterinary and Pharmaceutical Sciences Brno (Czech Republic) within a day. It was anesthetized to insensitiveness using isofluranum (Isofluran, Piramal Healthcare, UK), and then

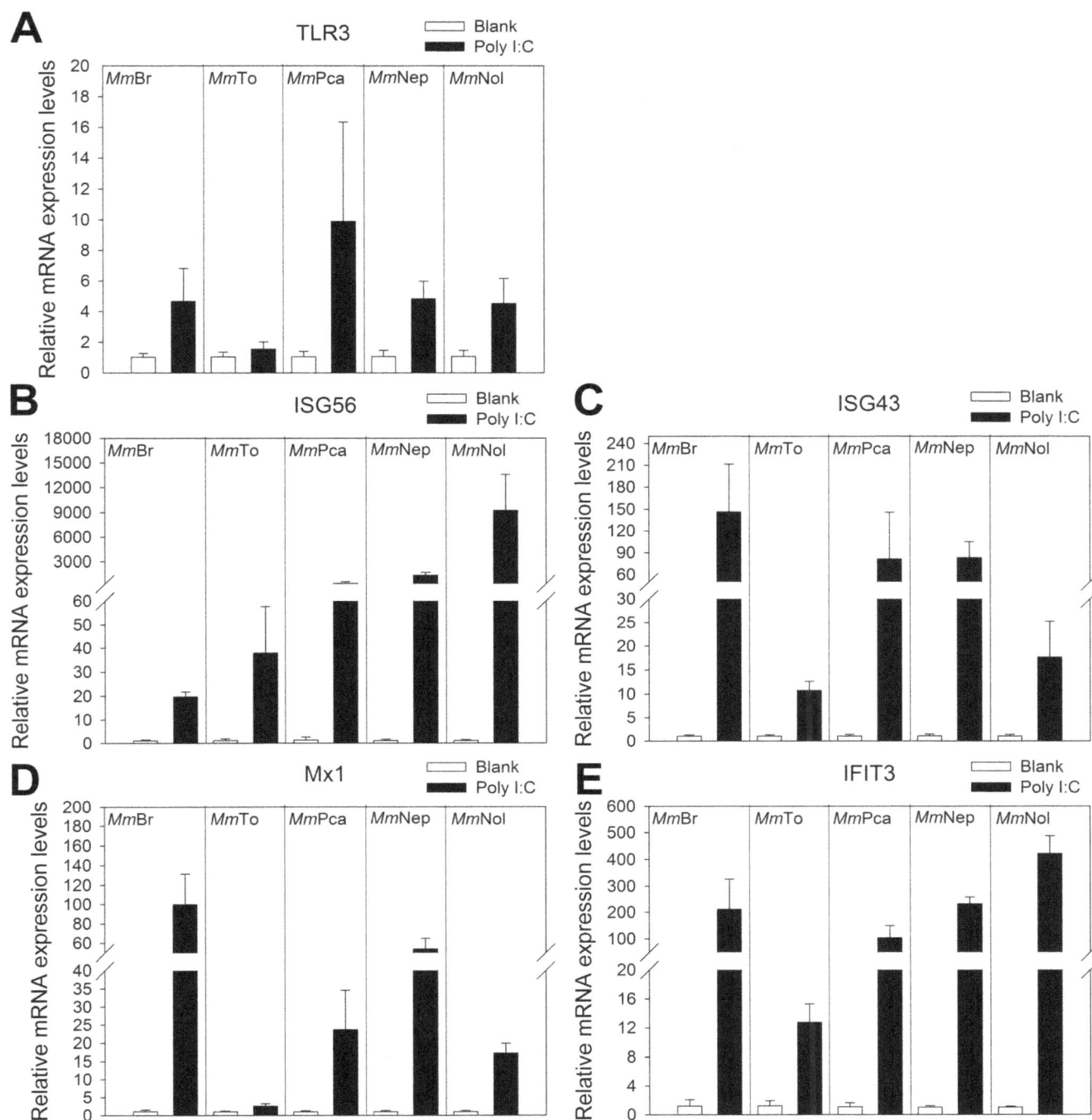

Figure 3. Comparative analysis of the expression patterns of antiviral molecules after poly I:C stimulation. The established immortalized *M. myotis* cell lines were transfected with poly I:C (10 μg/mL) by lipofectamine 2000. The unstimulated cells were used as blank control. Twenty four hours post transfection, the mRNA expression levels of TLR3, ISG56, ISG43, Mx1 and IFIT3 were measured by qRT-PCR (n = 3). The expression level was shown as a related fold and normalized against β-actin. The expression level of different molecules of blank group in individual cell line was presented as one fold.

euthanized by decapitation and subjected to necropsy in order to collect organs and tissues. Tissues were freshly isolated from the euthanized bat, and then minced and cultured in Dulbecco's Modified Eagle Medium (DMEM) supplemented with 10% fetal calf serum (FCS), penicillin 100 units/mL and streptomycin 100 mg/mL (Sigma). Primary cells were cultured in 6-well plates

till the confluence reaches 50–70%. Immortalization was done by transfection of pRSVAg1 plasmid expressing Simian Vacuolating Virus 40 large T antigen (SV40T) with lipofectamine 2000 according to the protocol (Invitrogen). Immortalized cells were expanded and stock frozen. After several passages, the mRNA expression of SV40T (in the established lines) was tested by reverse

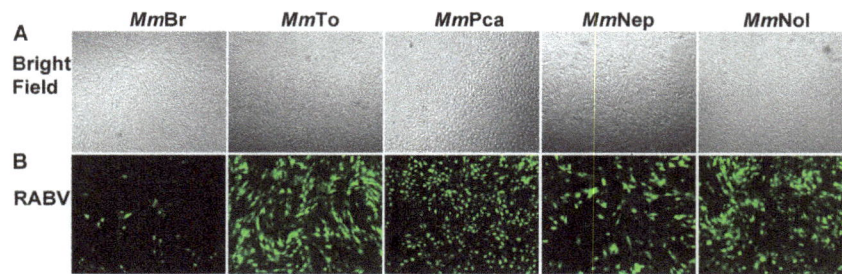

Figure 4. Susceptibility of *M. myotis* cell lines to lyssavirus infection. Immortalized cells (third passage, morphology and cell density visualized in bright filed) were infected with GFP fused RABV at a MOI of 10, and the propagation of RABV was visualized by fluorescence microscope at 24 hpi. Note the low amount of viral antigen positive cells in *Mm*Br in contrast to the other 4 cell lines.

transcription PCR (RT-PCR) using SV40T specific primers [35]. The protein expression was controlled by the immunofluorescence and western blot as described below. Briefly, cells were first fixed with 3% paraformaldehyde and permeabilized with 0.5% triton X. After washing with PBS, cells were stained with mouse anti-SV40T monoclonal antibody (Santa Cruz Biotechnology) and goat anti-mouse IgG Alexa Fluor (Invitrogen) as second antibody and visualized by fluorescence microscope. For western blot, the same mouse antibody was used as primary antibody and bound antibody was detected with goat anti-mouse IgG peroxidase (Sigma). Images were developed using the ECL kit (Thermo Scientific Pierce) according to the manufacturer's instructions.

Species confirmation of different cell lines by PCR

To confirm the identity of the established *M. myotis* cell lines derived from brain (cerebrum) (designated *Mm*Br), tonsil (*Mm*To), peritoneal cavity (*Mm*Pca), nasal epithelium (*Mm*Nep) and nervus olfactorius (*Mm*Nol), a *M. myotis*-specific PCR was developed. An NADH dehydrogenase subunit 1 (ND1) gene (Genbank accession number: DQ915043) from *M. myotis* was used as a species specific molecular marker. The genomic DNA from different cell lines was isolated by DNeasy Blood & Tissue Kit (Qiagen). The concentration and purity of genomic DNA were determined by Nanodrop (Thermo). PCR was performed using a specific primer pair ND1-F and ND1-R (Table 1) and genomic DNA as a template by GoTaq Flexi DNA Polymerase (Promega) to get the ND1 fragments. PCR products were cloned into PCR2.1 vector (Invitrogen) and transformed into *E. coli* competent cells. Plasmids were extracted from positive clones and sequenced by Applied Biosystems 3130 Genetic Analyzer (Life Technologies) at the Friedrich-Loeffler-Institute, Germany.

Poly I:C stimulation

To evaluate the IFN response of *M. myotis* cell lines and the induction of IFN mediated signaling, poly I:C was used to stimulate the cells. Different cell lines were seeded in 24-well plates at a density ranging from 1.2 to 2×10^5 cells/well, and cultured as described above. Around 20 hours after seeding, cells were transfected with poly I:C (Sigma) at a concentration of 10 μg/mL by lipofectamine 2000 (Invitrogen) following the manufacturer's instructions. Twenty four hours post stimulation, cells were harvested into RLT buffer (Qiagen) for RNA extraction by an RNeasy mini kit (Qiagen).

Lyssaviruses infection

Early after immortalization, the third passage immortalized cell lines were used to check the infectivity of RABV. Cells were infected with RABV (European fox isolate, fused with green

fluorescent protein, GFP) at a MOI of 10. Twenty four hours post infection (hpi), infected cells were fixed and permeabilized as described above and visualized by fluorescence microscope. *Mm*Br and *Mm*To cells that were infected with a serial MOI of 0.01, 0.1, and 1.0 were harvested at 24 hpi and used for RNA extraction. To confirm the infectivity in later passaged cells, different immortalized cell lines of more than 15 passages were infected with lyssaviruses RABV, EBLV-1 (*E. serotinus* isolate) and EBLV-2 (*M. daubentonii* isolate) at a MOI of 0.1. The infected cells were cultured as described above. Cells were collected for RNA extraction at 24 hpi and quantitative real-time PCR (qRT-PCR) was performed on the CFX96 TouchDetection System (Bio-Rad) using SensiFAST SYBR one-step kit (Bioline) according to the protocol. Immunofluorescence analysis was performed on fixed cells using FITC conjugated anti-rabies monoclonal antibody (SIFIN) at 72 hpi as described before [36]. To further confirm the susceptibility, *Mm*Br and *Mm*Nol cell lines were infected with EBLV-1 at a MOI of 0.01 to set the sensitivity at a Ct value of 22 for the inoculation dose. The viral supernatant was either changed or not changed with fresh medium at 1 hpi, and viral replication levels were measured by qRT-PCR over 72 hpi.

Quantitative real-time PCR

qRT-PCR was introduced to measure the mRNA expression levels of immune related molecules in response to poly I:C stimulation and virus infection. The selected molecules include IFN induced genes: IFN stimulated gene 56 (ISG56), ISG43, myxovirus resistance 1 (Mx1) and IFN induced protein with tetratricopeptide repeats 3 (IFIT3), and pattern recognition receptors (PRRs): toll-like receptor 3 (TLR3), retinoic acid-inducible gene 1 (RIG-1) and melanoma differentiation-associated protein 5 (MDA+5). All of these primers were designed based the sequence resources from our own un-published sequence database and public databases of bat species. The softwares for primers design include primer premier 5, online tools: http://bioinfo.ut.ee/primer3-0.4.0/ and http://www.ncbi.nlm.nih.gov/tools/primer-blast/. Primers of target genes and internal control β-actin were listed in Table 1. qRT-PCR was performed on the CFX96 TouchDetection System (Bio-Rad) using SensiFAST SYBR one-step kit (Bioline) according to the protocol. To assess the specificity of the PCR amplification, a melting curve analysis was performed at the end of the reaction. The relative expression levels of targets were calculated by $2^{-\Delta\Delta Ct}$ method [37].

Molecular characterization of the *Mm*BrBecause the *Mm*Br is derived from the CNS, the target of fatal infections by lyssaviruses, a further characterization of cell type of *Mm*Br is desired to improve the understanding of the antiviral defense in the CNS. The expressions of cluster of differentiation (CD) 68, a marker for

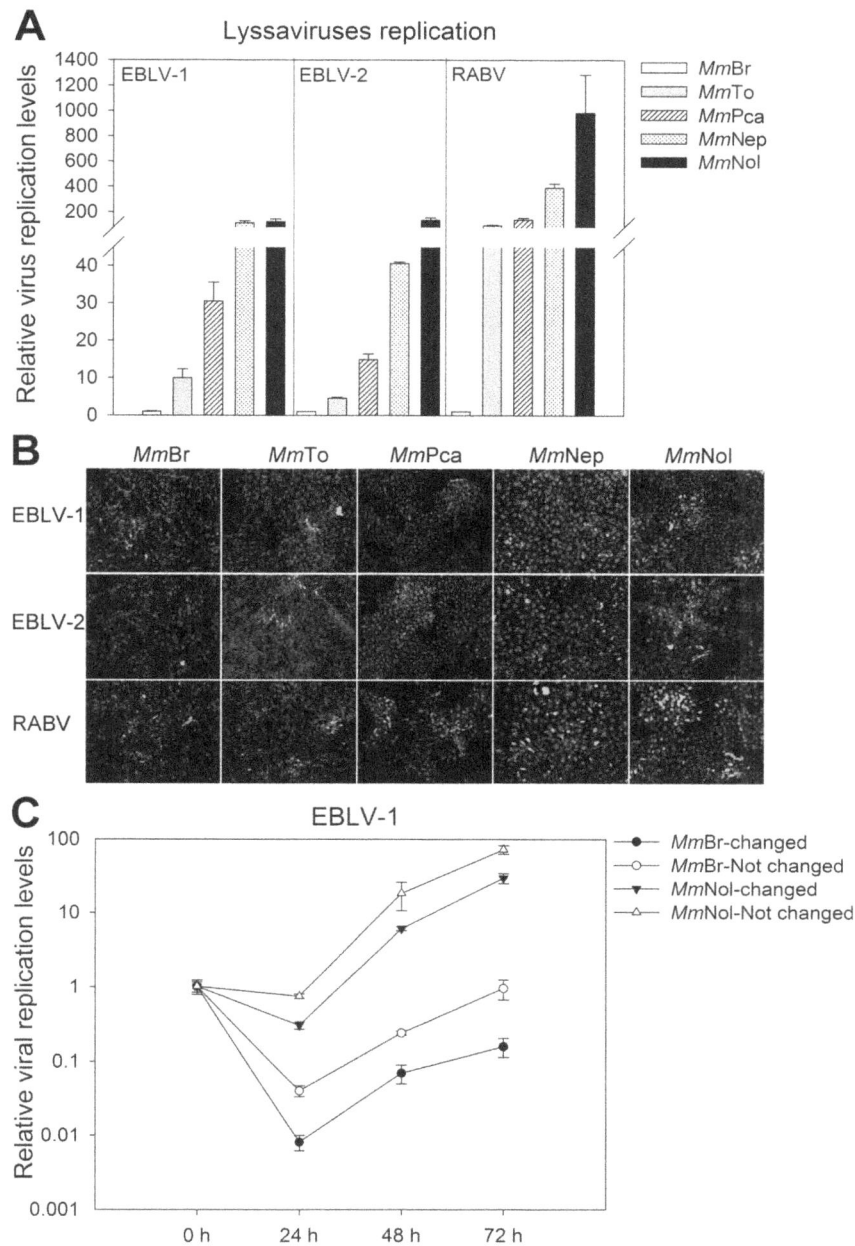

Figure 5. Susceptibility of the established immortalized *M. myotis* cell lines to lyssaviruses (EBLV-1, EBLV-2 and RABV) infection. (A) Cell lines were infected with lyssaviruses at a MOI of 0.1, and virus replication levels were measured by qRT-PCR at 24 hpi (n = 2). The viral RNA level was shown as a related fold and normalized against β-actin. The viral replication levels of EBLV-1, EBLV-2 and RABV were lowest in *Mm*Br and presented as one fold, respectively. **(B)** Viral growth was analysed by immunofluorescence using anti-rabies monoclonal antibody at 72 hpi. Green: lyssavirus infected cell, Blue: nuclei stained with DAPI. **(C)** To further confirm the susceptibility, *Mm*Br and *Mm*Nol cell lines were infected with EBLV-1 at a MOI of 0.01, and viral replication levels were measured by qRT-PCR over 72 hpi (n = 2). – changed: viral supernatant was changed with fresh medium at 1 hpi. – Not changed: viral supernatant was not changed during the whole infection process.

cells of macrophage lineage [38], and CD14, a marker expressed in activated microglia [39], were investigated by RT-PCR in different cell lines. Specific primers for CD14, CD68 and internal control β-actin were listed in Table 1. The RT-PCR was prepared according to the instructions of the one-step RT-PCR kit (Qiagen).

Statistical analysis

All data were presented as means ± S.D. Statistical significant differences were analysed by one-way ANOVA using the SPSS software package.

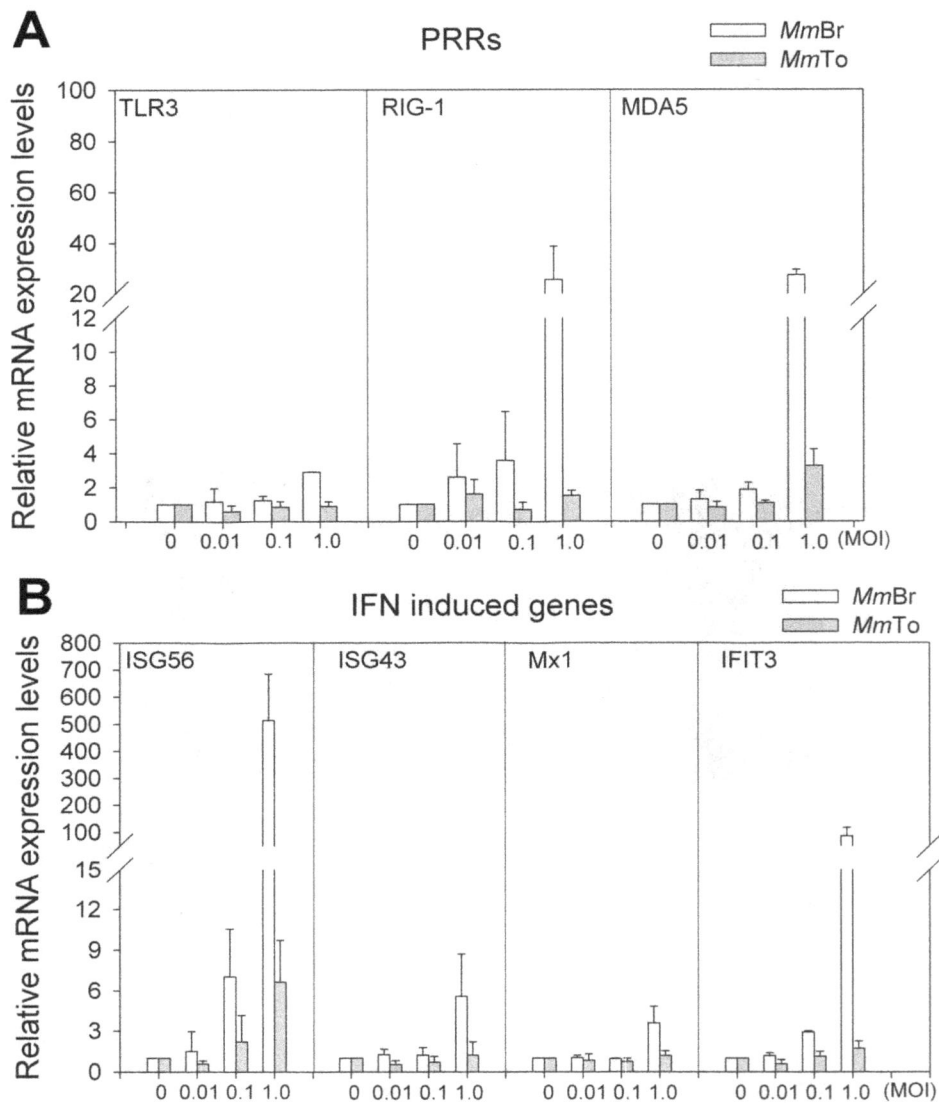

Figure 6. Comparative analysis of the expression patterns of antiviral molecules in *Mm*Br and *Mm*To during RABV infection. *Mm*Br and *Mm*To were infected with RABV at a serial MOI of 0.01, 0.1 and 1.0, respectively. (**A**) The expression patterns of PRRs: TLR3, RIG-1 and MDA5 in the infected cells were investigated by qRT-PCR at 24 hpi (n = 2). (**B**) The expression patterns of IFN induced genes: ISG56, ISG43, Mx1 and IFIT3 were measured by qRT-PCR at 24 hpi (n = 2). The expression level was shown as a related fold and normalized against β-actin. The expression level of different molecules in blank group (MOI: 0) in both cell lines *Mm*Br and *Mm*To was presented as one fold, respectively.

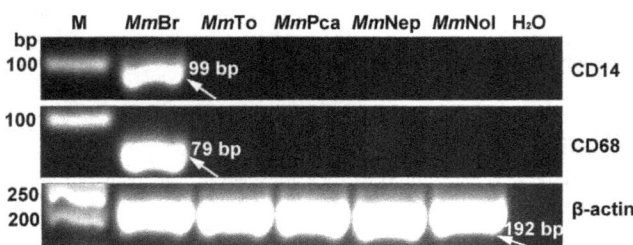

Figure 7. The mRNA expression patterns of CD14 and CD68 in immortalized *M. myotis* cell lines. Note the absence of these two monocyte lineage markers in four of the five cell lines.

Results

Permanent cell lines of different origin could be established after immortalisation

Five *M. myotis* cell lines brain (*Mm*Br), tonsil (*Mm*To), peritoneal cavity (*Mm*Pca), nasal epithelium (*Mm*Nep) and nervus olfactorius (*Mm*Nol) were successfully established by transformation with SV40T gene integrating into the chromosomal DNA. Varying cell morphologies were observed in the cell lines, with *Mm*Br, *Mm*To, *Mm*Nep and *Mm*Nol being fibroblastic-like, and *Mm*Pca being epithelial-like (Fig. 1A). The mRNA expression of SV40T antigen was detected in all cell lines (Fig. 1B). Protein level expression was confirmed in four of the five cell lines by immunofluorescence microscopy and western blot, respectively (Fig. 1C and D). In *Mm*To, the protein level of SV40 T antigen was under detectable because the transcriptional level is significantly low determined by RT-PCR (Fig. 1B). After immortalization, all five cell lines grew for more than 30 passages. The identity of the cell lines was validated by a *M. myotis* specific PCR using ND1 gene as a molecular marker. A predicted 515-bp fragment was obtained from genomic DNA of each cell line, and further confirmed by sequencing.

M. myotis permanent cell lines express major innate immune molecules

As the first step towards the characterization of the innate immune competence of different cell lines, the permanent or inducible expression of molecules involved in cell autonomous responses was examined. The PRRs: TLR3, RIG-1 and MDA5, display a various distribution pattern in different cell lines (Fig. 2). Of note, *Mm*Br has the lowest levels of TLR3, RIG-1 and MDA5 (Fig. 2). For TLR3, about 10-fold higher mRNA levels ($p<0.05$) were observed in *Mm*To, *Mm*Pca, *Mm*Nep and *Mm*Nol compared to *Mm*Br, respectively, while for MDA5 about 30-fold (*Mm*To; *Mm*Nep) or about 6-fold (*Mm*Pca; *Mm*Nol) ($p<0.05$) higher expression levels were measured (Fig. 2). Additionally, more than 200 times higher expression levels of RIG-1 were shown in other cell lines compared to *Mm*Br ($p<0.05$) (Fig. 2).

Further investigation focused on the expression of TLR3, ISG56, ISG43 and Mx1 induced by the poly I:C stimulation (Fig. 3). The obtained results indicate a 4-fold in *Mm*Br, *Mm*Nep and *Mm*Nol and 8-fold in *Mm*Pca ($p<0.05$) increase in the TLR3 expression, whilst no change in *Mm*To (Fig. 3A). All of the IFN induced genes were up-regulated to different extents in different cell lines (Fig. 3B, C, D and E). In detail, ISG56 expression increased from 19-fold in *Mm*Br to as high as more than 9000-fold in *Mm*Nol ($p<0.05$) (Fig. 3B). The expression of ISG43 ranged from 10 to 145 times more and Mx1 from 2 to 100 times more in *Mm*To and *Mm*Br, respectively (Fig. 3C and D). IFIT3 was up-regulated from 12 to 420 times more in *Mm*To and *Mm*Nol, respectively ($p<0.05$) (Fig. 3E).

M. myotis permanent cell lines display different susceptibility to lyssaviruses infection

Being a natural reservoir species, the main advantage of the permanent *M. myotis* cell lines is their susceptibility to lyssavirus infection. At an early stage of immortalization, cell lines displayed a significant susceptibility to RABV (MOI of 10 at 24 hpi) as demonstrated by the infection with GFP fused RABV. Notably, *Mm*Br exhibited considerably lower viral load compared to the other cell lines (Fig. 4). Later, all passaged immortalized cell lines showed susceptibility to EBLV-1, EBLV-2 and RABV in a different extent (Fig. 5A and B). Generally, the *Mm*Br cell line presented lower sensitivity to all three lyssaviruses (MOI of 0.1) than the other four cell lines measured by qRT-PCR at 24 hpi (Fig. 5A), and monitored by immunofluorescence at 72 hpi (Fig. 5B). Thus, the susceptibility could be ordered as *Mm*Nol and *Mm*Nep fully susceptible with a very high replication rate, *Mm*Pca and *Mm*To susceptible with a much less viral replication of EBLV-1 and 2, *Mm*Br susceptible for EBLV-1 and RABV with a very low viral replication and just single infected cells after EBLV-2 infection (Fig. 5B). The different susceptibility of the cell lines to lyssavirus infection was further confirmed by the growth kinetics of EBLV-1 in two representative models: *Mm*Br, much less susceptible and *Mm*Nol, highly susceptible (Fig. 5C).

M. myotis permanent cell lines respond differently to RABV infection

To further evaluate the cell line models for study of the different susceptibility between *Mm*Br and other cell lines, mRNA expressions of PRRs and IFN induced genes were investigated in *Mm*To and *Mm*Br after RABV infection (MOI 0.01 to 1.0). The expression of all three PRRs remained mostly unchanged in *Mm*To, while it was significantly regulated in *Mm*Br with 2-fold increased expression of TLR3, about 25-fold increased expression of RIG-1 and MDA5 at MOI of 1.0 ($p<0.05$) (Fig. 6A). A comparable expression pattern was observed for the ISG56, ISG43, Mx1 and IFIT3, which was nearly not up-regulated in *Mm*To but displayed a dose dependent increase in *Mm*Br along with the increase of MOI, especially for ISG56 and IFIT3 (Fig. 6B). ISG56 mRNA level increased from 6 to 513 times, IFIT3 from 2 to 85 times in the infected *Mm*Br ($p<0.05$).

The brain derived cell line *Mm*Br are microglia-like cells

Microglia are macrophage-like cells that are resident immune effector cells in the CNS [39]. They are activated in response to infection or injury and play a central role in immune surveillance and host defense [39]. The RT-PCR results showed that CD14 and CD68 are expressed only in *Mm*Br but not in the other four cell lines (Fig. 7). This suggested *Mm*Br is a microglia-derived cell line.

Discussion

Cell autonomous and innate immune mechanisms are the first line defenses against viral infections. This is mediated mainly by the PRRs and the machinery of the IFNs and IFN induced effector molecules [40–42]. Viral pathogens like lyssaviruses developed evasive strategies to escape these host defenses by counteracting the IFN mediated immune responses [43]. Co-evolution of the lyssaviral evading and bat's protective mechanisms resulted in an optimal balance, which protect bats as the 'natural host' from severe clinical symptoms or death. Bats, which changed very little over past 52 million years, illustrate this phenomenon very well by the resistance to lethal diseases caused by viruses in other mammals [11,21–24].To understand the specificity of host-pathogen interactions in 'natural host' like bats, studies in bats have to be performed. However, due to the strict protection of the endangered European bat species, *in vitro* models have to be used. In this study, we successfully established five *M. myotis* cell lines derived from neural and immune related tissues. To ensure the suitability of these cell lines to analyze virus-host cell interaction, the susceptibility to the infection as well as the presence of corresponding defensive pathways have to be confirmed.

First, the existence of the viral sensors TLR3, RIG-1 and MDA5 in these permanent cell lines suggests a capacity of these cell lines to sense a broad range of RNA viruses. The increased expression of dsRNA receptor TLR3 and IFN induced genes ISG56, ISG43, Mx1 and IFIT3 after stimulation with poly I:C mimicking a viral infection indicates that these cell lines can be used as effective *in vitro* models to study the bat's innate immune responses to virus infection [32,44]. Furthermore, to serve as valuable models would be a varying susceptibility of such cell lines to infection by lyssaviruses. In the present study, different susceptibility observed in different *M. myotis* cell lines using EBLV-1, EBLV-2 and RABV might be related to the different capacity of the cell lines to produce antiviral mediators and control the infection. Moreover, the strong difference in the susceptibility to RABV infection between *Mm*Br and other cell lines provides a unique opportunity for comparative investigations of cell autonomous and innate immune mechanisms in a reservoir host. In addition to the lyssaviruses, the other member from the *Rhabdoviridae* family, like vesicular stomatitis virus (VSV) can also be investigated by using these models in the future studies. Preliminary results indicate a correlation between the observed varying susceptibility and the ability to up-regulate the PRRs and the IFN induced genes. Emerging evidences have shown that PRRs play pivotal roles in antiviral immunity in the CNS [45]. In the brain derived cell line *Mm*Br, the high up-regulations of RIG-1 and MDA5 revealed activation of RIG-I-like receptor pathway during RABV infection. As previously reported, RIG-1 is a major PRR to induce IFN in the RABV infected cells, and MDA5 may function to sustain the IFN induction [46]. The increased expressions of IFN induced genes: ISG56, ISG43, Mx1 and IFIT3 in *Mm*Br indicate that the production of IFN was induced by activated RIG-1 and MDA5. In contrast, the low expression level of TLR3 implies a vague involvement of TLR3 in anti-RABV infection immunity or resistance. It was shown that TLR3 participated in and benefited the RABV pathogenesis in human neuron cells [47]. However, the roles of TLR3 during RABV infection in bats need further investigations. Importantly, the significant expression patterns of PRRs observed in presented cell line models provide an access to this issue *in vitro*. To reach a successful infection, the viruses must overcome the barriers of innate immune system. It was reported that IFN production and signaling pathways were antagonized in *P. alecto* cell lines under henipavirus infection [32]. Similarly, a recent study showed limited expressions of type I IFNs and IFN induced genes during lyssaviruses infection in an *E. serotinus* brain cell line [33]. A correlation between the low viral load and high expression levels of IFN induced genes in *Mm*Br contrasts to the high viral load and a silent expression pattern of antiviral effectors in *Mm*To, providing an evidence of a countermeasure to IFN system by lyssavirus in the peripheral tissue versus a protective mechanism to infection in the brain tissue of bats. Microglial cells are one of the major cell populations in the brain tissue. Additionally, comparing to neurons, they can be infected by different RABV strains to a lesser extent [48,49]. The presence of CD14 and CD68 as well as the anti-lyssavirus responses in *Mm*Br support a microglia-like

feature of *Mm*Br in the CNS. It was reported that a mouse microglia cell line can activate strong innate immunity during RABV infection [50]. The robust immune responses of the microglia-like *Mm*Br demonstrated a critical role of microglia in the anti-rabies defense in bat's CNS. In addition to the function of microglia, the clearance of infected viruses in the CNS requires systematical responses through the complex interactions of different brain resident cells. Herein, the establishment and identification of a microglia-like cell model is a first step towards understanding of the complex reactions of CNS in response to lyssavirus infection in the reservoir species. Overall, this preliminary study using established cell lines implies that immune mechanisms that control the virus replication are present in the CNS of bats. It seems that the ability to control the pathogenic RABV replication via IFN system in the CNS contributes to the asymptomatic outcome in bats.

In conclusion, the established immortalized cell lines from the European bat *M. myotis* displaying a variable susceptibility to different lyssaviruses will serve as a useful model to study virus-host interactions and antiviral resistance mechanisms in the 'natural' *Lyssavirus* host. This study provides a preliminary insight into the antiviral innate immunity correlated to CNS against neurotropic viruses infection in bats.

Acknowledgments

We would like to thank Dr. Thomas Müller and Dr. Conrad M. Freuling for the EBLV-1 and EBLV-2 strains, Dr. Stefan Finke for the RABV strain, Matthias Lenk for the pRSVAg1 plasmid, and Dr. Miroslav Kovarik from the Administration of the Moravian Karst Protected Landscape Area (Nature Conservation Agency of the Czech Republic) for cooperation in obtaining the *M. myotis* specimen.

Author Contributions

Conceived and designed the experiments: XH TK BK. Performed the experiments: XH YZ JP HB BK. Analyzed the data: XH YZ BK. Contributed reagents/materials/analysis tools: XH JP HB JZ. Wrote the paper: XH TK JP BK.

References

1. Teeling EC, Madsen O, Van den Bussche RA, de Jong WW, Stanhope MJ, et al. (2002) Microbat paraphyly and the convergent evolution of a key innovation in Old World rhinolophoid microbats. Proc Natl Acad Sci U S A 99: 1431–1436.
2. Simmons NB, Seymour KL, Habersetzer J, Gunnell GF (2008) Primitive Early Eocene bat from Wyoming and the evolution of flight and echolocation. Nature 451: 818–821.
3. Calisher CH, Childs JE, Field HE, Holmes KV, Schountz T (2006) Bats: important reservoir hosts of emerging viruses. Clin Microbiol Rev 19: 531–545.
4. Lau SK, Woo PC, Li KS, Huang Y, Tsoi HW, et al. (2005) Severe acute respiratory syndrome coronavirus-like virus in Chinese horseshoe bats. Proc Natl Acad Sci U S A 102: 14040–14045.
5. Halpin K, Hyatt AD, Plowright RK, Epstein JH, Daszak P, et al. (2007) Emerging viruses: coming in on a wrinkled wing and a prayer. Clin Infect Dis 44: 711–717.
6. Mackenzie JS, Field HE (2004) Emerging encephalitogenic viruses: lyssaviruses and henipaviruses transmitted by frugivorous bats. Arch Virol Suppl: 97–111.
7. Dobson AP (2005) Virology. What links bats to emerging infectious diseases? Science 310: 628–629.
8. Li W, Shi Z, Yu M, Ren W, Smith C, et al. (2005) Bats are natural reservoirs of SARS-like coronaviruses. Science 310: 676–679.
9. Muller MA, Paweska JT, Leman PA, Drosten C, Grywna K, et al. (2007) Coronavirus antibodies in African bat species. Emerg Infect Dis 13: 1367–1370.
10. Shi Z, Hu Z (2008) A review of studies on animal reservoirs of the SARS coronavirus. Virus Res 133: 74–87.
11. Leroy EM, Kumulungui B, Pourrut X, Rouquet P, Hassanin A, et al. (2005) Fruit bats as reservoirs of Ebola virus. Nature 438: 575–576.
12. McElhinney LM, Marston DA, Leech S, Freuling CM, van der Poel WH, et al. (2013) Molecular epidemiology of bat lyssaviruses in europe. Zoonoses Public Health 60: 35–45.
13. Arechiga Ceballos N, Vazquez Moron S, Berciano JM, Nicolas O, Aznar Lopez C, et al. (2013) Novel lyssavirus in bat, Spain. Emerg Infect Dis 19: 793–795.
14. Knobel DL, Cleaveland S, Coleman PG, Fevre EM, Meltzer MI, et al. (2005) Re-evaluating the burden of rabies in Africa and Asia. Bulletin of the World Health Organization 83: 360–368.
15. Lumio J, Hillbom M, Roine R, Ketonen L, Haltia M, et al. (1986) Human rabies of bat origin in Europe. Lancet 1: 378.
16. Johnson N, Vos A, Freuling C, Tordo N, Fooks AR, et al. (2010) Human rabies due to lyssavirus infection of bat origin. Veterinary microbiology 142: 151–159.
17. Stantic-Pavlinic M (2005) Public health concerns in bat rabies across Europe. Euro surveillance : bulletin europeen sur les maladies transmissibles = European communicable disease bulletin 10: 217–220.
18. Nathwani D, McIntyre PG, White K, Shearer AJ, Reynolds N, et al. (2003) Fatal human rabies caused by European bat Lyssavirus type 2a infection in Scotland. Clin Infect Dis 37: 598–601.
19. Wibbelt G, Moore MS, Schountz T, Voigt CC (2010) Emerging diseases in Chiroptera: why bats? Biol Lett 6: 438–440.
20. Harris SL, Brookes SM, Jones G, Hutson AM, Fooks AR (2006) Passive surveillance (1987 to 2004) of United Kingdom bats for European bat lyssaviruses. Vet Rec 159: 439–446.
21. Middleton DJ, Morrissy CJ, van der Heide BM, Russell GM, Braun MA, et al. (2007) Experimental Nipah virus infection in pteropid bats (Pteropus poliocephalus). J Comp Pathol 136: 266–272.
22. Williamson MM, Hooper PT, Selleck PW, Westbury HA, Slocombe RF (2000) Experimental hendra virus infectionin pregnant guinea-pigs and fruit Bats (Pteropus poliocephalus). J Comp Pathol 122: 201–207.

23. Baker ML, Schountz T, Wang LF (2013) Antiviral immune responses of bats: a review. Zoonoses Public Health 60: 104–116.
24. Williamson MM, Hooper PT, Selleck PW, Gleeson LJ, Daniels PW, et al. (1998) Transmission studies of Hendra virus (equine morbillivirus) in fruit bats, horses and cats. Aust Vet J 76: 813–818.
25. Johnson N, Vos A, Neubert L, Freuling C, Mansfield KL, et al. (2008) Experimental study of European bat lyssavirus type-2 infection in Daubenton's bats (Myotis daubentonii). J Gen Virol 89: 2662–2672.
26. Wynne JW, Wang LF (2013) Bats and viruses: friend or foe? PLoS Pathog 9: e1003651.
27. McColl KA, Chamberlain T, Lunt RA, Newberry KM, Middleton D, et al. (2002) Pathogenesis studies with Australian bat lyssavirus in grey-headed flying foxes (Pteropus poliocephalus). Aust Vet J 80: 636–641.
28. Crameri G, Todd S, Grimley S, McEachern JA, Marsh GA, et al. (2009) Establishment, immortalisation and characterisation of pteropid bat cell lines. PLoS One 4: e8266.
29. Biesold SE, Ritz D, Gloza-Rausch F, Wollny R, Drexler JF, et al. (2011) Type I interferon reaction to viral infection in interferon-competent, immortalized cell lines from the African fruit bat Eidolon helvum. PLoS One 6: e28131.
30. Mourya DT, Lakra RJ, Yadav PD, Tyagi P, Raut CG, et al. (2013) Establishment of cell line from embryonic tissue of Pipistrellus ceylonicus bat species from India & its susceptibility to different viruses. Indian J Med Res 138: 224–231.
31. Eckerle I, Ehlen L, Kallies R, Wollny R, Corman VM, et al. (2014) Bat airway epithelial cells: a novel tool for the study of zoonotic viruses. PLoS One 9: e84679.
32. Virtue ER, Marsh GA, Baker ML, Wang LF (2011) Interferon production and signaling pathways are antagonized during henipavirus infection of fruit bat cell lines. PLoS One 6: e22488.
33. He X, Korytar T, Schatz J, Freuling CM, Muller T, et al. (2014) Anti-Lyssaviral Activity of Interferons kappa and omega from the Serotine Bat, Eptesicus serotinus. J Virol 88: 5444–5454.
34. Zukal J, Bandouchova H, Bartonicka T, Berkova H, Brack V, et al. (2014) White-nose syndrome fungus: a generalist pathogen of hibernating bats. PLoS One 9: e97224.
35. Heinsohn S, Golta S, Kabisch H, zur Stadt U (2005) Standardized detection of Simian virus 40 by real-time quantitative polymerase chain reaction in pediatric malignancies. Haematologica 90: 94–99.
36. Schatz J, Freuling CM, Auer E, Goharriz H, Harbusch C, et al. (2014) Enhanced passive bat rabies surveillance in indigenous bat species from Germany - a retrospective study. PLoS Negl Trop Dis 8: e2835.
37. Livak KJ, Schmittgen TD (2001) Analysis of relative gene expression data using real-time quantitative PCR and the 2(-Delta Delta C(T)) Method. Methods 25: 402–408.
38. Holness CL, Simmons DL (1993) Molecular cloning of CD68, a human macrophage marker related to lysosomal glycoproteins. Blood 81: 1607–1613.
39. Rock RB, Gekker G, Hu S, Sheng WS, Cheeran M, et al. (2004) Role of microglia in central nervous system infections. Clin Microbiol Rev 17: 942–964, table of contents.
40. Takaoka A, Yanai H (2006) Interferon signalling network in innate defence. Cell Microbiol 8: 907–922.
41. Levy DE, Marie IJ, Durbin JE (2011) Induction and function of type I and III interferon in response to viral infection. Curr Opin Virol 1: 476–486.
42. Honda K, Yanai H, Takaoka A, Taniguchi T (2005) Regulation of the type I IFN induction: a current view. Int Immunol 17: 1367–1378.
43. Rieder M, Conzelmann KK (2009) Rhabdovirus evasion of the interferon system. J Interferon Cytokine Res 29: 499–509.
44. Zhou P, Cowled C, Todd S, Crameri G, Virtue ER, et al. (2011) Type III IFNs in pteropid bats: differential expression patterns provide evidence for distinct roles in antiviral immunity. J Immunol 186: 3138–3147.
45. Carty M, Reinert L, Paludan SR, Bowie AG (2013) Innate antiviral signalling in the central nervous system. Trends Immunol.
46. Rieder M, Conzelmann KK (2011) Interferon in rabies virus infection. Adv Virus Res 79: 91–114.
47. Menager P, Roux P, Megret F, Bourgeois JP, Le Sourd AM, et al. (2009) Toll-like receptor 3 (TLR3) plays a major role in the formation of rabies virus Negri Bodies. PLoS Pathog 5: e1000315.
48. Nakamichi K, Saiki M, Sawada M, Takayama-Ito M, Yamamuro Y, et al. (2005) Rabies virus-induced activation of mitogen-activated protein kinase and NF-kappaB signaling pathways regulates expression of CXC and CC chemokine ligands in microglia. J Virol 79: 11801–11812.
49. Ray NB, Power C, Lynch WP, Ewalt LC, Lodmell DL (1997) Rabies viruses infect primary cultures of murine, feline, and human microglia and astrocytes. Arch Virol 142: 1011–1019.
50. Zhao P, Yang Y, Feng H, Zhao L, Qin J, et al. (2013) Global gene expression changes in BV2 microglial cell line during rabies virus infection. Infect Genet Evol 20: 257–269.

14

Highly Adaptable Triple-Negative Breast Cancer Cells as a Functional Model for Testing Anticancer Agents

Balraj Singh, Anna Shamsnia[¤a], Milan R. Raythatha[¤a], Ryan D. Milligan, Amanda M. Cady[¤b], Simran Madan[¤c], Anthony Lucci*

Department of Surgical Oncology, and Morgan Welch Inflammatory Breast Cancer Research Program and Clinic, The University of Texas MD Anderson Cancer Center, Houston, Texas, United States of America

Abstract

A major obstacle in developing effective therapies against solid tumors stems from an inability to adequately model the rare subpopulation of panresistant cancer cells that may often drive the disease. We describe a strategy for optimally modeling highly abnormal and highly adaptable human triple-negative breast cancer cells, and evaluating therapies for their ability to eradicate such cells. To overcome the shortcomings often associated with cell culture models, we incorporated several features in our model including a selection of highly adaptable cancer cells based on their ability to survive a metabolic challenge. We have previously shown that metabolically adaptable cancer cells efficiently metastasize to multiple organs in nude mice. Here we show that the cancer cells modeled in our system feature an embryo-like gene expression and amplification of the fat mass and obesity associated gene *FTO*. We also provide evidence of upregulation of *ZEB1* and downregulation of *GRHL2* indicating increased epithelial to mesenchymal transition in metabolically adaptable cancer cells. Our results obtained with a variety of anticancer agents support the validity of the model of realistic panresistance and suggest that it could be used for developing anticancer agents that would overcome panresistance.

Editor: Gokul M. Das, Roswell Park Cancer Institute, United States of America

Funding: This study was supported by a State of Texas Rare and Aggressive Breast Cancer Research Program Grant, Breast Cancer Metastasis Research Program at MD Anderson Cancer Center, and CA16672 (Core) from the National Institutes of Health. The funders had no role in study design, data collection and analysis, decision to publish, or preparation of the manuscript.

Competing Interests: The authors have declared that no competing interests exist.

* Email: alucci@mdanderson.org

¤a Current address: Texas A&M Health Science Center College of Medicine, Bryan, Texas, United States of America
¤b Current address: UT Health Science Center School of Medicine, San Antonio, Texas, United States of America
¤c Current address: Baylor College of Medicine, Houston, Texas, United States of America

Introduction

Our understanding of cancer has advanced tremendously over the last four decades. However, translation of this knowledge into clinical applications to improve treatment outcomes has been slow, particularly for solid tumors. The difficulty stems in large part from the fact that only rare cancer cells (often representing as little as 0.001% of the total cell population) truly drive the disease, particularly metastasis [1–4]. These rare special cells are akin to Olympic decathlon winners; such cells may also be the cause of panresistance (resistance to all existing therapies), often seen in patients with advanced disease [5]. The difficulties in overcoming panresistance are best understood in the context of the mechanisms of tumor heterogeneity. Previous attempts to address the tumor heterogeneity problem by isolating important subpopulations of cancer cells using a variety of methods achieved various degrees of success. These methods include 1) *in vitro* selection based on the ability of cancer cells to invade the basement membrane, 2) *in vitro* selection based on the ability of cancer cells to grow in soft or hard agar, 3) selection of cancer cells based on their ability to colonize and grow at metastasis sites in nude mice, and 4) more recently, enrichment of cancer stem cells on the basis of specific cell surface markers [6–8]. Here, we describe a new strategy for delving deeper into the roots of cancer. Our strategy is based on the hypothesis that decathlon winner cancer cells/roots can resist severe metabolic challenges and this ability can be employed for selecting them. Increasingly, metabolic state is viewed not merely as a recipient of aberrant signaling in cancer but rather as an important driving factor in oncogenesis [e.g., see reference 9]. We applied this knowledge of a linkage between metabolic state and regulatory state to isolate rare cancer cells whose adaptability can drive metastasis.

Since current methods of preclinically evaluating new drug candidates poorly predict treatment response in cancer patients, we are developing a new strategy to test potential therapeutic agents with the goal of better predicting response in patients. The strategy consists of three elements, all aimed at improving the likelihood of accurately determining whether a test therapy can eradicate the roots of a therapy-resistant cancer: 1) selecting a cell line for optimal modeling of mutations and other features that drive therapy resistance in patients, 2) choice of body-like selection strategy to eliminate most cancer cells that would die under nutrient starvation, and 3) equally important, evaluating therapies in long-term assays to accurately predict response in the clinic.

Our strategy is focused on modeling cancer roots that are highly abnormal and highly adaptable. The rationale is that if a test therapy is effective against such roots, it is more likely to overcome therapy resistance and succeed in treating cancer patients. Here, we describe the application of this cell-based approach to triple-negative breast cancer (TNBC), which lacks expression of estrogen receptor, expression of progesterone receptor, and HER2 gene amplification. TNBC is an aggressive and heterogeneous breast cancer, with considerable overlap with basal-like breast cancers. The *TP53* tumor suppressor gene is commonly mutated in TNBC [10,11]. The germ line mutation in the BRCA1 gene is also associated with TNBC. These mutations indicate that TNBC is a disease of genome instability. TNBC-like molecular features have been observed in other cancers, e.g., high-grade serous ovarian carcinoma [11]. We chose to model the roots of TNBC with an intention to contribute to drug discovery efforts against cancers that are very heterogeneous and adaptable.

To model the roots of therapy-resistant TNBC, we chose the SUM149 cell line because it has the following characteristics: 1) origination from a very aggressive human triple-negative inflammatory breast cancer (TN-IBC), 2) several mutations that are often observed in TNBC, including a gain-of-function M237I mutation in the *TP53* gene that may affect protein stability and function similar to the hotspot R175H mutation [12,13], a defective RB pathway due to p16 deletion, a micro-scale genomic deletion in *PTEN*, 3) *BRCA1* mutation, making the cell line relevant for patients with a defective *BRCA1* gene [14,15], and 4) one of the highest rates of metastatic ability among TNBC cell lines, as indicated by the ability of the cells to reproducibly metastasize to distant organs from fat pad xenografts in nude mice [16,17]. It is important to stress that IBC resembles non-inflammatory breast cancers in terms of gene expression patterns; both TNBC and TN-IBC are heterogeneous subgroups [18,19]. Therefore, the TN-IBC cell line SUM149 is a good model for aggressive, therapy-resistant TNBC. After considering many approaches for isolating the TNBC cells responsible for therapy resistance, we hypothesized that a severe metabolic challenge would select rare cancer cells that can survive a variety of challenges and thus drive metastasis and panresistance. We reasoned that the therapy-resistant cancer cells selected by this approach may be more useful for testing new anticancer agents against the therapy-resistant disease than the resistant cancer cells selected to survive a specific chemotherapeutic or targeted agent. In support of this approach, we recently reported that lack of glutamine (Gln) in culture medium killed more than 99.99% of SUM149 cells. However, the rare cells (0.01%; termed MA for metabolically adaptable) that survive and grow are capable of surviving additional metabolic challenge, i.e., lack of glucose, they are resistant to chemotherapeutic drugs doxorubicin and paclitaxel, and they efficiently metastasize to multiple organs-lungs, liver, brain, and skin-from fat pad xenografts in nude mice [20]. Our results indicated that a robust *in vitro* selection method can be more useful and versatile for selecting cancer cells that drive metastasis than is the commonly used method of selecting cells from xenografts in nude mice, the latter often being limited by the tissue specificity of cancer cell colonization depending on the site of inoculation.

To determine the potential usefulness of *in vitro*-derived MA cells for testing anticancer agents, we investigated the cells' molecular and functional characteristics. Our data from gene expression arrays and from comparative genomic hybridization (CGH) arrays revealed several mechanisms in MA cells for generating tumor heterogeneity. These results also revealed that obesity-related molecular pathways, which have evolved to serve a beneficial role in helping organisms survive under harsh metabolic challenges, may be exploited by the rare cancer cells for survival under metabolic challenges. We evaluated the effects of a variety of potential anticancer compounds simultaneously on MA cells and the parental SUM149 cell line. The results validated the MA cells as a good model of panresistance, and that the model can be used for discovering effective anticancer agents.

Results and Discussion

Metabolic Adaptability of MA Cells

The SUM149 cell line is highly dependent on Gln for cell survival and growth. As reported previously, more than 99.99% of SUM149 cells die upon Gln withdrawal [20]. We obtained only 10–20 colonies after plating half a million cells into a 10-cm culture dish (see Fig. 1A for a representative result). In view of the tremendous cellular heterogeneity, it is likely that the 10–20 cells that survived a severe metabolic challenge (prolonged lack of Gln) and yielded colonies are not genomically and epigenomically identical. Significantly, once selected in this manner, MA cells could grow in a Gln-free medium indefinitely. Furthermore, even though initially selected for their ability to survive a lack of Gln, MA cells are also capable of surviving a prolonged lack of glucose

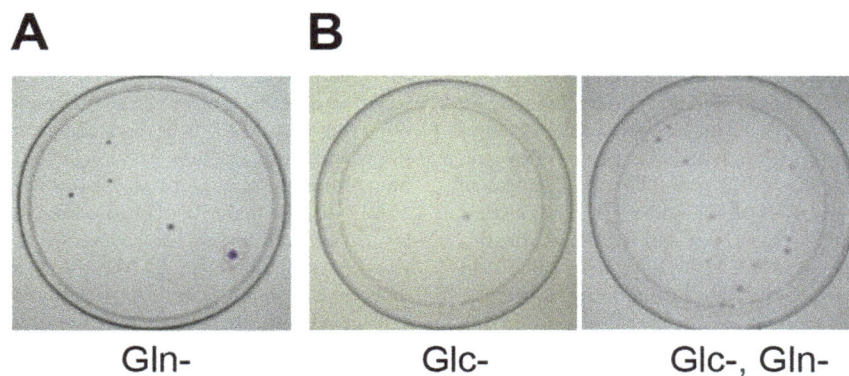

A   **B**

Gln-   Glc-   Glc-, Gln-

Figure 1. Selection of Rare Cancer Cells That Survive Lack of Glutamine and Glucose. (A) Half a million SUM149-Luc cells were plated in a 10-cm dish. The next day, the medium was changed to a medium containing dialyzed FBS and lacking glutamine. After growth for 34 days in the Gln-deficient medium, we stained the colonies with crystal violet. (B) Half a million SUM149-FP76 cells were plated in a 10-cm dish. The next day, the medium was changed to a medium containing dialyzed FBS lacking glucose (left dish) or lacking both glucose and glutamine (right dish). After 15 days, the media were replaced with complete medium. We stained the colonies with crystal violet after 18 days in complete medium.

[20]. To determine by an alternative approach whether the parental cell line contains rare cells that can survive a lack of both Gln and glucose, we incubated SUM149-FP76 cells (initially cultured from a fat pad xenograft in nude mouse) in Gln- and glucose-deficient medium for 15 days before switching them back to complete medium (all cells eventually die upon glucose withdrawal). We stained the resulting colonies after 18 days, and the staining showed that the number of colonies was not lower than the number of colonies obtained after withdrawal of only glucose (compare the 2 plates in Fig. 1B).

Although some SUM149-Luc cancer cells could survive without glucose or even without glucose and glutamine for approximately 21 days, the experiments involving a lack of glucose are technically more complicated than those involving a lack of glutamine. Since cells can only survive but not grow into colonies without glucose, we must provide glucose to determine the number of surviving cells that form colonies. The number (and presumably the adaptability properties) of surviving cells varies depending upon the number of days glucose was missing (data not shown). In contrast, selection of rare cells surviving the lack of glutamine is easy and reproducible in our system since the surviving cells form colonies within 4–5 weeks. Furthermore, the cells selected in this manner can indefinitely grow in a medium lacking glutamine [20]. Therefore, we chose to focus on SUM149-Luc cells selected in the absence of glutamine for this study.

Gene Expression and Chromosomal Analysis of MA Cells

To gain insight into the characteristic of MA cells that enables them to survive severe metabolic challenge, i.e., prolonged lack of glutamine and other challenges [20], we used gene expression microarrays and CGH arrays to compare these cells with the parental SUM149-Luc (luciferase-transfected) cells. We analyzed two independently selected cell populations, one from 0.5 million parental cells (designated MA1) and one from 1 million parental cells (designated MA2). We generated lists of 2843 gene probes for MA1 cells and 8521 gene probes for MA2 cells that detected significantly altered RNA expression (decreased or increased) relative to RNA expression in the parental cell line (some RNAs were detected with more than one probe) (Tables S1 and S2). Of the significant gene expression changes in MA1 cells, 61% were shared with MA2 cells (Fig. 2A). To gain insight into the significance of the gene expression changes, we performed an Ingenuity Systems pathway analysis. We chose to analyze data from MA2 cells because they had many more changes in gene expression than MA1 cells did. The gene expression profile of significantly altered genes in MA2 cells revealed several associated network functions and molecular and cellular functions that could potentially be important in TNBC (Table S3). Strikingly, network functions with the highest scores included embryonic development, indicating a linkage between metabolic adaptability and overall (embryo-like) adaptability in MA cells. The core analysis of gene expression in MA1 cells also yielded embryonic development as one of the top network functions (not shown). Based on this analysis, the other functions that are likely altered in MA2 cells include lipid metabolism, carbohydrate metabolism, molecular transport, cellular movement, organ development, etc. (Table S3).

We had previously observed that MA cells possess mesenchymal features, including the expression of cadherin 11 and vimentin [20]. On the basis of that finding, we searched for gene expression information for regulators of epithelial-to-mesenchymal transition (EMT) and found that MA cells significantly overexpress zinc finger E-box binding homeobox 1 (ZEB1), a key transcriptional regulator of the mesenchymal phenotype [21]. Furthermore, expression of grainy head-like 2 (GRHL2) RNA, which encodes a

transcriptional regulator of epithelial phenotype [21], in MA cells is significantly lower than in the parental cell line (Table S2). We determined by Western blotting that the protein levels of ZEB1 and GRHL2 are consistent with the gene expression data, i.e., ZEB1 protein is up-regulated and GRHL2 protein is down-regulated in MA cells (Fig. 2B). ZEB1 and GRHL2 repress each other, and final phenotype- epithelial or mesenchymal- is dependent on multiple inputs. These results suggest that MA cells possess characteristics of EMT. Cancer cells that have undergone EMT are enriched in cancer stem cell properties and they possess flexible epigenetic state [22,23]. A high degree of EMT would enable MA cells to generate embryo-like tumor heterogeneity, a prediction supported by the results of gene expression analysis (Table S3).

Our results support a model in which different subpopulations of cancer cells have different levels of metabolic adaptability-minor subpopulations of stem-like non-proliferating cells that can survive severe metabolic challenges and major subpopulations of proliferative, less adaptable non-stem-like cells (Fig. 2D). A severe metabolic challenge, such as the one used for selecting MA cells, which eliminated 99.99% of the cells, selects for highly adaptable, stem-like cancer cells (Fig. 2D). The severity of the metabolic challenge would increase the level of adaptability properties of the cancer cells.

We also examined whether MA cells have a drug-tolerant epigenetic state by measuring methylation and acetylation of histone H3. Significant reductions in both trimethylation at Lys-4 and acetylation at Lys-14 in histone H3 are associated with a drug-tolerant state in cancer [24]. We analyzed the protein modifications by western blotting and quantitated them by measuring band intensities and normalizing them based on equal histone H3 protein in samples. In this manner, we observed that both of these protein modifications were less common in MA cells growing without or with Gln than in the parental cell line (Fig. 2C). These results indicate that MA cells selected in the absence of Gln possess a drug tolerant epigenetic state. Furthermore, the drug tolerant epigenetic state is maintained after the MA cells are switched back to Gln-containing medium (Fig. 2C). These results further support our hypothesis that the adaptability of the rare MA cells may extend beyond metabolism, with drug tolerance being an integral component of this broad adaptability.

We analyzed chromosomal gains and losses in MA cells as compared to the parental SUM149-Luc cell line by CGH arrays. The CGH array analysis revealed a large number of deletions in MA cells, affecting all chromosomes, some more than others (see Fig. 3A for a graphic summary of chromosomal gains and losses in MA1 and MA2 variants). There were multiple amplifications as well, but amplifications were less frequent than deletions. We also observed that MA1 cells had significantly fewer chromosomal losses and gains than the MA2 cells did (compare Tables S4 and S5). Although some gains and losses were common to MA1 and MA2 cells, the majority of chromosomal changes were not shared by the two cell types (Tables S4–S6). It is not clear at this time whether this difference in MA1 and MA2 cells is due to the difference in input cells, e.g., 0.5 million versus 1 million cells, respectively, for selecting them. Alternatively, some differences could be due to passage number differences between MA1 and MA2 cultures. These results indicate a high degree of chromosomal changes associated with the selection of MA cells. They further suggest that the rare cells selected under a metabolic challenge are likely to be genetically heterogeneous.

Focusing on the chromosomal changes common to MA1 and MA2 cells, we found that the amplified regions involve only three chromosomes: chromosomes 11, 13 and 16 (Table S6). The

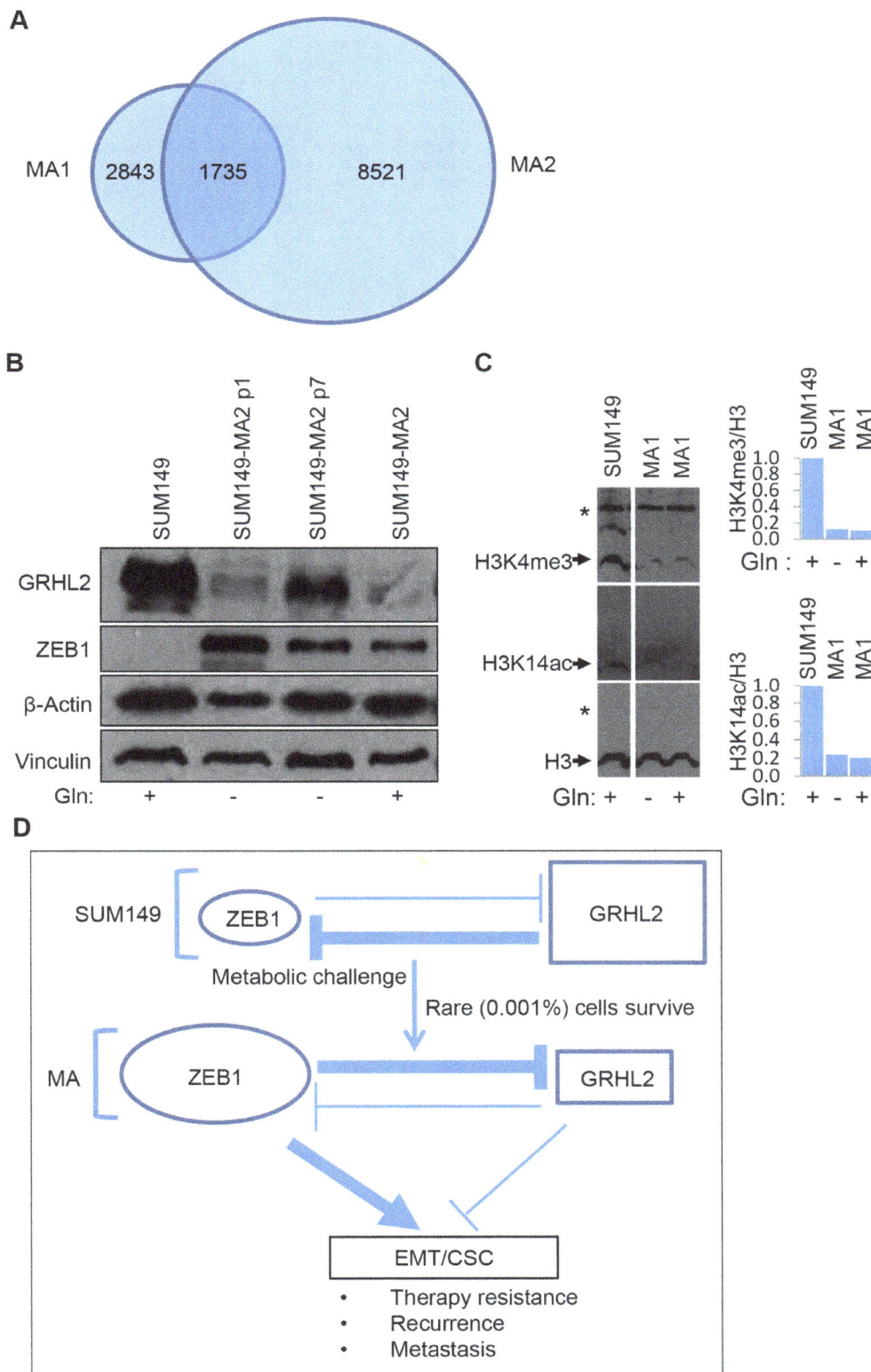

Figure 2. Evidence of a High EMT and a Drug-Tolerant State in MA Cells. (A) Overlap in gene expression changes between MA1 and MA2 variants. The Venn diagram depicts the number of gene probes that detected significantly higher or lower RNA levels in MA1 and MA2 cells than in the parental cell line. The common 1735 gene probes that detected the shared alterations between MA1 and MA2 cells were identified in Microsoft Excel by combining the spreadsheets (Tables S1 and S2) and then searching for duplicate primary sequence names. (B) We analyzed by Western blotting ZEB1 and GRHL2 proteins in MA2 variants and the parental cell line SUM149-Luc (labeled SUM149). After selection in Gln-free medium, MA2 variants were cultured in a medium without or with Gln as indicated at the bottom. p1 and p7 represent passages in Gln-free medium. The parental cell line was cultured in a medium with dialyzed FBS for six passages prior to preparation of lysates for Western blots to match the MA cells. For the

rightmost lane, MA2 cells were cultured in Gln-containing medium for four passages prior to preparation of the lysate. We blotted β-actin and vinculin as controls; β-actin is reduced in MA cells grown without Gln, as reported previously [20]. (C) We analyzed by western blotting trimethylation at lysine 4 of histone H3 and acetylation at lysine 14 of histone H3 in MA1 cells and the parental cell line SUM149-Luc (labeled SUM149). We blotted total histone H3 as a control. The star marks a background band that serves as an additional gel loading control in western blots. We cultured MA cells continuously in Gln- medium for 11 passages after selection (Gln-, middle lane), or we cultured MA cells in Gln- medium for 9 passages and then switched them back to Gln+ medium for 4 passages (Gln+, right lane). We measured the relative intensities of histone H3 bands detected as K4me3 and K14ac modified forms and normalized them by dividing with the intensity of total H3 band from the corresponding samples. The normalized data are shown in a graphical form on the right. (D) A model of therapy resistance in MA cells. According to our data, the balance of epithelial versus mesenchymal phenotypes, which is determined by the GRHL2/ZEB1 ratio, is shifted toward mesenchymal phenotype in MA variants. The sizes of ovals and rectangles and the widths of arrows indicate relative levels/strengths. See also Tables S1 and S2.

amplified regions harbor a total of 14 genes (*SBF2, MTUS2, ANKRD26P1, SHCBP1, CBLN1, C16orf78, ZNF423, TMEM188, HEATR3, PAPD5, RBL2, AKTIP, RPGRIP1L*, and *FTO*; see Tables S3–S5 for a list of all genes deleted or amplified in MA1 and MA2 cells and the gene alterations common to MA1 and MA2 cells). The gene that caught our attention in this analysis is *FTO*, one of the four genes (with *RBL2, AKTIP,* and *RPGRIP1L*) amplified on chromosome 16q12.2, for reasons described below.

FTO Gene Amplification in MA Cells

FTO stands for fat mass and obesity-associated protein. It is an alpha-ketoglutarate-dependent dioxygenase belonging to the AlkB family. RNA demethylase activity of FTO is likely to regulate the stability and function of mRNAs [25]. FTO plays an important role in nutrient sensing [26]. It coordinates food intake and consumption in the body. Lack of nutrients, as during starvation, leads to induction of FTO protein in multiple tissues [27]. Certain variants of the *FTO* gene correlate with obesity in humans. These observations suggest that FTO protein plays a key role in organismal survival in nutritionally harsh environments, and inappropriate activation of FTO-mediated protein networks in the absence of metabolic challenges contributes to obesity. Obesity is a commonly observed feature in breast cancer, especially in TNBC and TN-IBC, and may contribute to cancer progression and metastasis [28]. Importantly, genome wide single-nucleotide polymorphism studies strongly suggest that the *FTO* locus is associated with estrogen receptor-negative breast cancer including TNBC but not estrogen receptor-positive breast cancer [29].

Our gene expression microarray data showed that an increased copy number of the *FTO* gene correlated with significantly increased FTO mRNA in MA2 cells (Table S2); this correlation indicates a functional relevance of FTO to the survival of rare (0.01%) SUM149 cells under a severe metabolic challenge. To determine whether the increase in FTO mRNA observed in microarrays led to an increase in the FTO protein level, we

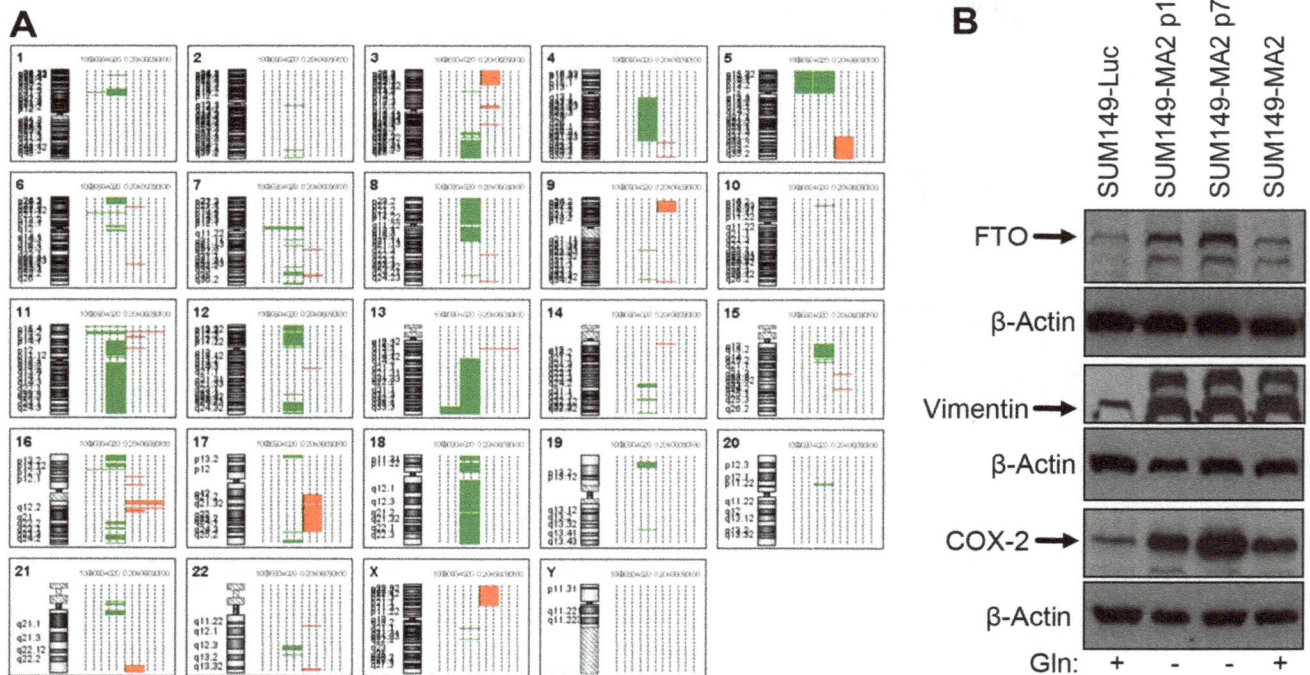

Figure 3. Amplification of the *FTO* Gene in MA Cells. (A) CGH array analysis was performed with MA cells compared to the SUM149-Luc cell line. Chromosomal gains (indicated in red) and losses (indicated in green) in MA1 and MA2 variants are presented as a composite graphic penetrance summary after removal of pseudoautosomal regions from the X and Y chromosomes. Chromosome numbers and the specific loci involved are indicated. Further details are provided in Tables S3 and S4. (B) FTO protein overexpression in MA cells. The cell lysates used in the analysis shown in Fig. 2 were used for these Western blots. The lysates were from the parental cell line, MA2 variants in Gln-free medium at passages 1 and 7, and MA2 variants after 4 passages in Gln-containing medium. We blotted vimentin and COX-2 as controls; we had previously detected their overexpression in MA1/Gln-independent variants [20]. See also Tables S4–S6.

analyzed the level of FTO protein by Western blotting. We found that MA cells produced a significantly higher amount of FTO protein than did the parental cell line (Fig. 3B). Interestingly, the level of FTO protein was higher in MA cells growing in a medium without glutamine than in MA cells growing in complete medium with Gln (compare lanes in Fig. 3B). These results are consistent with the results of starvation studies in animals showing that FTO protein is induced upon metabolic challenge. Although MA cells can grow without Gln (albeit slower than they grow in complete medium), we suspect that the cells sense Gln-deficient medium as a nutritionally challenging environment. Our results revealed how obesity and TNBC could be connected at the root level: inappropriate activation or overexpression of FTO could contribute to both obesity and cancer progression, with the latter being orchestrated through the enabling of small subpopulations of cancer cells to survive metabolic challenges in the body. We consider these results highly significant from the perspective of modeling the roots of therapy-resistance in TNBC since MA cells feature genomic abnormalities that are likely relevant in cancer evolution in patients.

Evidence of Panresistance in MA Cells

To determine whether MA cells are panresistant, we tested the cells' responses to a variety of therapeutic agents affecting many cellular processes. We did not solely base our choice of therapeutic agents on the levels of expression of specific targets in MA cells since such information may or may not apply to the roots of resistance (rare, highly adaptable cancer cells). In addition, to focus on the roots of resistance, we chose not to rely on cell proliferation assays. We tested the EGFR inhibitor erlotinib and the MEK inhibitor AZD6244; both targets, EGFR and MEK, are commonly active in TNBC and particularly in the SUM149 cell line [30]. We found that MA cells were more resistant to both of these inhibitors than the parental cell line was. Figure 4A shows representative data obtained with different doses and treatment times. Eleven-day treatment with 2 μM erlotinib or 1 μM AZD6244 was sufficient to eradicate most (>99%) cancer cells in the parental SUM149 cell line. In contrast, a significantly large number of MA cells survived these treatments, as was evident after the treated cells were allowed to recover in drug-free medium for 4 days (compare the stain in left and right panels in Fig. 4A). We also evaluated the effect of combining the two compounds. The combination inhibited MA cells significantly more than did either single agent; however, MA cells remained more resistant than the parental cell line (Fig. 4A, bottom panel). Although the difference in staining is not easily visible in Fig. 4A bottom panel, the staining of MA cells was significantly stronger than was the staining of the parental cell line. Additional experiments using longer recovery time clearly showed that MA cells are more resistant to combination therapy than is the parental cell line (data not shown). Regarding the choice of MA1 versus MA2 cell line for testing therapeutic agents, we have gradually shifted towards using MA2 since it represents a broader heterogeneity (Fig. 2). However, we have tested many therapies, including chemotherapeutic drugs doxorubicin and paclitaxel, on both cell lines and obtained similar results (not shown).

Crizotinib, which is used to treat EML4-ALK-positive non-small cell lung carcinoma, inhibits ALK, MET, and ROS1 tyrosine kinases. These kinases may play a role in breast cancer as well (e.g., see reference [31]). We treated cells with crizotinib for 4–5 days (the treatment killed 99% of the cells) and then allowed the surviving cells to recover and form colonies in drug-free medium for 14–16 days. We observed significantly more colonies in MA cells than in the parental cell line, indicating that MA cells are more resistant to crizotinib than are the parental cells (Fig. 4B).

In view of the fact that elevated PI3K and mTOR activities are common to many cancers including TNBC, BEZ235, a dual PI3K/mTOR inhibitor, is a promising therapeutic agent. A 0.4 μM dose of BEZ235 was ineffective against the very aggressive SUM149 cell line. Therefore, after first treating cells with BEZ235 for 14 days, we treated the cells with doxorubicin for 7 days thus killing >99% of the cells. We allowed the surviving cells to grow into colonies in a drug-free medium for 14 days, after which the cells were visualized by staining (Fig. 4C). MA cells were more resistant to BEZ235/doxorubicin treatment than was the parental cell line. The results indicated that BEZ235 treatment failed to sensitize MA cells to doxorubicin. Similarly, BEZ235 treatment also failed to sensitize MA cells to paclitaxel (Fig. 4C, bottom panel).

Some potentially therapeutic drugs have been identified that would be selective towards cancer stem cells. We reasoned that if there is a significant overlap between the cells selected by our approach and those selected by cancer stem cell approaches, the compounds that are effective against cancer stem cells may be effective against MA cells as well. Consequently, first we evaluated the potential cancer stem cell inhibitor salinomycin [32]. The top panels of Figure 4D show photographs of cells after 6 days of treatment with 4 μM salinomycin (this is how cells typically appear in our experiments after treatment with most test drugs). We found that treatment with 4 μM salinomycin for 7 days eradicated almost all cells in the parental cell line (with the exception of one cell that produced a small colony in drug-free medium after 12 days) , while approximately 40 colonies were observed for MA cells after the cells recovered in a drug-free medium for 12 days (bottom panel in Fig. 4D). This result suggests that the MA cells selected by our approach are resistant to salinomycin.

Thioridazine, a dopamine receptor antagonist that has been identified as an anti-cancer stem cell compound [33], was not very effective in killing SUM149 cells. MA cells were more resistant to 3-day treatment with 20 μM thioridazine than was the parental cell line, as observed after the cells were allowed to recover in a drug-free medium for 10 days (compare the left and right dishes in Fig. 4E). These results suggest that our stringent selection for metabolic adaptability (and co-selection of the embryo-like EMT phenotype) differs from selections that are used to identify cancer stem cells based on surface markers. Nonetheless, the results obtained with the cancer stem cell inhibitors- salinomycin and thioridazine- provide additional evidence of the panresistant nature of MA cells.

One strategy for targeting metastatic cancer cells relies on the observation that these cells suffer from high oxidative stress. It has been proposed that increasing oxidative stress further with therapeutic agents such as PEITC would kill cancer cells without damaging normal cells [34]. Since we had previously observed high oxidative stress in MA cells [20], we tested the efficacy of PEITC against MA cells. We treated cells with 5 μM PEITC for 3 days and allowed the surviving cells to recover and form colonies for 13 days in drug-free medium. We found that PEITC was not very effective against the SUM149 cell line and that it was significantly less effective against MA cells than against the parental cell line. MA cells yielded 10- to 20-fold more colonies than did parental cells (compare the left and right dishes in Fig. 4F), suggesting that the highly adaptable epigenetic state of MA cells enables them to survive various challenges, including high oxidative stress.

Itraconazole is a triazole antifungal drug that is being explored as an anticancer drug for several cancers, e.g., basal cell

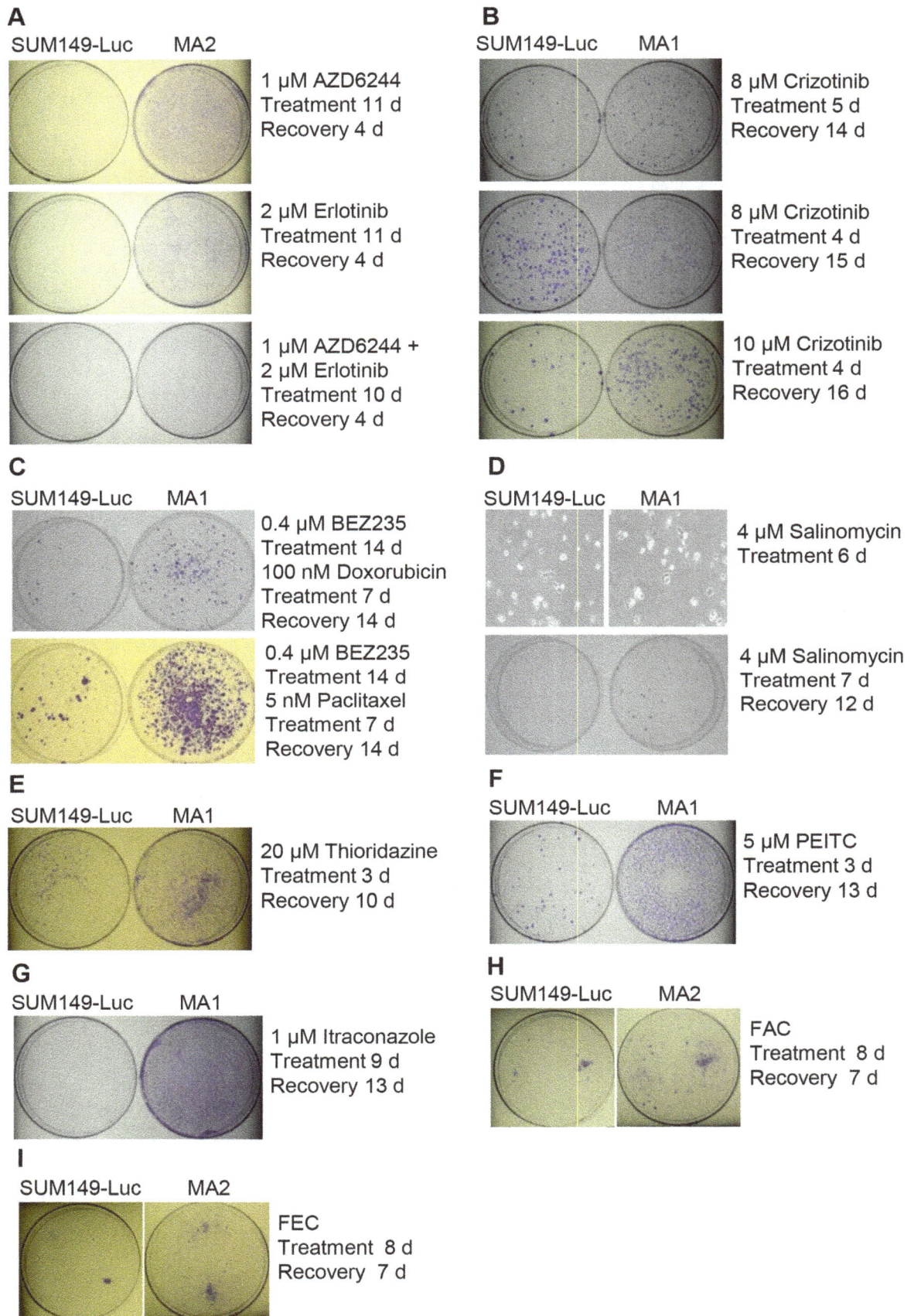

A

SUM149-Luc   MA2

1 μM AZD6244
Treatment 11 d
Recovery 4 d

2 μM Erlotinib
Treatment 11 d
Recovery 4 d

1 μM AZD6244 +
2 μM Erlotinib
Treatment 10 d
Recovery 4 d

B

SUM149-Luc   MA1

8 μM Crizotinib
Treatment 5 d
Recovery 14 d

8 μM Crizotinib
Treatment 4 d
Recovery 15 d

10 μM Crizotinib
Treatment 4 d
Recovery 16 d

C

SUM149-Luc   MA1

0.4 μM BEZ235
Treatment 14 d
100 nM Doxorubicin
Treatment 7 d
Recovery 14 d

0.4 μM BEZ235
Treatment 14 d
5 nM Paclitaxel
Treatment 7 d
Recovery 14 d

D

SUM149-Luc   MA1

4 μM Salinomycin
Treatment 6 d

4 μM Salinomycin
Treatment 7 d
Recovery 12 d

E

SUM149-Luc   MA1

20 μM Thioridazine
Treatment 3 d
Recovery 10 d

F

SUM149-Luc   MA1

5 μM PEITC
Treatment 3 d
Recovery 13 d

G

SUM149-Luc   MA1

1 μM Itraconazole
Treatment 9 d
Recovery 13 d

H

SUM149-Luc   MA2

FAC
Treatment 8 d
Recovery 7 d

I

SUM149-Luc   MA2

FEC
Treatment 8 d
Recovery 7 d

Figure 4. Evidence of an Elevated Panresistance in MA Cells. MA cells and the SUM149-Luc parental cell line were simultaneously treated with various anticancer agents as indicated to determine their relative level of resistance under the conditions that would inhibit most of the proliferating cells. (A) Increased resistance to EGFR and MEK inhibition in MA cells. The cells were treated with erlotinib or AZD6244 for 11 days (d) as indicated (each treatment killed most of the cells) and were allowed to recover in a drug-free medium before the colonies were stained. Combination treatment with the two drugs lasted 10 days, since it killed cells sooner than did the single-drug treatments; then, the cells were allowed to recover for 4 days. (B) Resistance to crizotinib in MA cells. (C) Resistance to PI3K/mTOR inhibition in MA cells. The cells were treated with BEZ235 for 14 days (d), passaged, treated with doxorubicin for 7 days (the treatment killed most of the cells), and allowed to recover and form colonies in a drug-free medium for before the colonies were stained (top). The bottom panel shows a similar experiment in which cells were treated with paclitaxel instead of doxorubicin. (D) Resistance to salinomycin in MA cells. The top panel shows bright-field photographs of representative cells (10X magnification) after 6 days of treatment with salinomycin. The cells were treated with salinomycin for 7 days (which killed most of the cells) and were allowed to recover and form colonies in a drug-free medium before the colonies were stained (bottom panel). (E) Resistance to thioridazine in MA cells. (F) Resistance to PEITC in MA cells. (G) Resistance to itraconazole in MA cells. The parental SUM149-Luc cell line and MA1 cells were treated in parallel with 1 µM itraconazole for 9 days (which killed most of the cells in the parental cell line) and were allowed to recover and grow in a drug-free medium for 5 days before being stained. Since itraconazole was ineffective in killing MA1 cells, the cells grew into a continuous monolayer rather than colonies. (H) Resistance to FAC combination chemotherapy agents in MA cells. (I) Resistance to FEC combination chemotherapy agents in MA cells.

carcinoma, non-small cell lung cancer, and prostate cancer (e.g., see reference [35]). It inhibits metastasis by antagonizing Smoothened (SMO), thus blocking Hedgehog signaling [36]. Our gene expression microarray analysis indicated SMO overexpression in MA cells. We found that itraconazole was not very effective in eradicating SUM149 cells (Fig. 4G). Furthermore, treatment with 1 µM itraconazole for 9 days followed by 5-day recovery in drug-free medium resulted in significantly more surviving MA cells than surviving parental cells (compare stain intensities, which are proportional to cell masses). The data in Fig. 4G represents a situation wherein treatment with a reasonable dose of a the drug kills most parental cells, but it does not kill MA cells sufficiently to bring down their number in a low range that would yield nicely separated colonies (instead we obtain a lot of cells growing to confluency). This result provides further evidence of the therapy-resistant nature of MA cells.

Finally, we tested combinations of chemotherapeutic agents, formulated as FAC and FEC, which are commonly prescribed for TNBC, particularly for IBC patients. We found that MA cells are more resistant to treatment with FAC (5-fluorouracil, adriamycin, and cyclophosphamide) or FEC (5-fluorouracil, epirubicin, and cyclophosphamide) than the parental cell line (Figs. 4H and 4I), further supporting the utility of MA cells as a cell-based model of therapy-resistant TNBC.

Strategies for Overcoming Panresistance

Although the question of what would eradicate the roots of the disease, exemplified by the rare MA cells, is difficult to answer, some general guidelines may help in addressing this core problem. Low-dose therapies that are well tolerated (and therefore can be taken for a long period) and that have a low risk of developing resistance to therapy would be preferable to high-dose therapies that are effective against the bulk of the disease (but that do not affect the roots or that may even enrich the roots and thereby lead to therapy resistance). The rationale is that dealing with the roots requires time. One promising approach for eradicating the roots may be the modification of chromatin with the use of agents that alter posttranslational modifications of histones and methylation of DNA. A tremendous amount of new knowledge is emerging regarding regulation of chromatin, leading to discovery of therapeutic agents that alter chromatin. The complexities of chromatin regulation and tumor heterogeneity make it difficult to predict which specific therapy would be optimal for eradicating the roots of TNBC, exemplified by the panresistant MA variants. In a proof-of-principle approach, we began with agents that have been extensively investigated as chromatin modifiers.

The goal was to determine whether long-term treatment with histone deacetylase (HDAC) inhibitors would sensitize MA

variants to chemotherapeutic drugs. HDAC inhibitors are being evaluated against a variety of cancers. Since we observed decreased acetylation at histone H3 lysine 14, which may be part of a drug-tolerant state [24], we evaluated sodium valproate and sodium butyrate, inhibitors of class I and class IIa HDACs [37]. We evaluated them in a low concentration range to minimize cytotoxic effects. We treated cells with various concentrations of the drugs for different periods in duplicate plates and then passaged the cells from subconfluent cultures by trypsinization. The results for cells treated with HDAC inhibitors for 7 days are presented in Fig. 5. After recovery for 1 day, we treated the cells with the chemotherapeutic drug doxorubicin or paclitaxel. When approximately 99% of the cells were killed (after 5–6 days) , we removed the drugs and allowed the cells to recover and form colonies. Representative data from these experiments show that MA cells yielded fewer colonies than did the parental cell line after either valproate-paclitaxel or valproate-doxorubicin treatment (compare plates in Fig. 5A; we obtained a similar results with butyrate-paclitaxel or butyrate-doxorubicin treatment (Fig. 5B). Given the therapy-resistant nature of MA cells, even equal numbers of colonies in MA cells and the parental cell line would indicate sensitization of MA cells to chemotherapy. Thus, our results indicate that both valproate and butyrate effectively sensitize MA cells to chemotherapeutic drugs.

A low-dose 5-azacytidine treatment leads to reduced DNA methylation. Azacytidine and its deoxy derivative are used for treating Myelodysplastic Syndromes, and they are being evaluated in clinical trials in various combination therapies for several additional cancers. Here, we investigated whether 5-azacytidine treatment can sensitize MA cells to chemotherapeutic drugs. Since reversal of epigenetic modifications may require multiple cell divisions, we treated cells with azacytidine for different periods ranging from 7 days (one passage) to 27 days (three passages). Although initially 1 µM azacytidine was not cytotoxic as evidenced by a lack of significant growth-inhibition, we noticed that after 27 days of treatment the parental cell line had approximately 50% fewer cells than did the MA2 cell line. At that time, we passaged the cells while continuing to treat them with azacytidine in combination with doxorubicin for 7 days; the combination treatment killed >99% of the cells. After allowing the surviving cells to recover and form colonies in a drug-free medium for 12 days, we stained the colonies (Fig. 5C). We found that azacytidine treatment resulted in significant sensitization of MA cells to doxorubicin: the number of doxorubicin-resistant colonies among MA cells was approximately the same as or lower than the number of doxorubicin-resistant colonies formed by the parental cell line (Fig. 5C). Given that without prior treatment, doxorubicin-treated MA cells yield significantly (5–10-fold) more colonies

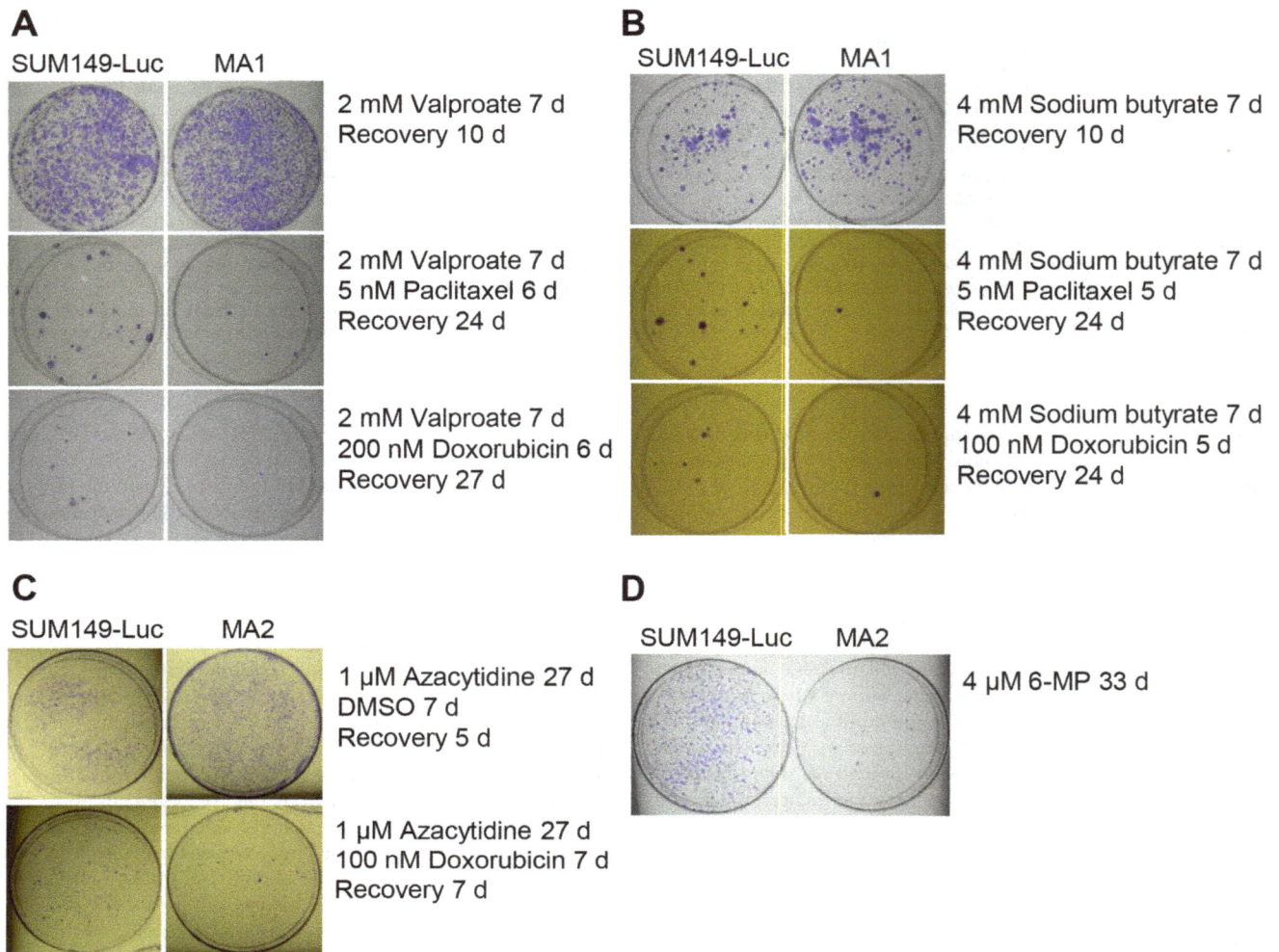

Figure 5. Testing Anticancer Agents for Their Ability to Overcome Panresistance in MA Cells. (A) Sensitization of MA cells to chemotherapeutic drugs by prior treatment with sodium valproate. The parental SUM149-Luc cell line and MA1 cells were treated in parallel with sodium valproate for 7 days (d), passaged, treated with paclitaxel or doxorubicin for 6 days (which killed most of the cells), and allowed to recover and form colonies in a drug-free medium before being stained. (B) Sensitization to chemotherapeutic drugs by prior treatment of MA cells with sodium butyrate. (C) Sensitization to doxorubicin by prior 27-day treatment of MA cells with azacytidine. The cells were treated with azacytidine for 27 days (d) (three passages in culture), passaged, and treated with doxorubicin along with for 7 days (which killed most of the cells), and allowed to recover and form colonies in a drug-free medium for 12 days before being stained (top). Also shown are cells that were treated with the DMSO solvent for 7 days as a control for doxorubicin treatment and that were allowed to recover in a drug-free medium for 5 days prior to being stained (bottom). (D) Eradication of MA cells by 33-day treatment with 6-mercaptopurine. The cells were treated with 6-mercaptopurine for 33 days (three passages in culture) and then stained.

than the parental cell line does [20], our results indicate that azacytidine may help overcome resistance to chemotherapeutic drugs in MA cells.

We hypothesized that 6-mercaptopurine, which is used to treat leukemia and several other diseases, including inflammatory bowel disease, may help overcome resistance of MA cells to anticancer agents. What makes 6-mercaptopurine attractive for our purpose is the safety and efficacy of low-dose 6-MP regimens administered over many years. We treated MA cells and the parental cell line in parallel with 4 μM 6-mercaptopurine for 33 days continuously, involving three passages in cell culture. 6-Mercaptopurine did not cause any appreciable cytotoxicity in either cell line during the first two passages. However, it had a dramatically greater effect on MA cells than on the parental cell line during the third passage. Colonies stained after 17 days of growth in passage 3 in the

presence of 6-mercaptopurine are shown in Fig. 5D. A significantly lower number of colonies from MA cells than the parental cell line indicates that long-term treatment with 6-mercaptopurine may have exhausted MA cells.

In view of the evolution-like nature of cancer, it is critical that we get better at predicting evolution of cancer as therapies are implemented. The therapies would influence cancer evolution and often rare clones that are relatively indolent at the time of diagnosis would assume an active role after therapeutic interventions [1,38,39]. Although it is not feasible to model the cancer evolution that occurs in individual patients, the strategy presented here can assist in modeling the roots of cancer that are likely to be relevant in continuously evolving cancer. An important lesson from our study is that to get most out of a cell culture model we must constantly ask what would be the likely fate of the specific subpopulations of cancer cells in cell culture when such cells are part of a heterogeneous tumor evolving in the body, particularly under currently offered therapeutic interventions. Addressing this question would be a basis of selecting adaptable cancer cells and for selecting/designing assays that inform about adaptable cells even though the cell culture conditions do not impose body-like selection pressures.

Based on our results, we believe that *FTO* gene amplification leading to higher FTO mRNA and protein level helped MA cells to survive a severe metabolic challenge in the form of lack of glutamine. Among other important elements in this selection may be the embryo-like nature of rare cells that survived. Based on work with whole animals and with cell lines, FTO is important in starvation response [27,40]. We found that the level of FTO protein begins to come down when cells are cultured in complete medium (Fig. 3B). We interpret these data to suggest that a higher level of FTO protein may not be required or may not be beneficial for cell growth in complete medium. By modulating the function of many mRNAs and non-coding RNAs through demethylation, FTO is an important part of circuitry that controls metabolism at cellular and organismal levels. Investigating the functional role of FTO in MA cells and how to therapeutically exploit it for cancer treatment may be fruitful. Besides the complexity of obesity networks involving FTO deregulation, the major interconnected problem in disease like TNBC is tumor heterogeneity and tumor adaptability.

There is significant interest in glutamine metabolism in cancer. Glutamine serves as an important carbon and nitrogen source for biosynthetic processes in cancer; it is also important in regulating redox status. As far as studies pertaining to metabolic challenges are concerned, a major effort has been to investigate the consequences of hypoxia in tumors. However, it is apparent that tumors not only lack oxygen in some regions, they also lack other nutrients, e.g., glutamine [41]. Such metabolic challenges are likely to influence tumor evolution. Studies have also been carried out to determine the effect of Gln withdrawal *in vitro*. However, unlike this study that aims to isolate metabolically adaptable rare cancer cells, those studies have mostly investigated the effect of Gln withdrawal on the major cancer cell population (e.g., see reference 41).

We have summarized our strategy for modeling the roots of TNBC, and for testing anticancer agents in Figure 6. Although our approach has unique features that make it useful for testing anticancer agents, we also recognize the limitations of the approach. Even though our strategy is good at modeling the roots of panresistance, all *in vitro* approaches including ours have a drawback of lacking suitable microenvironments that cancer cells encounter in the body. Although one can choose from a variety of techniques aimed at modeling a tumor microenvironment in cell culture, such techniques have severe limitations. Broadly speaking, modeling appropriate tumor environment is not easy since tumor cells face a variety of microenvironments in the body. In this study, our main goal was to investigate the most adaptable rare cells for their potential utility in anticancer drug testing. We recognize that all models of cancer have different sets of limitations. Therefore, unique strengths of different models, including that of ours, need to be utilized for developing effective therapies against cancer.

Conclusions

To summarize, we have a suitable model of hard-to-eradicate roots for testing anticancer therapies. Our approach can be used to test a growing list of potential therapeutic compounds that can target the majority of cancer cells through a variety of mechanisms. The cancer cells most responsible for panresistance

are heterogeneous, and the likelihood that a particular test therapy will affect these cells is difficult to predict. Our testing strategy is optimized to investigate the effects of potential therapeutic compounds on such cells. Our *in vitro* approach can complement other approaches, including therapy testing in genetically engineered mouse models and patient-derived cancer xenografts in mice. Since evaluation of combination therapies (drugs, doses, schedules, etc.) is much easier in cell lines than in mice, our *in vitro* strategy can be used to merge realistic function-based selection of cancer cells most responsible for therapy resistance with the common biomarker/driver-based drug discovery approach to develop therapies that can overcome resistance. Our approach can be used to test sensitizing agents (that may alter root cells when administered at a low dose over a long period) and inhibitors of molecular drivers of disease (whose range of target cells may broaden as a result of prior treatment with sensitizing agents) and to identify therapies that are effective in combating disease recurrence and metastasis and that are suitable for clinical trials. If aligned with the clinic, this approach can not only help choose superior therapies for clinical trials, but also suggest strategies for overcoming therapeutic resistance in a timely manner.

Materials and Methods

Cell Lines and Culture

The SUM149 IBC cell line, originally obtained from Stephen Ethier (Barbara Ann Karmanos Cancer Institute, Detroit, MI, USA), was grown in Ham's F-12 medium supplemented with 5% fetal bovine serum (FBS), 5 μg/ml of insulin, 1 μg/ml of hydrocortisone, 100 U/ml of penicillin, and 100 μg/ml of streptomycin in a humidified 5% CO_2 atmosphere. We previously described SUM149-Luc, a luciferase-transfected cell line [17]. SUM149-FP76 is another cell line recently developed in our laboratory by first culturing cells from a SUM149-Luc xenograft subcutaneously in a female nude mouse and then culturing cells from a fat pad xenograft [20].

Selection and Culture of MA Variants

We selected MA variants by plating 0.5 million or 1 million SUM149-Luc cells in a glutamine-deficient medium containing dialyzed FBS. We recently described selection of rare variants (13 colonies from 0.5 million cells) and initial characterization of a cell culture established from these colonies [20]. In the current study, we refer to this cell line as MA1, and we refer to another cell line that was similarly established by selecting variants from 1 million cells in a glutamine-deficient medium as MA2. Although MA1 and MA2 cells can be passaged indefinitely in glutamine-deficient medium, we cultured them in glutamine-containing medium for testing therapeutic agents. Both MA variants and the parental SUM149-Luc cell line were cultured identically in regular medium containing dialyzed FBS for this purpose. We performed all experiments with the cell lines before passage 10 in this medium. We also cultured MA cells in a glutamine-deficient medium in some experiments as indicated; the purpose of these experiments was to determine whether the continued absence of glutamine affected the phenotype under investigation.

Gene Expression Microarray Analysis

For gene expression microarray analysis, we cultured both MA variants and the parental cell line in a medium containing Gln and dialyzed FBS; we used dialyzed FBS since it was used for selecting the MA variants in Gln-free medium. RNA was isolated using standard RNA extraction protocols (NucleoSpin RNA II,

Strategy for Testing Anticancer Agents

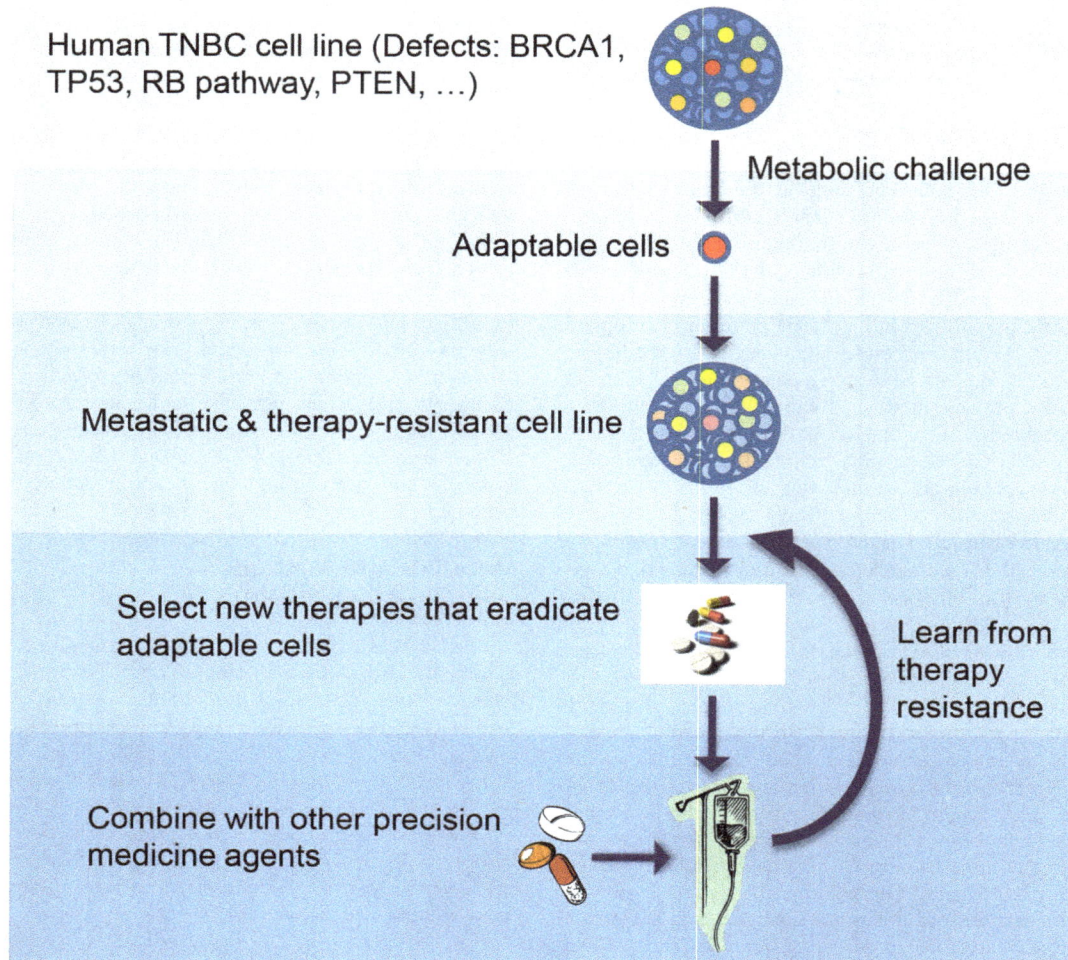

Figure 6. A Strategy for Testing Anticancer Agents. The strategy described in this paper is depicted as a flow diagram. See text for further details.

Macherey-Nagel, Bethlehem, PA) and quality checked with the 2100 Bioanalyzer (Agilent Technologies). To produce cyanine 3-labeled complementary RNA, the 100-ng RNA samples were amplified by linear T7 polymerase-based amplification and labeled using the Agilent Low Input Quick Amp Labeling Kit according to the manufacturer's protocol. Hybridization was performed according to the Agilent 60-mer oligo microarray processing protocol using the Agilent Gene Expression Hybridization Kit. Briefly, 600 ng of cyanine 3-labeled fragmented complementary RNA in hybridization buffer was hybridized overnight (17 hours, 65°C) to Agilent Whole Human Genome Oligo 8×60K Microarrays. The fluorescence signals of the hybridized microarrays were detected using the Agilent Micro-array Scanner System. Agilent G2567AA Feature Extraction Software version 10.7.3.1 was used to read out and process the microarray image files. The software determines feature intensities (including background subtraction), rejects outliers, and calculates statistical confidences. To determine differential gene expression, Feature Extraction Software-derived output data files were further analyzed using the Rosetta Resolver gene expression data analysis

system (Rosetta Biosoftware). Putative candidate genes with fold changes >2 and p-values <0.01 are summarized in the preselected candidate genes lists.

CGH Array Analysis

For CGH array analysis, cells were cultured in parallel with those used for the gene expression microarray. Genomic DNA was purified using the Gentra Puregene Cell kit (Qiagen). The integrity of the genomic DNA samples was assessed by agarose gel electrophoresis. A 2.5-µg quantity of each DNA sample was digested with the restriction enzymes *Alu*I and *Rsa*I. The digested DNA was used as template for a genomic DNA labeling reaction using the SureTag DNA Labeling Kit (Agilent Technologies) according to the manufacturer's protocol. The yields of labeled DNA and the dye incorporation rate after Klenow labeling were determined with the ND-1000 Spectrophotometer (NanoDrop Technologies). Hybridization was performed according to the Agilent Oligonucleotide Array-Based CGH for Genomic DNA Analysis protocol version 7.1 using the SureTag DNA Labeling Kit. The corresponding cyanine 3- and cyanine 5-labeled DNAs were combined and hybridized at 65°C for 40 hours to an Agilent Human Genome CGH 1M Microarray. The fluorescence signals of the hybridized Agilent Human Genome CGH microarrays were detected using the Agilent DNA microarray scanner. Agilent Feature Extraction Software was used to read out and process the microarray image files. Further analysis and visualization of the hybridization results were performed with the Agilent Genomic Workbench Lite 6.5. We used the following stringent filter settings for aberrations: (i) minimum number of probes present in an aberrant region = 5, (ii) minimum absolute average log2 ratio for region = 0.4 (corresponds to a fold change of 1.32), and (iii) threshold = 5. The statistical analysis of aberrant regions was based on the aberration detection method ADM-2. The ADM-2 algorithm identifies all aberrant intervals in a genome with consistently high or low log ratios based on the statistical score.

Western Blotting

We separated proteins by sodium dodecyl sulfate-polyacryl-amide gel electrophoresis and detected various proteins by Western blotting, as described previously [42]. The following primary antibodies were used for detection: anti-COX-2 mono-clonal antibody (Cayman Chemical, Ann Arbor, MI), anti-GRHL2 (Sigma-Aldrich, St. Louis, MO), anti-FTO (EMD Millipore, Billerica, MA), anti-ZEB1, anti-histone H3, anti-histone H3 tri-methyl Lys-4, anti-histone H3 acetyl Lys-14, and anti-vimentin antibodies (Cell Signaling, Danvers, MA). We used the ECL prime blocking agent (GE Healthcare Life Sciences, Piscataway, NJ) for blocking and Lumigen TMA-6 reagents for detection (Lumigen, Inc., Southfield, MI). The filters were stripped by incubating the membrane in 0.5% Triton X-100 and were re-probed with a monoclonal β-actin antibody (Sigma-Aldrich, St. Louis, MO) or with vinculin antibody (Abcam, Cambridge, MA), which served as gel-loading controls. To detect β-actin or vinculin, we used 2% non-fat dry milk (Bio-Rad, Hercules, CA) for blocking and ECL prime reagent for detection (GE Healthcare Life Sciences, Piscataway, NJ). We performed each western blot at least twice; the representative blots are shown. We quantified the protein bands on x-ray films by using the ImageJ image processing program (National Institutes of Health).

Metabolic Adaptability Assay

The basis of the metabolic adaptability assay is that prolonged extreme glucose deficiency is likely to select cells with a highly adaptable metabolic state. Cancer cells are highly addicted to glucose, so the majority of them die within a few days in the absence of glucose; in fact, a prolonged lack of glucose eventually kills all cancer cells in culture. To determine the metabolic adaptability of MA cells, we deprived them of glucose for 2 to 4 weeks and assessed whether a higher number of cells survived than the number of parental cells that survived glucose deprivation. We plated a half million cells on a 10-cm dish, switched them to glucose-free medium (custom medium from AthenaES, Baltimore, MD) the next day for 2 to 4 weeks, and then switched them back to complete medium for 2–4 weeks until the surviving cells yielded colonies. We stained the colonies with crystal violet and photographed them. In some experiments, we deprived cells of both glucose and glutamine and compared the surviving cells with those selected by glucose deprivation alone.

Test Drugs

We purchased the drugs from the commercial sources: paclitaxel, doxorubicin/adriamycin, 5-fluorouracil, epirubicin, cyclophosphamide, crizotinib, salinomycin, thioridazine, phe-nethyl isothiocyanate (PEITC), sodium valproate, sodium buty-

rate, 5-azacytidine, and 6-mercaptopurine (Sigma-Aldrich, St. Louis, MO), NVP-BEZ235 (Cayman Chemical, Ann Arbor, MI), and itraconazole (Selleckchem, Houston, TX). Naoto Ueno kindly provided AZD6244 (AstraZeneca, Wilmington, DE) and erlotinib (ChemieTek, Indianapolis, IN). We dissolved BEZ235, PEITC, itraconazole, crizotinib, salinomycin, doxorubicin, paclitaxel, AZD6244, and erlotinib in dimethyl sulfoxide (DMSO), thiorid-azine, sodium valproate, and sodium butyrate in water, 6-mercaptopurine in 0.1 M NaOH, and 5-azacytidine in dulbecco's phosphate buffered saline. To prepare FAC (5-fluorouracil, adriamycin, cyclophosphamide) and FEC (5-fluorouracil, epirubi-cin, cyclophosphamide) chemotherapy combinations, drugs were measured individually and dissolved into DMSO and combined. Final concentrations of drugs were 250 nM 5-fluorouracil, 25 nM adriamycin, and 250 nM cyclophosphamide for FAC and 250 nM 5-fluorouracil, 50 nM epirubicin, and 250 nM cyclophosphamide for FEC. We added equal volume of the solvent in all dishes including the control dishes without drugs. DMSO volume was ≤ 0.04% of the volume of the culture medium.

Evaluation of Test Drugs *in vitro*

In a comparative long-term evaluation of therapeutic agents, in which >99% of the cells in the parental cell line were killed, we determined the number of MA cells that survived and yielded colonies. Since it is often not easy to determine that therapeutic agents has killed more than 99% of the cells in a culture dish, we used different drug concentrations and treatment times to improve the likelihood of detecting this result. We typically plated 0.5 million cells per 10-cm dish in duplicate in culture medium with glutamine. After 24 hours, we added a test drug at four different concentrations (1X, 2X, 4X, and 8X) based on literature in a preliminary experiment to determine the drug concentration range that would eradicate >99% cells in about a week. The cells were photographed, and plates were monitored for different periods until most cells (>99%) were killed. To compare the cell survival rates in the MA cells and the SUM149 cell line, drugs were removed by rinsing twice with phosphate-buffered saline, and cells were maintained in a drug-free medium for 3–4 weeks until colonies were visible to the naked eye. Colonies were stained with crystal violet and photographed.

Because some therapeutic agents that are ineffective at killing cancer cells may nevertheless sensitize the cells to other chemotherapeutic drugs, we evaluated combination therapies, wherein two agents were added concurrently or sequentially to MA cells or the parental cell line. In evaluating therapies for their potential to sensitize MA cells to chemotherapeutic drugs, some experiments involved treatment that lasted several weeks, which made it necessary to passage the cell cultures.

Accession Numbers

The gene expression microarray data and CGH array data discussed in this publication have been deposited in NCBI's Gene Expression Omnibus and are accessible through GEO Series accession number GSE60017 (http://www.ncbi.nlm.nih.gov/geo/query/acc.cgi?acc=GSE60017).

Supporting Information

Table S1 Genes Encoding Significantly Altered RNA Expression in MA1 Cells, Related to Figure 2. We generated a list of genes with significantly altered RNA levels in MA1 cells relative to the RNA levels in the parental cell line as described in Extended Experimental Procedures. The genes are listed in order from the most reduced RNA to the most increased

RNA, with fold changes in RNA level and p-values indicated. The MA1 cells were passaged 11 times in glutamine-free medium followed by 5 times in glutamine-containing medium prior to this analysis.

Table S2 Genes Encoding Significantly Altered RNA Expression in MA2 Cells, Related to Figure 2. We generated a list of genes with significantly altered RNA levels in MA2 cells relative to the RNA levels in the parental cell line as described in Extended Experimental Procedures. The genes are listed in order from the most reduced RNA to the most increased RNA, with fold changes in RNA level and p-values indicated. The MA2 cells were passaged 10 times in glutamine-free medium followed by 3 times in glutamine-containing medium prior to this analysis.

Table S3 Molecular Alterations Affecting Several Networks in MA2 cells, Related to Figure 2. The significant alterations in gene expression in MA2 cells, which are listed in Table S2, were subjected to core analysis in the Ingenuity Pathway Analysis software. Significantly up-regulated or down-regulated molecules are grouped according to the diseases and functions they may impact. The analysis is composed of 25 networks.

Table S4 Chromosomal Gains and Losses in MA1 Cells, Related to Figure 3. The 42 aberrations detected under stringent settings are listed. The list includes all the genes located in the affected chromosomal regions.

Table S5 Chromosomal Gains and Losses in MA2 Cells, Related to Figure 3. The 283 aberrations detected under stringent settings are listed. The list includes all the genes located in the affected chromosomal regions.

Table S6 Chromosomal Gains and Losses Shared by MA1 and MA2 Cells, Related to Figure 3. The 13 aberrations (5 amplifications and 8 deletions) detected under stringent settings in both MA1 cells and MA2 cells are listed. The list includes all the genes located in the affected chromosomal regions. We extracted this information manually in Microsoft Excel from Tables S4 and S5.

Acknowledgments

We thank Ralph Arlinghaus for critical reading of the manuscript and Arthur Gelmis for editorial corrections. We thank Silvia Rüberg (Miltenyi Biotec GmbH) for help with the analysis of gene expression microarray and CGH array data. We thank Naoto Ueno for providing AZD6244 and erlotinib.

Author Contributions

Conceived and designed the experiments: BS AL. Performed the experiments: BS AS MR RM AC SM. Analyzed the data: BS AS MR RM AC SM AL. Contributed reagents/materials/analysis tools: BS AS MR RM AC SM AL. Contributed to the writing of the manuscript: BS AL.

References

1. Nowell PC (1976) The clonal evolution of tumor cell populations. Science 194: 23–28.
2. Fidler IJ, Kripke ML (1977) Metastasis results from preexisting variant cells within a malignant tumor. Science 197: 893–895.
3. Talmadge JE, Wolman SR, Fidler IJ (1982) Evidence for the clonal origin of spontaneous metastases. Science 217: 361–363.
4. Chambers AF, Groom AC, MacDonald IC (2002) Dissemination and growth of cancer cells in metastatic sites. Nat Rev Cancer 2: 563–572.
5. Talmadge JE, Fidler IJ (2010) AACR centennial series: the biology of cancer metastasis: historical perspective. Cancer Res 70: 5649–5669.
6. Cifone MA, Fidler IJ (1980) Correlation of patterns of anchorage-independent growth with in vivo behavior of cells from a murine fibrosarcoma. Proc Natl Acad Sci U S A 77: 1039–1043.
7. Al-Hajj M, Wicha MS, Benito-Hernandez A, Morrison SJ, Clarke MF (2003) Prospective identification of tumorigenic breast cancer cells. Proc Natl Acad Sci U S A. 100: 3983–3988.
8. Guo L, Fan D, Zhang F, Price JE, Lee JS, et al. (2011) Selection of brain metastasis-initiating breast cancer cells determined by growth on hard agar. Am J Pathol 178: 2357–2366.
9. McKnight SL (2010) On getting there from here. Science 330: 1338–1339.
10. Shah SP, Roth A, Goya R, Oloumi A, Ha G, et al. (2012) The clonal and mutational evolution spectrum of primary triple-negative breast cancers. Nature 486: 395–399.
11. The Cancer Genome Atlas Network (2012) Comprehensive molecular portraits of human breast tumours. Nature 490: 61–70.
12. Zhang Q1, Liu Y, Zhou J, Chen W, Zhang Y, et al. (2007) Wild-type p53 reduces radiation hypermutability in p53-mutated human lymphoblast cells. Mutagenesis. 22: 329–334.
13. Bullock AN, Henckel J, Fersht AR. (2000) Quantitative analysis of residual folding and DNA binding in mutant p53 core domain: definition of mutant states for rescue in cancer therapy. Oncogene 19: 1245–1256.
14. Chao HH, He X, Parker JS, Zhao W, Perou CM (2012) Micro-scale genomic DNA copy number aberrations as another means of mutagenesis in breast cancer. PLoS One. 7: e51719.
15. Barnabas N, Cohen D (2013) Phenotypic and molecular characterization of MCF10DCIS and SUM breast cancer cell lines. Int J Breast Cancer 2013: 872743.
16. Pan Q, Bao LW, Merajver SD (2003) Tetrathiomolybdate inhibits angiogenesis and metastasis through suppression of the NFkappaB signaling cascade. Mol Cancer Res 1: 701–706.
17. Singh B, Cook KR, Martin C, Huang EH, Mosalpuria K, et al. (2010) Evaluation of a CXCR4 antagonist in a xenograft mouse model of inflammatory breast cancer. Clin Exp Metastasis 27: 233–240.
18. Van Laere SJ, Ueno NT, Finetti P, Vermeulen P, Lucci A, et al. (2013) Uncovering the molecular secrets of inflammatory breast cancer biology: an integrated analysis of three distinct affymetrix gene expression datasets. Clin Cancer Res. 19: 4685–4696.
19. Masuda H, Baggerly KA, Wang Y, Iwamoto T, Brewer T, et al. (2013) Comparison of molecular subtype distribution in triple-negative inflammatory and non-inflammatory breast cancers. Breast Cancer Res. 15: R112.
20. Singh B, Tai K, Madan S, Raythatha MR, Cady AM, et al. (2012) Selection of metastatic breast cancer cells based on adaptability of their metabolic state. PLoS ONE 7: e36510.
21. Cieply B, Farris J, Denvir J, Ford HL, Frisch SM (2013) Epithelial-mesenchymal transition and tumor suppression are controlled by a reciprocal feedback loop between ZEB1 and Grainyhead-like-2. Cancer Res. 73: 6299–6309.
22. Mani SA, Guo W, Liao MJ, Eaton EN, Ayyanan A, et al. (2008) The epithelial-mesenchymal transition generates cells with properties of stem cells. Cell. 133: 704–715.
23. Tam WL1, Weinberg RA (2013). The epigenetics of epithelial-mesenchymal plasticity in cancer. Nat Med. 19: 1438–1449.
24. Sharma SV, Lee DY, Li B, Quinlan MP, Takahashi F, et al. (2010) A chromatin-mediated reversible drug-tolerant state in cancer cell subpopulations. Cell 141: 69–80.
25. Gerken T, Girard CA, Tung YC, Webby CJ, Saudek V, et al. (2007) The obesity-associated FTO gene encodes a 2-oxoglutarate-dependent nucleic acid demethylase. Science 318: 1469–1472.
26. Gulati P, Cheung MK, Antrobus R, Church CD, Harding HP, et al. (2013) Role for the obesity-related FTO gene in the cellular sensing of amino acids. Proc Natl Acad Sci U S A. 110: 2557–2562.
27. Fredriksson R, Hägglund M, Olszewski PK, Stephansson O, Jacobsson JA, et al. (2008) The obesity gene, FTO, is of ancient origin, up-regulated during food deprivation and expressed in neurons of feeding-related nuclei of the brain. Endocrinology. 149: 2062–2071.
28. Jain R, Strickler HD, Fine E, Sparano JA (2013). Clinical studies examining the impact of obesity on breast cancer risk and prognosis. J. Mammary Gland Biol. Neoplasia 18: 257–266.
29. Garcia-Closas M, Couch FJ, Lindstrom S, Michailidou K, Schmidt MK, et al. (2013) Genome-wide association studies identify four ER negative-specific breast cancer risk loci. Nat Genet. 45: 392–398.
30. Zhang D, LaFortune TA, Krishnamurthy S, Esteva FJ, Cristofanilli M, et al. (2009) Epidermal growth factor receptor tyrosine kinase inhibitor reverses

mesenchymal to epithelial phenotype and inhibits metastasis in inflammatory breast cancer. Clin Cancer Res 15: 6639–6648.

31. Raghav KP, Wang W, Liu S, Chavez-MacGregor M, Meng X, et al. (2012) cMET and phospho-cMET protein levels in breast cancers and survival outcomes. Clin. Cancer Res. 18: 2269–2277.

32. Gupta PB, Onder TT, Jiang G, Tao K, Kuperwasser C, et al. (2009) Identification of selective inhibitors of cancer stem cells by high-throughput screening. Cell 138: 645–659.

33. Sachlos E, Risueño RM, Laronde S, Shapovalova Z, Lee JH, et al. (2012) Identification of drugs including a dopamine receptor antagonist that selectively target cancer stem cells. Cell 149: 1284–1297.

34. Trachootham D, Zhou Y, Zhang H, Demizu Y, Chen Z, et al. (2006) Selective killing of oncogenically transformed cells through a ROS-mediated mechanism by beta-phenylethyl isothiocyanate. Cancer Cell 10: 241–252.

35. Rudin CM, Brahmer JR, Juergens RA, Hann CL, Ettinger DS, et al. (2013) Phase 2 study of pemetrexed and itraconazole as second-line therapy for metastatic nonsquamous non-small-cell lung cancer. J. Thorac. Oncol. 8: 619–623.

36. Kim J, Tang JY, Gong R, Kim J, Lee JJ, et al. (2010) Itraconazole, a commonly used antifungal that inhibits Hedgehog pathway activity and cancer growth. Cancer Cell 17: 388–399.

37. Xu WS, Parmigiani RB, Marks PA (2007) Histone deacetylase inhibitors: molecular mechanisms of action. Oncogene 26: 5541–5552.

38. Landau DA, Carter SL, Stojanov P, McKenna A, Stevenson K, et al. (2013) Evolution and impact of subclonal mutations in chronic lymphocytic leukemia. Cell 152: 714–726.

39. Wang Y, Waters J, Leung ML, Unruh A, Roh W, et al. (2014) Clonal evolution in breast cancer revealed by single nucleus genome sequencing. Nature Jul 30. doi: 10.1038/nature13600. [Epub ahead of print].

40. Berulava T, Ziehe M, Klein-Hitpass L, Mladenov E, Thomale J, et al. (2013) FTO levels affect RNA modification and the transcriptome. Eur J Hum Genet. 21: 317–323.

41. Reid MA, Wang WI, Rosales KR, Welliver MX, Pan M, et al. (2013) The B55α subunit of PP2A drives a p53-dependent metabolic adaptation to glutamine deprivation. Mol Cell 50: 200–211.

42. Singh B, Berry JA, Shoher A, Ayers GD, Wei C, et al. (2007) COX-2 involvement in breast cancer metastasis to bone. Oncogene 26: 3789–3796.

15

ASPP2 Links the Apical Lateral Polarity Complex to the Regulation of YAP Activity in Epithelial Cells

Christophe Royer[1,2]*, Sofia Koch[1], Xiao Qin[1], Jaroslav Zak[1], Ludovico Buti[1], Ewa Dudziec[1], Shan Zhong[1], Indrika Ratnayaka[1], Shankar Srinivas[2], Xin Lu[1]*

1 Ludwig Institute for Cancer Research, Nuffield Department of Clinical Medicine, University of Oxford, Oxford, United Kingdom, 2 Department of Physiology, Anatomy and Genetics, University of Oxford, Oxford, United Kingdom

Abstract

The Hippo pathway, by tightly controlling the phosphorylation state and activity of the transcription cofactors YAP and TAZ is essential during development and tissue homeostasis whereas its deregulation may lead to cancer. Recent studies have linked the apicobasal polarity machinery in epithelial cells to components of the Hippo pathway and YAP and TAZ themselves. However the molecular mechanism by which the junctional pool of YAP proteins is released and activated in epithelial cells remains unknown. Here we report that the tumour suppressor ASPP2 forms an apical-lateral polarity complex at the level of tight junctions in polarised epithelial cells, acting as a scaffold for protein phosphatase 1 (PP1) and junctional YAP via dedicated binding domains. ASPP2 thereby directly induces the dephosphorylation and activation of junctional YAP. Collectively, this study unearths a novel mechanistic paradigm revealing the critical role of the apical-lateral polarity complex in activating this localised pool of YAP *in vitro*, in epithelial cells, and *in vivo*, in the murine colonic epithelium. We propose that this mechanism may commonly control YAP functions in epithelial tissues.

Editor: Hong Wanjin, Institute of Molecular and Cell Biology, Biopolis, United States of America

Funding: CR was supported by an Oxford Stem Cell Institute Fellowship from the Oxford Martin School (http://www.oxfordmartin.ox.ac.uk/institutes/stem_cells/). XL was supported by the Medical Research Council (MR/J000930/1). The funders had no role in study design, data collection and analysis, decision to publish, or preparation of the manuscript.

Competing Interests: The authors have declared that no competing interests exist.

* Email: royer@dpag.ox.ac.uk (CR); xin.lu@ludwig.ox.ac.uk (XL)

Introduction

The Hippo pathway, by regulating the phosphorylation state and activity of YAP and TAZ (YAP/TAZ), has emerged as a crucial regulator of cell number and differentiation [1]. This is illustrated by recent findings that the Hippo pathway is central to tissue regeneration and the control of organ size. The Hippo pathway must therefore be able to sense global changes in tissue architecture. Interestingly, at the cellular level, the Hippo pathway can integrate signals through several routes, including GPCR signalling [2], mechanical forces [3] and apicobasal polarity [4]. The Hippo pathway is therefore at an ideal position to sense structural changes *in vivo* and consequently regulate these biological processes.

Originally uncovered in Drosophila, the core hippo signalling cascade in mammals is well established [1]: MST1/2 kinases phosphorylate Lats1/2 which subsequently phosphorylate the transcription cofactors YAP/TAZ to induce their cytoplasmic retention or degradation. Thus phosphorylated cytoplasmic YAP/TAZ are unable to bind and activate a variety of transcription factors including TEAD1-4. Interestingly however, several components of the apicobasal polarity machinery, including the Crumb complex, have been shown to interact with YAP/TAZ [5]. In fact an extended list of Hippo pathway regulators contains proteins involved in apicobasal polarity, planar cell polarity and the formation of cell-cell junctions [1]. With the advent of the Hippo interactome, the connection between YAP and junctional and polarity complexes is emerging [6–8]. Additionally, in epithelial cells, several positive regulators of the Hippo pathway, such as angiomotin [9], and Hippo pathway components themselves, such as Lats2 [10], co-localise with YAP/TAZ at the apical-lateral domain corresponding to tight junctions. However, how the junctional pool of YAP/TAZ is mobilised and activated by proteins within polarity complexes remains an open question. Despite the wealth of information about the kinases phosphorylating and repressing YAP/TAZ, much less is known about the proteins acting as positive switches. The mammalian Hippo interactome suggests that serine/threonine phosphatases play a key role in regulating the Hippo pathway [6]. Consistently, recent studies showed that protein phosphatase 1 (PP1) dephosphorylates YAP/TAZ *in vitro* [11,12]. PP1 is also known to interact with the polarity protein Par3 and induce its dephosphorylation *in vitro* [13]. Interestingly, recent evidence suggests that ASPP2, an interacting partner of Par3, YAP and TAZ, that also binds PP1 via its RVxF motif, may promote YAP activity, suggesting that it may link apicobasal polarity to the dephosphorylation of YAP/TAZ [7,11,14–17].

ASPP2 was originally characterised as an activator of the p53 family of proteins [18,19]. Recently it was also identified as an important regulator of cell polarity during central nervous system (CNS) development [14]. Mechanistically, ASPP2 regulates apicobasal polarity by interacting with Par3 via its N-terminal coiled-coiled region to form the apical-lateral polarity complex at tight junctions in epithelial cells [14,15]. The PPxY motif of ASPP2 was reported to interact with the WW domain of YAP *in vitro* [17] and the ASPP2/YAP interaction was confirmed *in vitro* by the recently reported mammalian Hippo interactome studies [6–8]. Despite this body of evidence, how ASPP2 mechanistically regulates YAP and the importance of the ASPP2/YAP interaction *in vivo* remain unknown.

To understand how YAP is activated *in vitro* and *in vivo*, in the context of polarised epithelial cells and tissues, we analysed the precise mechanism by which ASPP2 regulates junctional YAP, in parallel in epithelial cell lines and in the colon. Our results reveal the importance of ASPP2 within the apical-lateral polarity complex in linking apicobasal polarity to YAP activity via the recruitment of protein phosphatase 1 and the direct dephosphorylation of junctional YAP.

Results

ASPP2 and YAP form a complex at tight junctions

To understand how ASPP2 may control both apicobasal polarity and the activity of YAP, we examined their subcellular localisation in Caco-2 cells, a colorectal cancer cell line exhibiting strong epithelial characteristics and retaining the ability to polarise. Similar to previous observations in epithelial cells and tissues, ASPP2 co-localised with Par3 at tight junctions. Interestingly, in addition to its nuclear localisation, YAP was found to co-localise with ASPP2 at tight junctions, suggesting that they may form a complex at this level (**Figure 1A**). A similar observation could be made in polarised MDCK cells, as YAP co-localised with ASPP2 at the level of the apical-lateral domain where tight junctions reside (**Figure 1B and Figure S1A**). In addition, in colonic crypt cells, YAP was also expressed apically towards the lumen, in a localisation pattern reminiscent of ASPP2's, suggesting that, ASPP2 and YAP may also interact at tight junctions *in vivo* (**Figure 1C**). Importantly, reduced junctional and nuclear YAP signal following YAP knockdown could be observed, demonstrating the specificity of the antibody used. Of note, YAP depletion did not affect the localisation of ASPP2 at tight junctions, suggesting that YAP is not important for its subcellular localisation pattern (**Figure S1B**).

Sequence alignments of ASPP1, ASPP2 and iASPP reveal that only ASPP2 contains a WW domain binding PPxY motif within its proline-rich region (**Figure S1C**). Accordingly, ASPP1 was shown not to interact with YAP and neither ASPP1 or iASPP were identified as YAP-binding partners in the Hippo interactome studies [6,7,20]. In epithelial cells, ASPP2 co-immunoprecipitated with both Par3 and YAP, further suggesting that endogenous YAP and ASPP2 physically interact at tight junctions, where the majority of ASPP2 resides (**Figure 1D and Figure S1D**). Since YAP is regulated in a cell density-dependent manner, we tested whether the ASPP2/YAP interaction could be affected in a similar way. However, ASPP2/YAP complexes could be detected at all densities, suggesting that the ASPP2/YAP interaction is not regulated by cell contact inhibition (**Figure 1E and Figure S1D**). Collectively, these results define an ASPP2/YAP complex at tight junctions in epithelial cells and tissues, indicating that the apical-lateral domain may be the primary site of YAP regulation by ASPP2.

YAP phosphorylation at serine 127 (pYAP S127) is crucial in regulating its subcellular localisation [21,22]. Interestingly, some pYAP S127 co-localised with ASPP2 at cell-cell junctions (**Figure S1E**) and co-immunoprecipitated with ASPP2 in both Caco-2 and MDCK cells (**Figure 1E and Figure S1D**). To test the importance of YAP phosphorylation at S127 in regulating the ASPP2/YAP complex, we used Caco-2 cells stably expressing either wild type (hYAP-myc) or YAP mutated at S127 (hYAP-S127A-Flag). Exogenous wild type YAP adopted a localisation pattern reminiscent of endogenous YAP, partly co-localising with ASPP2 at tight junctions. However, hYAP-S127A-Flag was exclusively nuclear (**Figure 1F**) and was not able to co-immunoprecipitate with ASPP2 (**Figure 1G**). Collectively, these results reveal the importance of YAP phosphorylation at S127 in regulating its interaction with ASPP2 at tight junctions.

ASPP2 promotes the dephosphorylation of YAP

Since YAP phosphorylation can affect YAP localisation or stability [21,23], we investigated the impact of ASPP2 on YAP phosphorylation at S127. In Caco-2 cells, ASPP2 knockdown resulted in a marked increase in pYAP S127 without changes in YAP protein expression level (**Figure 2A**). This was observed in multiple cell lines, including MDCK cells (**Figure S2A**) and cancer cell lines of various epithelial origins (data not shown). Protein expression levels of ASPP1 and iASPP were not altered, suggesting that no compensatory effect occurred (**Figure 2A**). In addition, Par3 knockdown did not influence pYAP S127, further confirming the intrinsic role of ASPP2 (**Figure 2B**). Since YAP is directly phosphorylated by Lats kinases at S127 [21,24,25], we tested whether they mediate the increase in pYAP S127 following ASPP2 depletion (**Figure 2C**). As expected, Lats$_{1/2}$ knockdown resulted in the decrease of the ratio between pYAP S127 and total YAP protein level, whereas ASPP2 depletion resulted in its increase. However, when Lats$_{1/2}$ were depleted, ASPP2 depletion did not lead to an increased pYAPS127/YAP ratio, suggesting that this effect is Lats-dependent and that ASPP2 inhibits the Hippo pathway by antagonising Lats-mediated YAP phosphorylation.

Consistently, in the colon of $ASPP2^{\Delta exon3}$ mice, we observed increased pYAP S127, accompanied by a small decrease in total YAP protein level (**Figure S2B**). We next used $ASPP2^{\Delta 3loxP\text{-}CreER}$ mice to deplete ASPP2 expression in an acute manner using intraperitoneal injections of tamoxifen. Tamoxifen treatments resulted in a significant increase in the pYAPS127/YAP ratio, suggesting that ASPP2 also controls the phosphorylation of YAP at S127 *in vivo* (**Figure 2D and 2E**). Interestingly, as seen in Caco-2 cells four days following ASPP2 knockdown (**Figure 2B and 2C**), this was accompanied by a 2-fold decrease in total YAP protein level, confirming that the regulation of pYAP S127 by ASPP2 may have repercussions on the stability of YAP (**Figure 2D and 2F**).

The degradation of YAP is controlled by its phosphorylation status at several key serine residues [23]. To investigate whether ASPP2 controls additional YAP phosphorylation sites, we used several EJ stable cell lines in a Phos-Tag assay (**Figure 2G**). EJ cells stably expressing wild type YAP displayed up to five bands on a Phos-Tag gel whereas cells expressing a mutant YAP that cannot be phosphorylated [23] only displayed one. Following ASPP2 knockdown in wild type YAP-expressing EJ cells, we observed a shift from fast towards slowly migrating bands, corresponding to an increase in highly phosphorylated forms of YAP. These results indicate that ASPP2 controls the phosphorylation of YAP not only at S127 but also at other serine residues. Together, our results firmly establish ASPP2 as a repressor of YAP phosphorylation and the Hippo pathway.

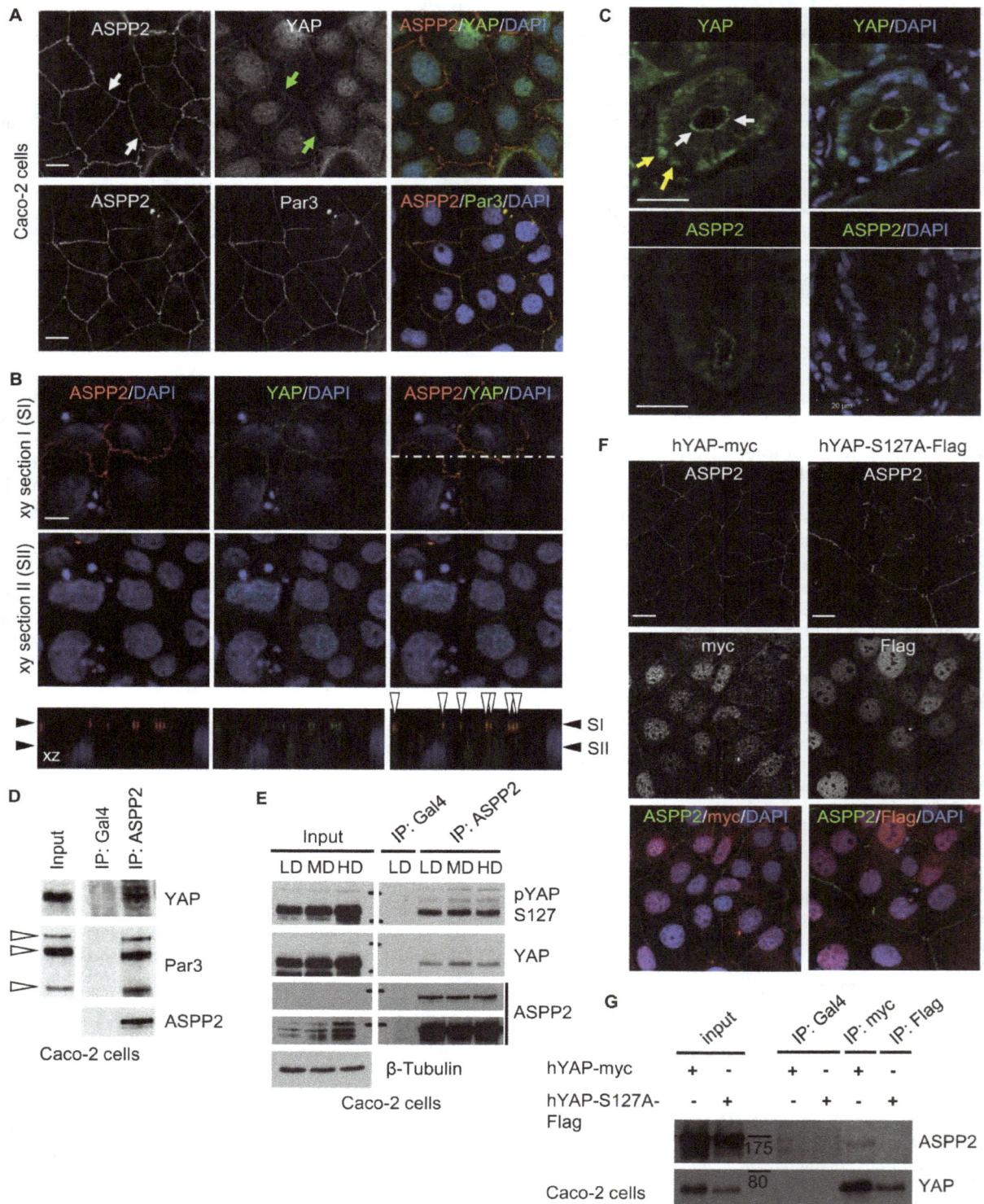

Figure 1. ASPP2 and YAP form a junctional complex in epithelial cells. (**A**) Immunostaining of ASPP2, Par3 and YAP in confluent monolayers of Caco-2 cells. White and green arrows point to tight junction localised ASPP2 and YAP respectively. Scale bars: 20 μm. (**B**) YAP and ASPP2 immunostaining in polarised MDCK cells. Sections SI and SII represent xy optical sections going through the apical-lateral domain and the middle of nuclei respectively. The bottom panel represents the xz section corresponding to the dashed line. SI and SII are shown with black arrowheads. White arrowheads show co-localisation of ASPP2 and YAP at the apical-lateral domain. Nuclei are counterstained with DAPI. Scale bar: 8 μm. (**C**) The localisation of YAP and ASPP2 was analysed by immunostaining of frozen sections obtained from wild type mice. YAP was apical (white arrows) and nuclear (yellow arrows) in the epithelial cells of colonic crypts. Nuclei are counterstained with DAPI. Scale bar: 20 μm. (**D**) The interaction between ASPP2 and Par3 was tested in Caco-2 cells. Endogenous ASPP2 was immunoprecipitated with an anti-ASPP2 mouse monoclonal antibody (DX50.13) and an anti-Gal4 mouse monoclonal antibody was used as a negative control. ASPP2, Par-3 and YAP were detected by SDS-Page/immunoblotting.

White arrowheads point to different Par-3 isoforms. (**E**) ASPP2 and YAP co-immunoprecipitation in Caco-2 cells plated at different cell densities. Lysates were obtained from Caco-2 cells plated at various cell densities and endogenous ASPP2 was immunoprecipitated with an anti-ASPP2 mouse monoclonal antibody (DX50.13) and an anti-Gal4 mouse monoclonal antibody was used as a negative control. ASPP2, YAP and YAP phosphorylated at S127 were subsequently detected by SDS-Page/immunoblotting. Long and short exposures are shown for ASPP2. β-tubulin was used as loading control. LD: low density; MD: medium density; HD: high density. (**F-G**) The phosphorylation status of YAP regulates its subcellular localisation and interaction with ASPP2. Stable Caco-2 cells expressing either hYAP-myc or hYAP-S127A-Flag were used to test the requirement of YAP phosphorylation at serine 127 for its interaction with ASPP2 and its junctional localisation. (**F**) hYAP-myc, hYAP-S127A-Flag and endogenous ASPP2 were detected by immunostaining. DAPI was used to stain nuclei. Scale bars: 20 μm. (**G**) hYAP-myc and hYAP-S127A-Flag were immunoprecipitated using an anti-myc monoclonal (9E10) and anti-Flag monoclonal antibody respectively. ASPP2 and YAP were subsequently detected by SDS-Page/immunoblotting. Molecular markers are indicated in the figure (values in kDa).

ASPP2 binds and dephosphorylates YAP via the recruitment of PP1

Since ASPP2 can directly bind PP1 [16,26] and PP1 has been shown to localise at tight junctions in epithelial cells [13], we hypothesised that ASPP2 may dephosphorylate YAP via the recruitment of PP1. Interestingly, in MDCK cells stably expressing an shRNA against endogenous canine ASPP2, both wild type and a PP1 binding defective ASPP2 mutant (ASPP2 (RAKA)-V5) [16] displayed the same junctional pattern, suggesting that ASPP2's localisation is not dependent on its interaction with PP1 (**Figure S3A**). In agreement, both constructs could interact with endogenous Par3 (**Figure S3B**). When wild type ASPP2 and YAP were co-transfected into HEK293 cells, pYAP S127 levels decreased, whereas they remained unchanged in the presence of ASPP2 (RAKA)-V5 (**Figure 3A**). In addition, wild type ASPP2, as opposed to ASPP2 (RAKA)-V5, decreased the amount of highly phosphorylated forms of YAP (**Figure 3B**). Together, these results demonstrate that ASPP2 dephosphorylates YAP via the recruitment of PP1. Finally, a mutant of ASPP2 containing two mutations in its YAP-binding motif (Y869A/Y874A) could not interact with YAP and was consequently unable to dephosphorylate pYAP S127, indicating that ASPP2 operates via the recruitment of YAP to directly induce its dephosphorylation (**Figure 3C**). Furthermore, partially contradicting a previous report and agreeing with the lack of a WW binding motif in their proline-rich region, neither ASPP1 or iASPP could interact with YAP to induce its dephosphorylation at S127 (**Figure 3D**) [20]. Collectively, these results highlight the role of ASPP2 as a PP1 regulatory subunit by scaffolding both phosphatase and substrate to specifically dephosphorylate YAP.

ASPP2 enhances the role of YAP as a TEAD transcription cofactor

YAP phosphorylation at S127 induces its export from the nucleus to the cytoplasm via the binding to 14-3-3. In Caco-2 cells, following ASPP2 depletion, we could observe a discrete increase in cytoplasmic YAP at low cell density. However, at high cell density, ASPP2-depleted cells exhibited a striking decrease in nuclear YAP, suggesting that ASPP2, by dephosphorylating pYAP S127, may indeed promote its nuclear localisation (**Figure 4A**). To test whether this would have consequences on the function of YAP as a transcriptional co-regulator, we used a TEAD reporter in Caco-2 cells (**Figure 4B**) [3]. Interestingly, transfecting YAP or ASPP2 alone had little effect on the reporter construct, suggesting that the endogenous machinery was able to regulate the excess of exogenous YAP or ASPP2. Moreover, over-expression of ASPP2 alone, despite reducing cytoplasmic YAP expression, did not seem to increase YAP nuclear localisation (**Figure S4A**). However, co-expression of both YAP and ASPP2 lead to a robust increase in luciferase activity, suggesting that together they synergise to overcome endogenous repression. Strikingly, ASPP2 (RAKA)-V5 was less effective than wild type ASPP2, consistent with ASPP2

acting through PP1 to dephosphorylate and activate YAP. In addition to TEAD1-4, YAP regulates a wide variety of transcription factors, including p73 [27]. Furthermore, ASPP2 is known to promote the apoptotic function of p53, p63 and p73 [18,19]. However, as opposed to the TEAD reporter, ASPP2 and YAP could not modulate the activity of a Bax-reporter in Caco-2 cells, therefore suggesting that, under unstressed conditions, they regulate TEAD rather than p53 family members (**Figure S4B**). In addition, ASPP2 did not further induce the activity of p73 on the Bax-luciferase reporter and only had a moderate effect on p53. Mutations in the YAP or PP1 binding sites of ASPP2 did not modify its ability to induce p53, suggesting that this effect is YAP independent (**Figure S4C and S4D**).

Together, these data demonstrate the role of ASPP2 in promoting the nuclear localisation and transcriptional activity of YAP *in vitro*. Recent studies have shown the importance of YAP during colonic regeneration [28,29]. Since ASPP2 and YAP also potentially form a junctional complex in the colon (**Figure 1C**), we investigated whether ASPP2 could influence YAP activity by analysing its subcellular localisation in the colonic crypts of wild type and $ASPP2^{\Delta exon3}$ mice (**Figure 4C**). As previously described, YAP was predominantly nuclear in the epithelial cells of wild type colonic crypts [28]. However, in $ASPP2^{\Delta exon3}$ crypts, nuclei exhibited reduced YAP expression, suggesting that ASPP2 promotes the nuclear localisation of YAP *in vivo*. Interestingly, CTGF mRNA level, an established endogenous readout of YAP/TAZ activity, was markedly reduced in $ASPP2^{\Delta exon3}$ colonic crypts, suggesting that ASPP2 stimulates the transcriptional function of YAP *in vivo* (**Figure 4D**). Collectively our results indicate that ASPP2, at the level of the apical-lateral domain, via the recruitment of PP1, induces the dephosphorylation of junctional YAP which subsequently leads to its activation *in vitro* and *in vivo*, in colonic crypts (**Figure 4E**).

Discussion

Our findings highlight how ASPP2, in addition to its role in regulating apicobasal polarity, acts as a scaffold for YAP and PP1 at the level of the apical-lateral domain in epithelial cells, *in vitro* and *in vivo*, therefore allowing PP1 to dephosphorylate junctional YAP.

Several regulators of apicobasal polarity such as Crumbs and AMOT have previously been associated with the regulation of YAP activity [5,9,30]. These proteins repress YAP by promoting the activity of the Hippo pathway. Likewise, ASPP2 can regulate apicobasal polarity by interacting with Par3 and regulating its localisation [14,15]. However, as opposed to AMOT and Crumbs, ASPP2 represents a positive regulator of YAP and may serve to counterbalance their activity. The fact that positive and negative regulators of YAP play a role in the establishment of apicobasal polarity and are localised at tight junctions underlies the importance of the apical lateral domain in epithelial cells as a key YAP-regulating hub. The recent unveiling of the Hippo

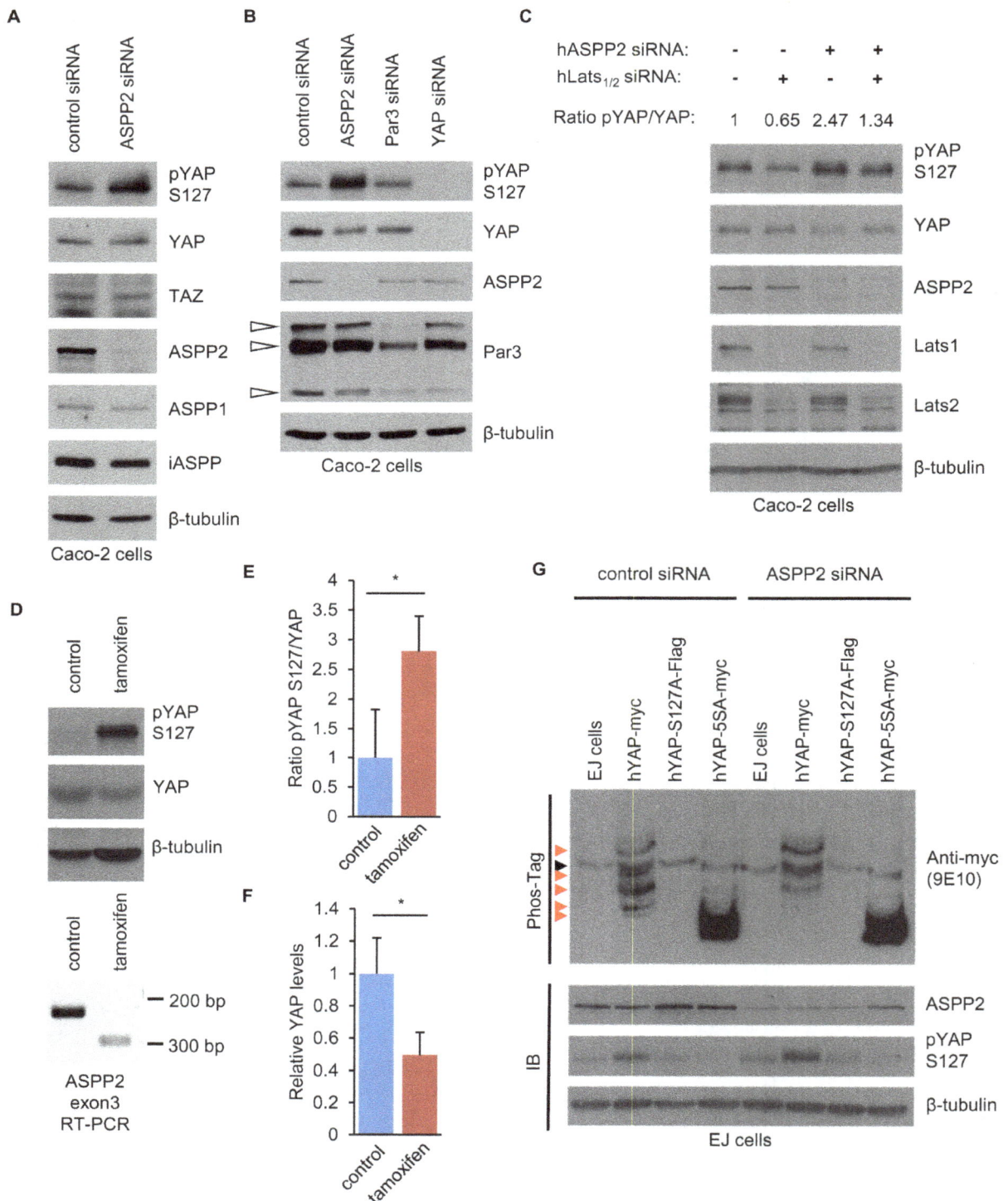

Figure 2. ASPP2 regulates YAP phosphorylation level *in vitro* and *in vivo*. (A) ASPP2 depletion leads to increased phosphorylated YAP at serine 127. ASPP2 was depleted in Caco-2 cells using siRNA and SDS-Page/immunoblotting was subsequently performed to analyse the expression level of the indicated proteins. An antibody specifically recognising YAP phosphorylated at serine 127 was used to analyse phosphorylation levels at this particular residue (A-E and G). **(B)** Par3 depletion does not affect the phosphorylation of YAP at S127. SiRNA knockdown was performed to deplete the indicated proteins and their expression was analysed by SDS-Page/immunoblotting. White arrowheads point to different Par-3 isoforms. **(C)** ASPP2 negatively regulates Lats-dependent phosphorylation of YAP at S127. SiRNA against ASPP2 and a combination of siRNA targeting Lats1 and lats2 were used as indicated in the figure. Four days following siRNA knockdown, protein levels were analysed by SDS-Page/immunoblotting. The ratios between pYAPS127 and YAP levels are indicated in the figure. **(D-F)** ASPP2 promotes the dephosphorylation of YAP at S127 in the colon. Exon 3 of *ASPP2* was deleted by injection of tamoxifen in *ASPP2*$^{\Delta3loxP\text{-}CreER}$ mice. **(D)** Upper panel: YAP and pYAP S127 levels were analysed by SDS-Page/immunoblotting using lysates obtained from control and tamoxifen-injected mice (5 controls and 3 tamoxifen-injected animals). β-tubulin was used as loading control. Lower panel: RT-PCR showing deletion of exon3 of ASPP2 following tamoxifen treatment using RNA isolated from the colon of the

same mouse. (E-F) Bar graphs representing the ratio between pYAP S127 and total YAP protein levels (E) and YAP protein expression levels normalised by β-tubulin (F). Data was normalised to the control. Error bars indicate standard deviation (*: p<0.05). (G) ASPP2 potentially induces the dephosphorylation of YAP at several serine residues. Following ASPP2 knockdown, lysates obtained from EJ cells stably expressing the indicated constructs were analysed by SDS-Page/immunoblotting using Phos-Tag (upper panel) or not (lower panel). Red arrowheads indicate differentially phosphorylated YAP proteins. The black arrowhead points to a non-specific band recognised by the anti-myc antibody (9E10).

interactome strongly reinforces this view by outlining extensive links between core Hippo pathway components and regulators of apicobasal polarity including ASPP2 [6–8]. Our study provides mechanistic insight into how one such interaction modulates YAP activity in a biologically significant context like the colon crypt.

The localisation of ASPP2, and potentially a whole group of other Hippo regulators, at the apical-lateral domain in epithelial cells is unlikely to be coincidental. It is tempting to speculate that its position may be important to integrate upstream information via the modulation of YAP activity which would consequently translate into an adapted transcriptional program. Our data suggests that in the colon, without ASPP2, YAP signalling is repressed, suggesting there exists a strong relationship between apicobasal polarity and YAP activity. This raises the possibility that ASPP2 may be required to translate polarity cues into an adapted transcriptional program via the activation of YAP. Considering the importance of YAP during colonic regeneration, the regulation of YAP by ASPP2 may be of particular relevance during tissue regeneration in order to coordinate correct tissue architecture and growth control.

Our study explores a function of ASPP2 that may in fact play roles in the control of the architecture, growth and homeostasis of other epithelial tissues. Born $ASPP2^{\Delta exon3}$ mice are smaller than their wild type litter mates which would agree with ASPP2 contributing to the control of organ size [31]. Moreover, although the mechanism by which ASPP2 controls apicobasal polarity during CNS development is relatively well-understood, how it controls the proliferation of neural progenitors remains unknown [14]. YAP has been shown to control the proliferation of neural progenitors and, as a result, it is not impossible that the proliferation phenotype observed in $ASPP2^{\Delta exon3}$ mice may be linked to abnormal YAP activity [32].

In conclusion, ASPP2, by linking apicobasal polarity to YAP activity, represents an exciting regulator of the Hippo pathway and future studies will define whether it may become an attractive target to modulate the activity of YAP in a disease context.

Materials and Methods

Ethics statement

All work involving animals has been carried out under a licence issued by the Home Office and have been approved by the Home Office.

Histology and tissue section immunostaining

The colons of 3 month old mice were dissected and washed in PBS. Tissues were kept in 10% formalin overnight. For frozen sections, the tissues were treated with 20% sucrose overnight, embedded in OCT and snap frozen in dry ice. Serial 15µm sections were obtained. Sections were then incubated in primary antibody diluted in 5% normal goat serum in PBS overnight at 4°C followed by incubation with Alexa Fluor secondary antibodies for 30 minutes. Nuclei were counterstained with DAPI (Roche). Slides were finally mounted onto coverslips.

Immunoprecipitation and SDS-PAGE/Immunoblotting

For immunoprecipitation experiments, HEK293 cells were transfected using Lipofectamine 2000 (Life Technologies) according to the manufacturer's protocol. Caco-2, EJ, MDCK, transfected HEK293 cells and colon samples were lysed in RIPA 1% IGEPAL for SDS-PAGE/Immunoblotting or a buffer containing 50 mM Tris-HCl at pH 8, 150 mM NaCl, 1 mM EDTA, Complete Protease Inhibitor Cocktail (Roche) and 1% Triton X-100 for immunoprecipitation. Lysates were then subjected to immunoprecipitation and SDS-PAGE/Immunoblotting as previously described (Yap et al., 2000). Densitometry analysis was performed using ImageJ. Phos-Tag gels were prepared according to the manufacturer's protocol.

Luciferase assay

Caco-2 cells were plated in 24-well plates and the next day transfected using Lipofectamine 2000 (Life Technologies) with the indicated constructs. The following amounts of DNA were used per well in the assay: Luciferase reporter, 200 ng; Renilla-luciferase, 10 ng; hYAP-myc, 50 to 200 ng; ASPP2-V5 and mutants, 800 ng; p73, 50 ng. PCDNA3.1 was used to keep DNA total amounts constant from one condition to another. The assay was performed 24 hours after transfection using the Dual Glo luciferase assay system (Promega) according to the manufacturer's protocol.

Real-Time Quantitative PCR

Real-time quantitative PCR (qRT-PCR) was performed as previously described with the 7500 real-time PCR system (Applied Biosystems) using the QuantiTect SYBR Green PCR kit (Qiagen) [33]. In short, each reaction was performed in triplicate using 1 µL of cDNA in a final volume of 25 µL. The expression level of CTGF was analyzed based on the ΔΔCt method, with GAPDH as an internal control. The following thermal cycle was used for all samples: 15 min at 95°C; 45 cycles of 15 s at 94°C, 30 s primer-specific annealing temperatures, 1 min at 72°C. For each experiment, the threshold was set to cross a point at which realtime PCR amplification was linear.

siRNA knockdown

siRNA oligos against human ASPP2, Lats1, Lats2, Par3, YAP and RISC-Free siRNA were purchased from Dharmacon. Cells were transfected with the indicated siRNA oligos at a final concentration of 35 nM using Dharmafect 1 reagent (Dharmacon) for 3 to 5 days, according to the manufacturer's instructions.

Immunocytochemistry

Caco-2 cells were seeded onto coverglasses. Lipofectamine 2000 was used for transient expression of ASPP2-V5 in Caco-2 cells. MDCK cells were plated on 6.5 mm diameter Transwell filters (Corning) with a 0.4 mm pore size as previously described [14]. Once experimental procedures had been carried out, cells were fixed with 4% paraformaldehyde in PBS for 10 min and then permeabilized with 0.1% Triton X-100 in PBS for 4 min. PBS containing 2% BSA was used as a blocking solution for 20 minutes prior to incubation with primary antibodies. Primary antibodies were diluted in PBS containing 2% of BSA, and applied to cells for

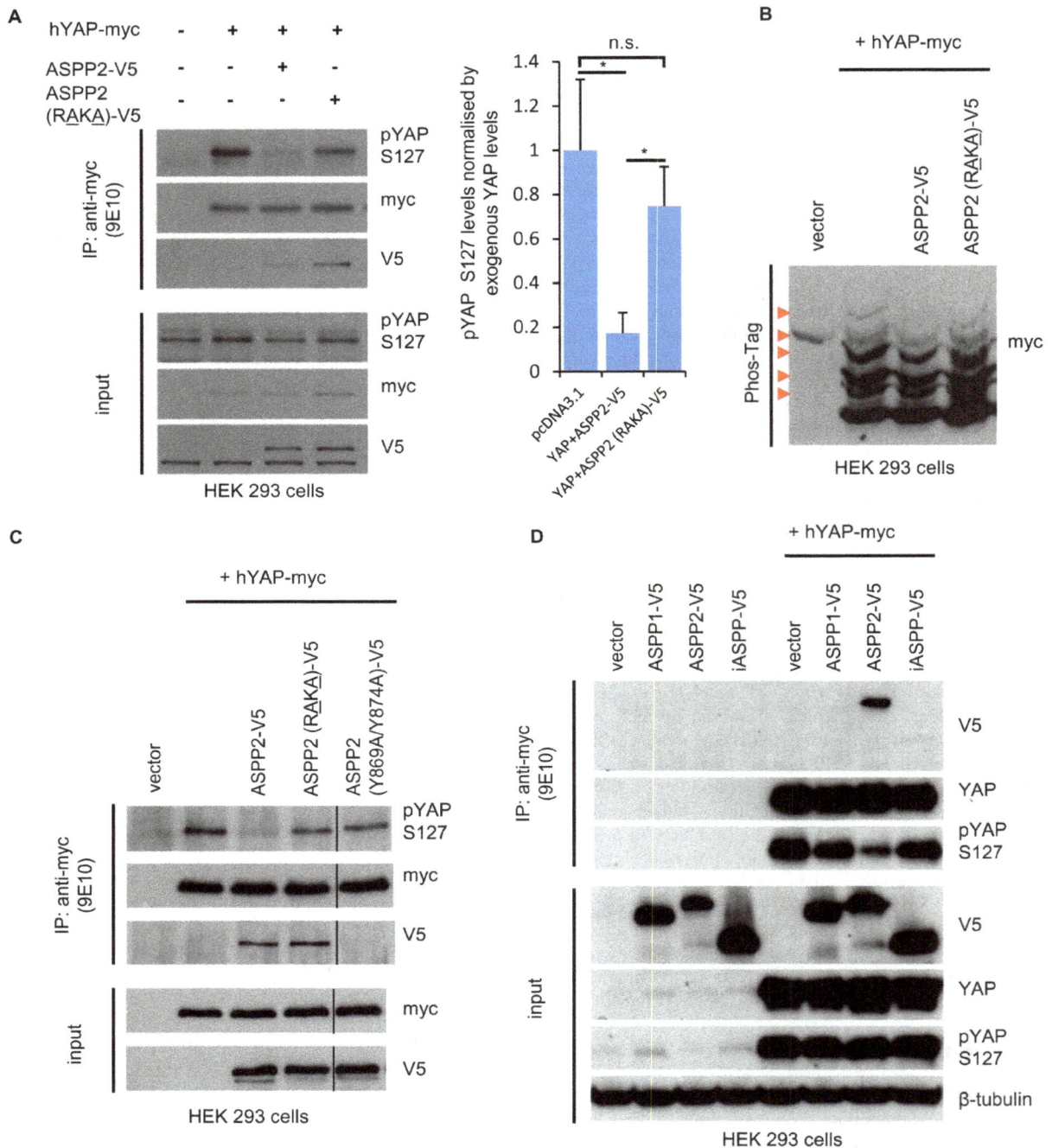

Figure 3. ASPP2 scaffolds PP1 to dephosphorylate YAP. (**A**) The importance of the ASPP2/PP1 interaction in dephosphorylating YAP was tested by transfecting the indicated constructs in HEK293 cells. Exogenous YAP was subsequently immunoprecipitated using an anti-myc mouse monoclonal antibody (9E10) and SDS-Page/immunoblotting was performed using the indicated antibodies. The bar graph represents the ratio between pYAP S127 and total YAP protein levels (n = 3; *: p<0.05; n.s.: non-significant). Error bars indicate standard deviation. (**B**) Lysates obtained from HEK293 cells transfected with the indicated constructs were analysed by SDS-PAGE/immunoblotting on a Phos-Tag gel. (**C**) The ability of ASPP2 (Y869A/Y874A)-V5 to dephosphorylate YAP was tested in HEK293 cells to investigate the importance of the YAP/ASPP2 interaction in this process. Following transfection with the indicated constructs in HEK293 cells, Exogenous YAP was immunoprecipitated using an anti-myc mouse monoclonal antibody (9E10) and SDS-Page/immunoblotting was performed using the indicated antibodies. (**D**) The ability of ASPP2, iASPP and ASPP1-V5 to interact with and dephosphorylate YAP was tested in HEK293 cells. Following transfection with the indicated constructs in HEK293 cells, Exogenous YAP was immunoprecipitated using an anti-myc mouse monoclonal antibody (9E10) and SDS-Page/immunoblotting was performed using the indicated antibodies.

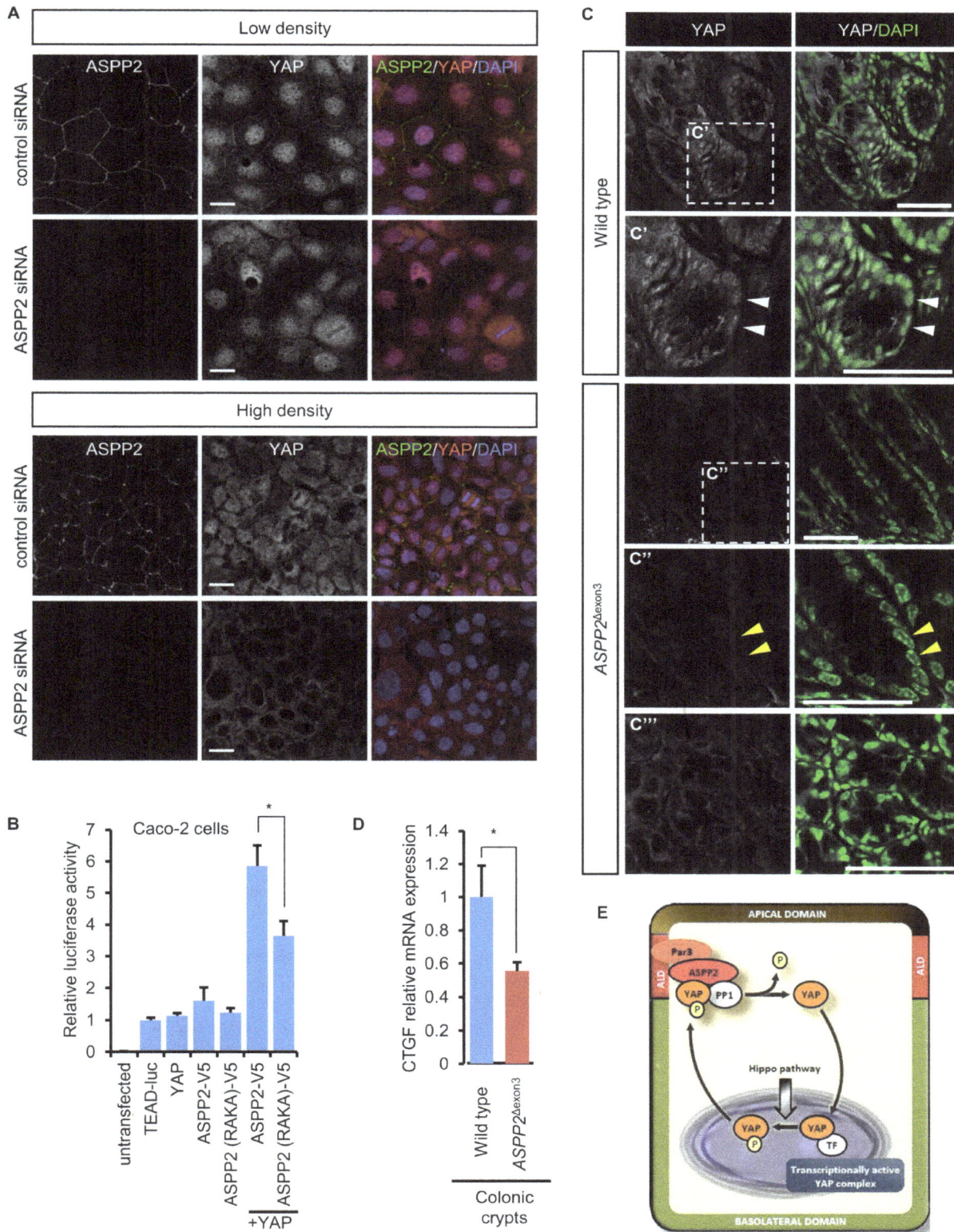

Figure 4. ASPP2 promotes the transcriptional activity of YAP. (**A**) Following transfection of Caco-2 cells with control or ASPP2 siRNA, the localisation of YAP and ASPP2 was analysed by immunostaining at low and high cell density. Scale bar: 20 μm. (**B**) The ability of wild type ASPP2 and ASPP2 (RAKA)-V5 to regulate TEAD-mediated transcription was analysed in a luciferase assay using the TEAD-luciferase reporter (8xGTIIC-luciferase) in Caco-2 cells. Values were obtained from three independent duplicate experiments and error bars indicate standard deviation (*: p<0.05). (**C**) YAP immunostaining of paraffin sections obtained from the colons of wild type and ASPP2$^{\Delta exon3}$ mice. Dashed white squares highlight magnified areas represented in the corresponding panels (C'-C'"). White arrowheads point to YAP positive nuclei in wild type crypts whereas yellow arrowheads point

to nuclei devoid of YAP in $ASPP2^{\Delta exon3}$ crypts. Nuclei were counterstained with DAPI. Scale bars: 50 µm. (**D**) *CTGF* mRNA levels were quantified by qRT-PCR using RNA obtained from the colons of wild type (n = 3) and $ASPP2^{\Delta exon3}$ mice (n = 3). Error bars represent standard deviation (*: p<0.05). (**E**) Diagram representing the regulation of YAP by ASPP2 in epithelial cells. Once YAP is phosphorylated at serine 127, it can interact with ASPP2 at the apical lateral domain where ASPP2 induces its dephosphorylation via the recruitment of PP1. YAP is consequently able to relocalise to the nucleus where it can modulate TEAD transcriptional activity.

40 minutes. Protein expression was detected using Alexa Fluor (1:400, Molecular Probes) for 20 min. DAPI (Invitrogen) was used to stain nuclei (1:2000).

Statistical analysis

The T-test was used to calculate the statistical significance between two measurements. Differences were considered significant at a value of p≤0.05

Supporting Information

Figure S1 (**A**) Maximum intensity projection of YAP and ASPP2 immunostaining in polarised MDCK cells. Nuclei are counterstained with DAPI. (**B**) YAP and ASPP2 immunostaining in Caco-2 cells transfected with control or YAP siRNA. Note the reduced junctional and nuclear YAP signal following YAP knockdown. (**C**) Protein sequence alignments performed using ClustalW. Left panel: amongst all ASPP family members, the YAP-binding motif is present in human ASPP2 only. Right panel: As opposed to the PP1-binding domain of ASPP2 that is conserved across evolution, the YAP-binding motif is only present in vertebrates. (**D**) Co-immunoprecipitation of ASPP2 and YAP in MDCK cells plated at different cell densities. Lysates were obtained from MDCK cells plated at various cell densities and endogenous ASPP2 was immunoprecipitated with an anti-ASPP2 mouse monoclonal antibody (DX50.13) and an anti-Gal4 mouse monoclonal antibody was used as a negative control. YAP and YAP phosphorylated at S127 were subsequently detected by SDS-Page/immunoblotting. β-tubulin was used as loading control. LD: low density; MD: medium density; HD: high density. (**E**) The localisation of YAP phosphorylated at S127 and ASPP2 was analysed by immunostaining in Caco-2 cells transfected with control or YAP siRNA. E′ and E″ are magnified views of the corresponding dashed areas. White arrowheads point to junctional YAP phosphorylated at S127.

Figure S2 (**A**) ASPP2 depletion leads to increased phosphorylated YAP at serine 127. ASPP2 was depleted in MDCK cells using an shRNA against canine ASPP2 and the levels of the indicated proteins were analysed by SDS-Page/immunoblotting. (**B**) SDS-Page/immunoblotting was performed on lysates obtained from the colons of wild type or $ASPP2^{\Delta exon3}$ mice to detect the expression levels of the indicated proteins. β-tubulin was used as a loading control.

Figure S3 (**A**) The localisation of ASPP2-V5 and ASPP2 (R<u>A</u>K<u>A</u>)-V5 was examined by immunostaining in MDCK cells depleted of endogenous canine ASPP2. Note the distribution of both constructs at cell-cell junctions. (**B**) The ability of iASPP-V5, ASPP2-V5 and ASPP2 (R<u>A</u>K<u>A</u>)-V5 to interact with endogenous Par3 was tested in U2OS cells. An anti-V5 mouse monoclonal antibody was used to immunoprecipitate these constructs and SDS-Page/immunoblotting was subsequently performed using the indicated antibodies. Black arrowheads point to ASPP2-V5 and iASPP-V5 respectively. White arrowheads point to different Par3 isoforms. Note that, as opposed to iASPP-V5, both ASPP2-V5 and ASPP2 (R<u>A</u>K<u>A</u>)-V5 could co-immunoprecipitate with endogenous Par3. As previously described, ASPP2 (R<u>A</u>K<u>A</u>)-V5 did not interact with endogenous $PP1_\alpha$.

Figure S4 (**A**) ASPP2-V5 was transfected into Caco-2 cells and its effect on endogenous YAP localisation was subsequently analysed by immunostaining. (**B**) The ability of ASPP2 and YAP to regulate transcriptional events mediated by TEAD or Bax was tested in a luciferase assay in Caco-2 cells. (**C-D**) The ability of ASPP2 to regulate the transcriptional function of p73 (B) and p53 (C) was analysed in Caco-2 cells using a Bax-luciferase reporter. Of note, ASPP2 (RAKA)-V5 and ASPP2 (Y869A/Y874A)-V5 behaved similarly to wild type ASPP2-V5 when co-expressed with p73 or p53.

File S1 Information on mouse models and reagents. This file contains detailed information on mouse colonies, cell lines, primary antibodies, plasmids and primers used in the study.

Acknowledgments

We thank Professor Colin Goding and Dr Paola Falletta for assistance on the Phos-Tag experiments and reagents.

Author Contributions

Conceived and designed the experiments: CR XL. Performed the experiments: CR SK JZ XQ LB ED. Analyzed the data: CR SK LB SS XL. Contributed reagents/materials/analysis tools: JZ LB SZ IR. Wrote the paper: CR SS XL.

References

1. Yu F, Guan K (2013) The Hippo pathway: regulators and regulations. 1: 355–371. doi:10.1101/gad.210773.112.a.

2. Yu F-X, Zhao B, Panupinthu N, Jewell JL, Lian I, et al. (2012) Regulation of the Hippo-YAP pathway by G-protein-coupled receptor signaling. Cell 150: 780–791. Available: http://www.pubmedcentral.nih.gov/articlerender.fcgi?artid=3433174&tool=pmcentrez&rendertype=abstract. Accessed 2013 Nov 6.

3. Dupont S, Morsut L, Aragona M, Enzo E, Giulitti S, et al. (2011) Role of YAP/TAZ in mechanotransduction. Nature 474: 179–183. Available: http://www.ncbi.nlm.nih.gov/pubmed/21654799. Accessed 2013 Mar 4.

4. Genevet A, Tapon N (2011) The Hippo pathway and apico-basal cell polarity. Biochem J 436: 213–224. Available: http://www.ncbi.nlm.nih.gov/pubmed/21568941. Accessed 2013 Mar 22.

5. Varelas X, Samavarchi-Tehrani P, Narimatsu M, Weiss A, Cockburn K, et al. (2010) The Crumbs complex couples cell density sensing to Hippo-dependent control of the TGF-β-SMAD pathway. Dev Cell 19: 831–844. Available: http://www.ncbi.nlm.nih.gov/pubmed/21145499. Accessed 2013 Feb 28.

6. Couzens a L, Knight JDR, Kean MJ, Teo G, Weiss a., et al. (2013) Protein Interaction Network of the Mammalian Hippo Pathway Reveals Mechanisms of Kinase-Phosphatase Interactions. Sci Signal 6: rs15–rs15. Available: http://stke.sciencemag.org/cgi/doi/10.1126/scisignal.2004712. Accessed 2013 Nov 20.

7. Hauri S, Wepf A, van Drogen A, Varjosalo M, Tapon N, et al. (2013) Interaction proteome of human Hippo signaling: modular control of the co-activator YAP1. Mol Syst Biol 9: 713. Available: http://www.pubmedcentral.nih.gov/articlerender.fcgi?artid=4019981&tool=pmcentrez&rendertype=abstract.

8. Wang W, Li X, Huang J, Feng L, Dolinta KG, et al. (2014) Defining the protein-protein interaction network of the human hippo pathway. Mol Cell Proteomics 13: 119–131. Available: http://www.ncbi.nlm.nih.gov/pubmed/24126142. Accessed 2014 Mar 24.

9. Zhao B, Li L, Lu Q, Wang LH, Liu C-Y, et al. (2011) Angiomotin is a novel Hippo pathway component that inhibits YAP oncoprotein. Genes Dev 25: 51–63. Available: http://www.pubmedcentral.nih.gov/articlerender.fcgi?artid=3012936&tool=pmcentrez&rendertype=abstract. Accessed 2013 May 27.

10. Paramasivam M, Sarkeshik A, Yates JR, Fernandes MJG, McCollum D (2011) Angiomotin family proteins are novel activators of the LATS2 kinase tumor suppressor. Mol Biol Cell 22: 3725–3733. doi:10.1091/mbc.E11-04-0300.

11. Liu C-Y, Lv X, Li T, Xu Y, Zhou X, et al. (2011) PP1 cooperates with ASPP2 to dephosphorylate and activate TAZ. J Biol Chem 286: 5558–5566. Available: http://www.pubmedcentral.nih.gov/articlerender.fcgi?artid=3037669&tool=pmcentrez&rendertype=abstract. Accessed 2013 Jun 3.

12. Wang P, Bai Y, Song B, Wang Y, Liu D, et al. (2011) PP1A-mediated dephosphorylation positively regulates YAP2 activity. PLoS One 6: e24288. Available: http://www.pubmedcentral.nih.gov/articlerender.fcgi?artid=3164728&tool=pmcentrez&rendertype=abstract. Accessed 2013 Apr 19.

13. Traweger A, Wiggin G, Taylor L, Tate SA, Metalnikov P, et al. (2008) Protein phosphatase 1 regulates the phosphorylation state of the polarity scaffold Par-3. Proc Natl Acad Sci U S A 105: 10402–10407. Available: http://www.pubmedcentral.nih.gov/articlerender.fcgi?artid=2475498&tool=pmcentrez&rendertype=abstract.

14. Sottocornola R, Royer C, Vives V, Tordella L, Zhong S, et al. (2010) ASPP2 binds Par-3 and controls the polarity and proliferation of neural progenitors during CNS development. Dev Cell 19: 126–137. Available: http://www.ncbi.nlm.nih.gov/pubmed/20619750. Accessed 2013 Aug 7.

15. Cong W, Hirose T, Harita Y, Yamashita A, Mizuno K, et al. (2010) ASPP2 regulates epithelial cell polarity through the PAR complex. Curr Biol 20: 1408–1414. Available: http://www.ncbi.nlm.nih.gov/pubmed/20619648. Accessed 2013 Aug 19.

16. Llanos S, Royer C, Lu M, Bergamaschi D, Lee WH, et al. (2011) Inhibitory member of the apoptosis-stimulating proteins of the p53 family (iASPP) interacts with protein phosphatase 1 via a noncanonical binding motif. J Biol Chem 286: 43039–43044. Available: http://www.pubmedcentral.nih.gov/articlerender.fcgi?artid=3234852&tool=pmcentrez&rendertype=abstract. Accessed 2013 Jun 4.

17. Espanel X, Sudol M (2001) Yes-associated protein and p53-binding protein-2 interact through their WW and SH3 domains. J Biol Chem 276: 14514–14523. Available: http://www.ncbi.nlm.nih.gov/pubmed/11278422. Accessed 2013 May 27.

18. Samuels-Lev Y, O'Connor DJ, Bergamaschi D, Trigiante G, Hsieh JK, et al. (2001) ASPP proteins specifically stimulate the apoptotic function of p53. Mol Cell 8: 781–794. Available: http://www.ncbi.nlm.nih.gov/pubmed/11684014.

19. Bergamaschi D, Samuels Y, Jin B, Crook T, Lu X, et al. (2004) ASPP1 and ASPP2: Common Activators of p53 Family Members ASPP1 and ASPP2: Common Activators of p53 Family Members. 24. doi:10.1128/MCB.24.3.1341.

20. Vigneron AM, Ludwig RL, Vousden KH (2010) Cytoplasmic ASPP1 inhibits apoptosis through the control of YAP. Genes Dev 24: 2430–2439. Available: http://www.pubmedcentral.nih.gov/articlerender.fcgi?artid=2964753&tool=pmcentrez&rendertype=abstract. Accessed 2013 Jun 3.

21. Zhao B, Wei X, Li W, Udan RS, Yang Q, et al. (2007) Inactivation of YAP oncoprotein by the Hippo pathway is involved in cell contact inhibition and tissue growth control. Genes Dev 21: 2747–2761. Available: http://www.pubmedcentral.nih.gov/articlerender.fcgi?artid=2045129&tool=pmcentrez&rendertype=abstract. Accessed 2013 May 27.

22. Schlegelmilch K, Mohseni M, Kirak O, Pruszak J, Rodriguez JR, et al. (2011) Yap1 acts downstream of α-catenin to control epidermal proliferation. Cell 144: 782–795. Available: http://www.pubmedcentral.nih.gov/articlerender.fcgi?artid=3237196&tool=pmcentrez&rendertype=abstract. Accessed 2013 May 30.

23. Zhao B, Li L, Tumaneng K, Wang C-Y, Guan K-L (2010) A coordinated phosphorylation by Lats and CK1 regulates YAP stability through SCF(beta-TRCP). Genes Dev 24: 72–85. Available: http://www.pubmedcentral.nih.gov/articlerender.fcgi?artid=2802193&tool=pmcentrez&rendertype=abstract. Accessed 2013 Jun 3.

24. Huang J, Wu S, Barrera J, Matthews K, Pan D (2005) The Hippo signaling pathway coordinately regulates cell proliferation and apoptosis by inactivating Yorkie, the Drosophila Homolog of YAP. Cell 122: 421–434. Available: http://www.ncbi.nlm.nih.gov/pubmed/16096061. Accessed 2013 May 28.

25. Oh H, Irvine KD (2008) In vivo regulation of Yorkie phosphorylation and localization. Development 135: 1081–1088. Available: http://www.pubmedcentral.nih.gov/articlerender.fcgi?artid=2387210&tool=pmcentrez&rendertype=abstract. Accessed 2013 May 28.

26. Skene-Arnold TD, Luu HA, Uhrig RG, De Wever V, Nimick M, et al. (2013) Molecular mechanisms underlying the interaction of protein phosphatase-1c with ASPP proteins. Biochem J 449: 649–659. Available: http://www.ncbi.nlm.nih.gov/pubmed/23088536. Accessed 2013 Jun 4.

27. Strano S, Monti O, Pediconi N, Baccarini A, Fontemaggi G, et al. (2005) The transcriptional coactivator Yes-associated protein drives p73 gene-target specificity in response to DNA Damage. Mol Cell 18: 447–459. Available: http://www.ncbi.nlm.nih.gov/pubmed/15893728. Accessed 2013 May 29.

28. Cai J, Zhang N, Zheng Y, de Wilde RF, Maitra A, et al. (2010) The Hippo signaling pathway restricts the oncogenic potential of an intestinal regeneration program. Genes Dev 24: 2383–2388. Available: http://www.pubmedcentral.nih.gov/articlerender.fcgi?artid=2964748&tool=pmcentrez&rendertype=abstract. Accessed 2013 Jun 5.

29. Barry ER, Morikawa T, Butler BL, Shrestha K, de la Rosa R, et al. (2013) Restriction of intestinal stem cell expansion and the regenerative response by YAP. Nature 493: 106–110. Available: http://www.ncbi.nlm.nih.gov/pubmed/23178811. Accessed 2013 Mar 1.

30. Wells CD, Fawcett JP, Traweger A, Yamanaka Y, Goudreault M, et al. (2006) A Rich1/Amot complex regulates the Cdc42 GTPase and apical-polarity proteins in epithelial cells. Cell 125: 535–548. Available: http://www.ncbi.nlm.nih.gov/pubmed/16678097. Accessed 2013 Dec 16.

31. Vives V, Su J, Zhong S, Ratnayaka I, Slee E, et al. (2006) ASPP2 is a haploinsufficient tumor suppressor that cooperates with p53 to suppress tumor growth. Genes Dev 20: 1262–1267. Available: http://www.pubmedcentral.nih.gov/articlerender.fcgi?artid=1472901&tool=pmcentrez&rendertype=abstract. Accessed 2013 Jun 5.

32. Cao X, Pfaff SL, Gage FH (2008) YAP regulates neural progenitor cell number via the TEA domain transcription factor. Genes Dev 22: 3320–3334. Available: http://www.pubmedcentral.nih.gov/articlerender.fcgi?artid=2600760&tool=pmcentrez&rendertype=abstract. Accessed 2013 Mar 10.

33. Tordella L, Koch S, Salter V, Pagotto A, Doondeea JB, et al. (2013) ASPP2 suppresses squamous cell carcinoma via RelA/p65-mediated repression of p63. Proc Natl Acad Sci U S A 110: 17969–17974. Available: http://www.ncbi.nlm.nih.gov/pubmed/24127607. Accessed 2013 Nov 8.

16

Inhibition of Tyrosine Kinase Receptor Tie2 Reverts HCV-Induced Hepatic Stellate Cell Activation

Samuel Martín-Vílchez[1♦], **Yolanda Rodríguez-Muñoz**[2,3♦], **Rosario López-Rodríguez**[2,3], **Ángel Hernández-Bartolomé**[2], **María Jesús Borque-Iñurrita**[4], **Francisca Molina-Jiménez**[4], **Luisa García-Buey**[2,3], **Ricardo Moreno-Otero**[2,3], **Paloma Sanz-Cameno**[2,3*]

1 Department of Cell Biology, University of Virginia School of Medicine, Charlottesville, Virginia, United States of America, **2** Unidad de Hepatología, Hospital Universitario de la Princesa, Instituto de Investigación Sanitaria Princesa (IIS-IP), Madrid, Spain, **3** Centro de Investigación Biomédica en Red de Enfermedades Hepáticas y Digestivas (CIBER-ehd), Instituto de Salud Carlos III (ISCIII), Madrid, Spain, **4** Unidad de Biología Molecular, Hospital Universitario de la Princesa, Instituto de Investigación Sanitaria Princesa (IIS-IP), Madrid, Spain

Abstract

Background: Hepatitis C virus (HCV) infection is a major cause of chronic liver disease (CLD) and is frequently linked to intrahepatic microvascular disorders. Activation of hepatic stellate cells (HSC) is a central event in liver damage, due to their contribution to hepatic renewal and to the development of fibrosis and hepatocarcinoma. During the progression of CLDs, HSC attempt to restore injured tissue by stimulating repair processes, such as fibrosis and angiogenesis. Because HSC express the key vascular receptor Tie2, among other angiogenic receptors and mediators, we analyzed its involvement in the development of CLD.

Methods: Tie2 expression was monitored in HSC cultures that were exposed to media from HCV-expressing cells (replicons). The effects of Tie2 blockade on HSC activation by either neutralizing antibody or specific signaling inhibitors were also examined.

Results: Media from HCV-replicons enhanced HSC activation and invasion and upregulated Tie2 expression. Notably, the blockade of Tie2 receptor (by a specific neutralizing antibody) or signaling (by selective AKT and MAPK inhibitors) significantly reduced alpha-smooth muscle actin (α-SMA) expression and the invasive potential of HCV-conditioned HSC.

Conclusions: These findings ascribe a novel profibrogenic function to Tie2 receptor in the progression of chronic hepatitis C, highlighting the significance of its dysregulation in the evolution of CLDs and its potential as a novel therapeutic target.

Editor: Gulam Waris, Rosalind Franklin University of Medicine and Science, United States of America

Funding: This work was supported in part by grants: 1) CIBER-ehd to Dr. R. Moreno-Otero, 2) SAF2010/21805 from Ministerio de Ciencia e Innovación to RM-O, 3) AIO Asociación Española Contra el Cáncer to PS-C, 4) Mutua Madrileña to RM-O, and 5) Mutua Madrileña to PS-C. The funders had no role in study design, data collection and analysis, decision to publish, or preparation of the manuscript.

Competing Interests: The authors have declared that no competing interests exist.

* Email: paloma_march@hotmail.com

♦ These authors contributed equally to this work.

Introduction

Hepatitis C virus (HCV) infection is a major cause of chronic liver disease (CLD) in developed countries, including chronic hepatitis C (CHC), fibrosis, cirrhosis and hepatocellular carcinoma (HCC) [1,2]. Unresolved chronic HCV infection triggers the persistent stimulation of immune responses and tissue repair mechanisms, which propel the progression of CHC toward cirrhosis and hepatocarcinoma (HCC) through incessant activation of fibrogenic and angiogenic processes [3,4,5].

Liver fibrosis is often observed in chronic HCV infections and is sustained primarily by liver-specific cells, called hepatic stellate cells (HSC). HSC are major injury-sensing cells in the liver, and their overactivation is considered the central event in the development of fibrosis and, ultimately, cirrhosis [6,7]. Once activated, HSC become highly proliferative and contractile,

increase their migratory abilities, and secrete extracellular matrix compounds, such as collagen and extracellular matrix (ECM) proteins [8,9,10,11]. In addition, HSC secrete several growth factors, such as vascular endothelial growth factor (VEGF), connective tissue growth factor (CTGF), and platelet-derived growth factor (PDGF), which promote the differentiation of mesenchymal cells and endothelial activation, migration, and proliferation [6,12].

This sequence of events effects the accumulation of ECM substances and endothelial and myofibroblast-like cells, which occlude sinusoidal fenestrations, altering the proper interchange of metabolites and oxygen between hepatocytes and blood. This process, termed sinusoidal capillarization, results in increased intrahepatic resistance to blood flow and oxygen delivery, to which HSC respond by increasing their expression of angiogenic factors, such as VEGF and angiopoietin-1 (Ang1), as well as the respective

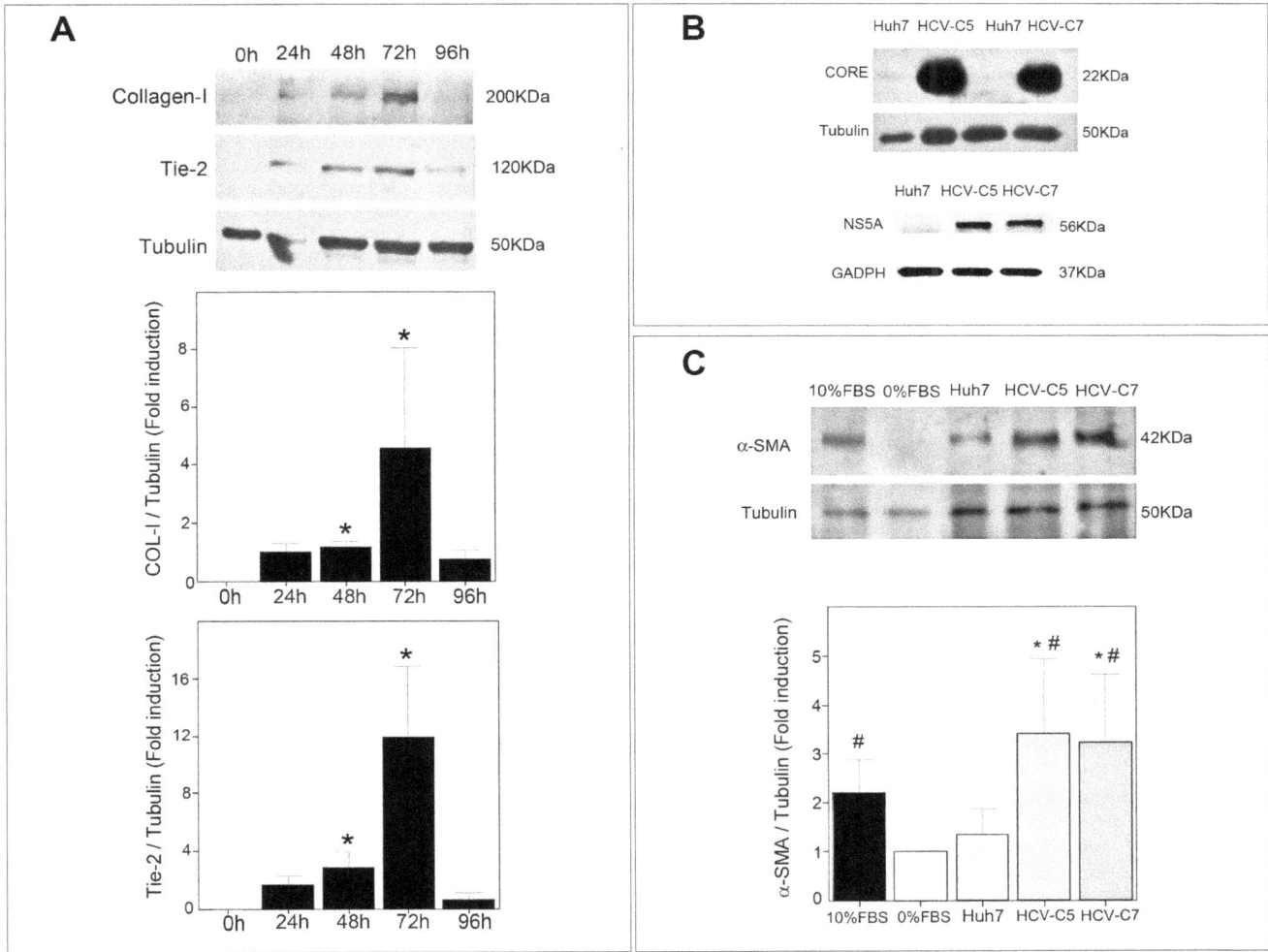

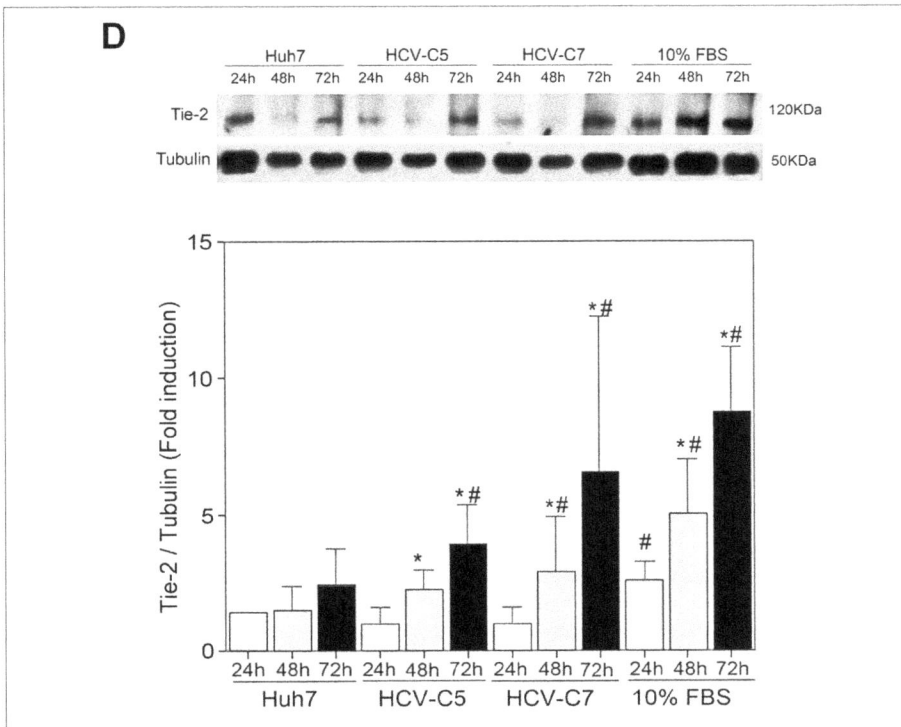

Figure 1. Activation of HSC increases Tie2 expression. (A) Kinetics of Tie2 expression during *in vitro* activation of HSC. HSC, plated at 50,000 cells/cm2 density and grown in 2% DMEM during 24, 48, 72 and 96 hours, were lysed in Laemmli buffer and loaded in 7% acrylamide gels (20 µL of total protein extract per well). SDS-PAGE resolved proteins were transferred to nitrocellulose membranes and probed with respective antibodies against COL-I (H-197, 1:2000), Tie2 (AF313, 1:200) or tubulin (DM1A, 1:5000). Quantitative analysis of Tie2 or COL-1 chemiluminescence in relation to tubulin (FUJIFILM Science Lab Image Gauge) is shown in respective graphs. *$p < 0.05$ *versus* 24 h (mean +SD, 3 independent experiments). (B) Expression of HCV proteins by hepatic cell lines (Huh7 or HCV replicons). Protein lysates from hepatic cells (20 µL each) were SDS-PAGE resolved and probed with antibodies against Core (C7–50, 1:500), NS5A (6F3, 1:1000) or tubulin (DM1A, 1:5000). (C) Expression of α-SMA in HSC exposed to conditioned media (CM) from hepatic cell lines (Huh7 and HCV replicons) compared to that expressed by HSC incubated with 0% FBS DMEM. Expression α-SMA by HSC grown in 10% FBS DMEM was used as positive control of HSC activation. CM from hepatic cells, plated at equal densities and cultured during 24 h in 0% FBS DMEM, were used to grow HSC deprived of serum 24 h before. After HSC lysis in Laemmli, equal volumes (20 µL) of protein extracts were analyzed for α-SMA expression by western blotting (1A4, 1:1000). Tubulin expression (DM1A, 1:5000) was also evaluated in order to control protein loading and relative α-SMA/tubulin data are shown in the graph (mean +SD of 3 independent experiments). #$p < 0.05$: statistic differences *versus* cultures in 0% FBS DMEM; *$p < 0.05$: statistic differences *versus* Huh7. (D) HSC conditioned during different times with media from hepatic cells (Huh7 or HCV replicons) or 10% FBS DMEM. HSC, plated at same density in 0% FBS DMEM 24 h before, were exposed during 24, 48 or 72 h to 10% FBS DMEM or CM from hepatic cell lines and the expression of Tie2 (AF313, 1:200), normalized with tubulin (DM1A, 1:5000), was analyzed. Graph shows quantitative densitometric analysis of western blot bands (mean +SD of 3 independent experiments). #$p < 0.05$: statistic differences *versus* Huh7 at same times; *$p < 0.05$, same CM *versus* 24 h.

receptors, VEGFR-2 and Tie2, exacerbating the pathology by enhancing cellular proliferation, migration, and deposition of ECM compounds [13].

Neoangiogenesis is a common feature of many CLD [14,15]; particularly, CHC is notably characterized by the development of an abnormal angioarchitecture in the liver, which is strongly linked with the fibrogenic progression of the disease. Accordingly, considerable alterations in systemic levels of diverse angiogenic factors have been reported in patients with CHC, being angiopoietin 2 (Ang2) significantly related to the fibrosis stage [16,17]. Due to HSC express angiopoietin's receptor Tie2 [18], a central regulator of physiological and pathological angiogenesis, we aimed to study the fibrogenic role of HCV-infected hepatocytes on HSC activation via Angiopoietin/Tie2 signaling axis. With that aim, we studied the expression of Tie2 receptor throughout the *in vitro* and HCV-induced activation of HSC mainly focused on investigating the effects of Tie2 inhibition on HSC behaviour as potential antifibrogenic target.

Results demonstrated that the tyrosine kinase Tie2 receptor is upregulated during HSC activation. This phenomenon was enhanced by conditioned media from HCV-expressing cells and mediated the activation and migration of HSC. Consistent with these findings, Tie2 blockade by a neutralizing antibody reduced HSC activation with regard to alpha-smooth muscle actin (α-SMA) expression and their migratory and invasive capacity. Inhibition of the key Angiopoietin/Tie2 signaling pathways PI3K/AKT and MAPK [19] notably diminished Tie2 expression on HSC and their activated phenotype. These findings reveal the significance of Tie2 in CHC progression and its related fibrogenesis, highlighting this signaling route as a valuable pharmacological target for CLD intervention.

Materials and Methods

Ethics statement

This study was approved by the Ethical Committee of Hospital Universitario de La Princesa and conducted per the Declaration of Helsinki.

Cell lines and culture conditions

The human hepatic stellate cell line LX-2 [20], plated at 50,000 cells/cm^2, was grown in Dulbecco's modified Eagle's medium (DMEM) that was supplemented with 10% fetal bovine serum (FBS), 2 mM glutamine, and 100 U/ml penicillin for 24 hours until 100% adherence and shifted to 2% FBS DMEM.

The HCV replicons HCV-C5 and HCV-C7, kindly provided by Dr. Aldabe (Hospital Universitario de Navarra, CIMA), express

the complete genome of HCV [21,22] and were generated in the Huh7 cell line from the genotype 1b HCV strain.

Hepatocyte cell lines were grown in DMEM with 10% FBS, 2 mM glutamine, and 100 U/ml penicillin, and HCV-expressing cells (C5 and C7) were selected specifically with 500 µg/ml G418 (Geneticin; Life Technologies GmbH). When hepatocyte cultures reached 60% confluence, they were shifted to 0% FBS DMEM without G418; the supernatants that were collected at 24, 48, 72, and 96 hours were used as conditioned media (CM) for further experiments.

Before the experiments, HSC were serum-starved in 0% FBS DMEM for 24 hours until their exposure to CM from HCV replicons or Huh7 cells. HSC that were grown in 0% and 10% FBS were used as negative and positive activation controls, respectively.

Blockade of Tie2 signaling

To determine the Tie2-mediating effects on HSC activation and to analyze the signaling pathways that are involved, 8 µg/ml of neutralizing anti-Tie2 (AF313, R&D Systems, Minneapolis, MN), 8 µg/ml of an isotype control antibody (BD 340473, Becton Dickinson, Mountain View, CA), 25 µmol/mL of PI3-K inhibitor or 25 µmol/mL of MAPK inhibitor (LY294002 and PD98059, respectively, Calbiochem, La Jolla, CA) were added to the appropriate cell cultures.

Western blot and zymography

Whole HCV replicons and HSC extracts were obtained with Laemmli buffer and sonicated (Soniprep 150, MSE, UK); the insoluble debris was removed by centrifugation at 9500 *g* for 5 minutes. Total protein extracts were resolved by SDS-PAGE and transferred to a nitrocellulose membrane (Bio-Rad). The membranes were blocked with a 7.5% solution of nonfat dry milk in Tris-HCl-buffered solution (TBS, pH 7.5) and incubated with antibodies against COL-I (H-197, Santa Cruz Biotechnology, Santa Cruz, CA, at 1:2000), Tie2 (AF313, R&D Systems, 1:200), matrix metalloproteinase-2 (AF902, R&D Systems, 0.2 µg/mL), α-SMA, (1A4, Sigma-Aldrich, St. Louis, MO, 1:1000), Tubulin (DM1A, 1:500, Sigma-Aldrich, St. Louis, MO, 1:5000), HCV core (C7–50, Santa Cruz Biotechnology, Santa Cruz, CA, 1:500) or NS5A (6F3, ViroStat, Portland, ME, 1:1000).

After extensive washes, the membranes were incubated with blocking buffer, containing horseradish peroxidase-conjugated goat anti-mouse or goat anti-rabbit (Santa Cruz Biotechnology, Santa Cruz, CA) at 1:1000. Proteins were detected by chemoluminescence (SuperSignal West Pico, Thermo Fisher Scientific Inc, Rockford, IL USA), and band intensities were quantitated using FUJIFILM Science Lab Image Gauge Ver. 4.0; expression of

A

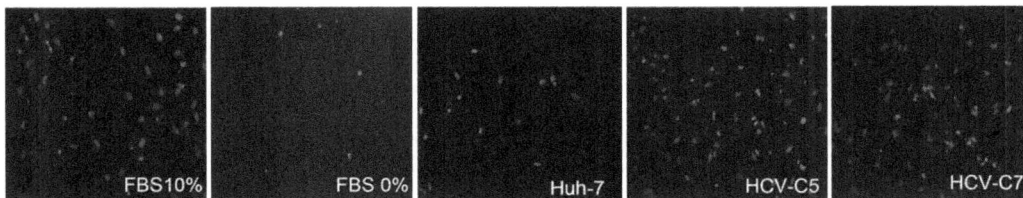

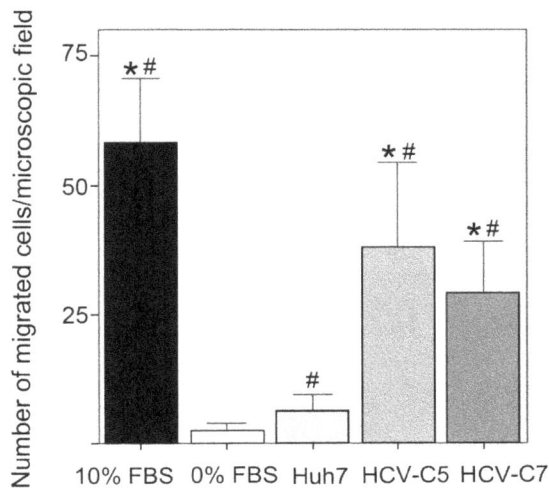

B

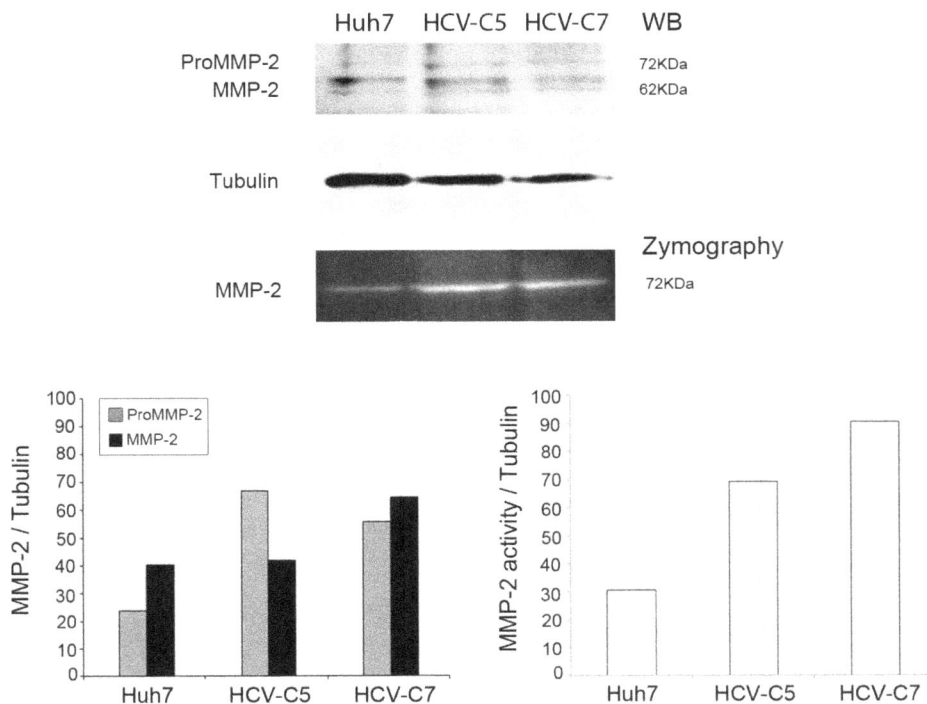

Figure 2. Conditioned media from HCV-expressing cells enhance invasive potential of HSC. (A) Effect of different CM on HSC invasion through transwell inserts precoated with collagen. Same amount of serum starved HSC (5×10^4 cells/100 μL) were seeded at the upper chambers of transwells and exposed during 24 h to different CM (hepatic-derived CM, 0% FBS DMEM or 10% FBS DMEM) dispensed at the lower compartments. At the end of migration, the upper surface of the membrane was washed and HSC adhered to the lower surface were fixed in methanol, stained with DAPI and counted in 5 randomly chosen microscopic fields (400x) in an epifluorescence microscope (Leica, Wetzlar, Germany). Data from 4

independent experiments are shown as mean +SD. #p<0.05, HSC cultured with CM from HCV replicons or DMEM 10% FBS *versus* HSC cultured in DMEM 0% FBS; *p<0.05, HSC cultured with CM from HCV-expressing cells or 10% FBS *versus* HSC cultured with CM from Huh7. (B) ProMMP-2 and MMP-2 expression (AF902, 0.2 µg/mL) and activity were respectively examined by western blotting (WB) and zymography and further quantified in protein extracts (20 µL) of HSC cultured during 24 h with CM from Huh7 or HCV-expressing cells. Bars represent the mean of densitometric analysis from two independent experiments.

proteins was normalized to that of Tubulin for each lysate. Zymographies were performed essentially as described [23] by separating equal amounts of total HSC extracts on a 10% polyacrylamide gel that contained 1% gelatin.

Cell migration assay

The migratory and invasive abilities of HCS were analyzed in a 24-well transwell, containing 8-µm pore inserts (Corning, Corning, NY), that was precoated with collagen (Pure Col, Advanced BioMatrix, California, EE.UU.). CM from Huh7 cells or HCV-derived replicons (600 µL) were placed in the bottom compartment of the chamber. After overnight serum starvation, 100 µL of LX-2 cell suspension (500 cells/µL) was added to the upper compartment. The chambers were incubated at 37°C with 5% CO2 for 24 hours. Then, the culture medium was removed, and adherent cells were fixed with 4% paraformaldehyde and stained with DAPI solution. The number of nuclei from LX-2 cells on the lower surface of the filters was counted in 5 randomly chosen microscopic fields for each specimen on an epifluorescence Leica microscope (Leica, Wetzlar, Germany).

Statistical analysis

Data were expressed as mean ± standard deviation (SD). Statistical analysis of the results was performed using unpaired student's *t* test. Two-tailed P values below 0.05 were considered significant (SPSS 16.0, SPSS, Chicago, IL).

Results

Tie2 expression is upregulated in HSC cultures in a time-dependent manner

During activation of HSC by culture, the expression of Tie2 increased progressively, peaking at 72 hours of incubation. As Figure 1A shows, the rise in Tie2 was linked to the enhancement of COL-I, a key marker of HSC activation (p<0.05). Further, the upregulation of these molecules correlated significantly at 48 and 72 hours (data not shown).

HCV replicons activate HSC

HCV-C5 and HCV-C7, previously characterized as described [22], were grown in serum-starved media at various times and used to obtain CM. The expression of viral proteins in HCV-expressing hepatocytes was confirmed by assessing the expression of core and NS5A proteins in the lysates of HCV replicons (Figure 1B). As figure 1C shows, CM from HCV-expressing cells upregulated the activation marker α-SMA in HSC compared with CM from Huh7.

In addition, Tie2 expression rose on HSC that were exposed to HCV-C5 and HCV-C7 CM in contrast to HSC that were conditioned with CM from Huh7 cells (Figure 1D, p<0.05). CM from HCV replicons increased Tie2 expression on HSC at 48 and 72 hours compared with 24 hours (Figure 1D, p<0.05). However, HSC failed to significantly enhance Tie2 expression when exposed to CM from Huh7 cells at any time.

HCV replicons promote HSC invasion

HCV replicons significantly stimulated the invasive potential of HSC. As Figure 2A illustrates, HSC that were exposed to HCV-C5 or HCV-C7 CM enhanced their migration and invaded the collagen matrix to a greater extent than HSC that were cocultured with Huh7 or 0% FBS control media (Figure 2A, p<0.05 all). Notably, the invasion of HSC that were exposed to media from HCV replicons was accompanied by an increase in MMP-2 expression and activity, as shown by western blot and zymography (Figure 2B). However, this effect was not so evident in HSC that were exposed to CM from Huh7 cells.

Effect of Tie2 blockade on HSC physiology

Pretreatment of HSC with the specific Tie2-neutralizing antibody (AF313, R&D Systems, Minneapolis, MN) downregulated α-SMA expression, as shown in Figure 3A, but this blockade was not effective in HSC that were conditioned with media from Huh7 cultures. This effect was not observed when a control isotype antibody was added instead anti-Tie2 neutralizing antibody (Figure S1), which argues in favour of Tie2 specificity. Similarly, Tie2 blockade by the same neutralizing antibody markedly reduced the invasion of HSC that were conditioned with media from HCV-C5 and HCV-C7 replicons through transwell inserts (Figure 3B).

Additionally, because PI3K/AKT and MAPK signaling mediates the key effects of angiopoietin/Tie2, including a wide range of downstream targets that regulate tumor-associated processes, such as cell growth, cell cycle progression, survival, migration, epithelial-mesenchymal transition, and angiogenesis, we examined the influence of selective PI3K/AKT and MAPK neutralization on Tie2-mediated HSC activation and migration. Administration of PI3-K/AKT or MAPK inhibitors (LY294002 and PD98059, respectively) to HSC cultures that were activated with CM from HCV replicons downregulated α-SMA and Tie2 expression (Figure 3C), which was followed by a notable decrease in their migration rate (Figure 3D).

Discussion

Liver fibrosis and angiogenesis have been suggested to contribute significantly to the progression of CLD [8,24]. Earlier results from our group and other laboratories have reported a link between the dysregulation of intrahepatic vascular homeostasis and the stage of liver disease. Further, the upregulation of several angiogenic mediators, such as Ang2 and VEGF, in diverse chronic inflammatory diseases and CLD [15,25,26,27,28,29] prompted us to implicate them as noninvasive surrogate markers of CHC evolution [16,17,30].

In chronic liver injury, the self-limiting inflammation, necrosis, and regeneration of the hepatic parenchyma are perturbed by the persistence of injury. Quiescent HSC experience constant activation, which entails significant phenotypic and functional changes in cultured human and rat HSC [6,31]. These changes, comprising the acquisition of a contractile, proliferative, and profibrogenic phenotype by HSC, effect the overproduction of ECM compounds, which ultimately results in architectural and functional alterations of the liver [9,11,12,32,33,34,35].

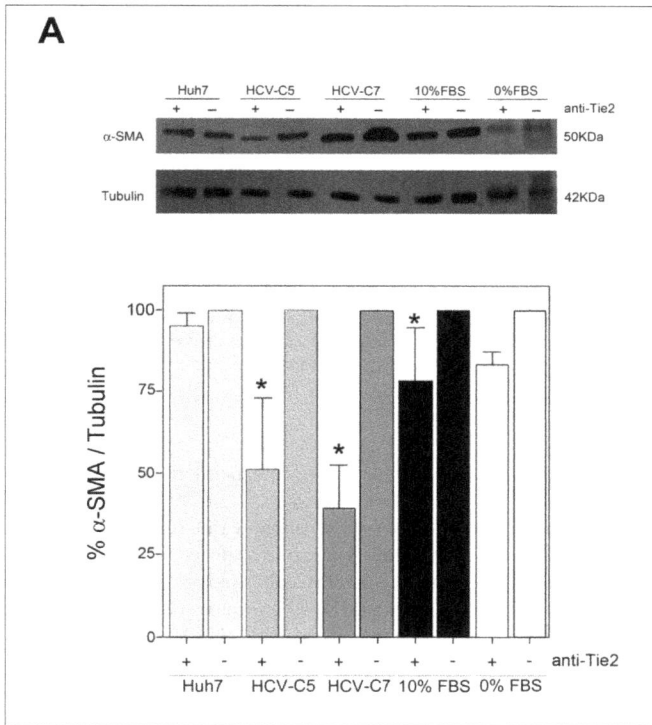

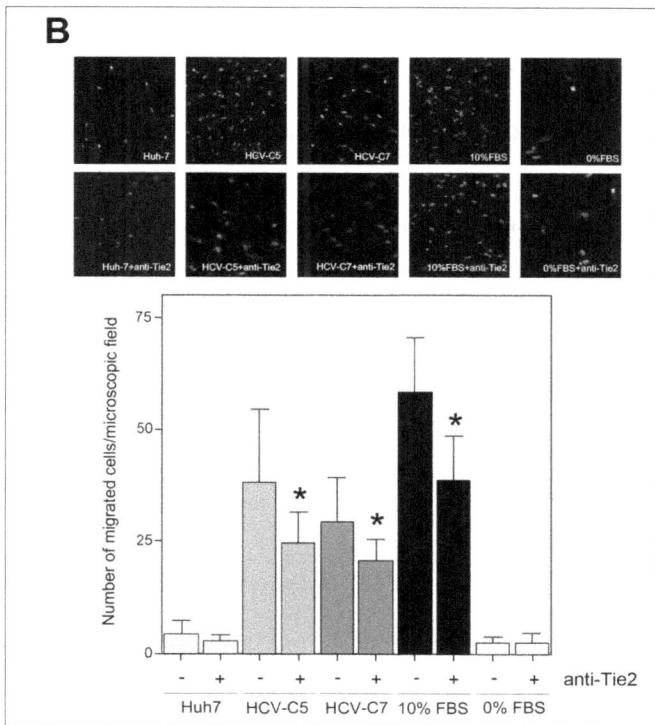

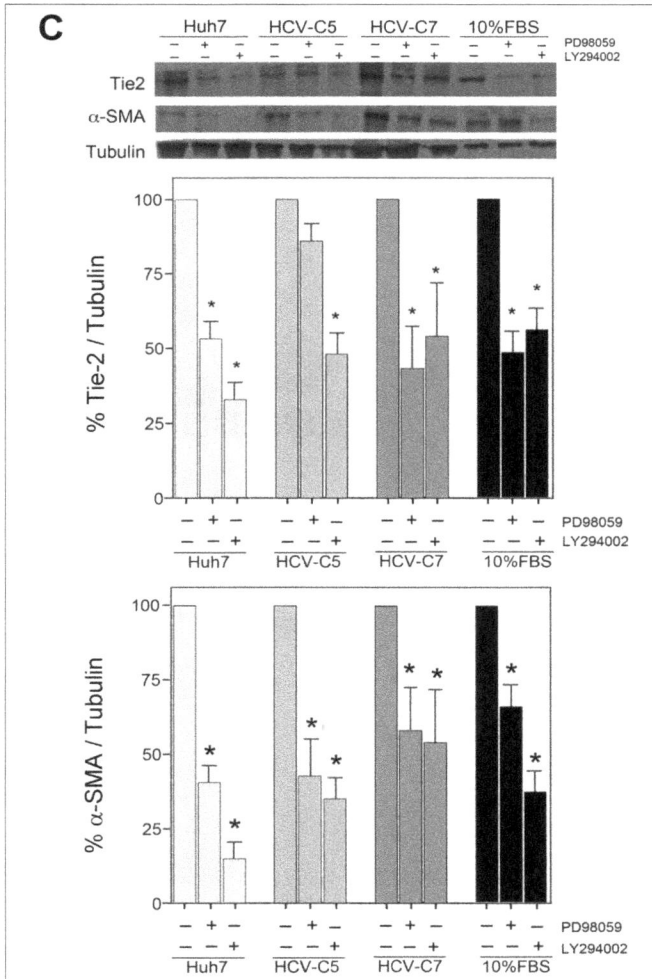

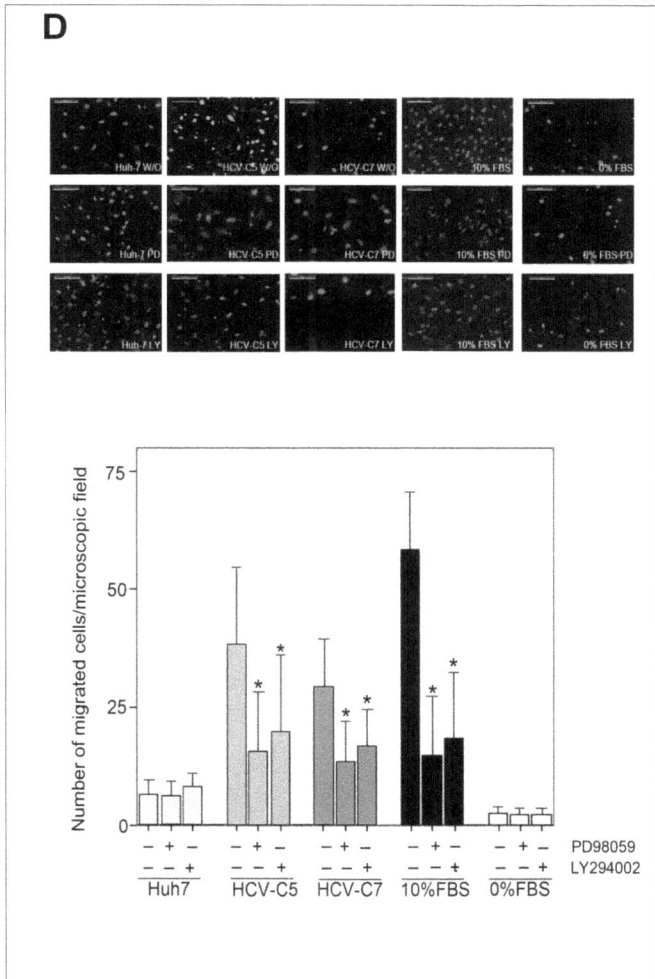

Figure 3. The activation and invasive potential of HSC are prevented by a Tie2 neutralizing antibody or by the inhibition of Akt/PI3k and MAPK signaling pathways. (A) α-SMA expression was assessed by western blotting (1A4, 1:1000) in 20 μL of Laemmli lysates from HSC exposed during 24 h to different culture conditions (hepatic-derived CM, 0% FBS DMEM or 10% FBS DMEM) in presence (+) or absence (-) anti-Tie2 blocking antibody (AF313, 8 μg/ml). Quantitative analysis of α-SMA/tubulin bands in presence of neutralizing antibody is displayed in the graph as percentage of expression observed without anti-Tie2 (100%) for each experimental condition. Bars show mean +SD of 3 independent experiments. *$p<0.05$. (B) HSC cells (5×10^4 cells/100 μL) cultured on upper transwell chambers able to invade collagen under different stimuli (hepatic-derived CM, 0% FBS DMEM or 10% FBS DMEM) with or without AF313 anti-Tie2 antibody (8 μg/ml) dispensed at bottom compartments of transwell are illustrated in the representative epifluorescence pictures after DAPI staining (400x). 5 randomly selected microscope fields were quantified per experiment (3 independent). Bars show the mean +SD average of migrating cells per field from the different experiments. *$p<0.05$ indicates statistical differences owing to the presence of anti-Tie2 neutralizing antibody. (C) The influence of different CM in presence or absence of LY294002 and PD98059 at 25 μmol/ml each (Akt/PI3k and MAPK inhibitors, respectively) on the expression of Tie2 (AF313, 1:200) and α-SMA (1A4, 1:1000) by HSC is illustrated. Respective western blots were analyzed in relation to tubulin to normalize total protein loading and the expression in presence of inhibitors was displayed as percentage of the expression without the respective compound (100%) in the same experimental conditions. Data from 3 experiments in duplicate are shown. *$p<0.05$. (D) Invasive potential of HSC (5×10^4 cells/100 μL) under the influence of LY294002 and PD98059 (25 μmol/ml both) at different experimental conditions (hepatic-derived CM, 0% FBS DMEM or 10% FBS DMEM) is shown by fluorescence images (400x). Graph depicts the mean +SD of average migrating HSC/field (5 microscope fields) per experiment (3 independent) of denoted HSC culture conditions. *$p<0.05$: statistical difference of presence *versus* absence of inhibitors.

Angiogenesis is also stimulated during fibrotic development in an attempt to restore the intrahepatic-blood interchange of cells and metabolites that is impaired by the progressive sinusoidal deposition of ECM. The fibrotic background promotes the upregulation of MCP1, VEGF, and Ang1 by HSC in response to surrounding stimuli (eg, hypoxia and leptins) [13,36,37,38], accelerating the development of a disorganized and dysfunctional intrahepatic vasculature [3].

Consistent with the upregulation of the VEGF receptors Flt-1 and Flk-1 and Tie2 by activated HSC in areas of active fibrogenesis [18,38], we noted time-dependent upregulation of Tie2 during *in vitro* HSC activation.

Despite previous studies have pointed out some discrepancies concerning the potential paracrine effects of hepatic cells expressing HCV proteins on HSC physiology, main results from stable hepatic transfectants of HCV proteins, HCV genomic and subgenomic replicons or infectious JFH1 HCV cultures, account for the substantial profibrogenic effects of HCV-expressing cells towards the activation of HSC. It has been described that recombinant core [39], as well as core-expressing cells, stimulate HSC activation as detected by the augmented expression of

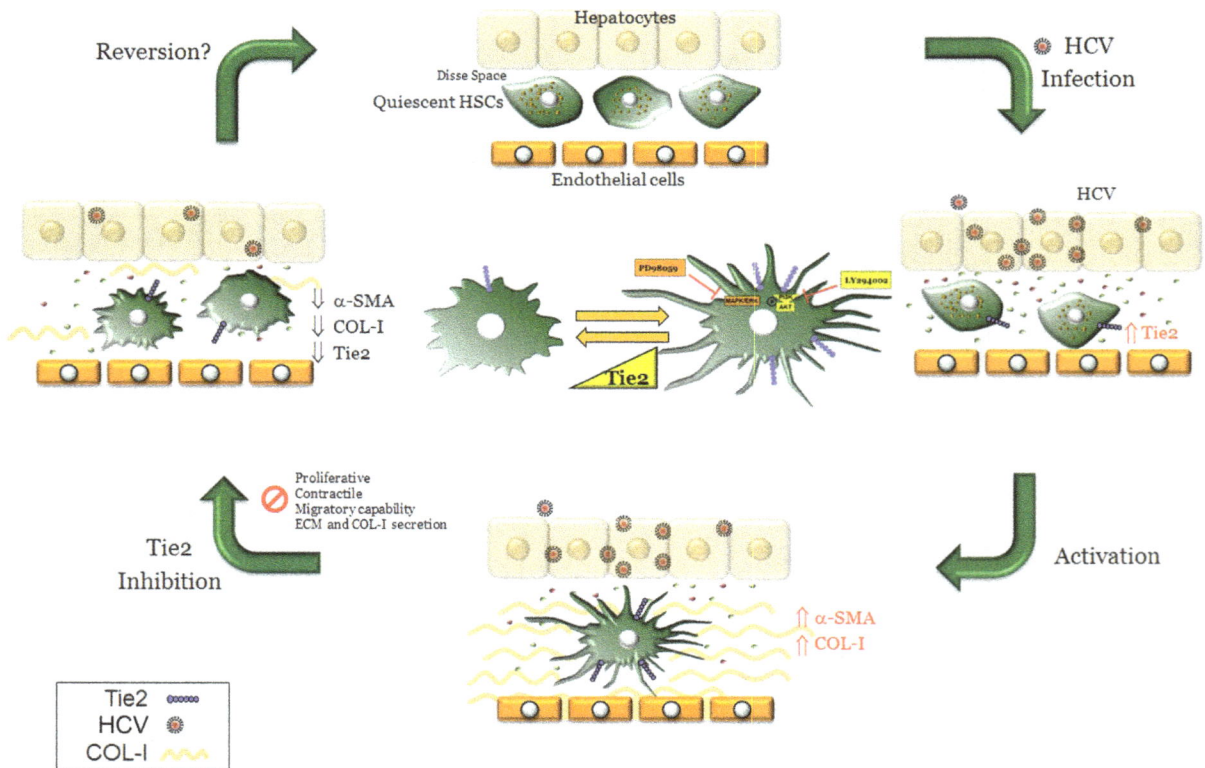

Figure 4. Phenotypic and functional changes in HSC during chronic HCV infection. After liver injury, hepatic stellate cells undergo activation, which entails the transition of quiescent cells into proliferative, fibrogenic, and contractile myofibroblasts. HCV-infected hepatocytes release factors that induce paracrine activation of HSC, triggering the expression of angiogenic Tie2 receptor, COL-I, and α-SMA and their invasive and migratory capacity. Blockade of Tie2 receptor or Akt/PI3k and MAPK signaling reverts the fibrogenic features of HSC during HCV infection. This phenomenon might facilitate the regression to quiescent HSC, raising the possibility of resolution of liver injury.

α-SMA [40,41]. In addition, nonstructural genes of HCV promote the expression of profibrogenic factors by hepatic cells leading to progression of liver fibrosis [42]. HCV core protein may assist hepatic fibrogenesis via up-regulation of CTGF and TGF-beta1 [39]. In addition, comparable significant increases of CTGF and TGF-β1 in a stable E2-expressing Huh7 cell line were also observed [43]. Conversely, the subgenomic replicon expressing HCV nonstructural proteins (NS3-5B) in Huh7 cells did not show considerable effects on proliferation and migration of HSC [44] which suggests higher paracrine effects of HCV structural proteins on such events. However, a differential regulation of core and nonstructural proteins on HSC biologic functions has been reported: whereas the expression of core protein increases cell proliferation in a Ras/ERK and PI3K/AKT dependent manner, NS3-5B protein expression mainly induce proinflammatory processes through the NF-kappa B and c-Jun N-terminal kinase pathways [40]. All these findings indicate both direct and indirect actions of HCV proteins on different aspects of HSC physiology, finally contributing to HCV-induced liver fibrosis.

In addition, HSC that were exposed to HCV replicons became highly invasive and had greater MMP2 expression and activity. Core and nonstructural HCV proteins affect a wide range of hepatocyte processes such as proliferation, adhesion, autophagy, and chemokine secretion [45,46,47], regulating the fibrogenic properties of HSC [48,49,50]. Further, HCV RNA replication alters the expression of extracellular matrix-related molecules in HSC [50]. These findings demonstrate that certain factors that are released by HCV-infected hepatocytes modify HSC behavior, enhancing their migration and enzymatic activities, ultimately effecting the fibrogenic progression of CHC.

Despite fibrogenic effects of HCV-expressing hepatocytes via TGF-β1 have been established, the observed induction of Tie2 throughout HSC activation suggests the possible paracrine role of Angiopoietins/Tie2 axis on liver fibrogenesis. Accordingly, it has been described that HBV/HCV could enhance Ang2 promoter expression through mitogen-activated protein kinase (MAPK) pathways [51] and other authors have reported that pericytes express a functionally active Tie2 receptor which may be significantly upregulated by both Ang1 and Ang2, concluding that the Angiopoietin/Tie2 axis influences the activation state and recruitment of pericytes during angiogenesis [52]. Hence, these findings might indicate the relevant role of Tie2 receptor on HCV-induced fibrogenesis. However, the molecular basis by which HCV affects Ang/Tie2 signaling requires further investigation.

In addition, we noted that blockade of Tie2 signaling by a specific neutralizing antibody significantly reduced α-SMA expression, primarily on activated HSC (exposed to 10% FCS and HCV replicon CM), impairing their invasive potential. Our finding that Tie2 blockade counteracts the activation of HSC by HCV replicons suggests that Tie2 signaling is critical in promoting and sustaining the profibrotic features of HSC during HCV infection.

As described, Tie2 receptor primarily involves the PI3K/Akt and MAPK signaling—essential pathways that have been implicated in cell survival, migration, and invasion [19,53,54,55] and modulates important functions in HSC [19,56,57,58,59].

Accordingly, blockade of the Akt/PI3k and MAPK pathways by selective inhibitors decreased the activation and migration of HSC, accompanied by the downregulation of Tie2. These findings suggest a regulatory feedback loop of Tie2 signaling that modulates HSC functions.

Paradoxically, VEGF-induced shedding of Tie2 by AKT signaling has been recently described in endothelial cells, and AKT-induced Ang2 promotes the endocytosis of Tie2 [60,61,62].

Yet, like other RTK receptors, Tie2 regulation is cell type- and context-dependent and tightly regulated over time and by location; thus, depending on cell type and conditions (resting or stimulated), Tie2 interacts with various coreceptors (integrins, Tie1) and ligands (Ang1 and Ang2, at different multimeric orders), leading to many effects [63,64,65]. Consequently, the modulation of cell-matrix interactions by Tie2 ligand (Ang1/Ang2) or Tie2-coreceptor complexes (Tie1, integrins), based on subcellular location, effects many responses by altering localized signaling routes [66,67], explaining the conditional agonistic versus antagonistic function of Ang2. Thus, the effects of HCV on HSC physiology implicate a novel profibrogenic mechanism that promotes the progression of CHC via paracrine upregulation of Tie2.

The profibrogenic and proinflammatory nature of activated HSC has been characterized extensively, but their function in pathological angiogenesis remains poorly understood. HSC are an important source of angiogenic cytokines under hypoxia [2,13,38] and in acute and chronic liver injury [2,15,68,69,70]. In addition, recent studies have identified HSC as important targets of VEGF and Ang1, which stimulate their proliferation, chemotaxis, and synthesis of type I collagen [69,70]. These data, with our findings regarding the implication of Tie2 receptor in modulating HSC function by HCV replicons, suggest that the dysregulation of Tie2 signaling orchestrates key events in HSC physiology, leading to CHC pathogenesis through neoangiogenesis, inflammation, and fibrogenesis.

Inhibition of Tie2 signaling might have antifibrotic effects through impaired activation and recruitment of these fibrogenic cells to injured tissues, preventing the release and accumulation of ECM compounds and their pathological effects on sinusoidal architecture and physiology (Figure 4). Thus, the reversion of HSC activation by inhibition of Tie2 signaling implicates the angiopoietin-Tie2 axis as a potential therapeutic target.

In summary, our results demonstrate that factors that are released by HCV-expressing hepatocytes effect the acquisition of an invasive profibrotic phenotype in HSC via the Tie2 receptor. Notably, Tie2 receptor blockade and Akt/PI3k and MAPK inhibitors significantly impair crucial fibrogenic events that are induced by HCV, such as HSC activation and migration. Our data indicate that the angiogenic Tie2 receptor regulates HSC physiology in HCV infection. An in-depth study of the complex regulation of Tie2 should increase our understanding of the molecular mechanisms of the progression of CLD, providing novel and valuable therapeutic strategies for antifibrotic pharmacological interventions.

Supporting Information

Figure S1 Anti-Tie2 neutralizing antibody reduces HSC activation. The expression of α-SMA by HSC exposed during 24 h to different conditioned media (0% FBS DMEM, Huh7, HCV-C5, HCV-C7 and 10% FBS DMEM) was assessed by western blotting in presence (+) or absence (−) of anti-Tie2 blocking antibody (AF313, 8 μg/ml) or isotype control antibody (BD-340473, 8 μg/ml). Quantitative analysis of α-SMA/GAPDH bands for each experimental condition is displayed in the graph. Bars show the mean of 2 independent experiments.

Acknowledgments

We are grateful to the Gastroenterology Service, Hospital Universitario de la Princesa, Madrid, Spain.

Author Contributions

Conceived and designed the experiments: SM-V PS-C. Performed the experiments: SM-V YR-M MJB-I. Analyzed the data: SM-V YR-M PS-C.

Contributed reagents/materials/analysis tools: RL-R AH-B FM-J LG-B. Wrote the paper: SM-V YR-M RM-O PS-C.

References

1. Chisari FV (2005) Unscrambling hepatitis C virus-host interactions. Nature 436: 930–932.
2. Medina J, Arroyo AG, Sanchez-Madrid F, Moreno-Otero R (2004) Angiogenesis in chronic inflammatory liver disease. Hepatology 39: 1185–1195.
3. Sanz-Cameno P, Trapero-Marugan M, Chaparro M, Jones EA, Moreno-Otero R (2010) Angiogenesis: from chronic liver inflammation to hepatocellular carcinoma. J Oncol 2010: 272170.
4. Schuppan D, Krebs A, Bauer M, Hahn EG (2003) Hepatitis C and liver fibrosis. Cell Death Differ 10 Suppl 1: S59–67.
5. Poynard T, Ratziu V, Benhamou Y, Opolon P, Cacoub P, et al. (2000) Natural history of HCV infection. Baillieres Best Pract Res Clin Gastroenterol 14: 211–228.
6. Friedman SL (2008) Hepatic stellate cells: protean, multifunctional, and enigmatic cells of the liver. Physiol Rev 88: 125–172.
7. Friedman SL (2008) Mechanisms of hepatic fibrogenesis. Gastroenterology 134: 1655–1669.
8. Pinzani M, Rombouts K (2004) Liver fibrosis: from the bench to clinical targets. Dig Liver Dis 36: 231–242.
9. Friedman SL (2003) Liver fibrosis — from bench to bedside. J Hepatol 38 Suppl 1: S38–53.
10. Novo E, di Bonzo LV, Cannito S, Colombatto S, Parola M (2009) Hepatic myofibroblasts: a heterogeneous population of multifunctional cells in liver fibrogenesis. Int J Biochem Cell Biol 41: 2089–2093.
11. Bataller R, Brenner DA (2005) Liver fibrosis. J Clin Invest 115: 209–218.
12. Pinzani M, Marra F (2001) Cytokine receptors and signaling in hepatic stellate cells. Semin Liver Dis 21: 397–416.
13. Novo E, Cannito S, Zamara E, Valfre di Bonzo L, Caligiuri A, et al. (2007) Proangiogenic cytokines as hypoxia-dependent factors stimulating migration of human hepatic stellate cells. Am J Pathol 170: 1942–1953.
14. Fernandez M, Semela D, Bruix J, Colle I, Pinzani M, et al. (2009) Angiogenesis in liver disease. J Hepatol 50: 604–620.
15. Medina J, Sanz-Cameno P, Garcia-Buey L, Martin-Vilchez S, Lopez-Cabrera M, et al. (2005) Evidence of angiogenesis in primary biliary cirrhosis: an immunohistochemical descriptive study. J Hepatol 42: 124–131.
16. Hernandez-Bartolome A, Lopez-Rodriguez R, Rodriguez-Munoz Y, Martin-Vilchez S, Borque MJ, et al. (2013) Angiopoietin-2 Serum Levels Improve Noninvasive Fibrosis Staging in Chronic Hepatitis C: A Fibrogenic-Angiogenic Link. PLoS One 8: e66143.
17. Salcedo X, Medina J, Sanz-Cameno P, Garcia-Buey L, Martin-Vilchez S, et al. (2005) The potential of angiogenesis soluble markers in chronic hepatitis C. Hepatology 42: 696–701.
18. Taura K, De Minicis S, Seki E, Hatano E, Iwaisako K, et al. (2008) Hepatic stellate cells secrete angiopoietin 1 that induces angiogenesis in liver fibrosis. Gastroenterology 135: 1729–1738.
19. Huang H, Bhat A, Woodnutt G, Lappe R (2010) Targeting the ANGPT-TIE2 pathway in malignancy. Nat Rev Cancer 10: 575–585.
20. Xu L, Hui AY, Albanis E, Arthur MJ, O'Byrne SM, et al. (2005) Human hepatic stellate cell lines, LX-1 and LX-2: new tools for analysis of hepatic fibrosis. Gut 54: 142–151.
21. Bartenschlager R (2005) The hepatitis C virus replicon system: from basic research to clinical application. J Hepatol 43: 210–216.
22. Pietschmann T, Lohmann V, Kaul A, Krieger N, Rinck G, et al. (2002) Persistent and transient replication of full-length hepatitis C virus genomes in cell culture. J Virol 76: 4008–4021.
23. Martin-Vilchez S, Sanz-Cameno P, Rodriguez-Munoz Y, Majano PL, Molina-Jimenez F, et al. (2008) The hepatitis B virus X protein induces paracrine activation of human hepatic stellate cells. Hepatology 47: 1872–1883.
24. Valfre di Bonzo L, Novo E, Cannito S, Busletta C, Paternostro C, et al. (2009) Angiogenesis and liver fibrogenesis. Histol Histopathol 24: 1323–1341.
25. Cho YJ, Ma JE, Yun EY, Kim YE, Kim HC, et al. (2011) Serum angiopoietin-2 levels are elevated during acute exacerbations of COPD. Respirology 16: 284–290.
26. David S, John SG, Jefferies HJ, Sigrist MK, Kumpers P, et al. (2012) Angiopoietin-2 levels predict mortality in CKD patients. Nephrol Dial Transplant 27: 1867–1872.
27. Eleuteri E, Di Stefano A, Tarro Genta F, Vicari C, Gnemmi I, et al. (2011) Stepwise increase of angiopoietin-2 serum levels is related to haemodynamic and functional impairment in stable chronic heart failure. Eur J Cardiovasc Prev Rehabil 18: 607–614.
28. Medina J, Caveda L, Sanz-Cameno P, Arroyo AG, Martin-Vilchez S, et al. (2003) Hepatocyte growth factor activates endothelial proangiogenic mechanisms relevant in chronic hepatitis C-associated neoangiogenesis. J Hepatol 38: 660–667.
29. Sanz-Cameno P, Martin-Vilchez S, Lara-Pezzi E, Borque MJ, Salmeron J, et al. (2006) Hepatitis B virus promotes angiopoietin-2 expression in liver tissue: role of HBV x protein. Am J Pathol 169: 1215–1222.
30. Salcedo X, Medina J, Sanz-Cameno P, Garcia-Buey L, Martin-Vilchez S, et al. (2005) Review article: angiogenesis soluble factors as liver disease markers. Aliment Pharmacol Ther 22: 23–30.
31. Lee UE, Friedman SL (2011) Mechanisms of hepatic fibrogenesis. Best Pract Res Clin Gastroenterol 25: 195–206.
32. Cassiman D, Libbrecht L, Desmet V, Denef C, Roskams T (2002) Hepatic stellate cell/myofibroblast subpopulations in fibrotic human and rat livers. J Hepatol 36: 200–209.
33. Forbes SJ, Parola M (2011) Liver fibrogenic cells. Best Pract Res Clin Gastroenterol 25: 207–217.
34. Friedman SL (2004) Stellate cells: a moving target in hepatic fibrogenesis. Hepatology 40: 1041–1043.
35. Parola M, Marra F, Pinzani M (2008) Myofibroblast - like cells and liver fibrogenesis: Emerging concepts in a rapidly moving scenario. Mol Aspects Med 29: 58–66.
36. Aleffi S, Petrai I, Bertolani C, Parola M, Colombatto S, et al. (2005) Upregulation of proinflammatory and proangiogenic cytokines by leptin in human hepatic stellate cells. Hepatology 42: 1339–1348.
37. Ankoma-Sey V, Wang Y, Dai Z (2000) Hypoxic stimulation of vascular endothelial growth factor expression in activated rat hepatic stellate cells. Hepatology 31: 141–148.
38. Wang YQ, Luk JM, Ikeda K, Man K, Chu AC, et al. (2004) Regulatory role of vHL/HIF-1alpha in hypoxia-induced VEGF production in hepatic stellate cells. Biochem Biophys Res Commun 317: 358–362.
39. Shin JY, Hur W, Wang JS, Jang JW, Kim CW, et al. (2005) HCV core protein promotes liver fibrogenesis via up-regulation of CTGF with TGF-beta1. Exp Mol Med 37: 138–145.
40. Bataller R, Paik YH, Lindquist JN, Lemasters JJ, Brenner DA (2004) Hepatitis C virus core and nonstructural proteins induce fibrogenic effects in hepatic stellate cells. Gastroenterology 126: 529–540.
41. Clement S, Pascarella S, Conzelmann S, Gonelle-Gispert C, Guilloux K, et al. (2010) The hepatitis C virus core protein indirectly induces alpha-smooth muscle actin expression in hepatic stellate cells via interleukin-8. J Hepatol 52: 635–643.
42. Schulze-Krebs A, Preimel D, Popov Y, Bartenschlager R, Lohmann V, et al. (2005) Hepatitis C virus-replicating hepatocytes induce fibrogenic activation of hepatic stellate cells. Gastroenterology 129: 246–258.
43. Ming-Ju H, Yih-Shou H, Tzy-Yen C, Hui-Ling C (2011) Hepatitis C virus E2 protein induce reactive oxygen species (ROS)-related fibrogenesis in the HSC-T6 hepatic stellate cell line. J Cell Biochem 112: 233–243.
44. Sancho-Bru P, Juez E, Moreno M, Khurdayan V, Morales-Ruiz M, et al. (2010) Hepatocarcinoma cells stimulate the growth, migration and expression of pro-angiogenic genes in human hepatic stellate cells. Liver Int 30: 31–41.
45. Abdalla MY, Ahmad IM, Spitz DR, Schmidt WN, Britigan BE (2005) Hepatitis C virus-core and non structural proteins lead to different effects on cellular antioxidant defenses. J Med Virol 76: 489–497.
46. Chu VC, Bhattacharya S, Nomoto A, Lin J, Zaidi SK, et al. (2011) Persistent expression of hepatitis C virus non-structural proteins leads to increased autophagy and mitochondrial injury in human hepatoma cells. PLoS One 6: e28551.
47. Dolganiuc A, Kodys K, Kopasz A, Marshall C, Do T, et al. (2003) Hepatitis C virus core and nonstructural protein 3 proteins induce pro- and anti-inflammatory cytokines and inhibit dendritic cell differentiation. J Immunol 170: 5615–5624.
48. DeBusk LM, Hallahan DE, Lin PC (2004) Akt is a major angiogenic mediator downstream of the Ang1/Tie2 signaling pathway. Exp Cell Res 298: 167–177.
49. Shin JY, Yoon SK, Wang JS, Hur W, Ryu JS, et al. (2003) [The role of hepatitis C virus core protein in liver fibrogenesis: a study using an in vitro co-culture system]. Korean J Gastroenterol 42: 400–408.
50. Watanabe N, Aizaki H, Matsuura T, Kojima S, Wakita T, et al. (2011) Hepatitis C virus RNA replication in human stellate cells regulates gene expression of extracellular matrix-related molecules. Biochem Biophys Res Commun 407: 135–140.
51. Li Y, Chen J, Wu C, Wang L, Lu M, et al. (2010) Hepatitis B virus/hepatitis C virus upregulate angiopoietin-2 expression through mitogen-activated protein kinase pathway. Hepatol Res 40: 1022–1033.
52. Cai J, Kehoe O, Smith GM, Hykin P, Boulton ME (2008) The angiopoietin/Tie-2 system regulates pericyte survival and recruitment in diabetic retinopathy. Invest Ophthalmol Vis Sci 49: 2163–2171.
53. Fiedler U, Augustin HG (2006) Angiopoietins: a link between angiogenesis and inflammation. Trends Immunol 27: 552–558.
54. Qian Y, Zhong X, Flynn DC, Zheng JZ, Qiao M, et al. (2005) ILK mediates actin filament rearrangements and cell migration and invasion through PI3K/Akt/Rac1 signaling. Oncogene 24: 3154–3165.
55. Shih MC, Chen JY, Wu YC, Jan YH, Yang BM, et al. (2012) TOPK/PBK promotes cell migration via modulation of the PI3K/PTEN/AKT pathway and is associated with poor prognosis in lung cancer. Oncogene 31: 2389–2400.

56. Reif S, Lang A, Lindquist JN, Yata Y, Gabele E, et al. (2003) The role of focal adhesion kinase-phosphatidylinositol 3-kinase-akt signaling in hepatic stellate cell proliferation and type I collagen expression. J Biol Chem 278: 8083–8090.

57. Son G, Hines IN, Lindquist J, Schrum LW, Rippe RA (2009) Inhibition of phosphatidylinositol 3-kinase signaling in hepatic stellate cells blocks the progression of hepatic fibrosis. Hepatology 50: 1512–1523.

58. Uyama N, Iimuro Y, Kawada N, Reynaert H, Suzumura K, et al. (2012) Fascin, a novel marker of human hepatic stellate cells, may regulate their proliferation, migration, and collagen gene expression through the FAK-PI3K-Akt pathway. Lab Invest 92: 57–71.

59. Wang FP, Li L, Li J, Wang JY, Wang LY, et al. (2013) High mobility group box-1 promotes the proliferation and migration of hepatic stellate cells via TLR4-dependent signal pathways of PI3K/Akt and JNK. PLoS One 8: e64373.

60. Bogdanovic E, Nguyen VP, Dumont DJ (2006) Activation of Tie2 by angiopoietin-1 and angiopoietin-2 results in their release and receptor internalization. J Cell Sci 119: 3551–3560.

61. Findley CM, Cudmore MJ, Ahmed A, Kontos CD (2007) VEGF induces Tie2 shedding via a phosphoinositide 3-kinase/Akt dependent pathway to modulate Tie2 signaling. Arterioscler Thromb Vasc Biol 27: 2619–2626.

62. Phelps ED, Updike DL, Bullen EC, Grammas P, Howard EW (2006) Transcriptional and posttranscriptional regulation of angiopoietin-2 expression mediated by IGF and PDGF in vascular smooth muscle cells. Am J Physiol Cell Physiol 290: C352–361.

63. Fukuhara S, Sako K, Minami T, Noda K, Kim HZ, et al. (2008) Differential function of Tie2 at cell-cell contacts and cell-substratum contacts regulated by angiopoietin-1. Nat Cell Biol 10: 513–526.

64. Fukuhara S, Sako K, Noda K, Nagao K, Miura K, et al. (2009) Tie2 is tied at the cell-cell contacts and to extracellular matrix by angiopoietin-1. Exp Mol Med 41: 133–139.

65. Pietila R, Natynki M, Tammela T, Kangas J, Pulkki KH, et al. (2012) Ligand oligomerization state controls Tie2 receptor trafficking and angiopoietin-2-specific responses. J Cell Sci 125: 2212–2223.

66. Marron MB, Singh H, Tahir TA, Kavumkal J, Kim HZ, et al. (2007) Regulated proteolytic processing of Tie1 modulates ligand responsiveness of the receptor-tyrosine kinase Tie2. J Biol Chem 282: 30509–30517.

67. Singh H, Milner CS, Aguilar Hernandez MM, Patel N, Brindle NP (2009) Vascular endothelial growth factor activates the Tie family of receptor tyrosine kinases. Cell Signal 21: 1346–1350.

68. Ankoma-Sey V, Matli M, Chang KB, Lalazar A, Donner DB, et al. (1998) Coordinated induction of VEGF receptors in mesenchymal cell types during rat hepatic wound healing. Oncogene 17: 115–121.

69. Corpechot C, Barbu V, Wendum D, Kinnman N, Rey C, et al. (2002) Hypoxia-induced VEGF and collagen I expressions are associated with angiogenesis and fibrogenesis in experimental cirrhosis. Hepatology 35: 1010–1021.

70. Yoshiji H, Kuriyama S, Yoshii J, Ikenaka Y, Noguchi R, et al. (2003) Vascular endothelial growth factor and receptor interaction is a prerequisite for murine hepatic fibrogenesis. Gut 52: 1347–1354.

17

A Model for Cell Wall Dissolution in Mating Yeast Cells: Polarized Secretion and Restricted Diffusion of Cell Wall Remodeling Enzymes Induces Local Dissolution

Lori B. Huberman[1,2], Andrew W. Murray[1,2]*

1 Molecular and Cellular Biology, Harvard University, Cambridge, MA, United States of America, 2 Faculty of Arts and Sciences Center for Systems Biology, Harvard University, Cambridge, MA, United States of America

Abstract

Mating of the budding yeast, *Saccharomyces cerevisiae*, occurs when two haploid cells of opposite mating types signal using reciprocal pheromones and receptors, grow towards each other, and fuse to form a single diploid cell. To fuse, both cells dissolve their cell walls at the point of contact. This event must be carefully controlled because the osmotic pressure differential between the cytoplasm and extracellular environment causes cells with unprotected plasma membranes to lyse. If the cell wall-degrading enzymes diffuse through the cell wall, their concentration would rise when two cells touched each other, such as when two pheromone-stimulated cells adhere to each other via mating agglutinins. At the surfaces that touch, the enzymes must diffuse laterally through the wall before they can escape into the medium, increasing the time the enzymes spend in the cell wall, and thus raising their concentration at the point of attachment and restricting cell wall dissolution to points where cells touch each other. We tested this hypothesis by studying pheromone treated cells confined between two solid, impermeable surfaces. This confinement increases the frequency of pheromone-induced cell death, and this effect is diminished by reducing the osmotic pressure difference across the cell wall or by deleting putative cell wall glucanases and other genes necessary for efficient cell wall fusion. Our results support the model that pheromone-induced cell death is the result of a contact-driven increase in the local concentration of cell wall remodeling enzymes and suggest that this process plays an important role in regulating cell wall dissolution and fusion in mating cells.

Editor: Yanchang Wang, Florida State University, United States of America

Funding: This work was supported by National Institute of General Medical Sciences (http://www.nigms.nih.gov/) Grant P50GM068763 of the National Centers for Systems Biology Program and a National Science Foundation (http://www.nsf.gov/) Graduate Research Fellowship and Ashford Fellowship (http://www.gsas.harvard.edu/fellowships.php) to L.B.H. The funders had no role in study design, data collection and analysis, decision to publish, or preparation of the manuscript.

Competing Interests: The authors have declared that no competing interests exist.

* Email: amurray@mcb.harvard.edu

Introduction

Cell fusion is essential for sexual reproduction and plays an important role in the development of many organisms [1]. In mammals, cell fusion is involved in the formation of myoblasts [2], osteoclasts [3], giant cells [4], and placental cells [5]. It is also important in the development of *Caenorhabditis elegans* [6] and *Drosophila melanogaster* [7]. Perhaps the simplest and most well studied form of cell fusion is the mating of the budding yeast, *Saccharomyces cerevisiae* [8].

Budding yeast can exist in both a diploid and haploid state. In either state, cells can replicate asexually by budding, producing daughters that are genetically identical to their mothers [9]. Haploid cells can be one of two mating types, **a** or α, which are defined by two alternative alleles of a single locus, *MAT***a** or *MAT*α. These mating types express reciprocal pheromones and pheromone receptors, which they use to signal to each other. Exposing a *MAT***a** cell to α-factor, the pheromone secreted by *MAT*α cells, (or vice versa) induces a pheromone response that includes transcription of pheromone response genes, cell cycle arrest in G1, and polarization in the direction of highest pheromone concentration to form a mating projection known as a shmoo [10].

After *MAT***a** and *MAT*α cells have successfully communicated and grown towards each other, they must fuse [9]. The two cells initially bind to each other at their shmoo tips using mating agglutinins [11–13], but their plasma membranes are still separated by two, approximately 100nm thick, cell walls [14]. Before the mating partners can fuse, the cell wall that lies between the two membranes must be dissolved and the boundaries of the remaining cell walls, which surround the site of cell fusion, must fuse to form a single, continuous structure that will enclose the newly formed zygote [8]. The osmotic pressure differential between the cytoplasm and the extracellular environment makes this spatially regulated cell wall dissolution and fusion a dangerous task [15,16]. If the cell wall is opened at the wrong time or place, exposing the plasma membrane directly to the environment, there will be no elastic force to resist the turgor pressure of the cell, water will rush into the cell from the extracellular environment, and the cell will lyse [15,16].

Various studies have been done on the molecular basis for cell wall dissolution. In 1996, Brizzio *et al*. showed that high pheromone concentrations are required for efficient fusion and hypothesized that vesicles found at the shmoo tip might contain cell wall remodeling enzymes [17]. Later, Cappellaro *et al.* found several proteins with homology to known cell wall glucanases, including *SCW4*, whose deletion makes mating less efficient [18]. The promoter of another putative glucanase gene identified by Cappellaro *et al.* [18], *SCW11*, has a binding site for Ste12 [19], the transcription factor that induces genes in response to pheromone stimulation [20].

Several proteins that are required for efficient cell fusion play a role in delivering secretory vesicles to the shmoo tip [21–24]. A complex containing Rvs161, an amphiphysin-like protein that binds curved membranes [25,26], and Fus2 [22,27,28], is hypothesized to direct vesicle transport to the cell fusion zone [23]. Once the vesicles reach the plasma membrane, they are anchored by Fus1 [23], a membrane spanning protein [29] that interacts with the polarisome [30], a protein complex associated with polarized actin polymerization [31], presumably ensuring tight clustering of the secretory vesicles. Although these proteins direct vesicle secretion towards the shmoo tip, their roles do not explain how cell wall dissolution is limited to the site of contact with a polarized partner.

The problem of remodeling the cell wall is not unique to mating. Even when cells are growing isotropically, there must be a balance between cell wall synthesis and destruction to allow the continual increase in cell diameter and volume, which is accomplished through spatially uniform secretion of synthesizing and remodeling enzymes (Figure 1A) [32,33]. Polarized growth, such as that associated with budding and shmooing, is achieved through polarized secretion of these enzymes [33] (Figure 1B). Most cell wall synthesizing enzymes are attached to the plasma membrane, whereas most wall-degrading enzymes are free to diffuse through the cell wall [34]. Synthesis and destruction must be carefully balanced: an excess of synthesis over degradation will lead to an increased cell wall thickness and eventually to slow growth, whereas an excess of degradation will weaken the cell wall until it is unable to resist the osmotic pressure inside the cell [35].

We propose a simple model to explain how cell walls are dissolved at the point where two polarized mating partners contact each other. When cells are not stuck to each other by mating agglutinins, the degradative enzymes diffuse through the cell wall and are then lost into the medium (Figure 1C). But when two mating partners stick to each other, using agglutinins, the enzymes must take a much longer path to escape, and because distance diffused only rises as the square root of time, their concentration at the site of fusion must rise, leading to an excess of destruction over synthesis and the eventual dissolution of the cell wall (Figure 1D).

If our model is correct, it should be possible to cause pheromone-induced cell death by tightly apposing pheromone-treated cells to impermeable surfaces, thus, mimicking the attachment of two cells to each other during mating (Figure 1E). Although previous studies [36,37] have reported that pheromone treatment can cause cell death, they neither hypothesized a mechanism through which this process is regulated, nor carefully examined the effect of holding cells against impermeable surfaces. We therefore set out to test the idea that slowing the escape of cell wall-degrading enzymes would lead to cell wall dissolution and death.

We observed that the frequency of cell death increases as the amount of cell contact with an impermeable surface increases and as the osmotic pressure differential between a cell and its environment rises, whereas decreasing the osmotic pressure differential reduces cell death. Deleting Fus1 and Fus2, proteins important for cell wall fusion [24], as well as the putative cell wall

glucanases Scw4 and Scw11 [18], also decreases the frequency of cell death. Our evidence argues that the pheromone-induced cell death is due to a contact-dependent increase in the local concentration of cell wall remodeling enzymes, leading to the dissolution of the cell wall and eventual lysis of the cell. This mechanism may ensure safe and accurate cell wall fusion during mating.

Results

A model for pheromone-stimulated cell wall dissolution

We propose a simple model for cell wall dissolution: cell-cell contact increases the concentration of cell wall remodeling enzymes because they have to diffuse further within the cell wall from their site of secretion to reach the aqueous solution that surrounds the cells. We mathematically analyzed the distribution of cell wall remodeling enzymes in two situations: cells that are free in aqueous medium and those apposed to an impermeable surface. In both cases, we assume that the enzymes diffuse much more slowly through the cell wall than they do in the surrounding medium, and that this medium represents an infinite sink, allowing us to set the enzyme concentration outside the cell wall to zero. For unapposed cells, the enzymes need only diffuse through the thickness of the cell wall. At steady state, the flux through all points from the external surface of the plasma membrane to the external surface of the cell wall must be constant, implying a linear gradient in the enzymes' concentration. For apposed cells, we assume that they have a circular area of the cell wall pressed against an impermeable surface and that secreted enzymes must diffuse through the wall, parallel to the impermeable surface, before they can escape. At the center of the apposed region, the cell is secreting enzymes into the wall and the enzymes are diffusing away from the center of this region. In this region, the flux through the circumference of circles inscribed in the cell wall increases as the radius of the circles increases. Because the area of secretory activity increases with the square of the radius, whereas its circumference increases only linearly, the flux per unit length of the circumference increases, and thus the steepness of the gradient increases, moving outwards from the center of the secretory zone. Beyond this zone, no new enzyme secretion occurs, the flux through successively larger circles remains constant, and since their circumference increases, the radial concentration gradient becomes progressively shallower. If we assume that the cell wall is 0.1 μm thick, the radius of the secretory zone is 0.25 μm, and the radius of the apposed zone is 1 μm, the concentration of cell wall degrading enzymes at the center of the apposed zone is more than ten times the mean enzyme concentration in the wall of an unapposed cell. Details of this analysis are found below.

Mathematical analysis of pheromone-stimulated cell wall dissolution

First we consider an enzyme diffusing one dimensionally through a cell wall with diffusion coefficient D. If the radius of the secretion zone is substantially greater than the thickness of the cell wall, it is reasonable to treat the escape of enzymes secreted at the center of this zone as proceeding by one-dimensional diffusion through the thickness of the cell wall. As it diffuses through the wall, the enzyme's flux through a unit area of the cell wall, parallel to the surface of the cell, J, must be constant and is given by Fick's law

$$J = -D\frac{dC}{dx}$$

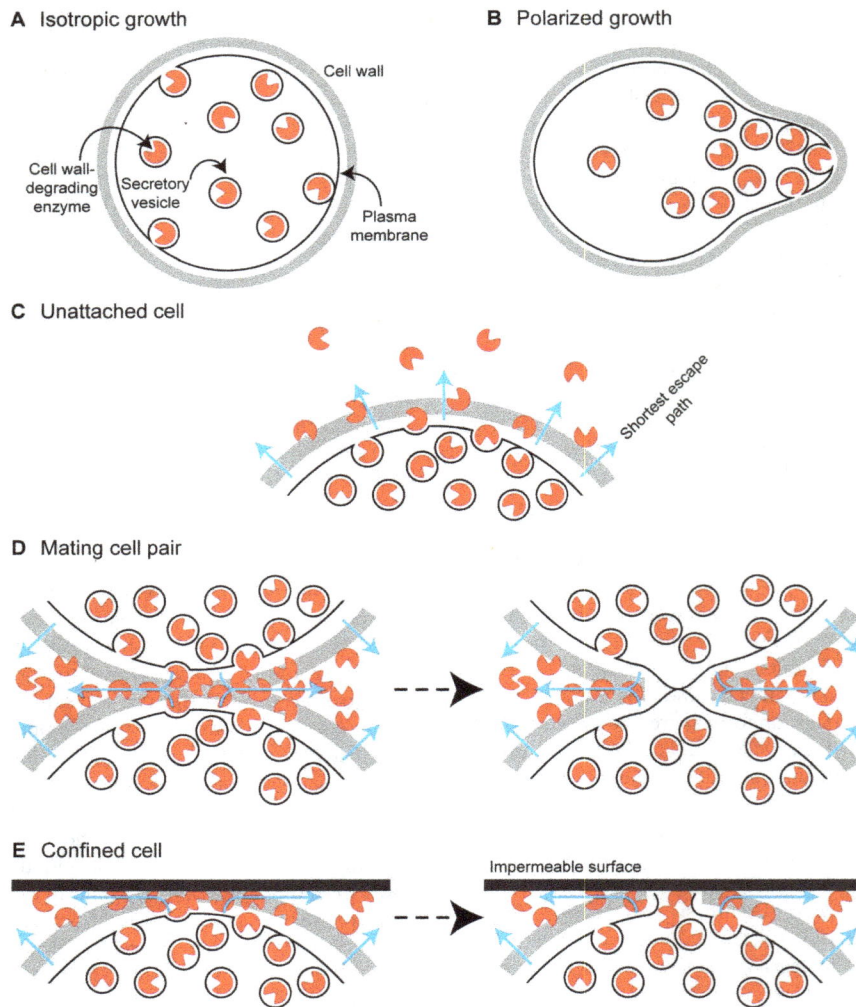

Figure 1. Model: Confining cell wall degrading enzymes in the fusion zone leads to cell wall destruction. A. Isotropically growing cells increase the size of their cell walls equally in all directions to grow larger while maintaining an ellipsoidal shape, so cell wall remodeling enzymes are secreted equally in all directions. **B.** Polarized cells grow anisotropically, so they polarize secretion of cell wall remodeling enzymes to expand their cell walls in the direction of polarization. **C.** When pheromone stimulated cells are unattached, the cell wall remodeling enzymes secreted from the shmoo tip exit the cell wall along the shortest path by traveling perpendicular to the plasma membrane. These enzymes break cell wall bonds as they diffuse through the wall to allow continual expansion of the shmoo up the pheromone gradient, but the wall is not breached. **D.** When two pheromone-stimulated cells are attached by mating agglutinins, the cell wall remodeling enzymes secreted into the future fusion zone must now travel further to exit the cell wall, traveling parallel to the plasma membrane until they reach the edge of the agglutinated zone, increasing the local concentration of cell wall remodeling enzymes in this zone. The cell wall remodeling enzymes dissolve the two cell walls at the point of contact while cell wall synthesizing enzymes simultaneously interlock them, allowing the plasma membranes of the two cells to contact one another and fuse without exposing the cell to osmotic lysis. **E.** We mimicked the attachment of two cells by tightly apposing single cells to impermeable surfaces, forcing cell wall remodeling enzymes to exit the wall by traveling parallel to the plasma membrane until they reached bulk solution and thus increasing the concentration of cell wall remodeling enzymes at the point of attachment to the impermeable surface. This causes a hole to form in the cell wall, exposing the plasma membrane to the extracellular environment and causing the cell to undergo osmotic lysis.

where x is the distance along the axis that runs perpendicular to the external surface of the plasma membrane to the outer surface of the cell wall. Since the flux is constant at all the points along this axis, the gradient must be the same at all points through the thickness of the wall, and thus the total concentration difference across the wall ΔC must increase linearly with the thickness of the wall, $\Delta C = x/D$. If we set D within the cell wall to be much lower than it is in solution, the concentration outside the wall will be close to zero, and C_0, the concentration at the site of secretion will be given by $C_0 = Jx/D$, and if we set the units of x to the thickness of the cell wall (roughly 100 nm), $C_0 = J/D$.

Now we consider an apposed cell, in which diffusion proceeds radially, in the plane of the cell wall, from the site of secretion to the edge of the apposed area. We consider two concentric regions within the opposed area, a central one where both secretion and diffusion occurs, and a peripheral one, where there is just diffusion. Remembering that we have set the unit of length equal to the thickness of the cell wall, within the region with secretion and diffusion, the flux that must leave an area of radius r is

$$\text{Flux} = J\pi r^2$$

where J is the rate of enzyme secretion per unit area. At any radius, r, within the secretory zone, the enzyme must pass through a ring whose area is $2\pi r$, specifying the radial concentration gradient, according to Fick's Law, and then, by integration, the concentration difference between the center (C_0) and any radial distance, r, within the secretion zone,

$$J\pi r^2 = -2\pi r D \frac{dC}{dr}$$

$$\frac{dC}{dr} = -\frac{J\pi r^2}{2\pi r D} = -\frac{Jr}{2D}$$

$$dC = -\frac{Jr}{2D}dr$$

$$C_r - C_0 = -\int_0^r \frac{Jr}{2D}dr = -\frac{Jr^2}{4D}$$

$$C_0 - C_r = \frac{Jr^2}{4D}$$

To get the concentration difference between the center and edge of this region, which is at r_s, the radius of the secreting region, we substitute r_s for r.

$$C_0 - C_{r_s} = \frac{Jr_s^2}{4D}$$

In the region where there is just diffusion, we can calculate the concentration difference from the inside edge to the outside edge of the region. Within the region, we must satisfy the condition that the total flux through each successive circumference is equal to the rate of enzyme production over the total producing region, which is $J\pi r_s^2$. Thus for all $r > r_s$,

$$J\pi r_s^2 = -2\pi r D \frac{dC}{dr}$$

We can rearrange and integrate to get the change in concentration between the concentration at r_s and r.

$$J\pi r_s^2 = -2\pi r D \frac{dC}{dr}$$

$$dC = -\frac{J\pi r_s^2}{2\pi r D}dr = -\frac{Jr_s^2}{2rD}dr$$

$$C_{r_s} - C_r = \int_{r_s}^r \frac{Jr_s^2}{2rD}dr$$

$$C_{r_s} - C_r = \frac{Jr_s^2}{2D}\log\left(\frac{r}{r_s}\right)$$

The overall concentration drop from the center of the secreting region to the edge of the apposed area is then given by adding the concentration drop from the center to the edge of the secreting region and the drop from the edge of the secreting region to edge of the apposed region, situated at r_{max}.

$$C_0 - C_{r\max} = (C_0 - C_{r_s}) + (C_{r_s} - C_{r\max}) =$$

$$\frac{Jr_s^2}{4D} + \frac{Jr_s^2}{2D}\log\left(\frac{r_{max}}{r_s}\right) = \frac{Jr_s^2}{4D}\left(1 + 2\log\left(\frac{r_{max}}{r_s}\right)\right)$$

If we set $r_s = 2.5$, and $r_{max} = 10$, corresponding to radii of secretion and apposition of 0.25 and 1 μm, and consider the concentration drop from the center of the secreted region to its edge, C_s, we get

$$C_s = 1.56 J/D$$

and for the drop from the outer edge of the secreted region to the edge of the apposed region, C_d, we get

$$C_d = 4.33 J/D$$

giving a total concentration drop, $C_{tot} = 5.89J/D$, which implies a maximum hydrolase concentration that is nearly six times that attained when a cell is not apposed to another cell.

The difference between the mean concentration at the center of the apposed region and the mean concentration in the wall of an unapposed cell is even higher. In unapposed cells, the mean concentration, felt half way through the cell wall, is $J/2D$, which is the average of a concentration J/D at the cell surface and 0 at the interface between the wall and solution. In the apposed cell however, the surface that the cell is exposed to acts as a reflecting barrier so that the concentration is constant across the thickness of the wall, and thus the mean concentration, in the scenario we have described is more than ten times higher for the apposed than for the unapposed cells (Figure 2).

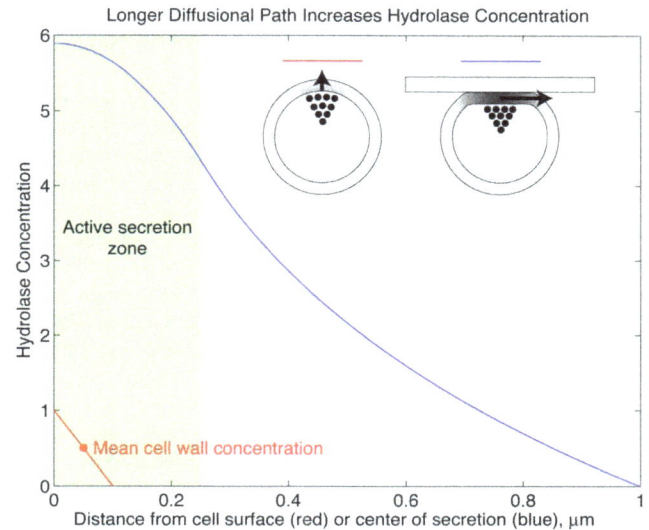

Figure 2. The role of radial diffusion through the cell wall of apposed cells in increasing the concentration of cell wall degrading enzymes. The graph shows analytical results for the relative enzyme concentration in two scenarios: *red*, diffusion through the cell wall, perpendicular to the cell surface, of a cell free in solution and not in contact with other cells or solid surfaces, and *blue*, diffusion through the cell wall, parallel to the cell surface, of a cell that is apposed to a solid surface, with a circular contact area whose radius is 1 μm.

Pheromone-stimulated cells die when attached to an impermeable surface

If contact with another cell leads to cell wall dissolution by increasing the local concentration of wall-degrading enzymes, we should be able to mimic the phenomenon by confining cells against an impermeable surface. We compared the response of pheromone-stimulated cells in environments where the cells were either free-floating, simulating cells in a mating mixture that are not attached to a fusion partner, or attached to an impermeable surface, simulating cells attached to a fusion partner via mating agglutinins. We observed cells in three different environments: bulk culture, attachment to a single, flat, impermeable surface, and confinement between two impermeable surfaces (Figure 3A). Cells expressing the α-factor protease, *BAR1* [38], are capable of decreasing the pheromone concentration at their surface [38,39], so we used *MAT**a** bar1Δ* cells for our investigations. We incubated *bar1Δ* cells in 50 nM α-factor in bulk culture for five hours and found that roughly 10% of the cells die (Figure 3B). Although cells grown in bulk culture have no enforced contacts with the other cells or the impermeable surface of the culture tube, it is difficult to control the physical interactions of cells when they are free-floating in liquid culture and possible that cells could stick either to each other, perhaps due to incomplete separation after budding, or to the surface of the culture tube.

To mimic the attachment of two cell walls via mating agglutinins, we attached cells to the impermeable surface of a glass coverslip using the lectin, concanavalin A (ConA), which binds to carbohydrates in the cell wall [40] (Figure 3A). In order to image the yeast cells for an extended period of time, we created a chamber several hundred times the diameter of a yeast cell. Cells were adhered to the ConA-coated coverslip, and the chamber was filled with medium containing 50 nM α-factor using capillary action and then sealed and observed over a period of five hours. We found that *MAT**a** bar1Δ* cells attached covalently to an impermeable surface were 1.6 times more likely to die than those in bulk culture, indicating that forced attachment to an impermeable surface increases the rate of cell death (Student's *t*-test, p = 0.01) (Figure 3B).

As cells attached to a single impermeable surface grow, they are free to expand away from the glass coverslip, resulting in a low proportion of the cell wall attached to an impermeable surface and making it likely that cells will polarize away from the impermeable surface. To address this problem we used a second technique to mimic the attachment of two cell walls via mating agglutinins. We trapped cells in a microfluidic chamber whose floor and ceiling are separated by the height of a single yeast cell and through which new medium is constantly perfused (Figure 3A). Cells are loaded into this device and then trapped between the two impermeable surfaces of a silicone ceiling and a glass floor. In addition, as the cells grow, the fraction of their surface that is pressed against the floor and ceiling rises, making them more likely to polarize towards an impermeable surface. Using an inverted microscope, it is possible to image cells over time through the glass floor as medium perfuses through the chamber. Once again we imaged *MAT**a** bar1Δ* cells in 50 nM α-factor for five hours (Figure 3C and Movies S1–S3). In the flow chamber, the rate of death of the *MAT**a** bar1Δ* cells was more than twice as high as in bulk culture and 1.5 times the rate of death when attached to ConA-coated coverslips, suggesting that a larger area of attachment to an impermeable surface causes increased cell death (Student's *t*-test, p<0.003) (Figure 3B).

Pheromone-induced cell death increases with increased cell polarization

Decreased pheromone production has been reported to cause decreased cell fusion [17]. Thus, if the cell death seen here is due to the same activities that normally promote cell fusion, we would expect to see a decrease in cell death with decreased pheromone concentration. We chose to assay the effect of decreased pheromone concentration in the flow chamber, where the highest percentage of cells died when exposed to 50 nM α-factor. As previously reported for cells in bulk culture [37], decreasing the α-factor concentration decreased the percentage of cells that died in the flow chamber. In 5 nM α-factor, 50-fold fewer cells died than when cells were exposed to 50 nM α-factor in the flow chamber (Student's *t*-test, p = 10^{-6}), and the fraction of dead cells increased as the concentration of α-factor was increased (Student's *t*-test, p< 0.02) (Figure 4A). Although shmoo formation occurs at 5 nM α-factor, as the pheromone concentration was increased, the cells became more tightly polarized, forming pointier shmoos (Figure 4B).

The flow chamber traps cells by wedging them into a space minutely smaller than a single cell in height. When cells are arrested, such as by pheromone stimulation, the cells increase in size as they continue to grow without dividing [41]. Because of this, it is possible that the increased frequency of cell death in the flow chamber, as compared to bulk culture and when cells are attached to ConA-coated coverslips, is not due to the accumulation of enzymes that would normally degrade the cell wall during cell fusion but rather because the physical strain put on the cell wall is too high, which could be increased by the modest pressure (14 kPa = 2 psi) applied to drive the flow of the perfused medium. We therefore used a different method that would arrest the cell cycle without interfering with cell growth. Like pheromone treatment, treating cells with benomyl, a drug that leads to microtubule depolymerization, causes cells to become larger without dividing, but unlike pheromone-arrest, benomyl-arrested cells are unpolarized and arrest in mitosis instead of G1 [42] (Figure 4C). If cells in the flow chamber die because they were squashed, a substantial percentage of benomyl-arrested cells should die in the flow chamber. Although it is possible to find the occasional, dead, benomyl-arrested cell, 60-fold fewer cells die during five hours of benomyl-arrest than during exposure to 50 nM α-factor, indicating that death in the flow chamber is specific to pheromone-arrest, where cells are polarized, and is not due to growth under physical confinement (Figure 4D).

Pheromone-induced cell death is due to osmotic lysis

Yeast cells require cell walls at least in part due to osmotic pressure. Since the osmolarity of the cytoplasm is higher than the typical extracellular environment, without the rigidity of a cell wall, water would rush into the cell and cause it to lyse [33], and previous studies have shown that cells that are unable to regulate the osmotic balance between the cytoplasm and the extracellular environment have a cell fusion defect [15]. One interpretation of the death of pheromone-treated cells pressed against an impermeable surface is that accumulation of cell wall-degrading enzymes causes the cells to digest part of their cell walls leading to membrane expansion through a hole in the cell wall and eventual lysis. If this interpretation is correct, it should be possible to affect the rate of death by manipulating the osmotic pressure differential between the cell and the medium [43].

We did two experiments to determine whether the pheromone-induced deaths are due to osmotic lysis: either increasing or decreasing the osmotic pressure differential between the cytoplasm and the extracellular environment. We first tested the effect of

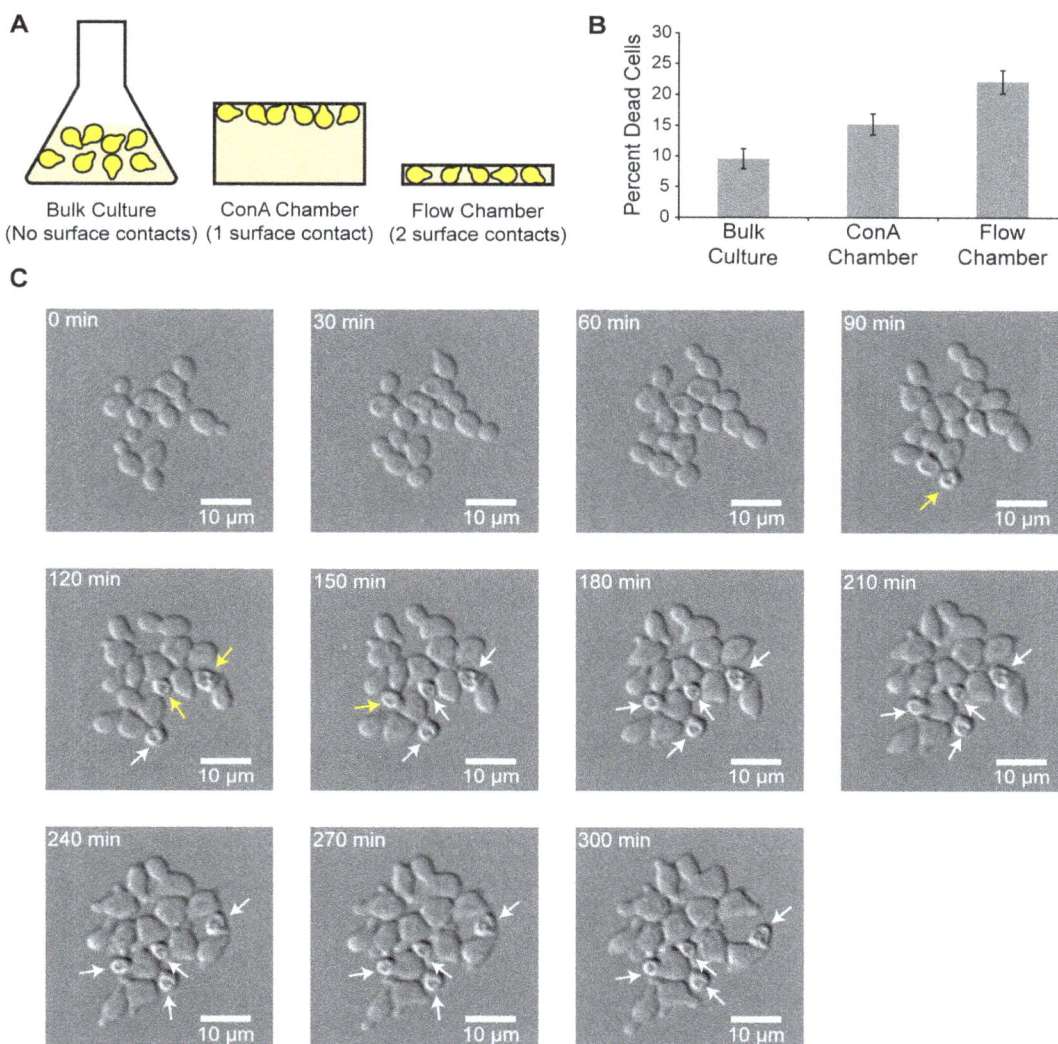

Figure 3. Pheromone-induced cell death increases with increasing attachments to an impermeable surface. A. Cells grown in bulk culture were incubated in test tubes on roller drums in liquid media without any enforced contact with impermeable surfaces. Cells grown in a concanavalin A (ConA) chamber were grown in a chamber whose depth was many times the diameter of a single yeast cell and attached to a single surface of the chamber (the ceiling provided by a glass coverslip) using the lectin, concanavalin A. For confinement, cells were loaded into a microfluidic chamber which traps cells between a ceiling and floor separated by the diameter of a single yeast cell, causing enforced contact with two surfaces. Medium is then constantly perfused through the chamber. **B.** Percent of *MAT**a** bar1Δ* cells that died after exposure to 50nM α-factor for five hours in three different physical environments. Error bars represent the standard deviation of at least three independent experiments. **C.** Time course of *MAT**a** bar1Δ* cells incubated in 50nM α-factor for the indicated amount of time in the flow chamber. Yellow arrows indicate cells that died since the previous time point. White arrows indicate cells that died earlier. The scale bar indicates 10 μm.

increasing the osmotic pressure differential between the cytoplasm and the extracellular environment. Cells were exposed to medium with 50 nM α-factor and 1M sorbitol, which increases the osmolarity of the medium, for five hours. When cells are exposed to high external osmolarity, they adapt to the osmotic stress by synthesizing glycerol, which can take place in a matter of minutes [44–47], thus increasing their internal osmolarity. Because of this partial restoration of the osmotic pressure gradient across the cell wall, we were not surprised to find that the fraction of cells that die when exposed to 50 nM α-factor and 1M sorbitol is similar to that of cells exposed to only 50 nM α-factor (Figure 5A). Nevertheless, we reasoned that some of the surviving cells would have holes in their cell walls that would be small enough to allow their survival until we increased the osmotic pressure difference between the

inside and outside of the cells. Thus, if we replace the sorbitol-containing medium with medium lacking sorbitol, we would expect to see rapid cell death due to the large pressure differential. To test this prediction, we waited until five hours after beginning pheromone treatment and then replaced medium containing 1M sorbitol and 50 nM α-factor with medium containing only 50 nM α-factor. Immediately following the sorbitol washout, the number of dead cells in the flow chamber more than doubled, supporting the idea that the cells in the chamber are dying due to a breach in their cell walls (Figure 5A and 5B and Movie S4).

To test the effect of decreasing the osmotic pressure differential between the cells and the extracellular environment, we exposed cells to 50 nM α-factor in the absence of 1M sorbitol for 80 minutes, at which point cells are just beginning to die

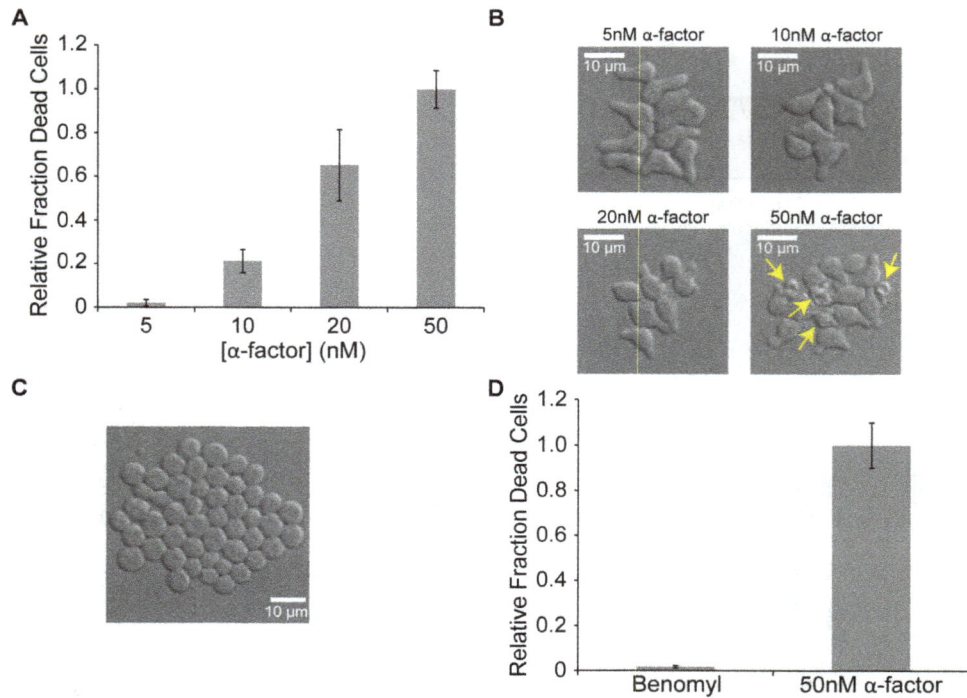

Figure 4. Pheromone-induced cell death increases with increased polarization. A. Fraction of *MATa bar1Δ* cells that died after five hours exposure to various concentrations of α-factor in the flow chamber relative to the fraction of *MATa bar1Δ* cells that died after five hours exposure to 50nM α-factor. Error bars represent the standard deviation of at least three independent experiments. **B.** *MATa bar1Δ* cells incubated in the indicated concentration of α-factor for five hours in the flow chamber. Yellow arrows indicate dead cells. The scale bar indicates 10 μm. **C.** *MATa bar1Δ* cells exposed to 0.1mM benomyl for five hours in the flow chamber. The scale bar indicates 10 μm. **D.** Fraction of *MATa bar1Δ* cells that died after five hours exposure to either 0.1mM benomyl or 50nM α-factor in the flow chamber relative to the fraction of *MATa bar1Δ* cells that died after five hours exposure to 50nM α-factor (Student's *t*-test, p<10^{-6}). Error bars represent the standard deviation of at least three independent experiments.

(Figure 3C and Movies S1–S3). We determined the percentage of dead cells at this point and then perfused the chamber with medium containing 1M sorbitol and 50 nM α-factor and observed the percentage of dead cells 60 minutes after the change of media. Since the sorbitol is washed in after the cells have begun to shmoo, the cells will have less time to induce the hyperosmotic response, and if the cell death is due to osmotic lysis, we should observe fewer cell deaths when 1M sorbitol is present in the medium. When we observe the fold change in cell death between 80 minutes and 140 minutes after α-factor addition in the absence of 1M sorbitol, there is an 8.9-fold increase in the fraction of dead cells (Figure 5C). However, when 1M sorbitol is added to the medium 80 minutes after α-factor addition, there is only a 1.4-fold increase in the fraction of dead cells between 80 and 140 minutes after α-factor addition, strengthening the evidence that pheromone-induced cell death is due to osmotic lysis (Figure 5C).

Proteins necessary for cell wall breakdown during mating are required for pheromone-induced cell death

We investigated the effects of deleting, *FUS1* and *FUS2*, two genes required for efficient cell fusion [24]. When *FUS1*, *FUS2*, or, both *FUS1* and *FUS2* are deleted in both mating partners, prezygotes, consisting of two shmoos bound to each other at their tips, are formed, but cells cannot dissolve their cell walls and thus fail to fuse [24,28]. Also, in *fus1* and *fus1fus2* mutants, the tightly polarized vesicles that are seen in the fusion zone of wild-type prezygotes and are hypothesized to contain cell wall remodeling

enzymes are fewer and more widely dispersed than in wild-type cells [28]. If cell death in the flow chamber is due to pheromone-stimulated cell wall breakdown, mutations known to impair cell wall fusion should reduce the frequency of pheromone-induced cell death events in the flow chamber. Corroborating previous results obtained in bulk cultures [37], deleting *FUS1* and *FUS2* alone and in combination caused more than a 14-fold reduction in cell death in the flow chamber when cells were exposed to 50 nM α-factor for five hours (Student's *t*-test, p<0.002) (Figure 6A).

If the pheromone-induced cell death in the flow chamber is due to holes formed in the cell wall from inappropriate cell wall dissolution, the deletion of cell wall remodeling enzymes should decrease the frequency of pheromone-induced cell death. We investigated the effects of two putative cell wall glucanases that have been implicated in mating: Scw11, a target of pheromone-induced gene expression [19], and its paralog, Scw4, whose deletion interferes with mating [18]. If the observed cell death is due to accumulation of cell-wall degrading enzymes and these glucanases are major contributors to cell wall remodeling during cell wall fusion, deleting them should reduce the frequency of pheromone-induced cell death in the flow chamber. To test this prediction, we incubated *MATa bar1Δ scw11Δ* cells in a flow chamber in medium containing 50 nM α-factor for five hours. Deleting *SCW11* caused a 20% reduction in cell death compared to *MATa bar1Δ* cells (Student's *t*-test, p = 8×10^{-4}), and removing both Scw11 and Scw4 caused a 40% reduction in cell death compared to *MATa bar1Δ* cells (Student's *t*-test, p = 3×10^{-5}) (Figure 6B).

A

B

C

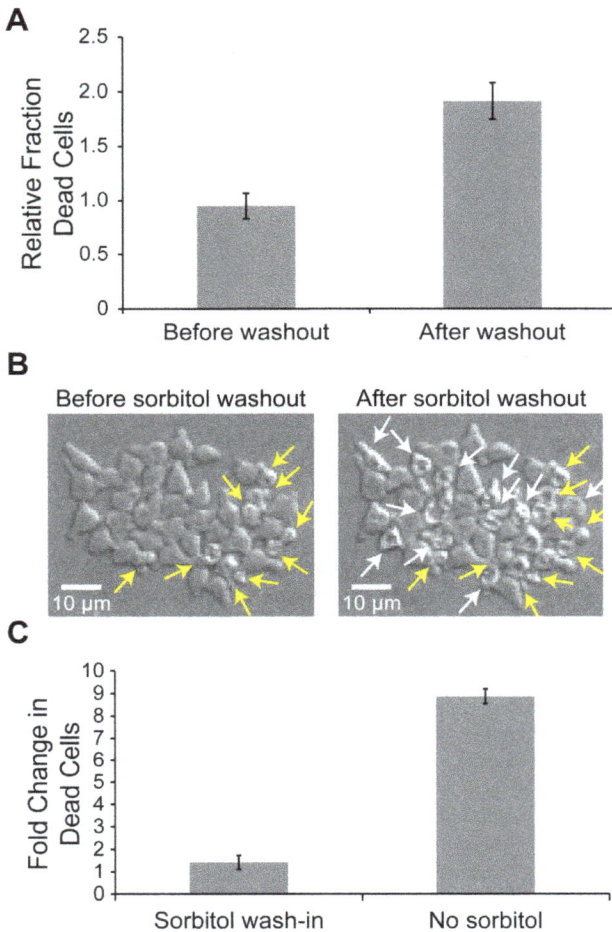

Figure 5. Pheromone-induced cell death is due to osmotic lysis.
A. *MATa bar1Δ* cells were grown in a flow chamber for five hours in medium with 50nM α-factor and 1M sorbitol. After five hours, the sorbitol was washed out, and the cells were incubated in medium with 50nM α-factor and no sorbitol. The fraction of dead cells 10 minutes before and 10 minutes after the 1M sorbitol was washed out relative to the fraction of cells that die when exposed to 50nM α-factor for five hours without the addition of sorbitol was determined (Student's t-test, $p = 2 \times 10^{-4}$). Error bars represent the standard deviation of at least three independent experiments. **B.** Cells imaged after 290 minutes in medium with 1M sorbitol and 50nM α-factor (Before sorbitol washout) and 10 minutes after the medium was replaced with medium with 50nM α-factor and no sorbitol (After sorbitol washout). Yellow arrows indicate the cells that died during the 290 minutes of pheromone treatment prior to the sorbitol washout. White arrows in the "After sorbitol washout" picture indicate cells that died during the twenty minute period that spanned the last 10 minutes with sorbitol and the first 10 minutes after the sorbitol washout. The scale bar indicates 10 μm. **C.** Cells were grown in the flow chamber for 80 minutes in medium with 50nM α-factor. After 80 minutes, 1M sorbitol was added to the medium such that the cells were incubated in medium with 1M sorbitol and 50nM α-factor. The fold change in the number of cells that died during the 80 minutes prior to and 60 minutes after the sorbitol wash-in was determined (Sorbitol wash-in). In control chambers (No sorbitol), no sorbitol was added to the medium, and the fold change in the number of cells that died in the two corresponding periods was determined (Student's t-test, $p = 9 \times 10^{-5}$). Error bars represent the standard deviation of at least three independent experiments.

Discussion

The mating of budding yeast is risky and elaborately choreographed. When two haploid yeast cells mate, they signal through reciprocal pheromones and receptors, stimulate each other to signal ever more strongly, arrest their cell cycles, use pheromone gradients to direct their polarization towards each other, and eventually fuse their cell walls, cell membranes, and nuclei to form a single diploid cell [8,10,48,49]. Although many aspects of yeast mating have been well studied, the mechanism by which cells dissolve their cell walls to allow fusion of their plasma membranes remains mysterious. Cell wall dissolution is a particularly dangerous step in yeast mating. The plasma membranes of the two partner cells cannot touch each other and fuse until the cell walls that lie between them have been dissolved [8]. Because the osmolarity inside a cell is so much higher than outside, the elasticity of the cell wall opposes the osmotic pressure difference between the cytoplasm and the environment, thus keeping water from rushing into the cell and causing it to lyse. A cell that dissolves any part of its cell wall that does not touch a closely apposed mating partner will die [15,16].

Pheromone-induced cell death was studied previously [36,37] and hypothesized to be due to inappropriate activation of cell fusion machinery, resulting in cell wall dissolution and eventual cell lysis [37]. Although it was observed that this lysis can be reduced by increasing cell wall integrity and deleting certain proteins involved in cell fusion, a hypothesis to explain why cells were dissolving their cell walls was not given [37]. Many hypotheses can be generated to explain how cell wall dissolution is regulated in time and space to promote mating and prevent accidental deaths. Most of them posit additional signaling systems in addition to the known mechanisms of pheromone signaling, but no additional signaling molecules have been uncovered, despite a variety of searches [8,28,50,51]. The failure of these attempts led us to propose a hypothesis that requires no new components and instead appeals to the physical differences between mating cell pairs and isolated, pheromone-stimulated cells.

In an isotropically growing cell, cell wall synthesizing and remodeling enzymes are secreted uniformly around the cell, whereas the polarized growth that accompanies both budding and shmooing requires similarly polarized secretion of these enzymes [33] (Figure 1A and 1B). Thus we hypothesize that cell wall remodeling enzymes, such as Scw4 and Scw11, are preferentially released at the shmoo tip, which locally weakens the cell wall, allowing the shmoo to grow continuously up the pheromone gradient. As a shmoo approaches a suitable partner, the concentration of pheromone increases, tightening the polarization, and increasing the concentration of cell wall remodeling enzymes in the part of the cell wall that has polarized towards its partner's site of maximum pheromone secretion [49,52–55]. If the remodeling enzymes are diffusible, the maximum concentration they can reach in a shmoo that has not bound to a partner is limited: even though the secretion rate of cell wall remodeling enzymes is high, the enzymes are able to diffuse through the cell wall, keeping their concentration in the range that is high enough to allow rapid remodeling of the growing shmoo but low enough to prevent cell wall rupture (Figure 1C). But when two shmoo tips are attached to each other via mating agglutinins, it takes longer for cell wall remodeling enzymes to diffuse out of the fusion zone because they must now travel laterally through the cell wall in order to escape, thus increasing the local concentration of the remodeling enzymes and leading to the gradual dissolution of the cell wall, exposing the two plasma membranes to each other and allowing their fusion to create a single, diploid cell (Figure 1D).

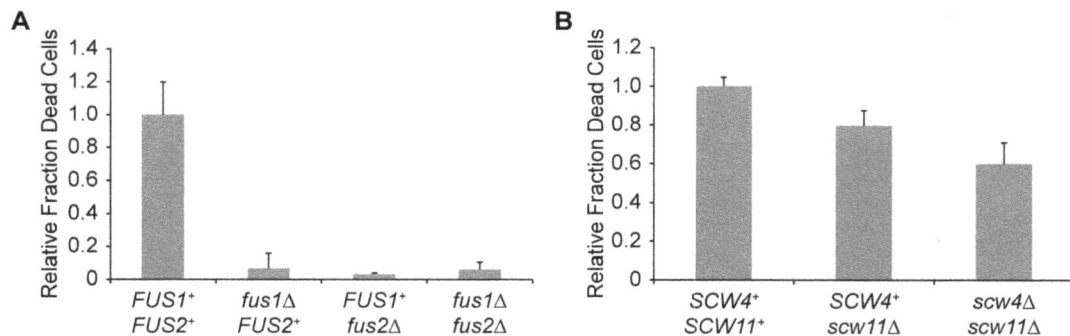

Figure 6. Pheromone-induced cell death is dependent on cell fusion proteins and putative glucanases. A. Fraction of dead *MATa bar1Δ* cells deleted for different combinations of *FUS1* and *FUS2* relative to the fraction of dead *MATa bar1Δ* cells incubated in 50nM α-factor for five hours in the flow chamber. Error bars represent the standard deviation of at least three independent experiments. **B.** Fraction of dead *MATa bar1Δ* cells deleted for different combinations of putative cell wall glucanases relative to the fraction of dead *MATa bar1Δ* cells incubated in 50nM α-factor for five hours in the flow chamber. Error bars represent the standard deviation of at least three independent experiments.

Our data supports the hypothesis that increased secretion of remodeling enzymes and the longer distances they have to diffuse when two polarization zones agglutinate to each other causes cell wall dissolution. We tightly apposed pheromone-treated cells to an impermeable surface, mimicking the effect of cell-cell attachment during mating while ensuring that the only signal these cells can receive is a uniformly high pheromone concentration (Figure 1E). The cell lysis events we observe are not due merely to the physical constraints of a flow chamber: cells also lyse when they are chemically attached to a glass coverslip with essentially infinite space to expand, and confined cells that grow larger isotropically while they are arrested in mitosis do not lyse [42]. By manipulating the presence of an osmoprotectant in the media, we show that the frequency of lysis events rises with increasing osmotic pressure difference between the osmolarity of the cell and its environment, implying that breaches in the cell wall lead to osmotically induced lysis. Lysis depends on Fus1 and Fus2, proteins that have been previously implicated in cell fusion [24], and the cell wall dissolution appears to be at least partially accomplished by two putative glucanases, Scw4 and Scw11.

Although a contact-driven increase in the concentration of cell wall remodeling enzymes is the simplest explanation for cell wall dissolution during mating, other viable hypotheses exist. One is that cells can only dissolve their cell walls in response to a high concentration of pheromone that they experience while attached to a mating partner. When cells are in the flow chamber, we do see an increase in cell lysis as we increase the pheromone concentration arguing that high pheromone concentrations promote cell wall dissolution; however, the increase in cell lysis in cells apposed to impermeable surfaces argues that factors other than pheromone concentration contribute to cell wall dissolution. Another hypothesis is that the cell is capable of detecting cell wall or cell membrane deformations that indicate that two shmoo tips are attached via mating agglutinins. Our experiments do not negate this possibility, since it is likely that contact with impermeable surfaces causes cell wall deformation, but the failure of previous attempts to identify additional signaling components [8,28,50,51] argues against this model. A third possibility is that cells respond to a direct signal from another cell, such as an additional, uncharacterized signaling mechanism similar to the G-protein coupled receptors involved in pheromone stimulation, or, perhaps, the oscillation in pheromone concentration that would occur if cells were close enough to detect the pulses of increased

pheromone concentration concomitant with the fusing of individual secretory vesicles with the plasma membrane. The fact that lysis occurs without the presence of a mating partner argues against any hypothesis that requires communication aside from that of the reciprocal pheromones and pheromone receptors between the two mating cells.

Taken together with previously published studies, our data supports a model that involves pheromone-induced, polarized secretion of cell wall remodeling enzymes. When cells are pheromone stimulated, a MAP kinase cascade activates transcription of pheromone-induced genes [10]. Along with many others, these genes include the expression of mating agglutinins and cell wall remodeling enzymes, which are packaged into vesicles for secretion into the extracellular environment [11–13,17–19]. Fus2 and Rvs161, a protein that binds to curved membranes [25,26] and is involved in cell fusion [51], bind to these vesicles and travel along actin cables to the site of polarization in a Myo2-dependent fashion [56] where they are anchored to the plasma membrane by Fus1 [23], which interacts with the polarisome [30]. Fus2 and Rvs161 in conjunction with Cdc42 may then facilitate the fusion of these vesicles with the plasma membrane [57].

When cells are weakly stimulated, they form broad shmoos (Figure 4B). Although these cells are polarized, the zone of polarization is relatively large, and presumably, the vesicles containing cell wall remodeling enzymes are released into a relatively large area. The enzymes cleave carbohydrate bonds as they diffuse through the cell wall matrix, weakening the cell wall and allowing for further expansion in the direction of highest pheromone concentration [34]. As a shmoo gets closer to a cell of the opposite mating type, the pheromone concentration increases and the shmoo tip becomes more tightly polarized [53,54]. This tighter polarization focuses the secretion of cell wall remodeling enzymes into a smaller fraction of the cell surface, increasing the concentration of cell wall remodeling enzymes in this zone. Although the concentration of cell wall remodeling enzymes in this zone has increased, it is not typically high enough to cause dissolution of the cell wall unless the shmoo tip is pressed against an impermeable barrier, forcing the enzymes to travel further to reach bulk solution. Since the time taken to diffuse a given distance rises with the square of the distance, the effective speed at which the enzymes move falls, and thus their concentration in the cell wall rises (Figure 1 and 2). Similarly, when the two polarized cells attach at their shmoo tips via mating agglutinins, the presence of a second cell membrane traps the remodeling enzymes in the cell wall by requiring them to move laterally along the cell surface to exit the cell wall, increasing their concentration and breaking down the cell wall (Figure 1D). As the wall dissolves, the two plasma membranes come into contact with one another, allowing membrane fusion to begin and pushing the Fus2-bound vesicles outward [23], which allows for the rest of the intervening cell wall to be dissolved and eventually full fusion of the newly formed zygote.

Understanding more about the cell fusion of budding yeast is an important step in understanding cell fusion in more complex organisms. Although animal cells do not have a cell wall, the extracellular matrix surrounding these cells must be dissolved prior to cell fusion. A similar local increase in enzyme concentration at points where cells are very close to each other could promote the digestion of the matrix and allow the plasma membranes of the two partners to touch each other.

Materials and Methods

Yeast strains and culturing

Strains used in this study are listed in Table 1. All strains were derived from the W303 background [58] (*ade2-1 can1-100 his3-11,15 leu2-112 trp1-1 ura3-1*) using standard genetic techniques. All media was prepared as described [59] and contained 2% wt/vol of glucose. Cells were either grown in Synthetic Complete medium (SC) or Yeast Extract Peptone Dextrose (YPD) at 30°C in culture tubes on roller drums or at room temperature (25°C) for timelapse microscopy. Bovine serum albumin (BSA) was used to reduce the non-specific absorption of α-factor to glass and plastic surfaces; it was made into 10% wt/vol stocks in deionized water and then diluted into media to 0.1% wt/vol. Synthetic α-factor (Biosynthesis, Lewisville, TX) was suspended in dimethyl sulfoxide (DMSO) and then diluted into YPD+0.1% BSA or SC+0.1% BSA at the appropriate concentration. When appropriate, 1M sorbitol was added to YPD by dissolving sorbitol into YPD. YPD containing 1-(butylcarbomoyl)-2-benzimidasolecarbamate (benomyl) was prepared by heating YPD to 65°C and adding 34 mM benomyl in DMSO dropwise to a final concentration of 0.1 mM. Yeast extract was obtained from EMD Millipore (Billerica, MA). Peptone and yeast nitrogen base were obtained from BD (Franklin Lakes, NJ). Bacto-agar was obtained from US Biological (Swampscott, MA). Unless otherwise noted, all chemicals were obtained from Sigma-Aldrich (St. Louis, MO).

Microscopy

Microscopy was done at room temperature using a Nikon Ti-E inverted microscope with a 20x Plan Apo VC 0.75NA air lens, and images were acquired with a Photometrics CoolSNAP HQ camera (Roper Scientific, AZ). Timelapse photography was done using Metamorph 7.7 (Molecular Devices, CA); pictures were acquired using differential interference contrast every 10 minutes with a 10 ms exposure.

Bulk culture lysis assay

Cells were grown to log phase ($\sim 5 \times 10^6$ cells/mL) at 30°C in YPD and counted using a Z2 Coulter counter (Beckman-Coulter, CA). Cells were washed in YPD+0.1% BSA and resuspended at 10^6 cells/mL into plastic 14 mL culture tubes (BD Falcon, MA) in YPD+0.1% BSA with 50 nM α-factor. These cultures were then incubated on a roller drum at 30°C for five hours. Cells were then put directly onto glass slides (Corning, NY) with uncoated coverslips (VWR, PA) and imaged at 20x magnification using differential interference contrast with a 10 ms exposure. Prior to the experiment, the plastic culture tubes were coated in BSA by incubating overnight at 4°C with phosphate buffered saline (PBS) with 2% wt/vol BSA. The PBS+2% BSA was poured out immediately prior to the addition of the cell cultures. To determine the percentage of cells that lysed, more than 50 cells were counted from each trial. Statistical significance was determined using Student's *t*-Test.

Concanavalin-A coated coverslip lysis assay

Coverslips (VWR, PA) were coated in concanavalin A (MP Biomedicals, OH) in a protocol modified from Joglekar *et al.* (2008) [60]. Briefly, coverslips were soaked in 1M NaOH for one hour at room temperature (25°C), rinsed five times with deionized, filtered water, and then incubated at room temperature for one hour in a solution of 10 mM Na_2HPO_4 pH 6.0 (Fisher Biotech, MA) +1 mM $CaCl_2$+0.5 mg/mL concanavalin A. Coverslips were then rinsed five times with deionized, filtered water and air-dried over a 100°C heat block. To make a chamber, strips of parafilm (American National Can, IL) were melted at 100°C on glass slides (Corning, NY) and concanavalin A-coated coverslips were placed on top of the strips. The parafilm was allowed to cool to room temperature, creating channels with a glass slide ceiling, concanavalin A-coated coverslip floor, and parafilm walls on two, parallel sides.

Cells were grown to log phase ($\sim 5 \times 10^6$ cells/mL) at 30°C in YPD and then washed in SC+0.1% BSA. 50 nM α-factor was added to the cells, and the cells were immediately injected into the chamber using capillary action. The cells were allowed to adhere to the concanavalin A-coated coverslip for 10 minutes, and then 200 μL of SC+0.1% BSA with 50 nM α-factor was flowed through the chamber using capillary action to wash off excess cells.

Table 1. Strains used in this study.

Strain Name	Genotype (all cells are in the W303 background)
LBHY52	*MATa bar1Δ::KanMX6 P_{ACT1}-yCerulean-HIS3MX6 @ P_{ACT1}*
LBHY77	*MATa bar1Δ::KanMX6 fus1Δ::NatMX4 P_{ACT1}-yCerulean-HISMX3 @ P_{ACT1}*
LBHY80	*MATa bar1Δ::KanMX6 fus2Δ::HphMX4 P_{ACT1}-yCerulean-HISMX3 @ P_{ACT1}*
LBHY84	*MATa bar1Δ::KanMX6 fus1Δ::NatMX4 fus2Δ::HphMX4 P_{ACT1}-yCerulean-HISMX3 @ P_{ACT1}*
LBHY136	*MATa bar1Δ::ADE2 SPA2-YFP:HIS3 scw11Δ::HphMX4 ade2-1 can1-100 his3-11,15 leu2-3,112 trp1-1 ura3-1*
LBHY153	*MATa bar1Δ::ADE2 SPA2-YFP:HIS3 scw4Δ::KanMX6 scw11Δ::HphMX4 ade2-1 can1-100 his3-11,15 leu2-3,112 trp1-1 ura3-1*
MP 384	*MATa bar1Δ::ADE2 SPA2-YFP:HIS3 ade2-1 can1-100 his3-11,15 leu2-3,112 trp1-1 ura3-1*

All strains are from this study except for MP 384, which is from M. Piel.

The chamber was sealed with candle wax and imaged at 20x magnification, using differential interference contrast with a 10 ms exposure, every 10 minutes for five hours from the time when the cells were exposed to α-factor-containing medium. To determine the percentage of cells that lysed, more than 400 cells were counted from each trial. Statistical significance was determined using Student's t-Test.

Flow chamber lysis assay

Cells were grown to log phase ($\sim 5 \times 10^6$ cells/mL) at 30°C in YPD and then washed in YPD+0.1% BSA. For experiments involving α-factor, the microfluidic chambers (CellAsic, Hayward, CA) [61] were pretreated by perfusing PBS+2% BSA through the chamber at 34 kPa (5 psi) for 10 minutes and then YPD+0.1% BSA through the chamber at 34 kPa (5 psi) for 10 minutes. After cells were loaded, YPD+0.1% BSA with the appropriate concentration of α-factor was perfused through the chamber at 14 kPa (2 psi), and pictures were taken at 20x magnification every 10 minutes for five hours using differential interference contrast with a 10 ms exposure.

For experiments involving benomyl, the microfluidic chambers were pretreated by perfusing YPD through the chamber at 34 kPa (5 psi) for 10 minutes. After the cells were loaded, YPD+0.1 mM benomyl was perfused through the chamber at 14 kPa (2 psi), and pictures were taken at 20x magnification every 10 minutes for five hours using differential interference contrast with a 10 ms exposure. To determine the percentage of cells that lysed, more than 250 cells were counted from each trial. Statistical significance was determined using Student's t-Test.

Sorbitol wash-out assay

Cells were grown to log phase ($\sim 5 \times 10^6$ cells/mL) at 30°C in YPD and then washed in YPD+0.1% BSA. The microfluidic chambers (CellAsic, Hayward, CA) [61] were pretreated by perfusing PBS+2% BSA through the chamber at 34 kPa (5 psi) for 10 minutes and then YPD+1M sorbitol +0.1% BSA through the chamber at 34 kPa (5 psi) for 10 minutes. After cells were loaded, YPD+1M sorbitol +0.1% BSA+50 nM α-factor was perfused through the chamber at 14 kPa (2 psi) for five hours. After five hours, the medium containing 1M sorbitol was washed out, and YPD+0.1% BSA+50 nM α-factor was perfused through the chamber at 14 kPa (2 psi) for two hours. Pictures were taken at 20x magnification every 10 minutes for seven hours using differential interference contrast with a 10 ms exposure. To determine the percentage of cells that lysed, more than 500 cells were counted from each trial. Statistical significance was determined using Student's t-Test.

Sorbitol wash-in assay

Cells were grown to log phase ($\sim 5 \times 10^6$ cells/mL) at 30°C in YPD and then washed in YPD+0.1% BSA. The microfluidic chambers (CellAsic, Hayward, CA) [61] were pretreated by perfusing PBS+2% BSA through the chamber at 34 kPa (5 psi) for 10 minutes and then YPD+0.1% BSA at 34 kPa (5 psi) through the chamber for 10 minutes. After cells were loaded, YPD+0.1%

BSA+50 nM α-factor was perfused through the chamber at 14 kPa (2 psi) for 80 minutes. At the end of 80 minutes, the YPD+ 0.1% BSA+50 nM α-factor was washed out, and YPD+1M sorbitol +0.1% BSA+50 nM α-factor was perfused through the chamber at 14 kPa (2 psi) for 60 minutes. Pictures were taken at 20x magnification every 10 minutes for 140 minutes using differential interference contrast with a 10 ms exposure. To determine the percentage of cells that lysed, more than 600 cells were counted from each trial. Statistical significance was determined using Student's t-Test.

Supporting Information

Movie S1 Pheromone-induced cell death in the flow chamber. *MAT***a** *bar1Δ* cells were incubated in medium containing 50 nM α-factor for five hours in the flow chamber. White arrows indicate cells that die during the movie. Cells were imaged every 10 minutes. The scale bar indicates 10 μm. Each movie is from an independent experiment.

Movie S2 Pheromone-induced cell death in the flow chamber. *MAT***a** *bar1Δ* cells were incubated in medium containing 50 nM α-factor for five hours in the flow chamber. White arrows indicate cells that die during the movie. Cells were imaged every 10 minutes. The scale bar indicates 10 μm. Each movie is from an independent experiment.

Movie S3 Pheromone-induced cell death in the flow chamber. *MAT***a** *bar1Δ* cells were incubated in medium containing 50nM α-factor for five hours in the flow chamber. White arrows indicate cells that die during the movie. Cells were imaged every 10 minutes. The scale bar indicates 10 μm. Each movie is from an independent experiment.

Movie S4 Pheromone-induced cell death is due to osmotic lysis. *MAT***a** *bar1Δ* cells were incubated in the flow chamber for five hours in medium containing 50 nM α-factor and 1M sorbitol. After five hours, the sorbitol was washed out, and the cells were incubated in medium containing 50 nM α-factor and no sorbitol. Cells were imaged every 10 minutes. The scale bar indicates 10 μm.

Acknowledgments

We thank V. Denic, R. Gaudet, C. Hunter, and members of the Murray laboratory for reading and commenting on the manuscript. We also thank M. Piel for strains.

Author Contributions

Conceived and designed the experiments: LBH AWM. Performed the experiments: LBH. Analyzed the data: LBH AWM. Wrote the paper: LBH AWM.

References

1. Chen EH, editor (2008) Cell fusion overviews and methods. New York, New York: Humana Press. 421 p.
2. Abmayr SM, Pavlath GK (2012) Myoblast fusion: Lessons from flies and mice. Development 139(4): 641–656.
3. Ishii M, Saeki Y (2008) Osteoclast cell fusion: Mechanisms and molecules. Mod Rheumatol 18(3): 220–227.
4. Galindo B, Lazdins J, Castillo R (1974) Fusion of normal rabbit alveolar macrophages induced by supernatant fluids from BCG-sensitized lymph node cells after elicitation by antigen. Infect Immun 9(2): 212–216.
5. Huppertz B, Borges M (2008) Placenta trophoblast fusion. Methods Mol Biol 475: 135–147.
6. Alper S, Podbilewicz B (2008) Cell fusion in *Caenorhabditis elegans*. Methods Mol Biol 475: 53–74.

7. Abmayr SM, Zhuang S, Geisbrecht ER (2008) Myoblast fusion in *drosophila*. Methods Mol Biol 475: 75–97.
8. Ydenberg CA, Rose MD (2008) Yeast mating: A model system for studying cell and nuclear fusion. Methods Mol Biol 475: 3–20.
9. Herskowitz I (1988) Life cycle of the budding yeast *Saccharomyces cerevisiae*. Microbiol Rev 52(4): 536–553.
10. Bardwell L (2005) A walk-through of the yeast mating pheromone response pathway. Peptides 26(9): 339–350.
11. Cappellaro C, Baldermann C, Rachel R, Tanner W (1994) Mating type-specific cell-cell recognition of *Saccharomyces cerevisiae*: Cell wall attachment and active sites of a- and alpha-agglutinin. EMBO J 13(20): 4737–4744.
12. Roy A, Lu C, Marykwas D, Lipke P, Kurjan J (1991) The AGA1 product is involved in cell surface attachment of the *Saccharomyces cerevisiae* cell adhesion glycoprotein a-agglutinin. Mol Cell Biol 11(8): 4196–4206.
13. Zhao H, Shen Z, Kahn PC, Lipke PN (2001) Interaction of α-agglutinin and a-agglutinin, *Saccharomyces cerevisiae* sexual cell adhesion molecules. J Bacteriol 183(9): 2874–2880.
14. Dupres V, Dufrêne YF, Heinisch JJ (2010) Measuring cell wall thickness in living yeast cells using single molecular rulers. ACS Nano 4(9): 5498–5504.
15. Philips J, Herskowitz I (1997) Osmotic balance regulates cell fusion during mating in *Saccharomyces cerevisiae*. J Cell Biol 138(5): 961–974.
16. de Nobel H, Ruiz C, Martin H, Morris W, Brul S, et al. (2000) Cell wall perturbation in yeast results in dual phosphorylation of the Slt2/Mpk1 MAP kinase and in an Slt2-mediated increase in FKS2-lacZ expression, glucanase resistance and thermotolerance. Microbiology 146 (Pt 9)(Pt 9): 2121–2132.
17. Brizzio V, Gammie AE, Nijbroek G, Michaelis S, Rose MD (1996) Cell fusion during yeast mating requires high levels of a-factor mating pheromone. J Cell Biol 135(6): 1727–1739.
18. Cappellaro C, Mrsa V, Tanner W (1998) New potential cell wall glucanases of *Saccharomyces cerevisiae* and their involvement in mating. J Bacteriol 180(19): 5030–5037.
19. Zeitlinger J, Simon I, Harbison CT, Hannett NM, Volkert TL, et al. (2003) Program-specific distribution of a transcription factor dependent on partner transcription factor and MAPK signaling. Cell 113(3): 395–404.
20. Dolan JW, Kirkman C, Fields S (1989) The yeast STE12 protein binds to the DNA sequence mediating pheromone induction. Proceedings of the National Academy of Sciences 86(15): 5703–5707.
21. Chenevert J, Valtz N, Herskowitz I (1994) Identification of genes required for normal pheromone-induced cell polarization in *Saccharomyces cerevisiae*. Genetics 136(4): 1287–1296.
22. Brizzio V, Gammie AE, Rose MD (1998) Rvs161p interacts with Fus2p to promote cell fusion in *Saccharomyces cerevisiae*. J Cell Biol 141(3): 567–584.
23. Paterson JM, Ydenberg CA, Rose MD (2008) Dynamic localization of yeast Fus2p to an expanding ring at the cell fusion junction during mating. J Cell Biol 181(4): 697–709.
24. Trueheart J, Boeke JD, Fink GR (1987) Two genes required for cell fusion during yeast conjugation: Evidence for a pheromone-induced surface protein. Mol Cell Biol 7(7): 2316–2328.
25. Friesen H, Humphries C, Ho Y, Schub O, Colwill K, et al. (2006) Characterization of the yeast amphiphysins Rvs161p and Rvs167p reveals roles for the rvs heterodimer in vivo. Mol Biol Cell 17(3): 1306–1321.
26. Peter BJ, Kent HM, Mills IG, Vallis Y, Butler PJG, et al. (2004) BAR domains as sensors of membrane curvature: The amphiphysin BAR structure. Science 303(5657): 495–499.
27. Bon E, Recordon-Navarro P, Durrens P, Iwase M, Toh-e A, et al. (2000) A network of proteins around Rvs167p and Rvs161p, two proteins related to the yeast actin cytoskeleton. Yeast 16(13): 1229–1241.
28. Gammie AE, Brizzio V, Rose MD (1998) Distinct morphological phenotypes of cell fusion mutants. Mol Biol Cell 9(6): 1395–1410.
29. Trueheart J, Fink GR (1989) The yeast cell fusion protein FUS1 is O-glycosylated and spans the plasma membrane. Proceedings of the National Academy of Sciences 86(24): 9916–9920.
30. Nelson B, Parsons AB, Evangelista M, Schaefer K, Kennedy K, et al. (2004) Fus1p interacts with components of the Hog1p mitogen-activated protein kinase and Cdc42p morphogenesis signaling pathways to control cell fusion during yeast mating. Genetics 166(1): 67–77.
31. Sheu Y, Barral Y, Snyder M (2000) Polarized growth controls cell shape and bipolar bud site selection in *Saccharomyces cerevisiae*. Mol Cell Biol 20(14): 5235–5247.
32. Klis FM (1994) Review: Cell wall assembly in yeast. Yeast 10(7): 851–869.
33. Cid VJ, Durán A, del Rey F, Snyder MP, Nombela C, et al. (1995) Molecular basis of cell integrity and morphogenesis in *Saccharomyces cerevisiae*. Microbiol Rev 59(3): 345–386.
34. Lesage G, Bussey H (2006) Cell wall assembly in *Saccharomyces cerevisiae*. Microbiol Mol Biol Rev 70(2): 317–343.
35. Lipke PN, Ovalle R (1998) Cell wall architecture in yeast: New structure and new challenges. J Bacteriol 180(15): 3735–3740.
36. Severin F, Hyman A (2002) Pheromone induces programmed cell death in S. *cerevisiae*. Current Biology: CB 12(7): R233.
37. Zhang N, Dudgeon DD, Paliwal S, Levchenko A, Grote E, et al. (2006) Multiple signaling pathways regulate yeast cell death during the response to mating pheromones. Mol Biol Cell 17(8): 3409–3422.
38. Sprague Jr GF, Herskowitz I (1981) Control of yeast cell type by the mating type locus: I. identification and control of expression of the a-specific gene BAR1. J Mol Biol 153(2): 305–321.
39. Chan RK, Otte CA (1982) Physiological characterization of *Saccharomyces cerevisiae* mutants supersensitive to G1 arrest by a factor and α factor pheromones. Mol Cell Biol 2(1): 21–29.
40. Biely P, Kratky Z, Bauer S (1976) Interaction of concanavalin A with external mannan-proteins of *Saccharomyces cerevisiae*. Glycoprotein nature of beta-glucanases. Eur J Biochem 70(1): 75–81.
41. Johnston G, Pringle J, Hartwell L (1977) Coordination of growth with cell division in the yeast *Saccharomyces cerevisiae*. Exp Cell Res 105(1): 79–98.
42. Li R, Murray AW (1991) Feedback control of mitosis in budding yeast. Cell 66(3): 519–531.
43. Hutchison HT, Hartwell LH (1967) Macromolecule synthesis in yeast spheroplasts. J Bacteriol 94(5): 1697–1705.
44. Brown A (1978) Compatible solutes and extreme water stress in eukaryotic micro-organisms. Adv Microb Physiol 17: 181–242.
45. Blomberg A, Adler L (1992) Physiology of osmotolerance in fungi. Adv Microb Physiol 33: 145.
46. Hirayarna T, Maeda T, Saito H, Shinozaki K (1995) Cloning and characterization of seven cDNAs for hyperosmolarity-responsive (HOR) genes of *Saccharomyces cerevisiae*. Molecular and General Genetics MGG 249(2): 127–138.
47. O'Rourke SM, Herskowitz I, O'Shea EK (2002) Yeast go the whole HOG for the hyperosmotic response. Trends Genet 18(8): 405–412.
48. Chang F, Herskowitz I (1990) Identification of a gene necessary for cell cycle arrest by a negative growth factor of yeast: FAR1 is an inhibitor of a G1 cyclin, CLN2. Cell 63(5): 999–1011.
49. Barkai N, Rose MD, Wingreen NS (1998) Protease helps yeast find mating partners. Nature 396(6710): 422–423.
50. Marsh L, Rose MD (1997) The pathway of cell and nuclear fusion during mating in *Saccharomyces cerevisiae*. In: Pringle JR, Broach JR, Jones EW, editors (1997) The Molecular and Cellular Biology of the Yeast *Saccharomyces* vol. 3, Cell Cycle and Cell Biology. New York: Cold Spring Harbor Laboratory Press. pp. 827–888.
51. Kurihara LJ, Beh CT, Latterich M, Schekman R, Rose MD (1994) Nuclear congression and membrane fusion: Two distinct events in the yeast karyogamy pathway. J Cell Biol 126(4): 911.
52. Jackson CL, Hartwell LH (1990) Courtship in S. *cerevisiae*: Both cell types choose mating partners by responding to the strongest pheromone signal. Cell 63(5): 1039–1051.
53. Bagnat M, Simons K (2002) Cell surface polarization during yeast mating. Proceedings of the National Academy of Sciences 99(22): 14183–14188.
54. Ayscough KR, Drubin DG (1998) A role for the yeast actin cytoskeleton in pheromone receptor clustering and signalling. Current Biology 8(16): 927–931.
55. Segall JE (1993) Polarization of yeast cells in spatial gradients of α mating factor. Proceedings of the National Academy of Sciences 90(18): 8332–8336.
56. Sheltzer JM, Rose MD (2009) The class V myosin Myo2p is required for Fus2p transport and actin polarization during the yeast mating response. Mol Biol Cell 20(12): 2909–2919.
57. Ydenberg CA, Stein RA, Rose MD (2012) Cdc42p and Fus2p act together late in yeast cell fusion. Mol Biol Cell 23(7): 1208–1218.
58. Thomas BJ, Rothstein R (1989) Elevated recombination rates in transcriptionally active DNA. Cell 56(4): 619–630.
59. Sherman F, Fink G, Lawrence C (1974) Methods in yeast genetics. New York: Cold Spring Harbor Laboratory Press. 73 p.
60. Joglekar AP, Salmon E, Bloom KS (2008) Counting kinetochore protein numbers in budding yeast using genetically encoded fluorescent proteins. Methods Cell Biol 85: 127–151.
61. Lee PJ, Helman NC, Lim WA, Hung PJ (2008) A microfluidic system for dynamic yeast cell imaging. BioTechniques 44: 91–95.

18

Clearance of Human IgG1-Sensitised Red Blood Cells *In Vivo* in Humans Relates to the *In Vitro* Properties of Antibodies from Alternative Cell Lines

Kathryn L. Armour[1]*, Cheryl S. Smith[1], Natasha C. Y. Ip[1¤a], Cara J. Ellison[1¤b], Christopher M. Kirton[1¤c], Anthony M. Wilkes[2], Lorna M. Williamson[3,4¤d], Michael R. Clark[1]

1 Department of Pathology, University of Cambridge, Cambridge, United Kingdom, 2 Bristol Institute for Transfusion Sciences, Bristol, United Kingdom, 3 National Health Service Blood and Transplant, Cambridge, United Kingdom, 4 Department of Haematology, University of Cambridge, Cambridge, United Kingdom

Abstract

We previously produced a recombinant version of the human anti-RhD antibody Fog-1 in the rat myeloma cell line, YB2/0. When human, autologous RhD-positive red blood cells (RBC) were sensitised with this IgG1 antibody and re-injected, they were cleared much more rapidly from the circulation than had been seen earlier with the original human-mouse heterohybridoma-produced Fog-1. Since the IgG have the same amino acid sequence, this disparity is likely to be due to alternative glycosylation that results from the rat and mouse cell lines. By comparing the *in vitro* properties of YB2/0-produced Fog-1 IgG1 and the same antibody produced in the mouse myeloma cell line NS0, we now have a unique opportunity to pinpoint the cause of the difference in ability to clear RBC *in vivo*. Using transfected cell lines that express single human FcγR, we showed that IgG1 made in YB2/0 and NS0 cell lines bound equally well to receptors of the FcγRI and FcγRII classes but that the YB2/0 antibody was superior in FcγRIII binding. When measuring complexed IgG binding, the difference was 45-fold for FcγRIIIa 158F, 20-fold for FcγRIIIa 158V and approximately 40-fold for FcγRIIIb. The dissimilarity was greater at 100-fold in monomeric IgG binding assays with FcγRIIIa. When used to sensitise RBC, the YB2/0 IgG1 generated 100-fold greater human NK cell antibody-dependent cell-mediated cytotoxicity and had a 10^3-fold advantage over the NS0 antibody in activating NK cells, as detected by CD54 levels. In assays of monocyte activation and macrophage adherence/phagocytosis, where FcγRI plays major roles, RBC sensitised with the two antibodies produced much more similar results. Thus, the alternative glycosylation profiles of the Fog-1 antibodies affect only FcγRIII binding and FcγRIII-mediated functions. Relating this to the *in vivo* studies confirms the importance of FcγRIII in RBC clearance.

Editor: Roberto Furlan, San Raffaele Scientific Institute, Italy

Funding: The work was supported by funding from the Department of Pathology, University of Cambridge through income that was derived from commercial exploitation of patented antibodies. The funders had no role in study design, data collection and analysis, decision to publish, or preparation of the manuscript.

Competing Interests: Kathryn L. Armour, Lorna M. Williamson and Michael R. Clark have filed patent applications ('Binding molecules derived from immunoglobulins which do not trigger complement mediated lysis' WO 99/58572) that are owned by the University of Cambridge and cover use of the mutant IgG constant region, G1Δnab. Antibodies including G1Δnab are included in some of the assays shown.

* Email: kla22@cam.ac.uk

¤a Current address: Lister Hospital, Stevenage, United Kingdom
¤b Current address: Medical Research Council Laboratory of Molecular Biology, Cambridge, United Kingdom
¤c Current address: Wellcome Trust Sanger Institute, Cambridge, United Kingdom
¤d Current address: National Health Service Blood and Transplant, Watford, United Kingdom

Introduction

For 40 years, human polyclonal anti-RhD antibodies have been used successfully in the prophylactic treatment of haemolytic disease of the foetus and newborn to prevent the immunisation of RhD-negative women by RhD-positive foetal RBC. The precise mechanisms by which the polyclonal anti-RhD IgG suppress immunisation against the RhD antigen are not fully understood but involve rapid, non-inflammatory, FcγR-mediated sequestration of the RhD-positive cells [1,2]. There is evidence that FcγRIIIa plays the major role in this clearance of sensitised RBC. Most notably, RBC clearance was slower following administration of an anti-FcγRIII monoclonal antibody to chimpanzees and to a patient [3,4]. Due to the problems implicit in the use of antibodies from hyperimmune plasma, there has been a drive to identify

effective monoclonal anti-RhD antibodies with which to replace polyclonal anti-RhD. As a result, monoclonal anti-RhD antibodies form perhaps the largest group of different antibodies against the same antigen that have been tested in humans. It appears that the most efficient antibodies for RBC clearance are those that give good antibody-dependent cell-mediated cytotoxicity (ADCC) with NK cells [5,6]. This does not necessarily imply that NK cells are involved in RBC clearance but that this assay is a good measure of ability to interact with FcγRIIIa. Phagocytosis by splenic macrophages is held to be the mechanism of IgG-sensitised RBC destruction but to achieve this by engagement of the high affinity IgG receptor, FcγRI, would require displacement of serum IgG, which occupies its binding site under physiological conditions. Strong binding of RBC-bound antibody to the intermediate affinity FcγRIIIa may allow rapid association of RBC and macrophages. This could both activate the macrophages directly and promote interactions via FcγRI molecules upon dissociation of non-specific IgG from their binding sites.

One of our interests lies in the development of mutated human IgG constant regions with different combinations of properties that can be tailored for therapeutic use. Combining these constant regions with the variable regions of the human anti-RhD IgG1 antibody Fog-1 [7] allowed measurement of their activity in various *in vitro* assays and offered the potential to study their effect on the intravascular survival of RBC in humans. Accordingly, aliquots of autologous RBC were labeled with different radionuclides and coated with either Fog-1 IgG1 antibody or a mutated version with reduced effector function (Fog-1 G1Δnab) before reinjection [8]. As anticipated, clearance of cells coated with Fog-1 G1Δnab from the circulation was significantly slower than the clearance of wild-type IgG1-coated cells. IgG1-mediated clearance was complete and irreversible, with accumulation in the spleen and liver and the appearance of radiolabel in plasma. Notably, the clearance mediated by our recombinant Fog-1 IgG1 was much more rapid than seen in a previous study that used the original Fog-1 antibody at comparable coating levels [9]. Monoclonal anti-RhD IgG do range widely in their ability to mediate RBC clearance and, whilst some of this variation results from the properties of the different variable regions and the choice of IgG1 or IgG3 constant regions, the cell line used for expression of the IgG appears to be crucial [5]. It is therefore relevant that the original Fog-1 was obtained from human-mouse heterohybridoma cells following fusion of Epstein-Barr virus-transformed B lymphocytes with the mouse myeloma line X63-Ag8.653 [10] whereas transfected YB2/0 rat myeloma cells were used for the production of both recombinant Fog-1 G1 and G1Δnab. The cell line influences the effector properties of an antibody sample by being responsible for its glycosylation profile.

IgG heavy chain carbohydrate moieties are linked to N297 of each chain, fill the space between the two CH2 domains and play roles in the stability and interactions of the Fc (reviewed [11]). Each oligosaccharide is of the complex biantennary type and consists of a basic heptasaccharide structure that can be enlarged by the presence of fucose on the primary N-acetylglucosamine (GlcNAc) residue, galactose (±sialic acid) on one or both of the terminal GlcNAc and a bisecting GlcNAc residue. Absence of carbohydrate results in a decrease in binding to all classes of Fc receptor whilst changing the oligosaccharide structure can modify binding. Serial truncation of Fc carbohydrate structures results in the movement of CH2 domains towards each other and conformational changes in the FcγR interface region that make receptor binding less favourable [12].

Apart from the Fog-1 studies, the only clinical investigation of anti-RhD antibodies that were produced from alternative cell lines involved a BRAD-5 and BRAD-3 mixture [13]. In contrast to the dissimilar RBC clearance rates of the two Fog-1 IgG1, comparable rates were mediated by the BRAD antibodies from EBV-immortalised human cell lines and CHO cells. We wished to understand the difference in potency between the Fog-1 IgG1 from YB2/0 and the original Fog-1 from human-mouse heterohybridoma cells by comparing the properties of IgG carrying glycosylation that is typical of the products of rat or mouse cell lines. We chose to compare the YB2/0-produced Fog-1 IgG1 with the same antibody produced in the mouse myeloma cell line NS0 since this is a line commonly used for therapeutic antibody production.

Comparison has previously been made between the activities of antibodies produced in YB2/0, NS0 and CHO cells [14]. When produced in YB2/0 cells, the humanised IgG1 antibody CAMPATH-1H was approximately 30-fold more efficient in ADCC assays than the same antibody made in either NS0 or CHO cells whilst the antibodies were equally active in monocyte killing assays. Oligosaccharide analysis showed that YB2/0-produced antibody contained less fucose (6–7% against 90% for CHO) and more bisecting GlcNAc residues than IgG made in CHO or NS0. These two properties are related since fucosylation prevents enzymatic addition of bisecting GlcNAc. The NS0- and CHO-derived antibodies contained carbohydrate of a similar structure but the NS0 IgG1 was significantly underglycosylated. Fractionation of the IgG showed that it was the lack of fucose, rather than the presence of bisecting GlcNAc, that led to YB2/0-produced IgG1 being more efficient at ADCC [15]. Furthermore, when YB2/0 cells were caused to over-express FUT8 mRNA to supplement their low levels α1,6-fucosyltransferase, this led to IgG1 with 81% fucosylation and 100-fold lower ADCC.

Since most of the literature that examines the effects of cell-specific glycosylation involves IgG produced in CHO cells, there has been no systematic comparison of YB2/0- and NS0-produced IgG. Although NS0 and CHO cell lines yield IgG with similar distributions of glycan structures, the presence of aglycosyl IgG in NS0 samples potentially reduces binding to all classes of FcγR. Here we evaluate the relative levels of binding of YB2/0- and NS0-produced Fog-1 IgG1 to a series of transfected cell lines that separately bear each type of human FcγR. We also test the antibodies' performance in ADCC, NK cell and monocyte activation and macrophage adherence and phagocytosis assays. So that the impact of altering the glycosylation profile of the antibody can be compared to the change in binding achieved through amino acid mutation, in some assays we include the Fog-1 IgG1 mutant with reduced activity, Fog-1 G1Δnab, made in YB2/0 or NS0 cells. As well as definitively assessing the relative levels of interaction of YB2/0- and NS0-produced IgG1 with the different FcγR, this work reinforces what is known about the mechanism of IgG-sensitised RBC clearance.

Materials and Methods

Antibody production and characterisation

The production of recombinant IgG1 and mutant G1Δab forms of Fog-1 in YB2/0 rat myeloma cells [16] and their subsequent characterisation has been described [17–19]. Fog-1 G1Δnab was produced by removing the G1m(1,17) allotypic residues from G1Δab, without effect on its properties [8]. Both wildtype and mutant Fog-1 antibodies were similarly produced from NS0 mouse myeloma cells [20]. The antibodies are denoted as Fog-1 G1 and Fog-1 G1Δnab followed by (YB2/0) or (NS0). The relative concentrations of the antibodies were confirmed by sandwich ELISA, using goat anti-human IgG, Fc-specific antibodies and HRPO-conjugated goat anti-human κ light chains antibodies (Sigma, Poole, UK).

Measurement of IgG binding to transfected cell lines bearing human FcγR

Cell lines transfected with appropriate cDNA expression vector constructs to express single human FcγR have been variously obtained. For FcγRI, the cell line was B2KA (S. Gorman and G. Hale, unpublished) and CHO cells expressing FcγRIIIb of allotypes NA1 and NA2 [21] were provided by J. Bux. CHO cell lines expressing FcγRIIIa of allotypes 158F and 158V as GPI-anchored receptors or the various FcγRII molecules as transmembrane proteins with native cytoplasmic domains have been constructed [22,23]. Continued and uniform expression of the appropriate FcγR was confirmed in each antibody binding assay by staining a sample of the cells for the receptor. Cells were incubated with CD64 (clone 10.1), CD32 (AT10) or CD16 (LNK16) monoclonal antibody (AbD Serotec, Kidlington, UK) and its binding detected with FITC-conjugated goat anti-mouse IgG antibodies (Sigma).

Binding of monomeric Fog-1 IgG to B2KA cells expressing FcγRI was measured as previously described using Fog-1 IgG2 antibody, produced in YB2/0, as a negative control [17]. Monomeric binding of IgG to FcγRIIIa was detected by the same protocol but using biotinylated goat F(ab')₂ anti-human κ (Rockland) followed by ExtrAvidin FITC (Sigma). Complexed Fog-1 antibody binding to FcγRII and III receptors was measured by pre-incubating the test antibodies with equimolar amounts of goat F(ab')₂ fragments that recognise human κ chain (Rockland) [19]. Human IgA1, κ purified myeloma protein (The Binding Site, Birmingham, UK) was used as a negative control test antibody. Complexes were detected using FITC-conjugated F(ab')₂ fragments of rabbit anti-goat IgG, F(ab')₂-specific antibodies (Jackson ImmunoResearch, Newmarket, UK) or FITC-conjugated donkey anti-goat IgG antibodies (Serotec). The mean fluorescence of cells from each sample was determined using a CyAn ADP flow cytometer and Summit v4.3 software (DakoCytomation, Ely, UK) or on a FACScan flow cytometer using LysisII software (Becton Dickinson, Oxford, UK).

Fold-differences in IgG binding were calculated as follows: Firstly, a curve was fitted to a subset of the mean fluorescence data of G1 (YB2/0), the higher-binding IgG. This was a logarithmic curve for the complexed IgG binding or a sigmoidal curve for the monomeric IgG binding. Using the mean fluorescence values of each of the three highest concentrations of the other IgG, the ratio of concentrations of the two IgG giving these mean fluorescence values was calculated. This was carried out for three or more independent experiments so that the fold-difference in binding could be expressed as the mean±sd of at least nine values.

Rosetting assays

O, RhD-positive RBC, which were shown to carry 9000 RhD sites/cell by SOL-ELISA [8], were incubated with dilutions of Fog-1 antibodies in V-bottom plates for 1 hour at room temperature. The cells were pelleted, washed three times in 150 μl/well PBS and resuspended in 100 μl RPMI (approximately 1% suspension). CHO+FcγRIIIa 158F or 158V cells were harvested using Cell Dissociation Buffer (Invitrogen), washed, resuspended at 4×10^6 cells/ml and 100 μl samples added to the sensitised RBC. The cells were pelleted together at $200 \times$ g for 2 min and incubated on ice for 1 h. 10 μl samples were transferred to slides with coverslips and representative images captured at $40 \times$ magnification.

Assays of functional responses to Fog-1 antibody-sensitised RBC

ADCC, macrophage adhesion and phagocytosis and monocyte activation assays were carried out using human cells as described previously [23]. NK cell activation was assessed using a method adapted from [24]. Peripheral blood mononuclear cells (PBMC) were prepared from FCGR3A- and FCGR2C-genotyped donors by adding 6 ml samples of EDTA-anti-coagulated blood to 45 ml samples of RBC lysis buffer (150 mM NH₄Cl, 10 mM KHCO₃, 1 mM EDTA) and incubating at room temperature for 15 min. White cells were collected by centrifugation, washed in RBC lysis buffer and resuspended in complete RPMI (RPMI containing 10% heat-inactivated FBS, 2 mM L-glutamine, 0.5 μg/ml amphotericin B, 100 U/ml penicillin, 0.1 mg/ml streptomycin).

Using the same donor as for the rosetting assays above, RBC were prepared by pelleting cells from 100 μl blood, washing twice in 1 ml RPMI and resuspending in complete RPMI. Samples of test antibody, 10^5 PBMC, 4×10^5 RBC were added to round-bottomed wells in 100 μl complete RPMI in triplicate and incubated at 37C in a humidified atmosphere of 5% CO₂ in air for 20 h. Control wells omitted test antibody or used 10 μg/ml IgG1 of irrelevant specificity. The surface expression of CD54 on NK cells (identified as CD3⁻CD56⁺) was determined by flow cytometry: The cells in each well were washed three times with FACS wash buffer (PBS containing 0.1% BSA, 0.1% NaN₃) then incubated in 100 μl FACS wash buffer containing PerCP/Cy5.5-conjugated CD3 (clone UCHT1), PE-conjugated CD56 (clone HCD56) and APC-conjugated CD54 (clone HA58, all from BioLegend, London, UK) for 45 minutes on ice. The cells were washed twice and fixed in 1% formaldehyde. Samples were analysed using a CyAn ADP flow cytometer and Summit v4.3 software with appropriate compensation settings. Mean APC fluorescence intensity was calculated for at least 2000 CD3⁻CD56⁺ cells from each well and plotted as mean±SD of the triplicate samples at each test antibody concentration. Samples containing control IgG1 antibody or RhD negative RBC showed no increase in APC fluorescence relative to samples containing no test antibody.

FCGR3A and FCGR2C genotyping

Typing for the FcγRIIIa 158F/V polymorphism was carried out on genomic DNA that had been purified using the QIAamp DNA Blood Mini Kit (Qiagen, Manchester, UK). A 235 bp section of DNA, which comprised parts of the 5th exon and following intron, was amplified using oligonucleotides (5′ CAT-ATTTACAGAATGGCAAAGG 3′, 5′ CAACTCAACTTCC-CAGTGTGAT 3′) that each mismatch the closely homologous FCGR3B gene at their 3′ nucleotide. The PCR products were directly sequenced using the second primer. Lack of FCGR3B contamination was confirmed by examining the electropherogram at two positions within the amplified DNA where the FCGR3A and FCGR3B sequences differ and the presence of the TTT (F) codon, GTT (V) codon or a mixture was determined.

High homology between the ectodomains of FcγRIIb and FcγRIIc means that FcγRIIc typing requires amplification of cDNA to enable use of a FcγRIIb/c-specific primer at the 5′ end and a FcγRIIa/c-specific primer at the 3′ end. Whole blood was diluted 10-fold with RBC lysis buffer and the white cells collected by centrifugation. These were lysed and stored in RNASafer (Omega bio-tek, Lutterworth, UK) before RNA was prepared using Tripure reagent (Roche, Burgess Hill, UK). First strand cDNA was synthesised from a FcγRIIa/c-specific primer (5′ AGCAAGTCTAGAGTATGACCACATGGCATAACGTTAC-TCTTTAG 3′) and was amplified by PCR using the same oligonucleotide in conjunction with a FcγRIIb/c-specific primer (5′ GACTGCTGTGCTCTGGGCGCCAGCTCGCTCCA 3′). Product was subjected to a second round of PCR using primers F (5′ AGGGAGTGATGGGAATCCTGTCATT 3′) and R (5′ CATAGTCATTGTTGGTTTCTTCAGG 3′). The nested PCR product was directly sequenced from primer F2 (5′ CAT-ATGCTTCTGTGGACAGCT 3′). For this cDNA segment, the FcγRIIc ORF and STP alleles differ at 3 positions in addition to the CAG (Q)/TAG (STP) codon corresponding to amino acid residue 13 [25].

Results

Comparison of binding of YB2/0 and NS0-generated antibodies to human FcγR

We used transfected cell lines, each expressing a single human FcγR, to pinpoint the effects on FcγR binding of changing the IgG production cell line. For FcγRIIa, FcγRIIIa and FcγRIIIb, functional polymorphisms are known [26] and testing was performed for two allotypes of each receptor. The IgG were titrated to obtain receptor-binding curves from which any differences in strengths of binding could be calculated. Such comparisons are traditionally made by relating the concentrations of each antibody required to give half-maximal binding but, for the IgG concentrations used here, either maximal binding was not achieved by the higher-binding IgG or the lower-binding IgG did not reach the half-maximal binding level. Therefore, comparison was made between the binding signals detected for the highest three concentrations of the lower-binding IgG and the binding curve of the higher-binding IgG as described in Materials and Methods.

Given the previous reports of FcγRIIIa importance in RBC clearance, we began by examining binding to this receptor. The binding of pre-complexed Fog-1 G1 (YB2/0) antibody to cells expressing FcγRIIIa of allotype 158F was 45-fold higher than that of pre-complexed G1 (NS0) (45 ± 17 fold over 5 experiments; Figure 1A). For the higher affinity allotype of the receptor, FcγRIIIa 158V, G1 (YB2/0) complexes bound 20-fold better than complexes of the NS0-produced antibody (20 ± 5 fold over 3 experiments; Figure 1B). This assay of complexed IgG binding to FcγRIIIa is sufficiently sensitive for binding of the Fog-1 IgG1 mutant with reduced effector function (Fog-1 G1Δnab) to be detected at the highest IgG concentrations used. Binding of G1Δnab (YB2/0) to the FcγRIIIa 158F and 158V molecules is 51-fold and 63-fold lower, respectively, than that of G1 (YB2/0) (158F: 51 ± 12 fold over 3 experiments; 158V: 63 ± 18 fold over only 2 experiments). Switching to NS0 for G1Δnab antibody production only gives a small additional decrease in binding.

FcγRIIIa is classed as a medium affinity Fc receptor and its interactions with monomeric antibody samples can also be measured. This was carried out to ensure that superiority of the YB2/0-derived antibody was not an artefact of the complexed antibody assay system. By this method, G1 (YB2/0) bound 102-fold and 97-fold more efficiently than G1 (NS0) to the 158F and 158V allotypes of the receptor respectively (158F: 102 ± 25 fold over 3 experiments; 158V: 97 ± 33 fold over 3 experiments; Figure 1C and D). The higher affinity of the 158V allotype of the receptor is clearly evident here since the binding curves for this allotype are displaced towards lower concentrations compared to the curves for the F allotype.

As an additional binding measurement, Fog-1 G1-sensitised RBC were tested for their ability to rosette the receptor-bearing cells. Figure 1, panels E - H show photographs from a representative experiment carried out with the FcγRIIIa 158V cell line. IgG with Fog-1 variable regions are known to give 100% saturation of RBC RhD sites (equivalent to 9000 IgG/RBC in these experiments) at a coating concentration of approximately 20 μg/ml and 50% saturation at 0.4 μg/ml (data not shown). A reduction in G1 (NS0) coating concentration from 100 μg/ml to 11 μg/ml reduced rosette formation even though this does not represent a large change in terms of IgG/RBC. Tight rosettes, similar to those seen at 100 μg/ml G1 (NS0), were formed by RBC coated with G1 (YB2/0) at 1.1 μg/ml or approximately 6000 IgG/RBC. The difference between the G1 (YB2/0) and G1 (NS0) coating concentrations that gave equivalent levels of

rosetting was greater in the case of the FcγRIIIa 158F cell line (data not shown).

We went on to compare the binding of monomeric YB2/0- and NS0-produced immunoglobulin to the high affinity FcγRI and complexed antibody binding to the remaining human FcγR, which are of low affinity. Very similar binding of the two IgG1 antibodies was observed for FcγRI, FcγRIIa (of allotypes 131R and 131H) and FcγRIIb (Figure 2, panels A–D). In contrast, for FcγRIIIb of NA1 and NA2 allotypes (Figure 2E and F), there were large differences in binding with the YB2/0-produced IgG1 being 39-fold and 38-fold better respectively (NA1: 39 ± 12 fold over 3 experiments; NA2: 38 ± 10 fold over 3 experiments). These differences are comparable to those measured for the FcγRIIIa molecules.

Measurement of functional cellular responses to Fog-1-sensitised RBC

Measurement of NK cell-mediated ADCC of Fog-1 IgG-sensitised RBC showed G1 (YB2/0) to cause lysis about 100-fold more efficiently than G1 (NS0) at sub-saturating concentrations (Figure 3A). However, the lysis mediated by G1 (NS0) was still higher than the background levels of lysis that were typical of G1Δnab (YB2/0).

We also examined the interaction with NK cells by using flow cytometry to measure the levels of the NK cell activation marker CD54 after overnight co-culture of PBMC and RBC in the presence of Fog-1 antibody. Results are shown for three donors of PBMC with different FCGR3A/FCGR2C genotypes (Figure 3, panels B–D). G1 (YB2/0) concentrations ≥1 ng/ml (Figure 3B), ≥10 ng/ml (Figure 3C) or ≥100 ng/ml (Figure 3D) produced CD54 levels that were significantly higher than in samples with no test antibody (p<0.05, Student's t-test). Even at 10 μg/ml, G1 (NS0) caused little increase in CD54 level with the fluorescence signals being similar to those seen in response to G1Δnab or non-specific IgG1 control and in samples without test antibody. Since 10 μg/ml G1 (NS0) generated a CD54 level that was similar to or lower than 10 ng/ml G1 (YB2/0), the two IgG appear to be at least 1000-fold different in their abilities to activate NK cells.

The activities of G1 (YB2/0) and G1 (NS0) were also compared using macrophages and monocytes as effector cells. Both of these cell types express FcγRI in addition to lower affinity receptors. Human macrophages were obtained from adherent mononuclear cells that had been cultured with M-CSF and differentiated with IFNγ and LPS and were CD64[+], CD32[+] and CD16[+] by flow cytometry (data not shown). The proportions of macrophages found to be interacting with Fog-1 antibody-saturated RBC following 1 h incubations were determined. For three different macrophage donors, there was no significant difference between the G1 (YB2/0) and G1 (NS0) samples in terms of the numbers of RBC interacting (p = 0.4, paired Student's t-test) or the proportion of these that had been phagocytosed (p = 0.3) (Figure 4A). Presence of the G1 antibodies resulted in higher macrophage/RBC interaction rates than G1Δnab and there were no instances of phagocytosis with the latter antibody.

Chemiluminescence assays of monocyte activation in response to Fog-1 antibody-sensitised RBC showed Fog-1 G1 (YB2/0) to be approximately 3-fold more efficient at activating monocytes than G1 (NS0) (Figure 4B). G1Δnab did not cause activation.

Discussion

Our comparison of the activities of IgG1 antibodies produced in YB2/0 and NS0 cell lines has shown heightened performance of the YB2/0 antibody in FcγRIII binding and FcγRIII-mediated

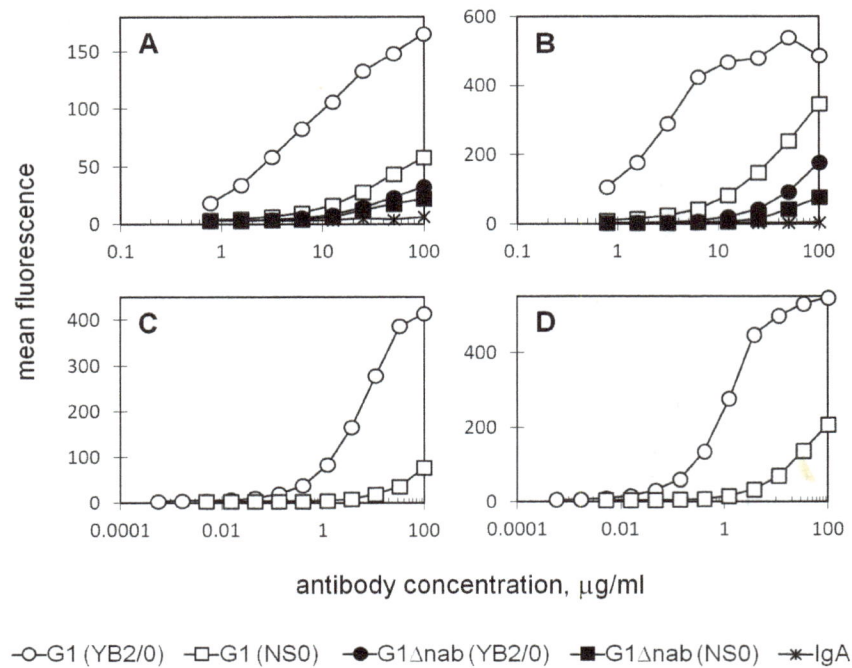

−◦−G1 (YB2/0) −□−G1 (NS0) −●−G1Δnab (YB2/0) −■−G1Δnab (NS0) −✳−IgA

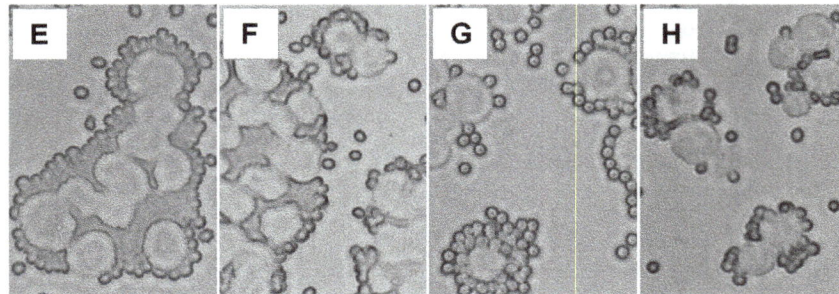

Figure 1. Binding interactions of YB2/0- and NS0-produced Fog-1 G1 and G1Δnab antibodies with human FcγRIIIa. A–D CHO cells expressing FcγRIIIa of allotypes 158F (A, C) and 158V (B, D) were incubated with pre-complexed (A, B) or monomeric (C, D) Fog-1 IgG and binding detected with fluorescent reagents and flow cytometry. Graphs show mean fluorescence of ≥12 000 cells at each antibody concentration and are typical of the results obtained in at least three experiments with each receptor. **E–H** Examples of the rosetting of FcγRIIIa 158V-expressing cells by RBC sensitised with Fog-1 G1 (YB2/0) at 10 μg/ml (E) and 1.1 μg/ml (F) or with Fog-1 G1 (NS0) at 100 μg/ml (G) and 11 μg/ml (H). Images are typical of eight independent experiments.

functions. With the amino acid sequence of the two IgG1 antibodies being identical, the difference in activity must be due to variations in glycosylation. Although we did not analyse the glycosylation profiles of the IgG tested here, previous studies have shown that NS0-produced antibody was underglycosylated compared to IgG produced in YB2/0, their glycan moieties contain more fucose and less bisecting GlcNAc residues [14] and that lower levels of fucose lead to greater efficiency in ADCC [15].

We were able to pinpoint through which receptors the differences in activity were generated by comparing the binding of YB2/0- and NS0-produced Fog-1 G1 to each type of human FcγR. The assays measured monomeric IgG binding to the high affinity FcγRI and medium affinity FcγRIIIa and complexed IgG binding to FcγRII and FcγRIII molecules. For each of the receptors FcγRI, FcγRIIa 131R and131H and FcγRIIb, G1 (YB2/0) and G1 (NS0) showed very similar binding to each other. In contrast, large inequalities in binding ability between G1 (YB2/0) and G1 (NS0) were observed in complexed IgG binding to both

FcγRIIIa and FcγRIIIb and in monomeric IgG binding to FcγRIIIa. In addition, Fog-1 G1 (YB2/0)-sensitised RBC readily formed rosettes with the FcγRIIIa transfectants at lower IgG coating concentrations than Fog-1 G1 (NS0)-sensitised RBC.

The superior FcγRIIIa binding of G1 (YB2/0) over G1 (NS0) translates into greater NK cell-mediated ADCC activity as might be expected since it is the only Fc receptor carried by NK cells in the majority of people. In some individuals, NK cells also express FcγRIIc, a low affinity, activating receptor with identical extracellular domains to FcγRIIb. This is due to a polymorphism, corresponding to amino acid 13 of the mature protein, where functional FcγRIIc depends on the presence of a Q codon, rather than stop codon [27]. In addition to the ADCC assays, we compared the abilities of the antibodies to cause NK cell activation when incubated with PBMC and RhD-positive RBC. Fog-1 G1 (YB2/0) was much more efficient in increasing levels of the activation marker CD54 than the NS0 antibody. Although our study was limited to one individual of each *FCGR3A* 158

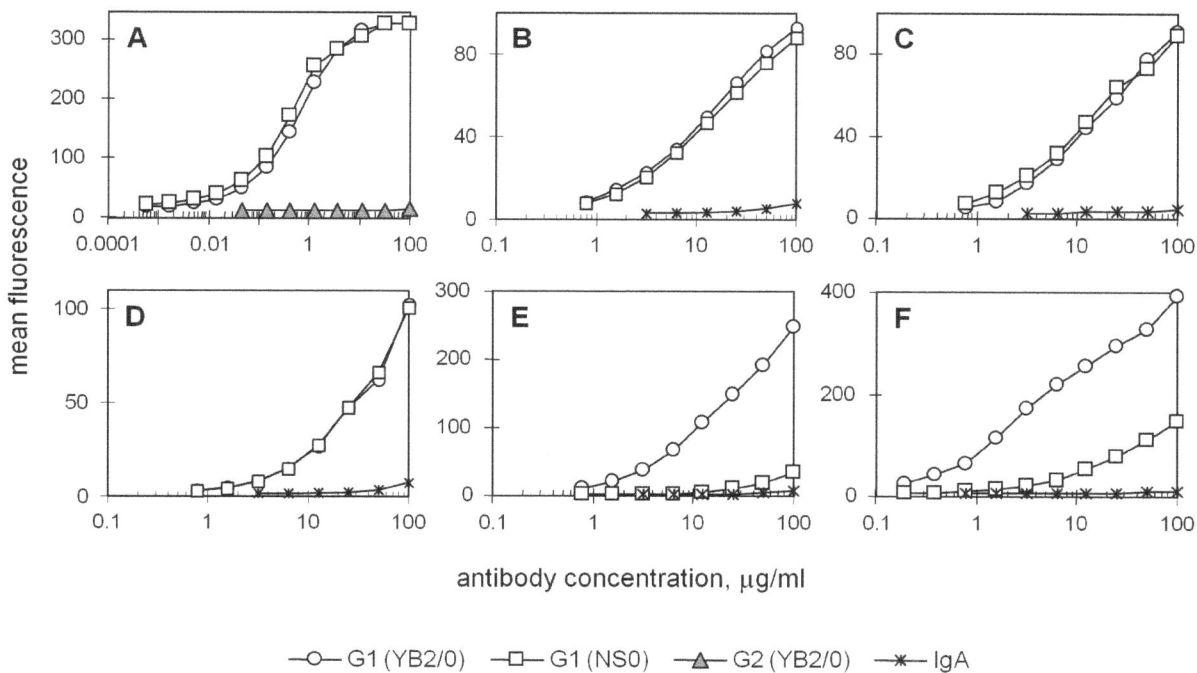

Figure 2. Binding of YB2/0- and NS0-produced Fog-1 G1 antibodies to human FcγR. A Binding of monomeric IgG was measured for FcγRI using the B2KA cell line and flow cytometry. Fog-1 IgG2 antibody, produced in the YB2/0 cell line, was used as the non-binding control antibody for this receptor. **B–F** Binding of pre-complexed IgG was measured using CHO cell lines expressing the low affinity receptors which were FcγRIIa, allotypes 131R (B) and 131H (C), FcγRIIb (D) and FcγRIIIb, allotypes NA1 (E) and NA2 (F). The level of background binding is given by the negative control antibody, IgA,κ. Graphs show mean fluorescence of ≥12 000 cells at each antibody concentration and are typical of the results obtained in at least three experiments with each receptor.

genotype, with only the VV donor being FcγRIIc-positive, we did see a trend in the lowest G1 (YB2/0) concentration at which significant activation was detected: VV<FV<FF. This is consistent with the activation being dependent on FcγRIIIa avidity for IgG.

In contrast to the large differences in activity seen in ADCC and NK cell activation experiments, G1 (YB2/0)- and G1 (NS0)-coated RBC interacted with macrophages to a similar extent and there was only a three-fold variation in ability to activate monocytes. In these two assays, the high affinity FcγRI, which binds the two antibodies equally efficiently, is present and able to play a dominant role in the absence of high levels of competing non-specific IgG. This is unlikely to be the case *in vivo* where the concentrations of IgG mean that FcγRI molecules will be occupied by antibody and macrophage FcγRIIIa becomes more important for RBC removal.

The difference in FcγRIIIa binding of IgG1 from YB2/0 and NS0 cell lines is similar to that between YB2/0- and CHO-derived IgG1 with the former being approximately 100-fold better at binding to soluble FcγRIIIa 158 F or V by ELISA [28]. A variant CHO line, Lec13, produces IgG molecules with 10% fucosylation and dimers of these IgG show up to 50-fold improved binding over normal CHO-produced IgG to FcγRIIIa α-chain in ELISAs [29]. Lower fucose gave less improvement when measuring the binding of trimers or mutated IgG with an intrinsically higher affinity for FcγRIIIa. This concurs with the results of our three measurements of FcγRIIIa binding where the greatest difference was observed when monomeric IgG was tested. These groups [28,29] showed essentially identical binding of their glycosylation variants to FcγRI and FcγRIIa, allotype 131H although the low fucose antibodies gave slightly higher signals with FcγRIIa, allotype 131R

and FcγRIIb. The difference in ADCC efficacy between G1 (YB2/0) and G1 (NS0) is also similar to that observed between YB2/0- and CHO-derived IgG1 in several studies. In ADCC assays using 20 different donors of PBMC, a YB2/0-produced version of the CD20 therapeutic antibody, rituximab, was 10–100 fold better than the original CHO-produced antibody [28]. YB2/0–produced IgG gave equivalent levels of ADCC to CHO-made antibody at lower antigen densities [30], at lower antibody concentrations or with lower numbers of effector cells [31].

The effect of switching from CHO- to YB2/0-produced IgG1 on the interaction with FcγRIIIa has also been analysed by surface plasmon resonance and isothermal titration calorimetry [32]. The enhancement of affinity arises mainly from an increased association rate that is a consequence of greater favourable enthalpy and implies additional non-covalent interactions. In line with the approximate 100-fold difference we saw between monomeric YB2/0- and NS0-produced IgG1 binding to cell-surface FcγRIIIa, low-fucose IgG1 was shown to have a 50-fold greater affinity than high-fucose IgG1 for soluble FcγRIIIa 158V by surface plasmon resonance [33]. Mutation of the receptor residue N162 reduced the affinity of low-fucose IgG1 by 13-fold whilst increasing that of high-fucose IgG1 by 3-fold. The authors infer that a high affinity interaction requires glycosylation at N162 of the receptor but that this carbohydrate can only have productive contacts with nonfucosylated IgG. The fucose residue, which protrudes from the carbohydrate core, may prevent close approach of the molecules. Although the crystal structure of an IgG1-Fc fragment in complex with FcγRIIIb was solved using aglycosyl receptor [34], it does indicate that a carbohydrate moiety attached to receptor residue N162 would be orientated towards the carbohydrate of the Fc. Only FcγRIIIa and FcγRIIIb of human FcγR

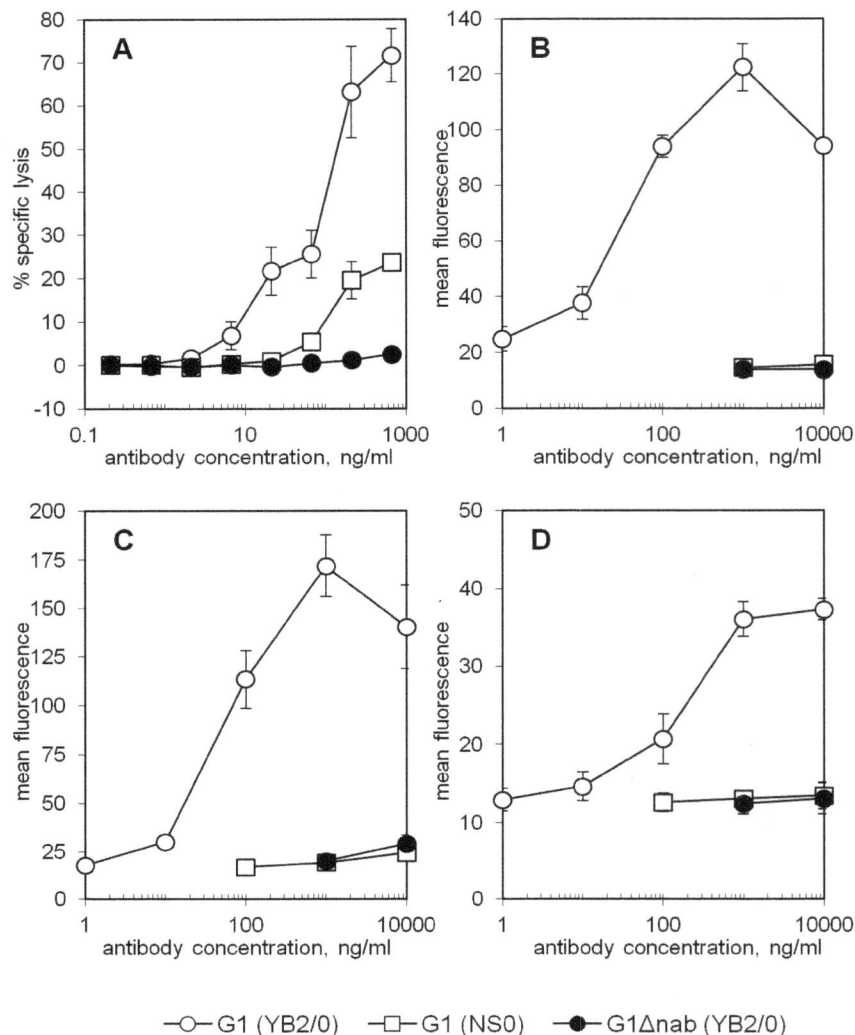

Figure 3. Functional responses of NK cells to RBC sensitised with Fog-1 antibodies. **A** The specific lysis of sensitised RBC by NK-cell mediated ADCC is presented as mean±SD of triplicate samples. This experiment used effector cells pooled from 6 donors but similar results were obtained in four experiments with individual donors of PBMCs. **B–D** The activation of NK cells in response to sensitised RBC as visualised by the level of CD54 on the surface of CD3$^-$, CD56$^+$ lymphocytes. Each graph shows the mean±SD of triplicate samples for each data point. Donors of PBMC were of the following genotypes: FcγRIIIa 158V/V, FcγRIIc 13Q/13Q (B), FcγRIIIa 158F/V, FcγRIIc 13STP/13STP (C) and FcγRIIIa 158F/F, FcγRIIc 13STP/13STP (D). CD54 signals for samples with no test antibody were 14.2±0.1 (B), 17.0±2.5 (C) and 13.0±1.1 (D). CD54 signals for samples incubated with irrelevant IgG1 were 13.3±0.4 (B), 18.7±4.1 (C) and 12.1±0.9 (D).

have a glycosylation site at position162, which accounts for the selectivity of the fucose effect.

Features of Fog-1 G1 (YB2/0)-mediated RBC clearance *in vivo*, namely the accumulation of RBC in the liver and the presence of radioactivity in the plasma [8], have raised concern about the use of YB2/0-produced IgG in the prevention of alloimmunisation [5]. The hepatic uptake could be indicative of YB2/0-derived antibodies having pro-inflammatory interactions via their carbohydrate residues with molecules of the innate immune system. Accumulation in the liver, in addition to the spleen, occurred in two subjects with >11000 IgG1 molecules/RBC (or more than 40 μg IgG1/ml packed cells) but not in a subject with 6800 molecules/cell. This accumulation must be FcγR-dependent since it did not occur in two subjects where RBC were coated with Fog-1 G1Δnab at >15000 molecules/cell. Few RBC clearance studies have included imaging but an investigation reported

hepatic uptake at high coating levels when RBC were sensitised with different amounts of polyclonal anti-Rh antibody from human serum [35]. Thus accumulation in the liver may be a consequence of FcγR binding of IgG that is presented at a high density on the RBC surface rather than interactions specifically with the glycan structure of the YB2/0-produced IgG. Most RBC survival studies use low coating levels as they are designed to discover if relatively low, prophylactic doses of the anti-RhD IgG would give sufficient RBC clearance. This might also explain the unusualness of detecting radioactivity in the plasma, which has been suggested to imply that some RBC were destroyed by a potentially pro-inflammatory extracellular cytotoxic mechanism rather than by phagocytosis [5]. The fate of radioisotopes following phagocytosis is unknown and cannot be assumed to be the same as for other products of cell destruction such as haemoglobin. The high rates of RBC destruction in our study may

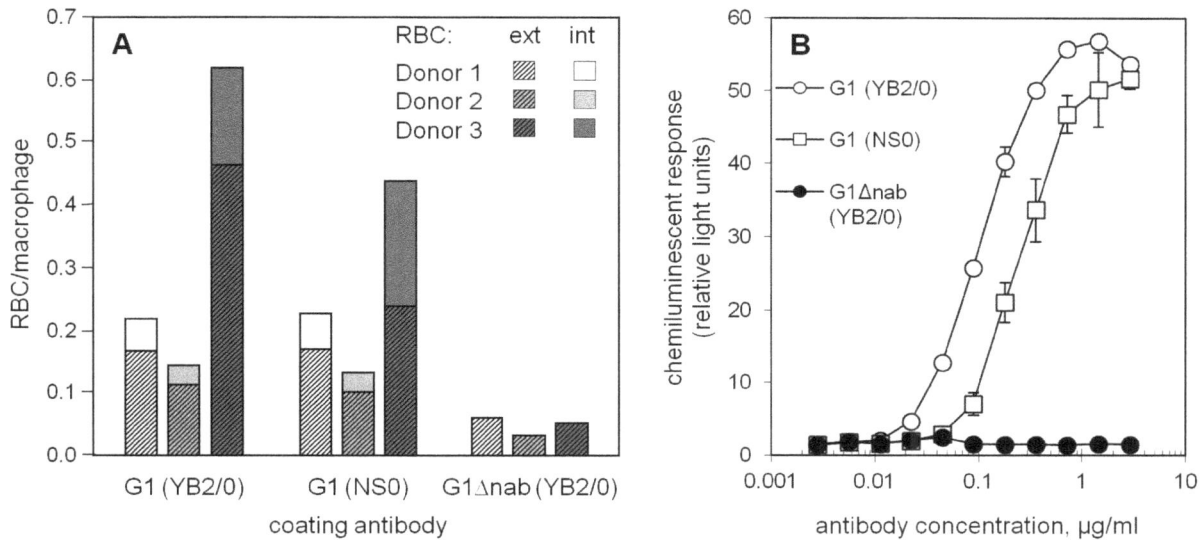

Figure 4. Interactions of Fog-1 IgG-sensitised RBC with monocytes and macrophages. A The numbers of adherent (ext) and phagocytosed (int) RBC per macrophage were determined for RBC sensitised with saturating concentrations of Fog-1 IgG. Results are shown for macrophages from three different donors. The numbers of unsensitised RBC associating with macrophages were typically 15- to 20-fold lower than the numbers of Fog-1 G1 (YB2/0)-sensitised RBC. **B** The mean chemiluminescent response of monocytes to sensitised RBC is plotted with the error bars indicating the range of the duplicate samples.

account for the levels of radiolabel in the plasma. These peaked at 4–6% injected dose 100 min post-injection, by which time more than 95% of the injected doses had disappeared from the cell fractions [8]. Mollison et al. [35] saw comparable levels of plasma radioactivity where there were rapid rates of destruction but most studies have not reported plasma radioactivities. In addition to Fog-1 IgG1, two other YB2/0-produced anti-RhD antibodies have been tested in humans. R297 was found to be at least as effective as commercially-available polyclonal anti-RhD in clearing RBC [36]. Its derivative, R593 (roledumab), is undergoing clinical trials [37] and has been shown to be well tolerated and have a similar pharmacokinetic profile to human polyclonal anti-RhD [38].

Superior ADCC activity, as afforded by YB2/0 production, can be achieved by alternative methods of controlling the level of antibody fucosylation (reviewed [39]). Ideally, all carbohydrate moieties within the therapeutic antibody sample should be without fucose since fucosylated molecules in mixtures can complete for antigenic sites on target cells and batch-to-batch variation in carbohydrate composition is a regulatory issue. A FUT8$^{-/-}$ CHO line that produces completely non-fucosylated antibodies but retains the growth characteristics of parent is attractive to the biopharmaceutical industry [40]. In ADCC, FUT8$^{-/-}$ CHO-produced chimeric anti-CD20 IgG1 was 100-fold more efficient than original rituximab and 2–3 fold better than YB2/0-produced antibody and enhancement was also seen for the other IgG subclasses [41]. As well as being an effective method of improving activity, removal of fucose should not result in immunogenicity since 10–20% of normal human IgG lacks fucose [42]. Clinical

trials have shown that non-fucosylated antibodies are tolerated and can give clinical effects at low doses [43,44].

This work has capitalised on a unique opportunity to compare the effects of alternative glycosylation profiles on the *in vivo* and *in vitro* properties of IgG molecules with the same amino acid sequence. Our comparison of YB2/0- and NS0-produced IgG1 antibodies has revealed that differences in Fc receptor binding are confined to FcγRIII and amount to 100-fold higher binding of YB2/0-produced IgG1 in the case of monomeric IgG binding to FcγRIIIa. The previously-reported underglycosylation of IgG from NS0 cells [14] would be expected to reduce binding to all FcγR but our results showed very similar binding to receptors of the FcγRI and FcγRII classes. The greater ability of Fog-1 IgG1 from YB2/0 cells, compared to the original Fog-1 from human-mouse heterohybridoma cells, to clear RBC *in vivo* therefore results from its improved FcγRIII binding. These pieces of evidence confirm the importance of FcγRIII in RBC clearance.

Acknowledgments

We thank Professor Jürgen Bux for providing the FcγRIIIb–bearing cell lines. We greatly appreciate the advice given by Dr James Robinson and Professor Ann Morgan regarding *FCGR2C* genotyping and Nishita Nigam's assistance with genotyping.

Author Contributions

Conceived and designed the experiments: KLA LMW MRC. Performed the experiments: KLA CSS NCYI CJE CMK AMW. Analyzed the data: KLA CSS NCYI CJE CMK AMW. Contributed reagents/materials/analysis tools: KLA CSS. Wrote the paper: KLA MRC.

References

1. Kumpel BM (2008) Lessons learnt from many years of experience using anti-D in humans for prevention of RhD immunization and haemolytic disease of the fetus and newborn. Clin Exp Immunol. 154: 1–5.

2. Brinc D, Lazarus AH (2009) Mechanisms of anti-D action in the prevention of hemolytic disease of the fetus and newborn. Hematology Am Soc Hematol Educ Program. 1:185–191.

3. Clarkson SB, Kimberly RP, Valinsky JE, Witmer MD, Bussel JB, et al. (1986) Blockade of clearance of immune complexes by an anti-Fcγ receptor monoclonal antibody. J Exp Med 164: 474–489.

4. Clarkson SB, Bussel JB, Kimberly RP, Valinsky JE, Nachman RL, et al. (1986) Treatment of refractory immune thrombocytopenic purpura with an anti-Fcγ-receptor antibody. N Engl J Med 314: 1236–1239.

5. Kumpel BM (2007) Efficacy of RhD monoclonal antibodies in clinical trials as replacement therapy for prophylactic anti-D immunoglobulin: more questions than answers. Vox Sang 93: 99–111.

6. Béliard R (2006) Monoclonal anti-D antibodies to prevent alloimmunization: lessons from clinical trials. Transfus Clin Biol 13: 58–64.

7. Bye JM, Carter C, Cui Y, Gorick BD, Songsivilai S, et al. (1992) Germline variable region gene segment derivation of human monoclonal anti-RhD antibodies. J Clin Invest 90: 2481–2490.

8. Armour KL, Parry-Jones DR, Beharry N, Ballinger JR, Mushens R, et al. (2006) Intravascular survival of red cells coated with a mutated human anti-D antibody engineered to lack destructive activity. Blood 107: 2619–2626.

9. Thomson A, Contreras M, Gorick B, Kumpel B, Chapman GE, et al. (1990) Clearance of RhD-positive red cells with monoclonal anti-D. Lancet 336: 1147–1150.

10. Melamed MD, Thompson KM, Gibson T, Hughes-Jones NC (1987) Requirements for the establishment of heterohybridomas secreting monoclonal human antibody to rhesus (D) blood group antigen. J Immunol Methods104: 245–251.

11. Jefferis R, Lund J (2002) Interaction sites on human IgG-Fc for FcγR: current models. Immunol Lett 82: 57–65.

12. Krapp S, Mimura Y, Jefferis R, Huber R, Sondermann P (2003) Structural analysis of human IgG-Fc glycoforms reveals a correlation between glycosylation and structural integrity. J Mol Biol 325: 979–989.

13. Chapman GE, Ballinger JR, Norton MJ, Parry-Jones DR, Beharry NA, et al. (2007) The clearance kinetics of autologous RhD-positive erythrocytes coated ex vivo with novel recombinant and monoclonal anti-D antibodies. Clin Exp Immunol 150: 30–41.

14. Lifely MR, Hale C, Boyce S, Keen MJ, Phillips J (1995) Glycosylation and biological activity of CAMPATH-1H expressed in different cell lines and grown under different culture conditions. Glycobiology 5: 813–822.

15. Shinkawa T, Nakamura K, Yamane N, Shoji-Hosaka E, Kanda Y, et al. (2003) The absence of fucose but not the presence of galactose or bisecting N-acetylglucosamine of human IgG1 complex-type oligosaccharides shows the critical role of enhancing antibody-dependent cellular cytotoxicity. J Biol Chem 278: 3466–3473.

16. Kilmartin JV, Wright B, Milstein C (1982) Rat monoclonal antitubulin antibodies derived by using a new nonsecreting rat cell line. J Cell Biol 93: 576–582.

17. Armour KL, Clark MR, Hadley AG, Williamson LM (1999) Recombinant human IgG molecules lacking Fcγ receptor I binding and monocyte triggering activities. Eur J Immunol 29: 2613–2624.

18. Armour KL, Williamson LM, Kumpel BM, Bux J, Clark MR (2000) Mutant IgG lacking FcγRIII binding and ADCC activities. Hematol J 1(Suppl.1): 27.

19. Armour KL, van de Winkel JGJ, Williamson LM, Clark MR (2003) Differential binding to human FcγRIIa and FcγRIIb receptors by human IgG wildtype and mutant antibodies. Mol Immunol 40: 585–593.

20. Clark MR, Milstein C (1981) Expression of spleen cell immunoglobulin phenotype in hybrids with myeloma cell lines. Somatic Cell Genet 7: 657–666.

21. Bux J, Kissel K, Hofmann C, Santoso S (1999) The use of allele-specific recombinant FcγRIIIb antigens for the detection of granulocyte antibodies. Blood 93: 357–362.

22. Armour KL, Smith CS, Clark MR (2010) Expression of human FcγRIIIa as a GPI-linked molecule on CHO cells to enable measurement of human IgG binding. J Immunol Methods 354: 20–33.

23. Armour KL, Smith CS, Turner CP, Kirton CM, Wilkes AM, et al. (2014) Low-affinity FcγR interactions can decide the fate of novel human IgG-sensitised red blood cells and platelets. Eur J Immunol 44: 905–914.

24. Bowles JA, Wang SY, Link BK, Allan B, Beuerlein G, et al. (2006) Anti-CD20 monoclonal antibody with enhanced affinity for CD16 activates NK cells at lower concentrations and more effectively than rituximab. Blood 108: 2648–2654.

25. Su K, Wu J, Edberg JC, McKenzie SE, Kimberly RP (2002) Genomic organization of classical human low-affinity Fcgamma receptor genes. Genes Immun 3 (Suppl 1): S51–56.

26. van Sorge NM, van der Pol WL, van de Winkel JG (2003) FcγR polymorphisms: Implications for function, disease susceptibility and immunotherapy. Tissue Antigens 61: 189–202.

27. Ernst LK, Metes D, Herberman RB, Morel PA (2002) Allelic polymorphisms in the FcγRIIC gene can influence its function on normal human natural killer cells. J Mol Med 80: 248–257.

28. Niwa R, Hatanaka S, Shoji-Hosaka E, Sakurada M, Kobayashi Y, et al. (2004) Enhancement of the antibody-dependent cellular cytotoxicity of low-fucose IgG1 is independent of FcγRIIIa functional polymorphism. Clin Cancer Res 10: 6248–6255.

29. Shields RL, Lai J, Keck R, O'Connell LY, Hong K, et al. (2002) Lack of fucose on human IgG1 N-linked oligosaccharide improves binding to human FcγRIII and antibody-dependent cellular toxicity. J Biol Chem 277: 26733–26740.

30. Niwa R, Sakurada M, Kobayashi Y, Uehara A, Matsushima K, et al. (2005) Enhanced natural killer cell binding and activation by low-fucose IgG1 antibody results in potent antibody-dependent cellular cytotoxicity induction at lower antigen density. Clin Cancer Res 11: 2327–2336.

31. Niwa R, Shoji-Hosaka E, Sakurada M, Shinkawa T, Uchida K, et al. (2004) Defucosylated chimeric anti-CC chemokine receptor 4 IgG1 with enhanced antibody-dependent cellular cytotoxicity shows potent therapeutic activity to T-cell leukemia and lymphoma. Cancer Res 64: 2127–2133.

32. Okazaki A, Shoji-Hosaka E, Nakamura K, Wakitani M, Uchida K, et al. (2004) Fucose depletion from human IgG1 oligosaccharide enhances binding enthalpy and association rate between IgG1 and FcγRIIIa. J Mol Biol 336: 1239–1249.

33. Ferrara C, Stuart F, Sondermann P, Brünker P, Umaña P (2006) The carbohydrate at FcγRIIIa Asn-162. An element required for high affinity binding to non-fucosylated IgG glycoforms. J Biol Chem 281: 5032–5036.

34. Sondermann P, Huber R, Oosthuizen V, Jacob U (2000) The 3.2-Å crystal structure of the human IgG1 Fc fragment-FcγRIII complex. Nature 406: 267–273.

35. Mollison PL, Crome P, Hughes-Jones NC, Rochna E (1965) Rate of removal from the circulation of red cells sensitized with different amounts of antibody. Br J Haematol 11: 461–470.

36. Béliard R, Waegemans T, Notelet D, Massad L, Dhainaut F, et al. (2008) A human anti-D monoclonal antibody selected for enhanced FcγRIII engagement clears RhD+ autologous red cells in human volunteers as efficiently as polyclonal anti-D antibodies. Br J Haematol 141: 109–119.

37. Quagliaroli D (2013) Do YB2/0 Cells Produce Alien Sugars? Biochemistry (Mosc) 78: 1371–1373.

38. Yver A, Homery MC, Fuseau E, Guemas E, Dhainaut F, et al (2012) Pharmacokinetics and safety of roledumab, a novel human recombinant monoclonal anti-RhD antibody with an optimized Fc for improved engagement of FcγRIII, in healthy volunteers. Vox Sang 103: 213–222.

39. Yamane-Ohnuki N, Satoh M (2009) Production of therapeutic antibodies with controlled fucosylation. MAbs 1: 230–236.

40. Yamane-Ohnuki N, Kinoshita S, Inoue-Urakubo M, Kusunoki M, Iida S, et al. (2004) Establishment of FUT8 knockout Chinese hamster ovary cells: an ideal host cell line for producing completely defucosylated antibodies with enhanced antibody-dependent cellular cytotoxicity. Biotechnol Bioeng 87: 614–622.

41. Niwa R, Natsume A, Uehara A, Wakitani M, Iida S, et al. (2005) IgG subclass-independent improvement of antibody-dependent cellular cytotoxicity by fucose removal from Asn297-linked oligosaccharides. J Immunol Methods 306: 151–160.

42. Jefferis R, Lund J, Mizutani H, Nakagawa H, Kawazoe Y, et al. (1990) A comparative study of the N-linked oligosaccharide structures of human IgG subclass proteins. Biochem J 268: 529–537.

43. Busse WW, Katial R, Gossage D, Sari S, Wang B, et al. (2010) Safety profile, pharmacokinetics, and biologic activity of MEDI-563, an anti-IL-5 receptor alpha antibody, in a phase I study of subjects with mild asthma. J Allergy Clin Immunol 125: 1237–1244.

44. Tobinai K, Takahashi T, Akinaga S (2012) Targeting chemokine receptor CCR4 in adult T-cell leukemia-lymphoma and other T-cell lymphomas. Curr Hematol Malig Rep 7: 235–40.

19

Benzalkonium Chloride Suppresses Rabbit Corneal Endothelium Intercellular Gap Junction Communication

Zhenhao Zhang[1], Yue Huang[2], Hui Xie[1], Juxin Pan[1], Fanfei Liu[1], Xuezhi Li[1], Wensheng Chen[1], Jiaoyue Hu[1]*, Zuguo Liu[1]*

1 Eye Institute and affiliated Xiamen Eye Center of Xiamen University; Fujian Provincial Key Laboratory of Ophthalmology and Vision Science, Xiamen, Fujian, China,
2 Institute of Stem Cell and Regenerative Medicine, Medical College, Xiamen University, Xiamen, Fujian, China

Abstract

Purpose: Gap junction intercellular communication (GJIC) plays a critical role in the maintenance of corneal endothelium homeostasis. We determined if benzalkonium chloride (BAK) alters GJIC activity in the rabbit corneal endothelium since it is commonly used as a drug preservative in ocular eyedrop preparations even though it can have cytotoxic effects.

Methods: Thirty-six adult New Zealand albino rabbits were randomly divided into three groups. BAK at 0.01%, 0.05%, and 0.1% was applied twice daily to one eye of each of the rabbits in one of the three groups for seven days. The contralateral untreated eyes were used as controls. Corneal endothelial morphological features were observed by *in vivo* confocal microscopy (IVCM). Immunofluorescent staining resolved changes in gap junction integrity and localization. Western blot analysis and RT-PCR evaluated changes in levels of connexin43 (Cx43) and tight junction zonula occludens-1 (ZO-1) gene and protein expression, respectively. Cx43 and ZO-1 physical interaction was detected by immunoprecipitation (IP). Primary rabbit corneal endothelial cells were cultured in Dulbecco's Modified Eagle Medium (DMEM) containing BAK for 24 hours. The scrape-loading dye transfer technique (SLDT) was used to assess GJIC activity.

Results: Topical administration of BAK (0.05%, 0.1%) dose dependently disrupted corneal endothelial cell morphology, altered Cx43 and ZO-1 distribution and reduced Cx43 expression. BAK also markedly induced increases in Cx43 phosphorylation status concomitant with decreases in the Cx43-ZO-1 protein-protein interaction. These changes were associated with marked declines in GJIC activity.

Conclusions: The dose dependent declines in rabbit corneal endothelial GJIC activity induced by BAK are associated with less Cx43-ZO-1 interaction possibly arising from increases in Cx43 phosphorylation and declines in its protein expression. These novel changes provide additional evidence that BAK containing eyedrop preparations should be used with caution to avoid declines in corneal transparency resulting from losses in GJIC activity and endothelial function.

Editor: Michael Koval, Emory University School of Medicine, United States of America

Funding: This work was supported by National Natural Science Foundation of China, Beijing, China (81100638, 81270978); Key Program of National Natural Science of China, Beijing, China (81330022) and Cross-Straits Science Foundation, Beijing. China (U1205025). The funders had no role in study design, data collection and analysis, decision on publish, or preparation of the manuscript.

Competing Interests: The authors have declared that no competing interests exist.

* Email: zuguoliu@xmu.edu.cn (ZL); mydear_22000@163.com (JH)

Introduction

Corneal endothelial functional activity and homeostasis are essential for this tissue layer to counter the natural tendency of the stroma to imbibe fluid and lose its transparency. [1] The maintenance of corneal deturgescence can be compromised by a variety of stresses that disrupt endothelial integrity. Some of the identified insults include intraocular surgical trauma [2,3] and corneal transplantation. [4] In addition, ocular surface diseases, topical application of drugs can cause corneal endothelial damage. [5,6] One reason for this vulnerability is that in humans *in vivo* corneal endothelial cell cannot proliferate since the cells are actively maintained in a non-proliferative, G1-phase–arrested state. [7] Losses in endothelial integrity caused by cell loss are instead compensated for through cell enlargement and stretching at sites proximal to the defect. Given this inability to undergo

proliferation, it is essential to characterize specific stress-induced cellular changes since such insight can help design strategies to hasten and improve restoration of transparency caused by an injury. [8]

Cell–cell communication is mediated mainly by gap junctions and is critical for tissue homeostasis. Gap junctions are formed by two hemichannels or connexons which consist of connexins in either homomeric or heteromeric configurations. They are needed for the transfer between cells of small molecules, ions, phosphorylated nucleotides, nutrients, and second messengers (<1KDa) such as cAMP, IP_3, and Ca^{2+} exchange. [9,10] Gap junction connectivity is required for numerous cellular processes that underlie the maintenance of tissue homeostasis. They include proliferation, differentiation, and embryonic development. [11] Disruption of gap junction complexes can change cell-cell communication and is associated with certain diseases such as cardio-cerebrovascular disease, [12,13] tumor, [14,15] a variety of skin diseases, [16] and congenital cataract. [17] There are approximately 21 human connexin genes and 20 mouse connexin genes which have been identified. [18] Connexin43 (Cx43) is widely expressed and ubiquitously present in a variety of cell types including macrophages. [19] It is also the most common connexin subtype expressed in the corneal endothelium of rat [20], rabbit [21] and human. [22] It has been shown that Cx43 gene knockdown accelerated rabbit corneal epithelial [23] and endothelial [24] wound healing. [25] The fact that losses in gap junctional communication (GJIC) has disparate effects on maintenance of tissue layer homeostasis and wound healing indicates the importance of delineating the effects of specific stresses on GJIC and tissue homeostasis.

Connexins undergo within 2–5 hours rapid turnover involving sequential transitions from synthesis, plaque clustering, junctional formation, final internalization leading to degradation. Cx43 repeatedly cycles through these aforementioned phases. [26–28] However, mechanisms that govern the dynamic patterning of gap junctions remain poorly defined. Immunoprecipitation experiments revealed that the carboxyl terminus of Cx43 interacts in other tissues with the second PDZ domain (PDZ2) of Zonula occludens-1 (ZO-1). [29] ZO-1 not only acts as a passive scaffold in organizing gap junction complexes including connexins and cytoskeletals, but also actively participates in the dynamic remodeling of gap junctions, including trafficking of Cx43, gap junction size regulation and Cx43 internalization. [30] Furthermore, gap junction plaques formation and cell communication maintenance are regulated through Cx43 phosphorylation status modulation. [31]

BAK is one of the most commonly used preservatives in ophthalmic preparations. [32] Our previous study demonstrated that BAK could affect changes in ZO-1 distribution and accordingly disrupt corneal endothelial tight junctional barrier function. [33] However, even though it is known that ZO-1 also interacts with the gap junctional component Cx43, the effect of BAK is unknown on corneal endothelial gap junctions. The purpose of this study was to evaluate the effects of BAK on corneal endothelial GJIC activity and on Cx43 gene and protein expression as well as protein-protein interaction between ZO-1 and Cx43.

Materials and Methods

Experiment animals

New Zealand albino rabbits (purchased from Shanghai Shilaike Laboratory Animal Center, Shanghai, China) weighing between 1.5 to 2.0 kg were used for this study. Rabbits were kept in a standard room with stable temperature and humidity. All procedures were performed in accordance with ARVO statement for the use of animals in ophthalmic and vision research and approved by the animal ethics committee of Medical College of Xiamen University.

BAK treatment

According to our previous experimental designs [33] which were slightly modified, thirty-six male adult New Zealand albino rabbits were randomly assigned into three groups. One eye of each rabbit was treated with BAK solution twice daily for seven days. BAK at different concentrations of 0.01%, 0.05%, and 0.1% was applied in different groups, respectively. The contralateral untreated eyes served as controls. At day 7, rabbits were sacrificed with an overdose of pentobarbital sodium, and corneas were removed carefully with ophthalmic surgical scissors.

Reagents and Antibodies

BAK, dimethyl sulfoxide (DMSO), and Triton X-100, collagenase I and Lucifer Yellow dye were purchased from Sigma Aldrich (St. Louis, MO); Protein A/G PLUS-Agarose Immunoprecipitation Reagent was from Santa Cruz Biotechnology (Santa Cruz, CA); PVDF Western Blotting Membrane was from Roche (Basel, Switzerland); pentobarbital sodium was from Abbott Laboratories (North Chicago, IL); Enhanced chemiluminescence (ECL) kit was obtained from GE Healthcare UK (Chalfont, UK); Mounting medium with 4, 6-diamidino-2-phenylindole (DAPI) and bovine serum albumin (BSA) were from Vector Laboratories (Burlingame CA); Dulbecco's modified Eagle's medium (DMEM) and fetal bovine serum (FBS) were purchased from Life Technologies (Carlsbad, CA); mouse-anti-rabbit ZO-1 antibody, Alexa488-conjugated donkey-anti-mouse IgG, and Alexa555-conjugated donkey-anti-goat IgG were from Life Technologies (Carlsbad, CA); goat polyclonal antibody for Cx43 and P-Cx43, Horseradish peroxidase (HRP)-conjugated donkey anti- goat IgG were from Santa Cruz Biotechnology (Santa Cruz, CA); mouse anti-rabbit β-actin antibody from Sigma Aldrich (St. Louis, MO); Horseradish peroxidase (HRP)-conjugated goat anti- mouse IgG from Merck (Darmstadt, Germany).

In Vivo Confocal Microscopy (IVCM)

After BAK treatment for 7 days, rabbits were anesthetized by intraperitoneal injection of sodium pentobarbital (20 mg/kg; Abbott Laboratories, North Chicago, IL) and intramuscular injection of xylazine (1 mg/kg body weight; Bayer, Shawnee Mission, KS). The Heidelberg Retina Tomograph III/Rostock Cornea Module (Heidelberg Engineer GmbH, Heidelberg, Germany) laser scanning in vivo confocal microscopy was used to examine corneal endothelial morphology. Before examination, a drop of carbomer gel (Alcon Laboratories, Fort Worth, TX) was applied to applanating lens. A diode laser was used as a light source at a wavelength of 670-nm. The microscope objective had an immersion lens covered by a polymethyl methacrylate cap (Olympus, Hamburg, Germany). Images consisted of 384×384 pixels, allowing a scanning area of 400 mm^2 with lateral and vertical resolutions of both 1 mm and a magnification of up to 800 times. The center of the cap was applanated onto the central cornea by adjusting the controller and the central corneal endothelium was examined. More than 10 in vivo digital images were captured on the computer. All measurements were performed by a single investigator masked to the specific experimental conditions.

Immunofluoresecence Staining

Rabbits were sacrificed with an intravenous overdose of pentobarbital sodium, eyes were enucleated and corneas were carefully excised around the limbal rim under a dissecting microscope (Model SZ40; Olympus, Tokyo, Japan). The remaining iris, lens, and retina were thrown away. The freshly isolated corneas were cut into four parts and two quarters were fixed with 4% paraformaldehyde in PBS for 5 minutes at room temperature (RT), followed by acetone for 3 minutes at −20°C. Subsequently, corneas were washed three times in PBS with 1% Triton X-100 and 1% dimethyl sulfoxide (TD buffer) followed by blocking with 2% BSA in PBS for 1 hour at RT. Then corneas were incubated with a polyclonal goat-anti-rabbit Cx43 antibody or mouse-anti-rabbit ZO-1 antibody diluted with 1% BSA overnight at 4°C. After washing three times with TD buffer, corneal tissues were incubated in secondary antibody (donkey-anti-goat IgG conjugated with Alexa Fluor 555 or donkey-anti-mouse IgG conjugated with Alexa Fluor 488) diluted with 1% BSA for 1 hour at RT.

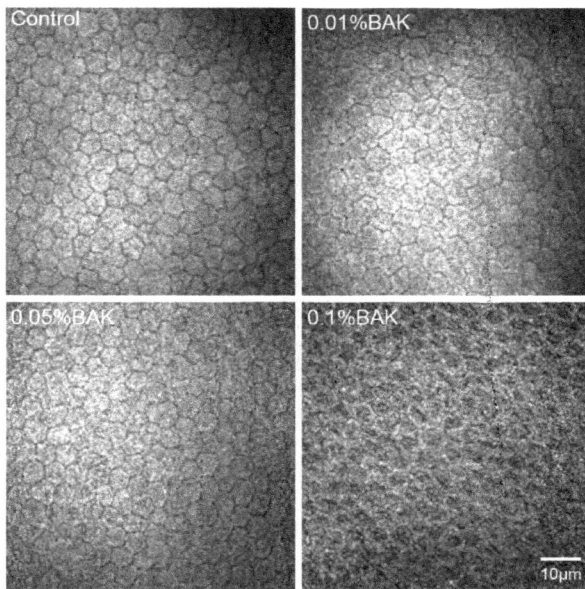

Figure 1. Effects of BAK on rabbit corneal endothelial cell morphology. BAK at 0.01%, 0.05%, and 0.1% was applied twice daily to one eye of each of rabbits for seven days. There was no significant difference of corneal endothelial cells morphology between 0.01% BAK-treated group and control group. In contrast, irregular hexagon cell morphology and blurry boundaries were apparent in 0.05% and 0.1% BAK treated groups. The number of rabbits is nine for three independent experiments (n = 9), three control corneas for IVCM and the remaining control eyes for the culture of corneal endothelial cells.

After that, tissues were mounted endothelial side up on a slide and stained with DAPI. Omission of primary antibody was used as a negative control. Then corneas were observed under a laser scanning confocal microscope (Olympus Fluoview 1000; Olympus, Japan), and images were captured and processed using Olympus Fluoview software.

RT-PCR analysis

Corneas were isolated as described above and cut into two parts. Corneal endothelial layer from one part was dissected. Total RNA was extracted with Trizol (Life Technologies, Carlsbad, CA) according to the manufacturer's instructions. 0.5 μg of the RNA were subjected to RT-PCR analysis (First strand cDNA synthesis kit, Fermentas EU) and was performed at 25°C for 5 minutes, 42°C for 2.5 hours, 70°C for 5 minutes, finally cooled to 4°C. After reverse transcription to cDNA, a PCR protocol was implemented to maintain amplification in the exponential phase. Sequence of the PCR primers were designed as follows: Cx43 sense, 5'-GCAAGCTCCTGGACAAAGTC-3'; Cx43 antisense, 5'- CGTTGACACCATCAGTTTGG-3'; ZO-1 sense, 5'-GTCTGCCATTACACGGTCCT-3'; ZO-1 antisense, 5'-GGTCTCTGCTGGCTTGTTTC-3'; Glyceraldehyde-3-phosphate dehydrogenase (GAPDH; internal control) sense, 5'-ACCACAGTCCACGCCATCAC-3'; and GAPDH antisense, 5'-TCCACCACCCTGTTGCTGTA-3'; RT and PCR incubation were performed with a PCR system (GeneAmp 2400-R; Perkin-Elmer, Foster City, CA) and the PCR cycle comprised of incubations at 94°C for 3 minutes, 94°C for 30 seconds, 55°C for 30 seconds, 68°C for 55 seconds, 68°C for 7 minutes. The reaction mixture was finally cooled to 4°C, and the products of amplification were fractionated by electrophoresis on a 2%

agarose gel and stained with ethidium bromide (EB). Band intensities were measured by Image Acquisition and Analysis System (UVP, Cambridge, UK). Band intensities were analyzed by image analysis software and those for Cx43 and ZO-1 were normalized by the corresponding value for GAPDH.

Western blot analysis

Endothelial layers from another part of the corneas was mechanically isolated and washed several times in PBS, then lysed in RIPA buffer containing 50 mM Tris-HCL (PH7.4), 150 mM NaCl, 1%NP-40, 0.5% deoxycholic acid, 0.1%SDS and 1% protease and phosphatase inhibitor cocktail (Thermo Scientific, Rockford, IL). Cell lysates were centrifuged at 15000×g for 25 minutes at 4°C and protein concentration was determined by Bio-Rad DC Protein Assay (Bio-Rad, Hercules, CA). The lysates (equal amounts of total protein) were subjected to SDS-PAGE on 10% (Cx43) and 8% (ZO-1) polyacrylamide gels. Proteins were transferred to PVDF membrane (Millipore) and blocked for 1 hour at RT with TBST (Tris-buffered saline with 0.05% Tween-20) containing 1% BSA. Then membranes were incubated with specific primary antibodies: goat-anti-rabbit Cx43, mouse-anti-rabbit ZO-1 and mouse-anti-rabbit β-actin diluted with blocking buffer (TBST containing 1% BSA) overnight at 4°C. After washing with TBST, membranes were incubated with secondary antibody (HRP conjugated donkey-anti-goat IgG and HRP conjugated goat-anti-mouse IgG) for 1 hour at RT. Immune complexes were detected with an ECL reagent. Band intensities were measured using Molecular Imager ChemiDoc XRS System (Bio-Rad, Hercules, CA) and analyzed with image analysis software.

Triton X-100 Fraction Isolation

Isolated half of corneal endothelial layer was washed in PBS and lysed in cell lysis buffer containing 1% Triton X-100, 1 mM EDTA and 1% protease and phosphatase inhibitor cocktail. Cell lysates were centrifuged at 15000 g for 25 minutes at 4°C. Supernatant was removed and 1% Triton X-100 insoluble pellet was further lysed in 1×SDS sample buffer by sonication. Both the insoluble and soluble fractions were subjected to western blot analysis using anti-Cx43 antibody separately. Band intensities were measured using Molecular Imager ChemiDoc XRS System and insoluble/soluble Cx43 ratios were calculated.

Immunoprecipitation

Isolated corneal endothelial layers were lysed in cold immunoprecipitation buffer (50 mM Tris-HCl (pH 7.5), 150 mM NaCl, 1 mM EDTA, 1% Nonidet P-40, 0.5% sodium deoxycholate, 0.1% SDS, 5 mM NaF, 1 mM Na_3VO_4, and 1% protease inhibitor cocktail) and solubilized in cell lysis buffer using a motor-driven tissue homogenate. Cell lysates were centrifuged at 15000 g for 15 minutes at 4°C and protein concentration was determined. Lysates were precleared with magnetic protein G beads (Millipore) for 1 hour at RT to remove nonspecific binding of cell proteins to beads. The beads were discarded and cleared lysates were incubated with goat-anti-rabbit Cx43 antibody and pre-blocked protein A/G agarose beads (Santa Cruz) with BSA overnight at 4°C. After centrifugation at 5000×g for 15minutes, the beads were washed three times in cell lysis buffer and heated at 100°C for 10 minutes in 1×loading buffer. Then proteins to bound beads were subjected to SDS-PAGE analysis using either anti-ZO-1 or anti-Cx43 antibodies as described above.

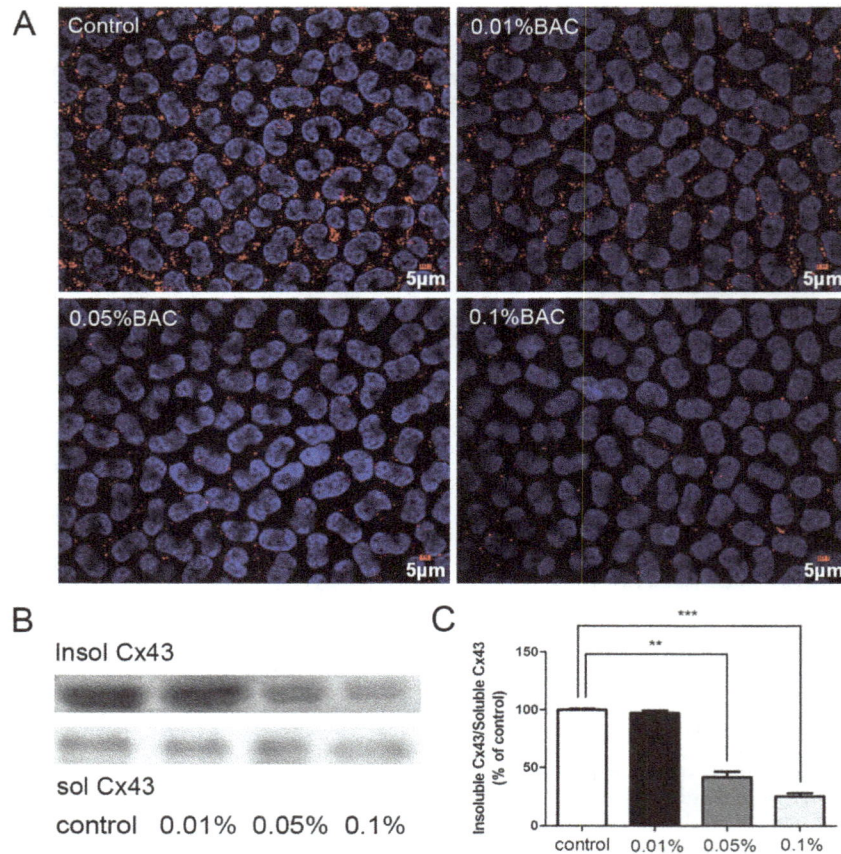

Figure 2. Effects of BAK on Cx43 distribution. In untreated cornea, Cx43 exhibited a great uniformity in appearance with numerous big gap junction plaque. The distribution of Cx43 became disorganized and large gap junction plaques were rarely noticeable in 0.05% and 0.1% BAK-treated group (A). Western blot analysis for 1% Triton X-100 insoluble Cx43 (B) and quantitative analysis of insoluble Cx43 abundance were shown (C). Data are means ± SE from three independent experiments. **P value<0.01, ***P value<0.001 for the indicated comparisons (ANOVA followed by Dunnett's tests).

Primary Cell Culture and GJIC activity

The scrape-loading dye transfer technique (SLDT) is a method to evaluate GJIC activity by calculating the number of cells containing the dye or measuring the distance of dye permeation through gap junctions. [34] Isolated corneal endothelial layers were digested into single cells in 10 mg/ml collagenase I at 37°C overnight and 0.25% trypsin-EDTA at RT for 2 minutes. Primary rabbit corneal endothelial cells were cultured in DMEM/F12 medium supplemented with 10% FBS and an antibiotic mixture (penicillin 100 U/ml and streptomycin 100 ug/ml) at 37°C in a humidified atmosphere containing 5% CO_2. BAK (0, 0.00001%, 0.00005% or 0.00025%) was added to the culture medium for 24 hours when cell densities reached about 70% confluence. After the cells were washed with PBS three times, 10 scrapes were made through confluent cells with a sterile pipette tip in the presence of warm PBS containing 1 mg/ml Lucifer Yellow. Cells were further incubated at 37°C and with 5% CO_2 for 5 minutes. Then Lucifer Yellow was removed, cells were washed with PBS three times and fixed with 4% paraformaldehyde in PBS. The number of cells containing Lucifer Yellow was counted under a laser confocal microscope and 20 sites were measured. GJIC activity was expressed as the mean number of cells containing the dye.

Statistical Analysis

Quantitative data are presented as mean ± SE and were analyzed with Dunnett's multiple comparison tests. A P value of < 0.05 was considered statistically significant.

Results

BAK Disrupts Endothelial Integrity

IVCM images of normal untreated corneal endothelial cells showed that they had hexagonal shapes and were organized in a regular array exhibiting bright cell bodies and dark cell borders. Following topical administration of 0.01% BAK, the cells were undamaged (Fig.1). In contrast, many of cells exposed to either 0.05% or 0.1% BAK lost their hexagonal shape and appeared damaged since their boundaries were distorted and ill defined.

BAK Treatment Altered Gap Junction Plaques Distribution

In untreated corneas, Cx43 was uniformly distributed within the cell plasma membrane in numerous big gap junction plaques. After 0.05% and 0.1% BAK treatment, the readily identifiable plaques along with Cx43 staining became progressively less evident along the endothelial cell borders (Fig. 2A). On the other hand with 0.01% BAK, these changes were less evident than at the

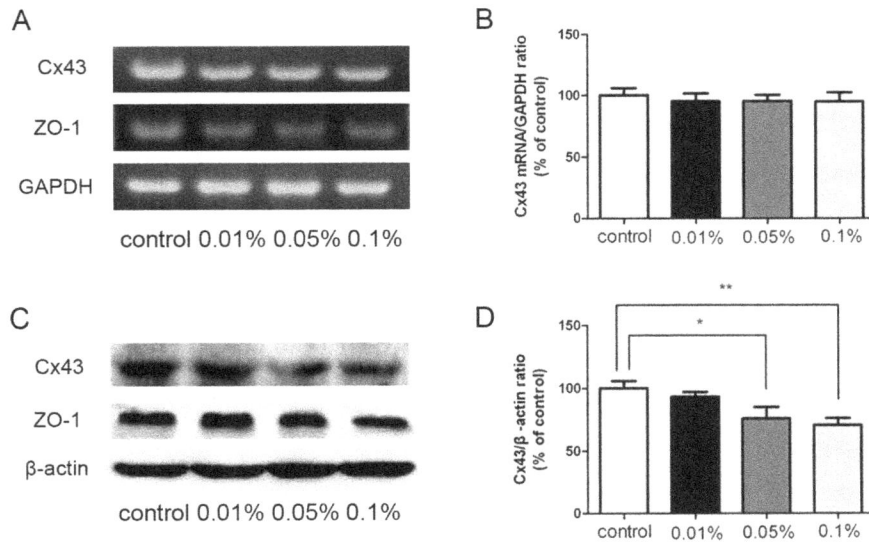

Figure 3. Effect of BAK treatment on Cx43 and ZO-1 expression. The significant decrease of Cx43 protein was induced by topical application of BAK at the concentration of 0.05% and 0.1% (C, D). In comparison, there was no significant decrease of Cx43 and ZO-1 mRNA in BAK treated group (A, B). Similarly, the expression of ZO-1 protein was consistent with mRNA level of ZO-1. Quantitative analysis of Cx43 mRNA and protein in corneal endothelium was shown in (B) and (D) respectively. Data are means ± SE from three independent experiments. *P value<0.05, **P value<0.01 for the indicated comparisons (ANOVA followed by Dunnett's tests).

two higher BAK concentrations. Associated with the progressive declines in punctate Cx43 staining, western blot analysis shown in Fig. 2B further indicates that gap junctional (1% Triton X-100 - insoluble) Cx43 also fell. Densitometric analysis of this change shown in Fig. 2C indicates that the insoluble/soluble Cx43 ratio reached a value following exposure to 0.1% BAK that was 25.38% of the control value. These declines indicate that BAK treatment led to functional Cx43 gap junction plaque disruption.

Differential Effects of BAK on Cx43 and ZO-1 Gene and Protein Expression

To determine if the losses in the Triton X-100 insoluble fraction are associated with declines in Cx43 and ZO-1 expression, we performed western blot analysis and RT-PCR, respectively (Fig 3). These approaches indeed revealed that Cx43 protein expression progressively declined during exposure to 0.05% and 0.1% BAK (Fig. 3C, D). On the other hand, RT-PCR analysis showed that in all BAK treated groups Cx43 mRNA expression was essentially invariant (Fig. 3A, B). The abundance of ZO-1 protein and mRNA was also unaffected in BAK treated groups compared with the control group (Fig. 3A and C).

BAK-Induced Declines in Cx43 and ZO-1 Co-localization and Interaction

We previously reported that exposing endothelial cells to BAK dose dependently disrupted their tight junctional barrier properties. ZO-1, a definitive tight junction marker, plays a critical role in maintaining corneal endothelial barrier function. Furthermore, the interactions between Cx43 and ZO-1 determine the gap junction size and GJIC activity. [30] To explore the effects of BAK on Cx43 and ZO-1 association, Cx43 and ZO-1 double immunostaining along with IP were used to assess BAK effects on gap junction plaque integrity. Immunofluorescence staining revealed that in the large gap junction plaques of normal cells, ZO-1 co-localization with Cx43 was very evident. On the other hand, it markedly declined in the 0.05% and 0.1% BAK treated

groups (Fig. 4A). These BAK effects agree with the IP result which revealed that the interaction between ZO-1 and Cx43 progressively decreased reaching a value at 0.1% BAK that was only 32.84% of the control group (Fig. 4B, C).

BAK-Induced Increases in Cx43 Phosphorylation

Cx43 Phosphorylation status modulation can regulate GJIC activity through a variety of mechanisms by affecting Cx43 plasma membrane transportation, assembly and degradation and modifying gap junctional gating. [35] Significant up-regulation of phosphorylated Cx43 was observed in the 0.05% and 0.1% BAK treated groups, while the 0.01% BAK treated group was not different from the control (Fig. 5A, B). These increases in Cx43 phosphorylation status could contribute to BAK-induced declines in GJIC activity.

BAK Treatment Suppresses GJIC Activity

Dye transfer is a commonly used method to evaluate GJIC activity by calculating the number of dye labeled cells or measuring gap junction dye permeation. To determine if the increases in Cx43 phosphorylation status are associated with changes in GJIC activity, we treated the cells for 24 hours with BAK (0, 0.00001%, 0.00005%, and 0.00025%, respectively) and evaluated dye transfer between neighboring cells. These lower BAK concentrations were chosen to minimize cytotoxic effects of prolonged BAK exposure. Gap junctional dye transfer declined at the two higher concentrations (0.00005% and 0.00025%) suggesting an association between increases in Cx43 phosphorylation status and declines in GJIC activity (Fig. 6A, B).

Discussion

Human corneal endothelial cells cannot replicate and their density declines at an average rate of about 0.6% per year throughout life. [36] Stresses resulting from allograft rejection, inflammation, surgical trauma and topical ophthalmic drug

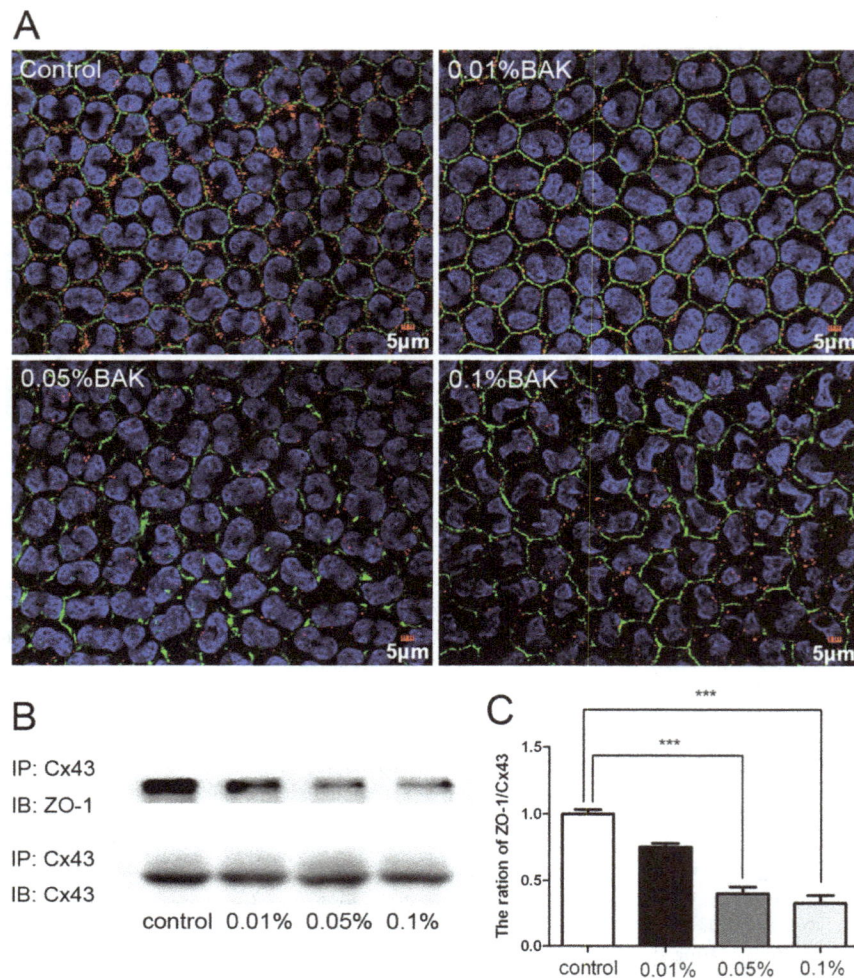

Figure 4. Effect of BAK treatment on co-localization and interaction of Cx43 and ZO-1. Cx43 (red) and ZO-1 (green) co-localization was observed under laser scanning confocal microscope (A). In control group, corneal endothelium presented large gap junction plaques and the co-localization of ZO-1 and Cx43 was obvious. Exposure to 0.05% and 0.1%BAK disturbed the distribution of Cx43 and ZO-1. IP was performed with goat polyclonal anti-Cx43 followed by immunoblotting with mouse monoclonal anti-ZO-1 (B). The ratio of ZO-1 to Cx43 which is normalized with control reduced significantly after BAK treatment especially in 0.05% and 0.1% BAK group (C). Data are means ± SE from three independent experiments. ***P value<0.001 for the indicated comparisons (ANOVA followed by Dunnett's tests).

application can induce additional increases in cell loss. As corneal transparency can compromised by BAK containing ophthalmic preparation, there is an extensive literature describing the potential cytotoxic effects on corneal function. We along with others reported that BAK topical application induced ocular surface discomfort, inflammation and disruption of corneal

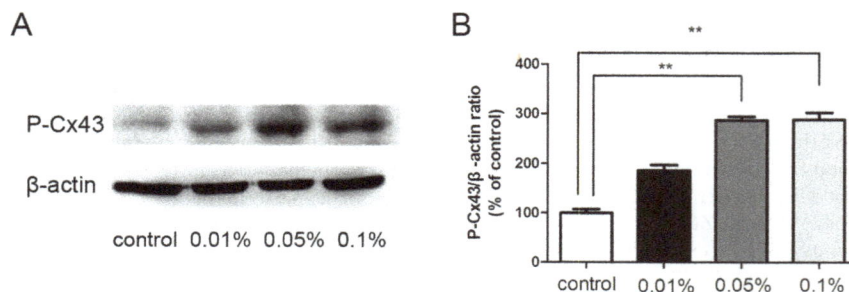

Figure 5. BAK treatment increased Cx43 phosphorylation status. Exposure to BAK induced the increase of phosphorylated Cx43 (A). In 0.05% and 0.1%BAK group, P-Cx43 increased significantly compared with control cornea. Quantitative analysis of P-Cx43 abundance was shown in (B). Data are means ± SE from three independent experiments. **P value<0.01 for the indicated comparisons (ANOVA followed by Dunnett's tests).

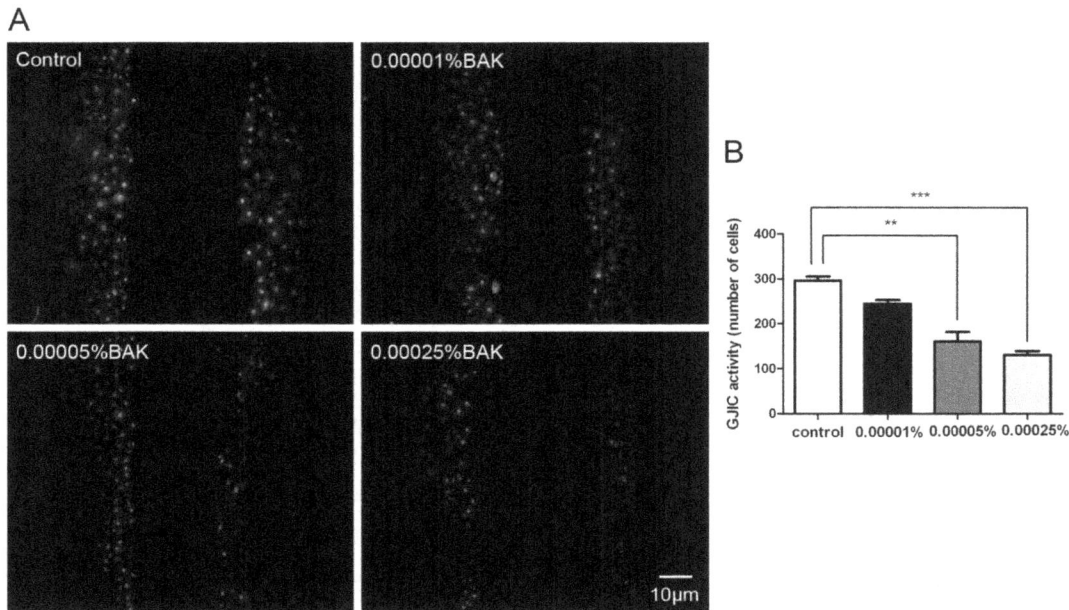

Figure 6. BAK treatment suppressed GJIC activity in primary rabbit corneal endothelial cells. Primary cultured corneal endothelial cells were treated with BAK for 24 hours, BAK was replaced with 1 mg/ml Lucifer Yellow and SLDT assay was performed as described before. Lucifer Yellow transfer through the gap junction was inhibited significantly after 0.00005% and 0.00025% BAK treatment (A). Quantitative analysis of GJIC activity of corneal endothelial cells was shown in (B). Data are means $\pm$ SE from three independent experiments. **P value<0.01, ***P value<0.001 for the indicated comparisons (ANOVA followed by Dunnett's tests).

epithelial and endothelial barrier function. [33] In the present study, we examined the effect of BAK on GJIC activity and its physical interaction with a tight junctional barrier protein ZO-1 in rabbit corneal endothelial cells. This was done since it is evident that preservation of GJIC activity and barrier properties is essential to the maintenance of corneal homeostasis. The BAK concentrations are the same that we used in our previous studies in which we described the destructive effects of BAK on rabbit corneal endothelial barrier function. [33]

We show here that topical application of BAK disrupted gap junctional Cx43 distribution (Fig. 2A), down-regulated Cx43 expression (Fig. 3) and suppressed GJIC activity (Fig. 6). Furthermore, these effects were accompanied with increases in Cx43 phosphorylation status (Fig. 5) and concomitant declines in Cx43 interaction with ZO-1 (Fig. 4). These findings suggest that BAK induced disruption of gap junction plaques formation in endothelial cells is associated with declines in Cx43 interaction with ZO-1 and declines in GJIC activity.

Cx43 is an important corneal endothelial gap junctional component which is stably accumulated into gap junctional plaques and its insertion is essential for regulating cell-cell communication under physiological conditions. We found that BAK dose dependently decreased gap junction plaque expression which was associated with declines in Cx43 protein expression and in 1% Triton X-100 extracts the insoluble Cx43 content also fell (Fig.2A–C). These results confirmed the destructive effects of BAK on gap junctions. On the other hand, even after exposure for 7 days to either 0.05% and 0.1% BAK, Cx43 mRNA expression remained unchanged whereas Cx43 protein decreased significantly (Fig.3). This difference suggests that the effects of BAK are restricted to a post-translational modification of Cx43. Many other studies in other tissues concur with this suggestion and indicate that such changes include alteration of Cx43 phosphorylation status and ubiquitination. [37] Declines in Cx43 interaction with

ZO-1 are solely due to Cx43 post translational modification since ZO-1 mRNA and protein expression were not affected by BAK treatment in our studies. Recently, *Thomas Tien* et al. found that Cx43 gene silencing with relevant siRNAs decreased ZO-1 and occludin protein expression and increased cell monolayer permeability in rat retinal endothelial cells. [38] It remains to be determined whether Cx43 can regulate ZO-1 expression.

Changes in Cx43 phosphorylation status affect gap junction plaque structure, function and degradation, [35] and regulate GJIC activity. [39,40] The Cx43 cytoplasmic carboxyl terminal domain is enriched in potential kinase phosphorylation target sites. [41] Gap junction plaques are dynamic cell plasma membrane structures with rapid turnover rates ranging between 2–5 hours. [26–28] Regulation of gap junction assembly and turnover controls intercellular communication in all kinds of tissues and cultured cells. [42] Many growth factors, hormones and inflammatory mediators can activate or inhibit intracellular signaling pathways and change connexin phosphorylation status. We found that Cx43 phosphorylation status increased significantly after BAK treatment, which was associated with declines in GJIC activity and Cx43 expression. A variety of kinases have been identified to induce Cx43 phosphorylation including tyrosine protein kinase, mitogen-activated protein (MAP) kinase, protein kinase C (PKC) and casein kinase I. Src is one of the non-receptor tyrosine kinases that can phosphorylate Cx43 carboxyl terminal at Tyr265 site accompanied by suppression of GJIC activity. [41] Further studies are needed to determine the precise signaling pathways that participate in this process.

The cytoplasmic carboxyl terminal domain of Cx43 contains the regulatory binding domains which can modulate Cx43 interactions with other cellular proteins. [41] ZO-1 can bind to the carboxyl terminal intracellular region of Cx43 and is associated with eliciting Cx43 endocytosis. This role for ZO-1 is based on results showing that Hexachlorocyclohexane (HCH), a non-genomic carcinogen that is known to be a potent inducer of Cx43 internalization, decreased the Cx43-ZO-1 association and thus accelerated endocytosis of gap junction plaques. [43] Conversely, *Danny S. Roh* et al. found that even though mitomycin C (MMC) also reduced the interaction of Cx43 and ZO-1, it instead increased the stability of gap junctional plaques in cultured bovine endothelial cells. [25] Our results are consistent with the notion that BAK-induced declines in ZO-1 and Cx43 interaction are associated with gap junctional plaque disruption and declines in GJIC activity. These effects of BAK may be sufficient to account for why at higher concentrations this preservative compromises functional rabbit corneal endothelial cell activity, which is needed to maintain corneal deturgescence and transparency.

After exposing the endothelial cells for 24 hours to Lucifer Yellow, BAK dose dependently reduced GJIC since the number of dye labeled cells declined. There are numerous other ophthalmic mediators that can also reduce GJIC activity. For example, multipurpose solutions used for hydrogel contact lens wear inhibited corneal keratocyte GJIC activity and simultaneously decreased Cx43 abundance. [44] In addition, EGF reduced bovine corneal endothelial cell GJIC activity. [25] So, our experimental results can be readily explained based on the fact that BAK, as a quaternary ammonium cationic surfactant, it disrupted GJIC activity by reducing Cx43expression.

BAK is most often used at a concentration of 0.01% (ranging from 0.004% to 0.025%) in ophthalmic preparations. Its cytotoxic effects are dependent on usage frequency and duration, active ingredients, etc. [32] In this study, we found that BAK is safe to use twice daily for 7 days at a concentration not higher than 0.01%. However, it is still prudent to be on alert for possible declines in endothelial cell GJIC activity if BAK containing ophthalmic preparations are used for longer duration and applied more frequently than twice per day.

In conclusion, topical BAK administration at concentrations above 0.01% induced gap junctional plaque disruption and reduced Cx43 protein expression, but this decline was not attributable to a decrease in Cx43 gene expression. The increases in Cx43phosphorylation status induced by BAK may contribute to observed declines in Cx43-ZO-1 interaction, which in any case are consistent with suppression by this preservative of GJIC activity. These effects of BAK make it even more evident that it is prudent to carefully monitor corneal function when using ophthalmic preparations containing this preservative since cytotoxic effects can severely compromise maintenance of corneal deturgescence and transparency.

Acknowledgments

The authors thank Dr. Peter S. Reinach from the State University of New York (SUNY) for his critical reading and careful revision of the manuscript.

Author Contributions

Conceived and designed the experiments: ZL JH ZZ. Performed the experiments: ZZ YH HX JP FL XL WC JH. Analyzed the data: ZL JH ZZ. Wrote the paper: JH ZZ.

References

1. Bourne WM (2010) Corneal endothelium—past, present, and future. Eye Contact Lens 36: 310–314.
2. Bourne WM, Kaufman HE (1976) Endothelial damage associated with intraocular lenses. Am J Ophthalmol 81: 482–485.
3. Bourne WM, Kaufman HE (1976) Cataract extraction and the corneal endothelium. Am J Ophthalmol 82: 44–47.
4. Laing RA, Sandstrom M, Berrospi AR, Leibowitz HM (1976) Morphological changes in corneal endothelial cells after penetrating keratoplasty. Am J Ophthalmol 82: 459–464.
5. Bourne WM, McLaren JW (2004) Clinical responses of the corneal endothelium. Exp Eye Res 78: 561–572.
6. Bourne WM (2003) Biology of the corneal endothelium in health and disease. Eye (Lond) 17: 912–918.
7. Joyce NC (2003) Proliferative capacity of the corneal endothelium. Prog Retin Eye Res 22: 359–389.
8. Gomes P, Srinivas SP, Vereecke J, Himpens B (2006) Gap junctional intercellular communication in bovine corneal endothelial cells. Exp Eye Res 83: 1225–1237.
9. Goodenough DA, Goliger JA, Paul DL (1996) Connexins, connexons, and intercellular communication. Annu Rev Biochem 65: 475–502.
10. Saez JC, Berthoud VM, Branes MC, Martinez AD, Beyer EC (2003) Plasma membrane channels formed by connexins: their regulation and functions. Physiol Rev 83: 1359–1400.
11. Loewenstein WR (1979) Junctional intercellular communication and the control of growth. Biochim Biophys Acta 560: 1–65.
12. Gros DB, Jongsma HJ (1996) Connexins in mammalian heart function. Bioessays 18: 719–730.
13. Miura T, Ohnuma Y, Kuno A, Tanno M, Ichikawa Y, et al. (2004) Protective role of gap junctions in preconditioning against myocardial infarction. Am J Physiol Heart Circ Physiol 286: H214–221.
14. Loewenstein WR, Kanno Y (1966) Intercellular communication and the control of tissue growth: lack of communication between cancer cells. Nature 209: 1248–1249.
15. Rivedal E, Opsahl H (2001) Role of PKC and MAP kinase in EGF- and TPA-induced connexin43 phosphorylation and inhibition of gap junction intercellular communication in rat liver epithelial cells. Carcinogenesis 22: 1543–1550.
16. Richard G (2005) Connexin disorders of the skin. Clin Dermatol 23: 23–32.
17. Gong X, Li E, Klier G, Huang Q, Wu Y, et al. (1997) Disruption of alpha3 connexin gene leads to proteolysis and cataractogenesis in mice. Cell 91: 833–843.
18. Oyamada M, Oyamada Y, Takamatsu T (2005) Regulation of connexin expression. Biochim Biophys Acta 1719: 6–23.
19. Alves LA, Coutinho-Silva R, Persechini PM, Spray DC, Savino W, et al. (1996) Are there functional gap junctions or junctional hemichannels in macrophages? Blood 88: 328–334.
20. Joyce NC, Harris DL, Zieske JD (1998) Mitotic inhibition of corneal endothelium in neonatal rats. Invest Ophthalmol Vis Sci 39: 2572–2583.
21. Williams KK, Watsky MA (2004) Bicarbonate promotes dye coupling in the epithelium and endothelium of the rabbit cornea. Curr Eye Res 28: 109–120.
22. Williams K, Watsky M (2002) Gap junctional communication in the human corneal endothelium and epithelium. Curr Eye Res 25: 29–36.
23. Grupcheva CN, Laux WT, Rupenthal ID, McGhee J, McGhee CN, et al. (2012) Improved corneal wound healing through modulation of gap junction communication using connexin43-specific antisense oligodeoxynucleotides. Invest Ophthalmol Vis Sci 53: 1130–1138.
24. Nakano Y, Oyamada M, Dai P, Nakagami T, Kinoshita S, et al. (2008) Connexin43 knockdown accelerates wound healing but inhibits mesenchymal transition after corneal endothelial injury in vivo. Invest Ophthalmol Vis Sci 49: 93–104.
25. Roh DS, Funderburgh JL (2011) Rapid changes in connexin-43 in response to genotoxic stress stabilize cell-cell communication in corneal endothelium. Invest Ophthalmol Vis Sci 52: 5174–5182.
26. Musil LS, Goodenough DA (1991) Biochemical analysis of connexin43 intracellular transport, phosphorylation, and assembly into gap junctional plaques. J Cell Biol 115: 1357–1374.
27. Laird DW, Puranam KL, Revel JP (1991) Turnover and phosphorylation dynamics of connexin43 gap junction protein in cultured cardiac myocytes. Biochem J 273(Pt 1): 67–72.
28. Laird DW, Castillo M, Kasprzak L (1995) Gap junction turnover, intracellular trafficking, and phosphorylation of connexin43 in brefeldin A-treated rat mammary tumor cells. J Cell Biol 131: 1193–1203.
29. Giepmans BN, Moolenaar WH (1998) The gap junction protein connexin43 interacts with the second PDZ domain of the zona occludens-1 protein. Curr Biol 8: 931–934.
30. Rhett JM, Jourdan J, Gourdie RG (2011) Connexin 43 connexon to gap junction transition is regulated by zonula occludens-1. Mol Biol Cell 22: 1516–1528.
31. Chen J, Pan L, Wei Z, Zhao Y, Zhang M (2008) Domain-swapped dimerization of ZO-1 PDZ2 generates specific and regulatory connexin43-binding sites. EMBO J 27: 2113–2123.
32. Baudouin C, Labbe A, Liang H, Pauly A, Brignole-Baudouin F (2010) Preservatives in eyedrops: the good, the bad and the ugly. Prog Retin Eye Res 29: 312–334.
33. Chen W, Li Z, Hu J, Zhang Z, Chen L, et al. (2011) Corneal alternations induced by topical application of benzalkonium chloride in rabbit. PLoS One 6: e26103.

34. el-Fouly MH, Trosko JE, Chang CC (1987) Scrape-loading and dye transfer. A rapid and simple technique to study gap junctional intercellular communication. Exp Cell Res 168: 422–430.

35. Solan JL, Lampe PD (2009) Connexin43 phosphorylation: structural changes and biological effects. Biochem J 419: 261–272.

36. Bourne WM, Nelson LR, Hodge DO (1997) Central corneal endothelial cell changes over a ten-year period. Invest Ophthalmol Vis Sci 38: 779–782.

37. Kimura K, Nishida T (2010) Role of the ubiquitin-proteasome pathway in downregulation of the gap-junction protein Connexin43 by TNF-{alpha} in human corneal fibroblasts. Invest Ophthalmol Vis Sci 51: 1943–1947.

38. Tien T, Barrette KF, Chronopoulos A, Roy S (2013) Effects of high glucose-induced Cx43 downregulation on occludin and ZO-1 expression and tight junction barrier function in retinal endothelial cells. Invest Ophthalmol Vis Sci 54: 6518–6525.

39. Solan JL, Lampe PD (2005) Connexin phosphorylation as a regulatory event linked to gap junction channel assembly. Biochim Biophys Acta 1711: 154–163.

40. Laird DW (2005) Connexin phosphorylation as a regulatory event linked to gap junction internalization and degradation. Biochim Biophys Acta 1711: 172–182.

41. Warn-Cramer BJ, Lau AF (2004) Regulation of gap junctions by tyrosine protein kinases. Biochim Biophys Acta 1662: 81–95.

42. Lampe PD, Lau AF (2000) Regulation of gap junctions by phosphorylation of connexins. Arch Biochem Biophys 384: 205–215.

43. Gilleron J, Fiorini C, Carette D, Avondet C, Falk MM, et al. (2008) Molecular reorganization of Cx43, Zo-1 and Src complexes during the endocytosis of gap junction plaques in response to a non-genomic carcinogen. J Cell Sci 121: 4069–4078.

44. Sumide T, Tsuchiya T (2003) Effects of multipurpose solutions (MPS) for hydrogel contact lenses on gap-junctional intercellular communication (GJIC) in rabbit corneal keratocytes. J Biomed Mater Res B Appl Biomater 64: 57–64.

Permissions

The contributors of this book come from diverse backgrounds, making this book a truly international effort. This book will bring forth new frontiers with its revolutionizing research information and detailed analysis of the nascent developments around the world.

We would like to thank all the contributing authors for lending their expertise to make the book truly unique. They have played a crucial role in the development of this book. Without their invaluable contributions this book wouldn't have been possible. They have made vital efforts to compile up to date information on the varied aspects of this subject to make this book a valuable addition to the collection of many professionals and students.

This book was conceptualized with the vision of imparting up-to-date information and advanced data in this field. To ensure the same, a matchless editorial board was set up. Every individual on the board went through rigorous rounds of assessment to prove their worth. After which they invested a large part of their time researching and compiling the most relevant data for our readers.

The editorial board has been involved in producing this book since its inception. They have spent rigorous hours researching and exploring the diverse topics which have resulted in the successful publishing of this book. They have passed on their knowledge of decades through this book. To expedite this challenging task, the publisher supported the team at every step. A small team of assistant editors was also appointed to further simplify the editing procedure and attain best results for the readers.

Apart from the editorial board, the designing team has also invested a significant amount of their time in understanding the subject and creating the most relevant covers. They scrutinized every image to scout for the most suitable representation of the subject and create an appropriate cover for the book.

The publishing team has been an ardent support to the editorial, designing and production team. Their endless efforts to recruit the best for this project, has resulted in the accomplishment of this book. They are a veteran in the field of academics and their pool of knowledge is as vast as their experience in printing. Their expertise and guidance has proved useful at every step. Their uncompromising quality standards have made this book an exceptional effort. Their encouragement from time to time has been an inspiration for everyone.

The publisher and the editorial board hope that this book will prove to be a valuable piece of knowledge for researchers, students, practitioners and scholars across the globe.

List of Contributors

Mónica González-Magaldi, Miguel A. Martín-Acebes and Francisco Sobrino
Centro de Biología Molecular Severo Ochoa, Consejo Superior de Investigaciones Científicas-Universidad Autónoma de Madrid, Madrid, Spain

Leonor Kremer
Centro Nacional de Biotecnología, Consejo Superior de Investigaciones Científicas, Madrid, Spain

Ci Fu, Jie Ao and Stephen J. Free
Department of Biological Sciences, SUNY University at Buffalo, Buffalo, New York, United States of America

Anne Dettmann
Institute for Biology II, Albert-Ludwigs University Freiburg, Freiburg, Germany

Stephan Seiler
Institute for Biology II, Albert-Ludwigs University Freiburg, Freiburg, Germany
Freiburg Institute for Advanced Studies (FRIAS), Albert-Ludwigs University Freiburg, Freiburg, Germany

Shuang Zhao, Xin-Pei Wang, Jing-Fei Jiang, Yu-Shuang Chai, Tian-Shi Feng, Fan Lei, Dong-Ming Xing and Li-Jun Du
MOE Key Laboratory of Protein Sciences, Laboratory of Molecular Pharmacology and Pharmaceutical Sciences, School of Life Sciences and School of Medicine, Tsinghua University, Beijing, China

Yu Tian and Yi Ding
Drug Discovery Facility, School of Life Sciences, Tsinghua University, Beijing, China

Jing Huang
Department of Chemistry, Virginia Polytechnic Institute and State University, Blacksburg, Virginia, United States of America

Maria E. Teves, Wei Li and Lauren van Reesema
Department of Obstetrics and Gynecology, Virginia Commonwealth University, Richmond, Virginia, United States of America

Jerome F. Strauss III and Zhibing Zhang
Department of Obstetrics and Gynecology, Virginia Commonwealth University, Richmond, Virginia, United States of America

Department of Biochemistry and Molecular Biology, Virginia Commonwealth University, Richmond, Virginia, United States of America

Patrick R. Sears
Cystic Fibrosis Center, University of North Carolina, Chapel Hill, North Carolina, United States of America

Zhengang Zhang
Department of Obstetrics and Gynecology, Virginia Commonwealth University, Richmond, Virginia, United States of America
Department of Infectious Diseases, Tongji Medical College, Huazhong University of Science and Technology, Wuhan, Hubei, China

Waixing Tang
Department of Otorhinolaryngology, University of Pennsylvania, Philadelphia, Pennsylvania, United States of America

Richard M. Costanzo
Department of Physiology and Biophysics, Virginia Commonwealth University, Richmond, Virginia, United States of America

C. William Davis and Michael R. Knowles
Department of Cell & Molecular Physiology of Medicine, University of North Carolina, Chapel Hill, North Carolina, United States of America

Kevin S. Jones
Biology Department, Howard University, Washington, DC, United States of America
Center for Neuroscience Research, Children's National Medical Center, Washington, DC, United States of America

Joshua G. Corbin
Center for Neuroscience Research, Children's National Medical Center, Washington, DC, United States of America

Molly M. Huntsman
Department of Pharmaceutical Sciences, Skaggs School of Pharmacy and Pharmaceutical Sciences, and Department of Pediatrics, School of Medicine, University of Colorado, Anschutz Medical Campus, Aurora, CO, United States of America

Sayonarah C. Rocha, Marco T. C. Pessoa, Luiza D. R. Neves, Herica L. Santos and Leandro A. Barbosa
Laboratório de Bioquímica Celular, Universidade Federal de São João del Rei, Campus Centro-Oeste Dona Lindú, Divinópolis, MG, Brazil

Silmara L. G. Alves and Jose A. F. P. Villar
Laboratório de Síntese Orgânica, Universidade Federal de São João del Rei, Campus Centro-Oeste Dona Lindú, Divinópolis, MG, Brazil

Soraya M. F. Oliveira, Alex G. Taranto and Moacyr Comar
Laboratório de Bioinformática, Universidade Federal de São João del Rei, Campus Centro-Oeste Dona Lindú, Divinópolis, MG, Brazil

Isabella V. Gomes and Fabio V. Santos
Laboratório de Biologia Celular e Mutagenicidade, Universidade Federal de São João del Rei, Campus Centro-Oeste Dona Lindú, Divinópolis, MG, Brazil

Fernando P. Varotti
Laboratório de Bioquímica de Parasitos, Universidade Federal de São João del Rei, Campus Centro-Oeste Dona Lindú, Divinópolis, MG, Brazil

Luciana M. Silva
Laboratório de Biologia Celular e Inovação Biotecnológica, Fundação Ezequiel Dias, Belo Horizonte, MG, Brazil

Natasha Paixão, Luis E. M. Quintas and François Noël
Laboratório de Farmacologia Bioquímica e Molecular, Instituto de Ciências Biomédicas, Universidade Federal do Rio de Janeiro, Rio de Janeiro, RJ, Brazil

Antonio F. Pereira
Laboratório de Bioquímica Microbiana, Instituto de Microbiologia Paulo Góes, Universidade Federal do Rio de Janeiro, Rio de Janeiro, RJ, Brazil

Ana C. S. C. Tessis
Laboratório de Bioquímica Microbiana, Instituto de Microbiologia Paulo Góes, Universidade Federal do Rio de Janeiro, Rio de Janeiro, RJ, Brazil
Instituto Federal de Educação, Ciência e Tecnologia do Rio de Janeiro (IFRJ), Rio de Janeiro, RJ, Brazil

Natalia L. S. Gomes and Otacilio C. Moreira
Laboratório de Biologia Molecular e Doenc͵as Endêmicas, Instituto Oswaldo Cruz/Fiocruz, Rio de Janeiro, RJ, Brazil

Ruth Rincon-Heredia and Rubén G. Contreras
Department of Physiology, Biophysics and Neurosciences, Center for Research and Advanced Studies (Cinvestav), Mexico City, Mexico

Gustavo Blanco
Department of Molecular and Integrative Physiology, University of Kansas Medical Center, Kansas City, Kansas, United States of America

Ildefonso M. De la Fuente
Institute of Parasitology and Biomedicine "López-Neyra", CSIC, Granada, Spain
Department of Mathematics, University of the Basque Country UPV/EHU, Leioa, Spain
Unit of Biophysics (CSIC, UPV/EHU), and Department of Biochemistry and Molecular Biology University of the Basque Country, Bilbao, Spain
Biocruces Health Research Institute, Hospital Universitario de Cruces, Barakaldo, Spain

Jesús M. Cortés
Biocruces Health Research Institute, Hospital Universitario de Cruces, Barakaldo, Spain
Ikerbasque: The Basque Foundation for Science, Bilbao, Basque Country, Spain

Edelmira Valero
Department of Physical Chemistry, School of Industrial Engineering, University of Castilla-La Mancha, Albacete, Spain

Mathieu Desroches
INRIA Paris-Rocquencourt Centre, Paris, France

Serafim Rodrigues
School of Computing and Mathematics, University of Plymouth, Plymouth, United Kingdom

Iker Malaina
Biocruces Health Research Institute, Hospital Universitario de Cruces, Barakaldo, Spain
Department of Physiology, University of the Basque Country UPV/EHU, Bilbao, Spain

Luis Martínez
Department of Mathematics, University of the Basque Country UPV/EHU, Leioa, Spain
Biocruces Health Research Institute, Hospital Universitario de Cruces, Barakaldo, Spain

Selina McHarg, Gemma Hopkins, Lusiana Lim and David Garrod
Faculty of Life Sciences, University of Manchester, Manchester, United Kingdom

Haiqiang Wu, Fang Zhang, , Liao Zhang, Zeqiu Liang, Jinyu Wang and Yizhi Zheng
College of Life Sciences, Shenzhen University, Shenzhen, China

Alan Tunnacliffe and Neil Williamson
Department of Chemical Engineering and Biotechnology, University of Cambridge, Cambridge, United Kingdom

Jie Jian
Department of Chemical Engineering and Biotechnology, University of Cambridge, Cambridge, United Kingdom
College of Pharmacy, Guilin Medical University, Guilin, China

Linkun An
School of Pharmaceutical Science, Sun Yat-sen University, Guangzhou, China

Siavash Hassanpour, Yongqiang Wang, Johannes W. P. Kuiper and Michael Glogauer
Matrix Dynamics Group, Faculty of Dentistry, University of Toronto, Toronto, Ontario, Canada

Hongwei Jiang
Matrix Dynamics Group, Faculty of Dentistry, University of Toronto, Toronto, Ontario, Canada
Department of Operative Dentistry and Endodontics, Guanghua School of Stomatology, Sun Yat-sen University, Guangdong Provincial Key Laboratory of Stomatology, Guangzhou, P. R. China

Dickson K. Kirui, Yeonju Lee and Weijia Zhang
Department of Nanomedicine, Houston Methodist Research Institute, Houston, Texas, United States of America

Eugene J. Koay
Department of Nanomedicine, Houston Methodist Research Institute, Houston, Texas, United States of America
Department of Radiation Oncology, M. D. Anderson Cancer Center, Houston, Texas, United States of America

Fazle Hussain
Department of Mechanical Engineering, Texas Tech University, Lubbock, Texas, United States of America

Lidong Qin and Haifa Shen
Department of Nanomedicine, Houston Methodist Research Institute, Houston, Texas, United States of America

Department of Cell and Developmental Biology, Weill Cornell Medical College, New York, New York, United States of America

Mauro Ferrari
Department of Nanomedicine, Houston Methodist Research Institute, Houston, Texas, United States of America
Department of Medicine, Weill Cornell Medical College, New York, New York, United States of America

Asma Kassas, Ivan C. Moura, Claudine Guérin-Marchand, Ulrich Blank, Renato C. Monteiro, Nicolas Charles and Marc Benhamou
INSERM U1149, Faculté de Médecine Xavier Bichat, Paris, France
University Paris-Diderot, Sorbonne Paris Cité, Laboratoire d'excellence INFLAMEX, DHU FIRE, Paris, France

Yumi Yamashita, Jorg Scheffel and Juan Rivera
Laboratory of Molecular Immunogenetics, Molecular Immunology and Inflammation Branch, NIAMSD, NIH, Bethesda, Maryland, United States of America

Peter J. Sims and Therese Wiedmer
Department of Medicine, University of Rochester School of Medicine and Dentistry, Rochester, New York, United States of America

Xiaocui He, Tomáš Korytář, Yaqing Zhu and Bernd Köllner
Institute of Immunology, Friedrich-Loeffler-Institute (FLI), Federal Research Institute for Animal Health, Greifswald- Insel Riems, Germany

Jiří Pikula and Hana Bandouchova
Department of Ecology and Diseases of Game, Fish and Bees, Faculty of Veterinary Hygiene and Ecology, University of Veterinary and Pharmaceutical Sciences Brno, Brno, Czech Republic

Jan Zukal
Institute of Vertebrate Biology, Academy of Sciences of the Czech Republic, Brno, Czech Republic
Department of Botany and Zoology, Masaryk University, Brno, Czech Republic

Balraj Singh, Anna Shamsnia, Milan R. Raythatha, Ryan D. Milligan, Amanda M. Cady, Simran Madan and Anthony Lucci
Department of Surgical Oncology, and Morgan Welch Inflammatory Breast Cancer Research Program and Clinic, The University of Texas MD Anderson Cancer Center, Houston, Texas, United States of America

Sofia Koch, Xiao Qin, Jaroslav Zak, Ludovico Buti, Ewa Dudziec, Shan Zhong, Xin Lu and
Indrika Ratnayaka
Ludwig Institute for Cancer Research, Nuffield Department of Clinical Medicine, University of Oxford, Oxford, United Kingdom

Shankar Srinivas
Department of Physiology, Anatomy and Genetics, University of Oxford, Oxford, United Kingdom

Christophe Royer
Ludwig Institute for Cancer Research, Nuffield Department of Clinical Medicine, University of Oxford, Oxford, United Kingdom
Department of Physiology, Anatomy and Genetics, University of Oxford, Oxford, United Kingdom

Samuel Martín-Vílchez
Department of Cell Biology, University of Virginia School of Medicine, Charlottesville, Virginia, United States of America

Àngel Hernández-Bartolomé
Unidad de Hepatología, Hospital Universitario de la Princesa, Instituto de Investigación Sanitaria Princesa (IIS-IP), Madrid, Spain

Yolanda Rodríguez-Muñoz, Rosario López-Rodríguez, Luisa García-Buey, Ricardo Moreno-Otero and Paloma Sanz-Cameno
Unidad de Hepatología, Hospital Universitario de la Princesa, Instituto de Investigación Sanitaria Princesa (IIS-IP), Madrid, Spain
Centro de Investigación Biomédica en Red de Enfermedades Hepáticas y Digestivas (CIBER-ehd), Instituto de Salud Carlos III (ISCIII), Madrid, Spain

María Jesús Borque-Iñurrita and Francisca Molina-Jiménez
Unidad de Biología Molecular, Hospital Universitario de la Princesa, Instituto de Investigación Sanitaria Princesa (IIS-IP), Madrid, Spain

Lori B. Huberman and Andrew W. Murray
Molecular and Cellular Biology, Harvard University, Cambridge, MA, United States of America
Faculty of Arts and Sciences Center for Systems Biology, Harvard University, Cambridge, MA, United States of America

Kathryn L. Armour, Cheryl S. Smith, Natasha C. Y. Ip, Cara J. Ellison, Christopher M. Kirton and Michael R. Clark
Department of Pathology, University of Cambridge, Cambridge, United Kingdom

Anthony M. Wilkes
Bristol Institute for Transfusion Sciences, Bristol, United Kingdom

Lorna M. Williamson
National Health Service Blood and Transplant, Cambridge, United Kingdom
Department of Haematology, University of Cambridge, Cambridge, United Kingdom

Zhenhao Zhang, Hui Xie, Juxin Pan, Fanfei Liu, Xuezhi Li, Wensheng Chen, Jiaoyue Hu and Zuguo Liu
Eye Institute and affiliated Xiamen Eye Center of Xiamen University; Fujian Provincial Key Laboratory of Ophthalmology and Vision Science, Xiamen, Fujian, China

Yue Huang
Institute of Stem Cell and Regenerative Medicine, Medical College, Xiamen University, Xiamen, Fujian, China

Index

www.ingramcontent.com/pod-product-compliance
Lightning Source LLC
Chambersburg PA
CBHW081710240326
41458CB00156B/4261